Resource Recovery and Reuse in Organic Solid Waste Management

Integrated Environmental Technology Series

The *Integrated Environmental Technology Series* addresses key themes and issues in the field of environmental technology from a multidisciplinary and integrated perspective.

An integrated approach is potentially the most viable solution to the major pollution issues that face the globe in the 21st century.

World experts are brought together to contribute to each volume, presenting a comprehensive blend of fundamental principles and applied technologies for each topic. Current practices and the state-of-the-art are reviewed, new developments in analytics, science and biotechnology are presented and, crucially, the theme of each volume is presented in relation to adjacent scientific, social and economic fields to provide solutions from a truly integrated perspective.

The *Integrated Environmental Technology Series* will form an invaluable and definitive resource in this rapidly evolving discipline.

Series Editor

Dr Ir Piet Lens, Sub-department of Environmental Technology, The University of Wageningen, P.O. Box 8129, 6700 EV Wageningen, The Netherlands. (piet.lens@wur.nl)

Published titles

Biofilms in Medicine, Industry and Environmental Biotechnology: *Characteristics, analysis and control*
Decentralised Sanitation and Reuse: *Concepts, systems and implementation*
Environmental Technologies to Treat Sulfur Pollution: *Principles and engineering*
Phosphorus in Environmental Technology: *Principles and Applications*
Water Recycling and Resource Recovery in Industries: *Analysis, technologies and implementation*

Forthcoming titles

Biofuels for Fuel Cells: *Biomass fermentation for use in fuel cells*
Pond Treatment Technology
Remediation of Soils and Sediments

www.iwapublishing.com

Resource Recovery and Reuse in Organic Solid Waste Management

Edited by

Piet Lens, Bert Hamelers,
Harry Hoitink and Werner Bidlingmaier

IWA Publishing

Published by IWA Publishing, Alliance House, 12 Caxton Street, London SW1H 0QS, UK

Telephone: +44 (0) 20 7654 5500; Fax: +44 (0) 20 7654 5555; Email: publications@iwap.co.uk
Web: www.iwapublishing.com

First published 2004

Printed by TJ International, Padstow, Cornwall, UK
Typeset by Gray Publishing, Tunbridge Wells, UK

British Library Cataloguing in Publication Data
A CIP catalogue record for this book is available from the British Library

Library of Congress Cataloging-in-Publication Data
A catalog record for this book is available from the Library of Congress

Front cover photographs
Top: Poultry manure composting at a bird layer facility in Ohio (USA). Photograph courtesy of Harry Hoitink and Frederick Michel Jr. (The Ohio State University, USA).
Bottom, left: Stirred tank anaerobic digester (ROS ROCA technology) of solid waste in Boden, Germany. Photograph courtesy of Joan Mata Alvares (University of Barcelona, Spain).
Bottom, right: Odour sampling from compost facility used to grow mushrooms. Photograph courtesy of Phil Hobbs (Institute of Grassland & Environmental Research, Devon, United Kingdom).

ISBN 1 84339 054 X

Contents

Preface

Environmental degradation continues to threaten and challenge societies across the world, independent of its level of development. Anthropogenic production and consumption has lead to changes in the environment. Examples are changes in concentrations of potentially toxic substances, in biogeochemical cycles, biodiversity and climate. Even subtle changes can have unforeseen impacts that disturb life-supporting systems. As a result, our society faces many undesirable and possibly irreversible consequences that result directly from socio-economic and technological developments. An important challenge is to address environmental degradation in such a way that alternative strategies allow the development of sustainable ecosystems and societies in an effective as well as efficient manner.

Although a general level of environmental awareness is now wide spread in much of the world, there is still an urgent need for the development of an adequate understanding of how to address and finally deal with many environmental issues. This need has arguably become more complex because early examples of success in environmental progress must now be followed by a period in which the more 'nasty' and seemingly 'invisible' environmental concerns must be addressed.

Solid waste management is an example an environmental issue that has become an integral part of every day life even though the public at large does not always fully realize the need for adequate environmental protection measures. Uncontrolled spread of waste materials often leads to health and environmental issues. To prevent these and other problems, infrastructure has been developed and implemented for waste management that oversees collection, treatment and disposal of wastes. Current management schemes are generally based on a hierarchy of three principles which include waste prevention or reduction, recycling/reuse and ultimate disposal.

Final or ultimate disposal is the least desirable option because it causes major emissions to the atmosphere, water bodies and the subsoil. For example, emission of methane to the atmosphere is an important source of greenhouse gases. Organic wastes therefore receive a lot of attention in waste management. In Europe it resulted in the issue of the Landfill Directive (99/31/EC) and of the Sewage Sludge Directive (86/278/EEC).

Utilization of proper options for treatment of organic wastes can turn this burden into an asset. Biological waste treatment in particular can lead to more effective resource management as well as the development of more sustainable practices. The following advantages may be listed:

- the greenhouse effect is tackled as methane emission from land filling is prevented,
- soil quality can be restored or enhanced by the use of composted wastes in agriculture,
- compost may replace peat in horticulture and home gardening and thus reduce greenhouse gas emissions and wetland exploitation,
- anaerobic digestion has the additional benefit of producing biogas that may be used as a source of fuel, and finally,
- pesticide use is reduced through proper utilization of disease suppressive properties of composted wastes.

Direct linkage of the field of waste management to the societal production system at large requires interactions with and involvement of many different disciplines, not only the natural sciences but also economics and the political and management disciplines in addition to product specialists, etc. Although most individuals involved in this overall process agree with these principles, fragmentation still is common in practice. For example, there is still insufficient exchange between agronomist working with compost utilization and technologist that address the composting process.

The purpose of this book volume is to 1) bridge these very different scientific fields and 2) stimulate the interest of stakeholders involved in the recovery and reuse of products generated from solid wastes. In order to design proper waste management strategies successfully, integrative aspects need to be addressed. For instance, the amount of solid waste produced highly depends on how production chains are set-up and how intake logistics of wasted products are organized. Reuse/recycling changes within production chains need to be identified, organized and implemented. Systems analysis is an indispensable tool in this process. The technical chapters in the volume present cutting-edge overviews of available treatment strategies for solid wastes as well as the utilization potential of the products. Specific reactor and design aspects of available physico-chemical and biotechnological treatment options for aerobic and anaerobic solid waste treatment are discussed. Theoretical aspects are illustrated with case studies. Finally, practical applications and constraints are presented for relevant solid wastes.

We wish to thank all the contributors for their enthusiastic support and timely submission of their manuscripts. This book is based on the Euro Summer School,

"Biotechnology in organic solid waste management: from solid waste disposal to resource recovery", held June 29th to July 4th 2003 in Wageningen, The Netherlands. This Euro Summer School was financially supported by the "Improving the Human Potential" Program of the EU (grant number HPCF-2002-00405). In addition to many of the oral presentations from the Euro Summer School, a number of invited contributions are included in this volume. We are grateful to Alan Click and Alan Peterson of IWA Publishing for their help and editorial support in realizing this book.

Piet Lens and Bert Hamelers
Wageningen, The Netherlands

Harry Hoitink
Ohio, USA

Werner Bidlingmaier
Weimar, Germany

July 2004

Contributors

Thorsten Ahrens
Bundesforschungsanstalt für Landwirtschaft (FAL), Institut für Technologie und Bioystemtechnik, Abteilung Technologie, Bundesallee 50, 38116 Braunschweig, Germany

Theodorita Al Seadi
University of Southern Denmark, Bioenergy Department, Niels Bohrs Vej 9, 6700 Esbjerg, Denmark

Kostas Angelopoulos
Department of Biology, University of Patras, Patras 26500, Greece

Heleen Bartelings
Aart, de Jong, Wilms & Goudriaan Public Economics bv, Lange Voorhout 94, 2514 EJ The Hague, The Netherlands

Josef Barth
European Compost Network ECN/ORBIT e.V., Am Landhagen 64a, 59302 Oelde, Germany

Werner Bidlingmaier
Department of Civil Engineering
Waste Management Group
Coudraystrasse 7
99423 Weimar, Germany

Reinhard Böhm
Institut für Umwelt und Tierhygiene sowie Tiermedizin mit Tierklinik 460b
Universität Hohenheim
D-70593 Stuttgart, Germany

Bernd Franke
ifeu-Institut Heidelberg GmbH
Wilckensstrasse 3
69120 Heidelberg, Germany

Rudi Braun
Environmental Biotechnology Division,
Department for Agrobiotechnology, IFA-Tulln, BOKU – University of Natural Resources and Applied Life Sciences,
Konrad Lorenzstrasse 20,
A-3430 Tulln, Austria

Jozef Coosemans
Laboratory of Phytopathology and Plant Protection, Katholieke Universiteit Leuven, W. de Croylaan 42, 3001 Leuven, Belgium

Deborah De Clercq
Laboratory of Phytopathology and Plant Protection, Katholieke Universiteit Leuven, W. de Croylaan 42, 3001 Leuven, Belgium

Rob Dellink
Environmental Economics and Natural Resources Group
The University of Wageningen
6700 EV Wageningen, The Netherlands

Harikleia N. Gavala
Department of Chemical Engineering, University of Patras, Patras 26500, Greece

Amaury de Guardia
Livestock and Municipal Waste Management Research Unit, Cemagref, 17 avenue de Cucillé, CS 64427, 35044 Rennes Cedex, France

Bert Hamelers
Sub-department Environmental Technology, University Wageningen, P.O. Box 8129, 6700 EV Wageningen, The Netherlands

Phil Hobbs
Institute of Grassland And Environmental Research, North Wyke, Okehampton, EX20 2SB Devon, UK

Dominic Hogg
Eunomia Research & Consulting Ltd., 120 Cotham Brow, Bristol BS6 6AR, UK

Harry A.J. Hoitink
Department of Plant Pathology and Env. Graduate Studies Program, The Ohio State University, 211 Selby Hall, 1680 Madison Avenue, Wooster, Ohio 44691-4096, USA

J.B. Holm-Nielsen
University of Southern Denmark
Bioenergy Department
Niels Bohrs Vej 9
6700 Esbjerg, Denmark

Ekko C. van Ierland
Environmental Economics Group
The University of Wageningen
P.O. Box 8130
6700 EV Wageningen, The Netherlands

Roland Kirchmayr
Department for Environmental Biotechnology
IFA-Tulln
BOKU – University of Natural Resources and Applied Life Sciences
Konrad Lorenz Straße 20
A-3430 Tulln, Austria

Florian Knappe
ifeu-Institut Heidelberg GmbH, Wilckensstrasse 3, 69120 Heidelberg, Germany

Heike Knicker
Lehrstuhl fur Bodenkunde, Technische Universität München, 85350 Freising Weihenstephan, Germany

Michael Kornaros
Department of Chemical Engineering, University of Patras, Patras 26500, Greece

Gustaf Landahl
Miljöförvaltningen, Environment and Health Protection Administration, Box 380 24, 100 64 Stockholm, Sweden

Paulo C. Lemos
Chemistry Department, CQFB/REQUIMTE, FCT/UNL, 2829-516 Caparica, Portugal

Gerasimos Lyberatos
Department of Chemical Engineering, University of Patras, Patras 26500, Greece

Sandra Macé
Dept. Enginyeria Química, Universitat de Barcelona, Martí i Franquès, 1, plata. 6. E-08028 Barcelona, Spain

Megan N. Marshall
Department of Biological and Agricultural Engineering, University of California, One Shields Ave., Davis, CA 95616, USA

José Martinez
Livestock and Municipal Waste Management Research Unit, Cemagref, 17 avenue de Cucillé, CS 64427, 35044 Rennes Cedex, France

Joan Mata-Alvarez
Dpt. Enginyeria Química. Universitat de Barcelona, Martí i Franquès, 1, plata. 6. E-08028 Barcelona, Spain

Frederick C. Michel Jr
Department of Food, Agricultural and Biological Engineering, OSU-OARDC, 1680 Madison Avenue, Wooster, OH 44691, USA

Henri C. Moll
Center for Energy and Environmental Studies IVEM, University of Groningen, Nijenborgh 4, 9747 AG Groningen, The Netherlands

Charlotte Plombin
Miljöförvaltningen, Environment and Health Protection Administration, Box 380 24, 100 64 Stockholm, Sweden

Ana M. Ramos
Chemistry Department, CQFB/REQUIMTE, FCT/UNL, 2829-516 Caparica, Portugal

Amitha P. Reddy
Department of Biological and Agricultural Engineering, University of California, One Shields Ave., Davis, CA 95616, USA

Maria A.M. Reis
Chemistry Department, CQFB/REQUIMTE, FCT/UNL, 2829-516 Caparica, Portugal

Tom L. Richard
Department of Agricultural and Biosystems Engineering, 3222 NSRIC, Iowa State University, Ames, IA 50011, USA

Wim H. Rulkens
Sub-department Environmental Technology, University Wageningen, P.O. Box 8129, 6700 EV Wageningen, The Netherlands

Jaak Ryckeboer
Laboratory of Phytopathology and Plant Protection, Katholieke Universiteit Leuven, W. de Croylaan 42, 3001 Leuven, Belgium

Luisa S. Serafim
Chemistry Department,
CQFB/REQUIMTE, FCT/UNL,
2829-516 Caparica, Portugal

Ioannis V. Skiadas
Department of Chemical Engineering,
University of Patras, Patras 26500,
Greece

Katerina Stamatelatou
Department of Chemical
Engineering, University of Patras,
Patras 26500, Greece

Jan-Olov Sundqvist
IVL Swedish Environmental
Research Institute, Box 210 60,
100 31 Stockholm, Sweden

Anne Tremier
SUEZ Environnement, Organic
Effluents Division, CIRSEE,
38 avenue du Président Wilson
78230 Le Pecq, France

Jean S. VanderGheynst
Department of Biological and
Agricultural Engineering,
University of California,
One Shields Ave., Davis,
CA 95616, USA

Lies Vandesteene
Laboratory of Phytopathology
and Plant Protection, Katholieke
Universiteit Leuven, W. de
Croylaan 42, 3001 Leuven,
Belgium

Kris (C.S.A.) Van Koppen
Environmental Policy, Wageningen
University, Hollandseweg 1,
6706 KN Wageningen,
The Netherlands

Adrie Veeken
Sub-department Environmental
Technology, University Wageningen,
P.O. Box 8129, 6700 EV Wageningen,
The Netherlands

Regine Vogt
ifeu-Institut Heidelberg GmbH
Wilckensstrasse 3
69120 Heidelberg, Germany

Peter Weiland
Bundesforschungsanstalt für
Landwirtschaft (FAL), Institut für
Technologie und Bioystemtechnik,
Abteilung Technologie, Bundesallee
50, 38116 Braunschweig,
Germany

PART ONE: Integral aspects of solid-waste management

1

Material metabolism through society and environment: a new view on waste management

Henri C. Moll

1.1 INTRODUCTION

An environmental science university course book (McKinney and Schoch 2003) describes three very different approaches for waste management: the "*disperse and dilute*" strategy, the "*concentrate and contain*" strategy, and the "*resource recovery*" strategy. In the course book description of these strategies, the "disperse and dilute" strategy is characterised to be historical (till the early industrial age). The "concentrate and contain" strategy was introduced in the beginning of the 20th century and has been dominant during this whole century. Presently the "resource recovery" strategy is getting increasingly attention and is expected to be the dominant approach for sustainable waste management in the next few decades.

The concept eco-efficiency draws highly the attention in modern business circles and is seen as an important principle to elaborate sustainable business solutions

 Resource Recovery and Reuse in Organic Solid Waste Management. Edited by Piet Lens, Bert Hamelers, Harry Hoitink, Werner Bidlingmaier. ISBN: 1 84339 054 X. Published by IWA Publishing, London, UK.

for the society and the environment. Eco-efficiency means (see DeSimone and Popoff 1997): less energy intensity, less material intensity and lower levels of environmental toxicity and risk. Eco-efficiency implies to go further than the pollution control approach (the so-called end-of-pipe approach) and to work on process integration, and further on higher scale levels on integral facility planning, industrial ecology on industrial sites, and on sustainable regions.

The quotations mentioned above demonstrate that waste management strategies have become interrelated with the material strategies in the production sectors. The outputs (i.e. waste) of the economic system should be related to the inputs (i.e. resources) of the economic system (see Ayres and Ayres 1998, 1999). Considering resources, materials, products and waste as parts of an integrated system enables to research the potential of the waste management strategy "resource recovery" as well as the material efficiency approach advocated presently in industry. In such a system, waste management and efficient material use may become the two sides of one valuable coin.

The quotations demonstrate also the importance of considering the history of waste management and material use. Analysis of past practice may give insight in the long-term effects of some waste handling practices and some past behaviour may give inspiring examples for the future. Also important lessons are derived from analysis of differences in the economic system and in the environmental conditions between present modern society and societies and cultures in the past.

In this chapter waste management is discussed from an environmental system analysis perspective. In this perspective substances are part of a cycle. Some cycles are mainly driven by natural processes, and some other cycles are driven mostly (or fully) by human activities. Each cycle has phases within the economic system as well as within environmental systems. Material use and waste management are both related to one or more phases of one or more substance cycles. New ways to manage waste and new approaches to recover resources from waste imply minor or possibly major changes in substance cycles and should be evaluated on their effects in the whole substance cycle involved. The overall cycle of a substance is named in this chapter the *material metabolism* of a substance.

The material metabolism patterns of the various substances differ mutually. Partly these differences are related to inherent properties of the substance itself, but also some differences are determined by the societal use of a substance. Effective waste management and resource recovery strategies should deal with the differences of the inherent substance properties and should also address or even change the usage characteristics of substances.

This chapter begins in Section 1.2 with a discussion on the history of waste and waste management, starting some millennia ago. In Section 1.3 the present-day status is discussed of waste treatment technologies and waste management approaches. In these sections all kinds of materials are considered with some specific focus on organic solid waste and organic wastewater. In Section 1.4 waste management is studied from the environmental system analysis perspective. Some substance cycles are discussed as well as the anthropogenic influence on these cycles. The elaboration of the cycle concept results in the identification of some environmentally sound waste management principles. These principles may form also the base of sustainable material and waste management, the issue addressed in Section 1.5. Especially increasing the use of resources recovered from consumer waste puts a high challenge to the present production and consumption system. In Section 1.6 some cases of material quality management, involving also changes in the usage characteristics, are reviewed. The road to sustainable waste management is examined in Section 1.7. In Section 1.8 the final conclusions are presented.

1.2 A BRIEF HISTORY OF WASTE MANAGEMENT

The generation of waste by humans is a very natural thing. As each living being man requires food to stay alive and produces urine and excrements to remove the digested remnants of food consumption. The generation of waste is also related to the structure and functioning of the society. Waste flows originate from the production, use and disposal of goods that are required to be an accepted member of a society. Although human waste generation has been omnipresent in history, waste management has not been always considered to be necessary. Two conditions should be met for the development of waste management approaches in a society: first, waste itself should be considered to be a problem, and second means should be available to manage this problem.

1.2.1 Waste in rural societies

In rural societies using mostly resources of biological origin harvested from the surrounding environment, the waste problem did not exist. Rest flows of agriculture, hunting and food preparation were used for other purposes like feeding domestic animals, clothing, construction, and fuelling fire. The excrements of humans and animals were used in many cases for fertilising agricultural land and pastures. Even when such flows became definite waste and were dumped in the close environment, these flows were decomposed quickly by natural processes. Materials like stone, sand and clay extracted from the surrounding environment used for building and the manufacture of goods like pottery, were reused often before they became waste disposed in the environment. After disposal aboveground or underground such waste generally did not cause environmental problems, because of its chemical and biological innocuous nature.

The substitution of resources derived directly from nature by mineral resources and fossil energy sources extracted from the (deep) underground has been a new cause of waste problems. The extraction and the processing of materials out of mineral resources generate often solid, fluid and gaseous substance flows, which may injure the local environment. Also when these materials become waste after their use in products, they may cause environmental damage. These substances and materials are not a part of the natural cycles of biological production and decomposition and may therefore interfere in a detrimental way with the functioning of natural systems.

1.2.2 Waste management in cities

The development of cities has become a source of waste problems (see Ponting 1992 for a discussion about the development of cities and the related environmental problems). Within a city a large number of people live closely together. Many of the resources and goods required to live within a city are not found and produced in the close surrounding of the city. So provision of materials and goods requires trade and transport. The generation of waste – human excrements, food waste and disposed goods – in the city causes problems. Neglecting the organisation of the removal and handling of city waste results in deterioration of the living conditions in the city: stench, polluted water and related illnesses and health problems.

Processing of natural and mineral resources and the production of goods has often taken place within the cities or in the close neighbourhood of cities. In pre-industrial times mostly on small scale, and since the start of industrialisation the average scale of process industry and manufacture enlarged substantially. Processing resources and the production of goods generate also waste flows that are not treated on the location of the industrial activity itself. These waste flows are transported and sometimes reused elsewhere but often also disposed into the water and soil resulting in environmental problems.

The large amounts of city waste generated by households as well as by industrial plants have created a strong pressure to develop management systems for city waste. City waste management systems are based on the collection of some waste flows, the transport of waste outside the city and some handling afterwards such as dumping or combusting the collected waste. Such systems were developed in most cases by the city authorities. Next to these removal-oriented systems also commercial systems developed based on the collection of specific waste flows with economic value because of their potential for reuse and material recovery.

1.3 PRESENT-DAY TECHNOLOGIES FOR WASTE MANAGEMENT

The increasing level of concentration of industrial production and consumption in the 20th century has increased the necessity of waste management. The technological development of the last century has enabled the possibility of efficient and effective waste management. The main waste handling technologies presently used are still based on practices applied already in the past: waste separation, landfilling, combustion or incineration, biological treatment, and recovery for use in agriculture and in industry. The present-day status of these technologies is discussed (see Stessel 1996 for an overview of waste processing technologies).

1.3.1 Some general waste handling practices

An element of all waste handling systems for specific kinds of materials is *waste separation* in order to obtain "pure" material flows in the waste handling system.

Waste separation improves the efficiency of the following processes and increases the quality of the resulting material. Separation of waste in pure material flows can be performed at the source of waste i.e. the household or the industrial plant. Instruction is required to realise successful separation at the source and the people involved should develop deliberate and consistent separation behaviour. Households are commonly able to produce pure material fraction for glass, paper and also organic waste. Also technology is available to separate waste in different material flows after collection. The material flows produced by separation after collection are generally more physically and chemically contaminated than the flows produced by separation at the source. But that does not pose problems for materials like metals that are easily cleaned afterwards.

Landfilling waste aboveground or underground at disposal sites has remained a very common practice to handle waste. But, the area occupied and the amount of waste stored on dumping sites is presently enormously larger than some centuries ago. Due to the possible contents of reactive and toxic substances in dumped waste, control technologies are being applied on modern waste deposit sites. The flow of rainwater and ground water through the deposed waste is limited by covering above and by application around with impermeable foils. Also waste deposit water treatment is applied regularly. The methane emitted by waste sites, which contain a large amount of organic digestible materials, is captured and mostly used as fuel to save energy. Sometimes restrictions are set on the types of waste that are disposed. For example, in many European countries it is not allowed anymore to dump on waste deposit site materials that are combustible or digestible.

At the beginning of the 21st century *waste incineration* plants demonstrate a large scale of operation (several hundred thousands tons of waste per year) and a high level of technological sophistication is available to recover useful energy and to remove substances that may cause air pollution, like dioxin, small size particulates (fly ash), sulphur oxide and nitrous oxides.

1.3.2 Treatment of organic waste: decomposition and incineration with potential of resource recovery

A variety of technologies are applied to treat the fluid waste flows of biological nature and to extract in some cases useful substances or materials from these flows. Some fluid organic waste flows like human excrements put in the sewage system are brought to sewage water treatment plants. In most cases the sewage water is *treated aerobically* with removal of nitrate and sometimes also of phosphate. Sewage water can also be *treated anaerobically* using bacterial methanogenesis, resulting in the production of biogas, usable for energy purposes. Sewage water cleaning results in the production of large amounts of sewage sludge potentially useful for agriculture. But in practice because of pollution by toxic substances from the sewage water, the sewage sludge is often too polluted for agricultural applications. Therefore sewage sludge is considered mostly as waste that should be treated further. Finally sewage sludge is disposed in specific depots or incinerated after a drying step to remove the largest part of water from the sludge. Also in industry many fluid organic process flows are generated that were in the past considered as waste and were emitted to surface water causing environmental problems like eutrophication and stench. Presently the content of these flows are mostly recovered and are converted to useful products. Sometimes aerobic and anaerobic wastewater treatment processes are also used in industry to remove the organic compounds from the wastewater flows.

Solid food waste flows occurring in food industry and in households are incinerated in some cases together with other waste flows despite its adverse effects on energy efficiency because of the substantial water content of solid food waste. Often food waste flows are treated together with private and public garden waste in specific processes. In large-scale plants these waste flows are composted in a few weeks under controlled aerobic circumstances. Depending on the (chemical) purity of the organic waste the quality of the compost is high enough for further use in agriculture and horticulture. Because of the large amount of compost produced presently in the Netherlands, problems exist to find sufficient markets for these amounts. Another way to treat these waste flows is anaerobic digestion. Digestion results in the production of methane containing biogas and sludge, comparable with the sludge produced in anaerobic wastewater treatment.

Solid organic waste flows like paper, wood and polymers produced from fossil energy sources are mostly treated in waste incineration plants or are directly fed in electricity production plants. Presently some other methods to handle these waste flows are also relevant. *Gasification* (at high temperature) and *pyrolysis* (at moderate temperature) of solid organic waste produce gaseous hydrocarbons that – after clean up of these gases – can be used for chemical synthesis purposes and very efficient electricity production in gas turbines. An important other application of these solid organic flows is *reuse and recycling*. Presently half (or more) of discarded paper is used as input in paper and cardboard production as substitute for virgin wood based pulp. Discarded wood is partly used for the production of chipboard and fibreboard. Discarded polymers are sometimes reused in the polymer production and the production of polymer products.

1.3.3 Treatment of mineral waste: reuse, recycling and landfilling

After the use phase, materials produced from mineral resources like metals, concrete, ceramics, bricks and stone are often disposed in waste dumping sites. Incineration of these materials does not result in a substantial reduction of the waste volume and requires energy instead of generating useful energy. Instead of becoming waste these materials are also often reused as secondary materials. Metals like steel, aluminium, copper and zinc are recycled after melting and purification in comparable applications as these metals in primary (i.e. gained from mineral resources) form. Materials like concrete, bricks, asphalt and stone regained from demolition of buildings and roads may also be reused to substitute new construction materials.

1.3.4 Waste prevention: a new challenge for waste management

Next to the instalment of waste treatment technologies, *waste prevention* has also become a part of the governmental waste management policy. Preventive approaches differ fundamentally from waste treatment. At consumer level waste prevention aims at responsible shopping by avoiding a lot of packaging material and by not buying goods with a short useful life, and at responsible use of products by good maintenance and repair, and by giving a product a second life instead of dumping it. At producer level waste prevention aims at the avoidance of substance losses in processes, recovery of material flows from process outputs and redesign of processes to increase the efficiency of use of the process inputs and to reduce uneconomic process outputs.

The waste management approaches discussed above are put together in a *hierarchy* in present waste management policy. *Prevention* is assigned the highest rank. At the second rank one finds *recycling, reuse and substance recovery*. *Waste incineration* gets the third rank, and *landfilling* has the lowest rank in this waste handling hierarchy. Present-day waste policy aims to bring waste handling on a higher level in accordance with this hierarchy.

1.4 ECONOMIC AND ENVIRONMENTAL EVALUATIONS OF WASTE MANAGEMENT

Waste is a form of matter, just as all materials used in the economic production system and the goods used by the consumers. So, what is the difference between waste materials and goods?

1.4.1 An economic viewpoint on waste management

From an economic perspective this difference is related to *economic valuation*. When a proprietor does not attribute anymore value to its possession and is not able to sell it to another person, that good or material has become waste according to economic thinking. Thus, waste is a matter with *no* value or even with a *negative* value. Therefore the owner wants to get rid of his waste as easily as possible. Disposing of waste entails costs or equivalently work, but the owner shall try to minimise his waste disposal costs.

Disposal of waste entails also *external costs*, i.e. costs not paid by the waste owner but inferred on other people or on the society. The space occupied by solid waste cannot be used for other economic relevant purposes. The organic waste emitted in surface water reduces the quality of the water and may cause the dying of fish. Atmospheric emission of exhaust gases may give rise to adverse health effects. The instalment of waste management systems aims at the mitigation of the external costs of waste disposal and at the compensation of inevitable external effects. So waste management may require a substantial amount of money or work. In present waste management policy the "*polluter pays principle*" is considered to be important. However, often the citizens or the taxpayers pay the costs of waste management instead of the polluter.

During the 20th century, the costs of waste management have increased substantially in western societies. Three general determinants of the increasing costs are identified: the increased amount of waste generated per person, the increased density of population, and the higher environmental standards. The technological development is a double-edged sword: technology enables the solution of environmental problems related to waste but it increases also the waste problem because of the increase of consumption levels accompanied with the technological development.

1.4.2 Environmental evaluation of waste

From an environmental perspective the difference between materials, goods and waste is illustrated by the quote of Mary Douglas (1966): "*Waste is matter out of place*." This quote of the anthropologist Douglas covers both the social implications and the ecological implications of waste. Waste disturbs social life and therefore one should remove waste to other places to render it harmless. In ecology, matter *per se* can never be problematic, but matter at certain places may disturb the proper functioning of ecological systems. To evaluate waste management from an environmental perspective, first the proper functioning of the environment should be considered and the potential disturbances of the environment by anthropogenic flows of matter. Therefore, the cycles relevant for organic matter produced by nature and the substances extracted by man from the deep underground are discussed below.

1.4.2.1 The natural cycles of organic matter

In the "natural" global environment some massive flows of matter are described with the *natural cycle* concept (see for a concise description of the most important natural cycles Smil 1997). Living matter consists mainly out of water, bound carbon and bound nitrogen (carbohydrates, fat and protein). So, very important natural cycles relevant for the living environment are carbon, water and nitrogen cycles. Both the carbon and water cycles are mainly driven by the solar input of radiation.

The main stages of the natural *carbon cycle* are the uptake of carbon dioxide by plants from the atmosphere together with the conversion of solar energy to chemical energy (production of organic matter) and the atmospheric release of carbon dioxide together with the digestion of organic matter. Between the formation of bound carbon by photosynthesis and the decomposition of bound carbon, it follows various pathways inside organisms and sometimes also between organisms.

The main stages of the natural *water cycle* are the evaporation by sunlight of water mainly from the sea, the rainfall of water on the land areas and the transport of water by gravity back to the sea. Between the rainfall and the return in the sea natural systems and humans extract a part of the water flow temporarily and discharge this elsewhere back in the water cycle.

The natural *nitrogen cycle* is very complex, therefore a rather simple description is given here. Several bacterial processes converting the nitrogen from one chemical form to another are the main drivers of this cycle. Bacteria convert atmospheric nitrogen to bound nitrogen compounds (ammonia, nitrate and nitrite). Plants extract this *bound nitrogen* from the soil and from water to produce proteins. The produced protein may cycle a period within an organism and between organisms but is afterwards released back to the soil and water in the form of human and animal excrements and of dead protein-containing matter. Bacteria digest this bound nitrogen-containing matter to soluble bound nitrogen compounds, which may be reused again by plants. Other bacteria convert a part of the soluble bound nitrogen compounds to gaseous forms of nitrogen (N_2 and N_2O).

The spatial scales of these cycles differ in size. The carbon cycle, dominated by the carbon dioxide in the global atmosphere and by dissolved carbonate in the oceans, functions mainly at *global scale*. The nitrogen cycle functions partly at global scale because of the global cycling of the (almost) stable atmospheric forms of nitrogen. The water cycle functions at the *fluvial scale* or the catchment's basin of continental rivers. The biological part of the nitrogen cycle of compounds bound to organic matter and solved in the (ground) water functions mainly at *regional and local scale*.

1.4.2.2 Anthropogenic impacts on the natural cycles

Human activities generate often directly and indirectly *disturbances of the natural cycles*. The combustion of energy carriers derived from fossil energy sources leads to an increased emission of carbon dioxide to the atmosphere resulting in an increased atmospheric carbon dioxide concentration. Human use of water disturbs the water flow patterns and the water availability and pollutes surface and ground water and the soil. For instance irrigation results into reduced water availability elsewhere and in salination of the soil on the long run. The nitrogen cycle is heavily influenced by energy use, agriculture and food trade. Combustion of energy carriers generates also emission of nitrous oxides that enter after atmospheric conversion processes the land and water compartments in the form of soluble bound nitrogen compounds. To improve the agricultural yield, large amounts of industrially produced nitrates are supplied to the soil. The surplus part of the fertiliser accumulates in the soil and may leak away to the ground water. The nitrate contained in human excrements is removed from the nitrogen cycle through sewage wastewater treatment. Export of food and fodder from other continents to western Europe results in losses from the nitrogen cycle in the export countries and in accumulation of nitrogen in the regional environment of western Europe (see van der Voet 1995 for a description of the bounded nitrogen flows in the European Union; see Galloway and Cowling 2002 for an analysis of present bound nitrogen flows at global level). Anthropogenic processes are presently dominant over the natural processes with regard to the bounded nitrogen flows at terrestrial level (see Vitousek and Mooney 1997).

Anthropogenic influence exerted on the natural cycles follows two main patterns. The first pattern results in *substance overload* in environmental compartments and *acceleration of substance cycling* because of the addition of new flows and the magnification of existing flows in the natural cycles (e.g. carbon dioxide emission by combustion of energy derived from fossil sources and the nitrogen fertilisation of agricultural land). The second pattern changes the *substance availability* in environmental systems because of the diversion or shifting away of flows (e.g. water flow diversion for irrigation and the nitrogen transport between continents by global exports of food and fodder). Moreover the *interaction* between the changes in the natural cycles may cause additional effects. The main effects of these disturbances in the natural cycles may be: changes in resource availability for nature and mankind, accumulation or depletion of substances in environmental compartments with undesired effects on economy and ecology, and the creation of new flows disturbing environmental systems.

1.4.2.3 The anthropogenically driven cycles of mineral matter and their environmental impacts

The use of mineral resources extracted from the (deep) underground, results in other material cycles in the global environment. These cycles are mainly anthropogenically driven, although natural processes may be also part of such cycles. Such a cycle starts with the "cradle" of a substance or material, the phase when the substance or material enters the economic system and obtains a positive economic value. After the "cradle" phase, the substance passes one or more processing and production phases and becomes part of an economic good usable for final consumption. After the usage phase of this good it is disposed and becomes waste. In the waste treatment phase the substance may be recovered and used again in the economic system. In many cases the substance is removed from the economic system and finds it "grave" in the environmental system.

These anthropogenically driven cycles differ from the natural cycles with regard to some important aspects. Firstly, these cycles are not closed (neglecting the timescale of very long-term geological processes), because the cradle differs from the grave for almost al substances. Metals produced from mineral ores and polymers produced from oil end their economic life on the waste deposit site and remain there forever. Human-made substances like CFCs and PCBs do not decompose at all (or at very slow rate) when they arrive after the usage phase in the environment. Secondly, the transition of substances from one phase to another phase of its lifecycle entails environmental effects. Especially the mining phase for ore and the processing phases of ores and oil for the production of substances and materials have severe environmental consequences. Also the waste treatment phase results often in considerable environmental effects. Thirdly, the substance itself may be toxic and detrimental to the environment.

1.4.3 Environmental evaluation of waste management strategies

The analysis of the material cycles and of the influence of human material use and material waste is helpful to develop environmentally sound ideas for waste management. Waste management for natural organic waste flows should aim at inclusion of the waste flow in the relevant natural cycles and at minimising the disturbance of the natural cycles. Prevention is also important. Potentially harmful flows that are not produced do not require management at all. Waste management addressing human-made and extracted mineral materials is more difficult than that for natural organic materials, because of the persistence of most humanly produced materials. Handling such waste does not solve the waste problem but shifts the waste problem to another place or another environmental compartment. However, the waste management related to humanly produced materials can follow about the same strategies as for the management of waste problems related to natural materials.

Starting from the idea "waste is matter out of place" the general principles to manage waste can be developed. Waste should be put back on its proper place (i.e. in accordance to its nature and to its position in the cycle). Following this line of thought a hierarchy of proper solutions is derived.

(1) Materials (or the goods containing these materials) should be maintained in their function as long as possible for instance by good maintenance, repair and second hand life of products.
(2) Materials losing their function should be given another function in the economic system (i.e. recycling). The use of waste flows substitutes new anthropogenic inputs in the economic system; the recycling of metals, wood, paper and organic fibres reduces the requirement of new materials for production and the use of biogas from digested organic waste reduces the use of fossil energy sources.
(3) Materials should be brought back in the natural cycle with minimal disturbance of these cycles and other environmental functions. Unpolluted and treated waste water should be disposed back on the spot the water was extracted and nitrogen compounds originating from food waste and excrements should brought back on agricultural land.
(4) Materials should be given an environmentally safe destination (a closed geological deposit or chemical decomposition in harmless substances). Anthropogenically produced carbon dioxide may be stored in new forests or underground in closed geological formations. Controlled waste deposit sites may also function as safe sinks. The nitrogen compounds in sewage and industrial wastewater may be chemically reduced in wastewater treatment plants.

At the last two levels in this hierarchy waste management infers substantial costs on the polluter and or on the society. Waste management at the first two levels is often also attractive from an economic point of view. At these two levels it is avoided that materials become waste. Accordingly a favourite slogan in environmental policy during the 1970s and 1980s was "pollution prevention pays". In this slogan pollution and waste can be used interchangeably. Waste prevention refers to a more fundamental approach aiming at reduction of material use and material losses in the whole material and product lifecycle. The design of processes in industry and of products should be directed to minimise the use of resources and the generation of uneconomic outputs (see for a discussion on waste prevention van Weenen 1991). Accordingly the slogan formulated in the 1990s is "pollution and waste are signs of inefficient resource use in the society". Extrapolating the challenge for the 21st century could be chosen as the transformation of waste into resources for the future.

1.5 SUSTAINABLE APPROACHES FOR MATERIAL AND WASTE MANAGEMENT

Aiming at sustainable development as proposed by Brundtland (WCED 1987) implies the application of an integral vision on society and the environment. Sustainable solutions should be economic viable, socially acceptable and ecological compatible. Looking for sustainable waste management strategies material policy and waste policy should be integrated. Put otherwise, waste management should be considered to be a part of resource management. Some of the arguments for this integrated perspective are:

- The growth of use of materials in the society generates on average also larger waste flows.
- The ways that materials are used to produce goods also determine the nature of waste flows and the possibilities to treat these flows in an ecological compatible way.
- Waste flows are potential sources for the production of materials but the present production practices determine whether this potential will be utilised.

1.5.1 Principles for sustainable management of material resources

Daly (1990) presented three criteria for sustainable resource management, which are useful for this integrated perspective on material use and waste:

(1) The flow rates through society of renewable resources should not exceed the natural renewal or regeneration capacity.
(2) The flow rates through society of non-renewable resources should be limited and should not exceed levels that can also be delivered in a sustainable way by renewable resources.
(3) The outflow rates of substances to the environment should not exceed the absorption and decomposition rates of the environment.

Renewable resources are produced by natural cycles and are therefore renewed and regenerated. *Non-renewable* resources are extracted or derived from mineral

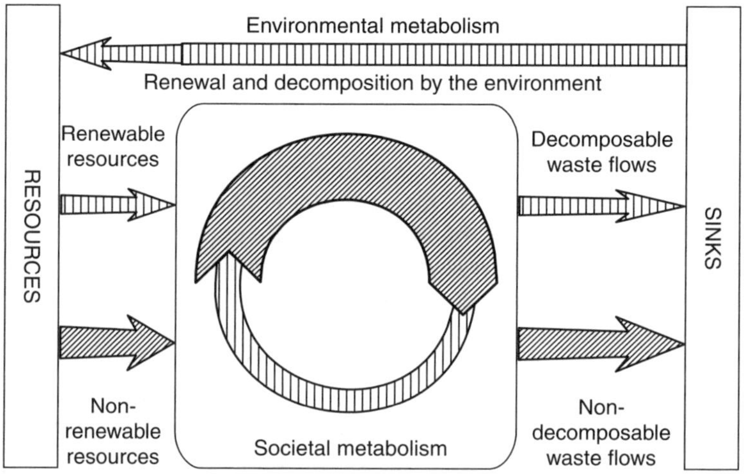

Figure 1.1 Environmental and societal metabolism of materials: resource flows from the environment entering the economic system, materials use and recycling within the economic system, and waste flows out of the economic system into the environment (*Source* Moll 1993).

resources that can be depleted. Waste materials that are absorbed and decomposed in the environment are included again in natural cycles.

This integrated perspective is summarised in the concept of the environmental and societal metabolism of materials (see Moll 1993). In Figure 1.1 the *material metabolism* is shown schematically. This figure demonstrates the natural cycles of yielding renewable resources and decomposition of the waste of these resources, the anthropogenic cycles (from cradle to grave) of non-renewable resources, and the recycling of materials within the society. Recycling is put in the centre of this scheme; increasing the recycling of materials serves two ends: the diminishment of the extraction of resources as well as the reduction of the amounts of waste.

Next to *preventive approaches* based on the renewal of the production and consumption system, the improvement of material recycling should form the base of sustainable approaches for material and waste management. Other approaches like material substitution and new waste technology may be helpful, but especially these approaches pose the risk of environmental problem shifting (for further discussions see Tromp 1995).

By the improvement of recycling in the economic system the natural cycles are partly imitated for anthropogenic produced material flows. The industrial analogy of natural cycles is named "*industrial ecology*". Economic waste becomes a resource for new production, according to this analogy, just as in nature biological waste is transformed in resources for the build up of new living matter (see for an elaboration of Industrial Ecology Chapter 2 by van Koppen; Graedel and Allenby (1995) offer an interesting overview on this subject).

1.5.2 Approaches to improve the conditions for recycling

In principle two approaches may be followed to improve material recycling: a *product-based approach* and a *substance-based approach*. The implications of both approaches will be discussed.

The *product-based strategy* aims at minimising the waste flows and maximising material recycling within the lifecycle of a product. The main focus of this strategy is the design phase of the product. Some elaborations of such an approach are *design to repair* (to maximise products' life), *design to dismantle* (to enable use of parts and materials of disposed products) and *design to recycle* (to construct products with materials which are easily recycled). The success of such approaches is primarily dependent on the efforts of the manufacturer of the product but also other actors in the product lifecycle (sellers, users, repairers, dismantlers, recyclers and so on) should behave themselves in accordance with the designed optimised product lifecycle. The producer (or in some cases the government) should take the responsibility to manage the product lifecycle to realise its functioning at an optimal level. A problem related to this strategy is the difficulty to close completely the material cycles within the product systems. Dismantled materials from products cannot fully be used again to manufacture the same products. So some material flows shall leave a product lifecycle for reuse elsewhere in the economic system. Then, the number of actors involved in the sustainable lifecycle of one product may become too big to be manageable.

Substance-based strategies depend not only on the nature of the substance concerned but also on the structure of production and consumption. The high requirements with regard to properties of materials demanded by producers and consumers influence substantially the possibilities for material recycling and waste reduction. In the opinion of producers, the quality of materials produced out of consumer waste flows is often too low to meet the product requirements. For many reasons it is difficult to recover materials and substances from consumer waste, which are physically pure and with a constant chemical composition. Presently recycling is often *downcycling*: materials derived from consumer waste flows are used for a small range of applications, with low requirements for purity and chemical composition.

Figure 1.2 shows a quality diagram of the material flows, representing the present way of material handling and recycling. Mineral ores are unfit for direct material use, but by extraction and processing materials are produced with a high quality (*fitness for use*) and put in several goods demanding materials at slightly different quality levels. After the usage phase of these goods due to mixing of materials derived from these goods, only materials with a low fitness for use in production are recovered and applied in the production of some goods with low demands with regard to material properties (downcycling). Also a considerable part of the material is disposed in the environment.

Remarkably the recycling of materials out of many producer waste flows is very fruitful. Metal waste of the metal machine and car production industry is sent back to the material producer and reprocessed to become the same metals or

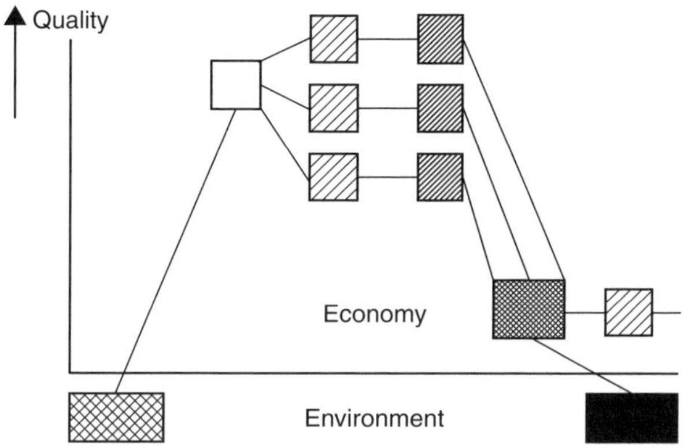

Figure 1.2 Quality diagram of present material flow pattern through the economy: material extraction from the environment, following material use, material recycling and finally disposal in the environment (*Source* Moll 1997).

metal alloys. Also plastic waste of the packaging material producers is reprocessed in the plastic production industry. The availability of the information about the material properties of industrial waste enables the recycling of the material at the same quality level.

1.5.3 Material quality management of substances in production and consumption

Improving the quality (i.e. the fitness for use) of the materials recovered from waste flows is very important. For instance, separate collection of waste flows, easily disassembled products, and materials' identification by using labels or marks will improve the potential for material recycling. However on the production side also changes are necessary. A wide variety of materials with a wide range of material properties are presently produced to realise very *high product functionality*. Instead of using standardised materials for a wide range of applications materials are developed and applied that are tailor-made for specific applications. The structure and composition of such *tailor-made* materials impedes often recovery and reuse of these materials after disposal of the products containing these materials. The high (functional) *fitness for use* of materials seems to entail a low *fitness for recycling* purposes of the same materials. *Material quality management* in production and in consumption is essential to raise the amounts of recycled materials applied in the production of consumer goods.

So it is important for sustainable material and waste management to diminish the difference between the high functional requirements on materials in production and the quality of recycled materials. A moderate lowering of the requirements on materials for production should be possible and will increase the potential to use recycled materials. *Standardisation* of materials for specific applications (car components like the car hood and the car frame, construction elements like window frames and reinforcement bars, and components in household appliances) is very useful for dismantling of high quality waste flows. Ideally all flows of a specific substance are ordered in a *cascade* based system. The cascade concept is also applied for the management of energy flows (cascading energy from high to low temperature applications) and of water flows on a production site (see Reijnders 1996 for a discussion of energy cascades and water cascades). The application of the material cascade concept implies that the primary substance is used for high material quality demanding applications; the waste flow of these products deliver the materials to meet the material demand at a (little) lower quality level, the lowest material quality demanding applications are produced with substance flows that already passed several times in the economic system, and only a tiny amount of this substance dissipates to the environment (Moll 1997). Thus, material flow management considering the material quality of these flows and prevention of large leaps in quality for use and for recycling are a prerequisite for sustainable material and waste management. Material quality management has different implications for different kinds of materials.

1.6 CASES OF IMPROVEMENT OF MATERIAL QUALITY MANAGEMENT

1.6.1 Cascade design for material quality management of polyolefines (polyethylene and polypropylene)

The most important ways to handle polyolefine waste flows are reuse of polyolefine containing products, polyolefine reuse after reprocessing in new products, reuse of polyolefines on lower quality levels substituting the use of other materials, incineration with energy recovery and dumping it on landfills (based on Moll 1977). In Figure 1.3 the distribution of the polyolefine waste flows is given for the Netherlands in 1988. From the viewpoint of material quality management these waste treatment methods can be ranked also from a material quality viewpoint. Product reuse and material reuse result in the conservation of the material quality of the material; material substitution and incineration with energy recovery extract some (but at a lower level) quality from the waste flow and results also in lowering the resource input of the economic system. Material dumping implies the full loss of the material quality of the disposed material. Figure 1.3 demonstrates high losses of material quality.

Designing a quality-based cascade for the use of polyolefines implies a reconsidering of the waste handling and the waste treatment as well as a reconsideration of the use of polyolefines in the economy itself. Therefore, the most important applications of polyolefines are evaluated with respect to improving the recycling potential. Product reuse is feasible e.g. for the frames of appliances and machines and for car parts. Material reuse can be realised for industrial and building waste, for a substantial part of packaging material and for disposed household

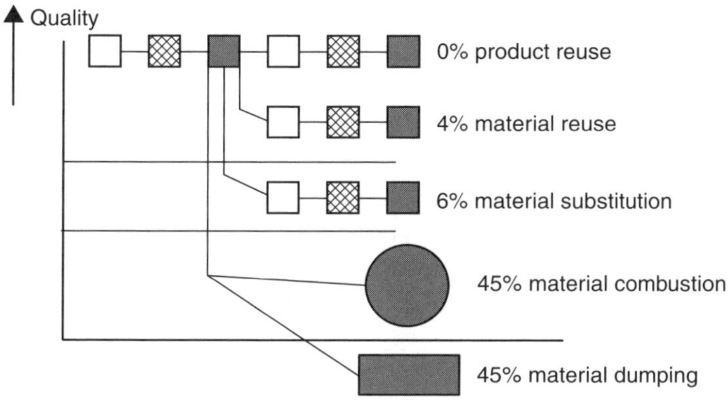

Figure 1.3 Quality diagram of polyolefine flows: primary use and treatment of disposed polyethylene in the Netherlands (1988) (*Source* Moll 1997).

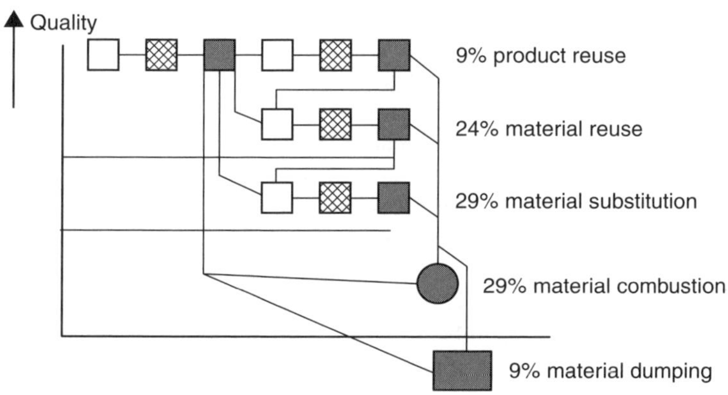

Figure 1.4 Quality diagram of polyolefine flows with an optimised cascade based on their use in the Netherlands (*Source* Moll 1997).

products like furnishing, buckets and baskets. These disposed materials are to be reused mainly in bags, containers and packaging materials. The remnant of the polyolefine waste can be burned separately with energy recovery, conserving energy. To realise the potential for recycling and incineration with energy recovery the strategies of standardisation and separate collection should be followed as far as possible.

Figure 1.4 demonstrates the designed cascade for polyolefines: about 60% reuse and less than 10% dumping. The material and energy saving results of this cascade of polyolefines amount to about 35% of the original material and energy requirements related to the production, use and disposal of polyolefines in the Netherlands.

1.6.2 Cascade design for material quality management for metals and natural and synthetic organic materials

For *metals* the chemical composition determines mainly their material quality (i.e. properties and fitness for use in production). Chemical contamination is the main cause for decline of material quality. Such contamination occurs by melting mixes of metals or mixes of different alloys of the same metal. For most metals also processes are available to remove unwanted contaminants of metals and metal alloys. However, such processes are often not economical and require high amounts of other resources like energy. Prevention of contamination is thus an important quality management strategy for metals.

For many *organic materials* like polymers, wood and paper, the structure (solidity and or plasticity) determines their material quality. The inherent structure is build up by the length of the polymer bonds (for plastics) and by the length of the fibres for wood and paper. Reprocessing these materials after disposal for new applications entails reduction of length of the polymer chains and of the fibres. Therefore, practically only a limited number of recycling turns is feasible because of the required minimal length of chains and fibres. Presently the recycling rate for these materials is further reduced by the disturbance of additives (fillers, colour agents, weakening agents, and UV protection agents) that are applied to improve the functional performance but also are limiting the potential for recycling because of this contamination. Prevention of contamination by additives is thus an important quality management strategy for organic materials. For these materials incineration with recovery of energy is the second best waste prevention strategy. Incineration of these materials implies loss of materials but is an important option to handle deteriorated and contaminated organic material waste. Gasification and pyrolysis of organic waste are alternative treatment methods that recover at least some useful substances from the waste. With these processes feedstock is recovered for chemical synthesis or fuel is produced for efficient electricity generation.

Also with regard to agriculture cascade design and management may be fruitful. Agricultural products are used for various purposes: human food, animal food, fibres, energy generation and many specific substances such as colouring, flavouring and fragrant agents but also substitutes for chemically produced substances. The harvesting and further processing of agricultural products often gives rise to large waste flows. Therefore material quality management principles can be applied to agriculture and the processing of these products and their waste. The hierarchy with regard to value per unit of mass is the following: the highest is the specific substances, second human food, third the fibres, fourth animal food and the lowest energy generation. The ordering describing the amount of the desired product per ton of harvested material is just contrariwise to the ordering according to economic value per kg product. The concentration of specific substances is low in the harvested material so, after the extraction of the specific substances, a large waste flow remains to be treated further. These waste flows should be used on lower levels of this hierarchy, but as high as possible. So it would be desirable to use the waste flows from the extraction processes and from food processing as source for fodder production instead as input for composting or digesting processes. Only agricultural waste flows that are unfit to produce animal feed should be used for these processes to recover some useful materials (compost) or energy carriers like biogas and (m)ethanol. This hierarchy demonstrates also that agriculture to produce energy crops is not a way to generate high value. Only land unfit for other types of agriculture should be used for energy crops production.

1.7 TOWARDS SUSTAINABLE WASTE MANAGEMENT

Concluding this chapter, the three approaches for waste management – discussed in Section 1.1 – are taken again into consideration: the "*disperse and dilute*" strategy, the "*concentrate and contain*" strategy, and the "*resource recovery*" strategy (McKinney and Schoch 2003). The historic material presented in this chapter demonstrates that the "disperse and dilute" strategy was not bad in ancient times, especially for waste flows which were treated by natural processes being a part of natural cycles. Since (early) industrial times this strategy has become ineffective and environmentally problematic because of the large flows of (waste) materials that are not part of natural cycles and the overload by anthropogenically driven flows of natural cycles. The explosive increase of waste management problems in the 19th and 20th century induced the development of a new waste management approach in the "concentrate and contain" strategy. With this strategy the uncontrolled pollution of water, air and the soil associated with severe environmental problems has largely been halted. However this strategy is not unproblematic; it requires high investments and large amounts of land area and it requires (almost) permanent control of the containment sites. Also the resources originally used in the production sectors to produce goods that finally become waste are also lost forever by this strategy. Although this strategy solves some pressing environmental problems it cannot be considered to be sustainable because of its high costs and its high consumption of resources. The "resource recovery" strategy has more potential to contribute to sustainability than the other waste management strategies. The "resource recovery" strategy strives at the imitation of the patterns of natural cycles and at prevention of the loss of material resources. The eco-efficiency approach proposed by the World Business Council of Sustainable Development (see DeSimone and Popoff 1997) is consistent with the "resource recovery" strategy.

Eco-efficiency and resource recovery are sustainable in the theory, but the practice is still very difficult and implementation is still underway. A successful implementation would require redesign of products and of production systems next to the development of new waste management technology and behaviour. Beek *et al.* (1990) spoke already about the challenge of industry to invent "reverse production": redesign products from a waste reduction perspective and redesign industrial processes to allow an optimal use of resources recovered from waste. For these redesign processes very specific knowledge is required about the material requirements for the manufacture of products, about the properties of materials during all phases of the material lifecycle, and about the technological systems to process materials in the phases of the material lifecycle. Studying the material metabolism of each specific substance will generate the knowledge to improve the eco-efficiency of material use and the recovery of resources out waste.

What are thus the implications of sustainable material use and sustainable waste management? First a reduction of the material requirements of production and consumption, next low extraction rates of resources from nature and high cycling rates of material within the economic system, and last safe sinks for waste flows that are not usable anymore within the economic system. The increase of the material cycling within the economic system is the most difficult challenge to realise. Material quality management within the production sectors and waste handling sector and the design of material cascades is an important tool in increasing the cycling of materials within the economic system.

1.8 CONCLUSIONS

Finally two conclusions are drawn:

(1) Waste management should not be developed only in a technical way disconnected from environmental and economic analysis. Waste flows are related to natural cycles or anthropogenic cycles and should be treated in accordance to the characteristics of the cycles they belong.

(2) Waste management and material use should be considered to be the two faces of the same coin. Recovering resources from waste flows and reusing these in the production sectors solve two problems at once: the resource depletion problem and the waste problem. Material quality management is an important approach to maximise the closure of material cycles within the economic system and to minimise the material losses to the environment.

REFERENCES

Ayres, R.U. and Ayres, L.W. (1998) *Accounting for Resources, Part 1: Economy-Wide Applications of Mass-Balance Principles to Materials and Waste*, Edward Elgar, Cheltenham, UK.

Ayres, R.U. and Ayres, L.W. (1999) *Accounting for Resources, Part 2: The Life Cycle of Materials*, Edward Elgar, Cheltenham, UK.

Beek, W., Boerstra, P. and Potma, T. (1990) De industrie van morgen: redder of ramp? In: *Het Milieu: Denkbeelden Voor De 21ste Eeuw*. Kerckebosch, Zeist, The Netherlands.

Daly, H.E. (1990) Towards some operational principles of sustainable development. *Ecological Economics*, vol. **2**, pp. 1–6.

DeSimone, L.D. and Popoff, F. (1997) *Eco-Efficiency; The Business Link to Sustainable Development*, MIT Press, Boston.

Douglas, M. (1966) *Purity and Danger: An Analysis of the Concepts of Pollution and Taboo*, Routledge & Kegan Paul, London and New York.

Galloway, J.N. and Cowling E.B. (2002) Reactive nitrogen and the world: 200 years of change. *Ambio* vol. **31**, pp. 64–71.

Graedel, T.E. and Allenby, B.R. (1995) *Industrial Ecology*, Prentice Hall, Englewoods Cliffs, USA.

Mc Kinney, M.L. and Schoch, R.M. (2003) *Environmental Science; Systems and Solutions* (Chapter 18, pp. 444–445) 3rd edn., Jones and Bartlett Publishers, Boston, USA.

Moll, H.C. (1993) *Energy Counts and Materials Matter in Models for Sustainable Development: Dynamic Lifecycle Modelling as a Tool for Design and Evaluation of Long-Term Environmental Strategies*, PhD. thesis, University of Groningen, The Netherlands.

Moll, H.C. (1997) The design of material cascades in the economic system. Theoretical considerations and some applications. In: *Regional and National Material Flow Accounting, From Paradigm to Practice of Sustainability* (eds. S. Bringezu, M. Fischer-Kowalski, R. Kleijn and V. Palm, Wuppertal Institute for Climate, Environment and Energy) Germany, pp. 215–220.

Ponting, C. (1992) *A Green History of the World*, Penguin Books, London, UK.

Reijnders, L. (1996) *Environmentally Improved Production Processes and Products: An Introduction*, Kluwer Academic Publishers, Dordrecht, The Netherlands.

Smil, V. (1997) *Cycles of Life. Civilisation and the Biosphere*, Scientific American Library, New York.

Stessel, R.I. (1996) *Recycling and Resource Recovery Engineering: Principles of Waste Processing*, Springer Verlag, Berlin, Germany.

Tromp, O-S. (1995) *Towards Sustainable Quality, A Methodological Principle for Sustainable Management of Material use*, PhD. thesis, University of Groningen, The Netherlands.

Van der Voet, E. (1995) *Substances from Cradle to Grave; Development of a Methodology for the Analysis of Substance Flows Through the Economy and the Environment of a Region, With Case Studies on Cadmium and Nitrogen Compounds*, PhD. thesis, University of Leiden, The Netherlands.

Van Weenen, J.C. (1991) *Waste Prevention, Theory and Practice*, PhD. thesis, University of Delft, The Netherlands.

Vitousek, P.M. and Mooney, H.A. (1997) Human domination of earth's ecosystems. *Science* vol. **277**, pp. 494–499.

WCED (Word Commission on Environment and Development) (1987) *Our Common Future*, Oxford University Press, UK.

2

Eco-industrial design of biomass cycles: flows, technologies, and actors

C.S.A. (Kris) Van Koppen

2.1 INTRODUCTION

In the past 10 years, industrial ecology (IE) has become a popular concept among environmental scientists, environmental consultants, and forerunners in industry. IE stands for an integrated and systemic approach to improving the environmental performance of production and consumption processes. Within this approach, the main emphasis has been on analyzing flows (e.g. Bringezu 2003). In this chapter it is argued that equal attention should be paid to technologies for environmental improvement, and to actors in the transformation process. Thus, the chapter deals with eco-industrial design from three angles: flows, technologies, and actors. In elaborating these angles, the emphasis will be on biomass cycles as one of the key areas of environmental concern. Before introducing the different angles and related tools for analysis and design, a brief overview of IE as a concept of environmental transformation is presented.

 Resource Recovery and Reuse in Organic Solid Waste Management. Edited by Piet Lens, Bert Hamelers, Harry Hoitink, Werner Bidlingmaier. ISBN: 1 84339 054 X. Published by IWA Publishing, London, UK.

2.2 IE, A BIRD'S EYE VIEW

2.2.1 Principles

Although its roots go back earlier (see Erkman 1997), the publication that marked the start of the current interest in IE was the article of Frosch and Gallopoulos in "Managing the Earth", a special issue of *Scientific American* (1989). In this article, the analysis and design of industrial flows is explicitly linked to the ecosystem concept. In the wording of the authors: "the traditional model of industrial activity – in which individual manufacturing processes take in raw materials and generate products to be sold plus waste to be disposed of – should be transformed into a more integrated model: an industrial ecosystem."

In the following years, many authors took up this challenge and a stream of projects and publications on IE – as well as on related concepts such as industrial metabolism – emerged. Key publications are, among others, Graedel and Allenby (1995), Ayres and Ayres (1996), Lowe *et al.* (1997), Côté and Cohen-Rosenthal (1998) and Bourg and Erkman (2003). Perhaps the most influential publication in the early days of IE was the brochure *Industrial Ecology: A New Environmental Agenda for Industry*, written by Tibbs, a consultant of Arthur D. Little (Tibbs 1991). The IE principles formulated by Tibbs aptly describe the central tenets of IE (see Table 2.1).

As is reflected in IE literature, the idea to approach industrial systems as ecosystems has three major aspects:

(1) An industrial system itself can be described to as a particular distribution of materials, energy and information flows, in analogy to an ecosystem.
(2) For its sustenance, an industrial system relies on resources and services provided by ecosystems.
(3) For the design of flows and technologies, the analogy with biological or ecological systems can serve a source of inspiration.

The first two aspects are generally acknowledged in IE research and closely correspond with the Tibb's principles of optimizing material and energy flows within the industrial systems, and between the industrial system and the surrounding natural ecosystem. Material and energy flows, in other words, should be analyzed and redesigned from a holistic and environmental point of view.

Table 2.1 Main principles of IE, according to Tibbs (1991).

Connect individual firms into industrial ecosystems: close loops, maximize materials and energy efficiency, minimize waste, and find markets for rest materials
Balance inputs and outputs to natural ecosystem capacities: reduce environmental impacts, design the industrial interface so that it is in balance with the natural world, avoid the hazardous waste transport
Re-engineer industrial use of energy and materials: redesign processes and products to reduce energy consumption and dissipative use of materials, and do more with less (dematerialization)
Align policy with a long-term perspective of industrial ecosystem evolution

The third aspect of the ecosystem analogy is more controversial. Several authors present design principles inspired by biology, such as using solar energy (like plants); useful application of wastes (like decomposers); and local synthesis of hazardous substances rather than long-distance transport (like snake venom). Most proponents of IE, however, would agree that though such analogies may be useful, their applicability is limited and the range of eco-industrial design principles has to be wider (Boons and Baas 1997; Lowe *et al.* 1997).

In fact, the methods for analysis and design in IE go far beyond biology and ecology as scientific disciplines, and encompass many tools and methods that were developed in other environmental fields. Examples of such methods and tools are pollution prevention (P2) methods, environmental management systems, environmental product design tools, and life-cycle assessment (LCA).

2.2.2 Practices

Turning from principles to practices, two main areas where the ideas of IE have been realized in concrete projects can be distinguished: eco-industrial parks (EIPs) and product life-cycle management.

2.2.2.1 EIPs

The viability of IE ideas has been demonstrated in several initiatives towards EIPs or industrial ecosystems. There are many definitions for EIPs; frequently quoted is the one from the US EPA Fieldbook for the development of EIPs: "An EIP is a community of manufacturing and service businesses, seeking enhanced environmental and economic performance through collaboration in managing environmental and resource issues." (Lowe *et al.* 1995). As the term "park" suggests, these companies are often situated together in one industrial area. However, organizations outside such a park can also be involved in co-operation and exchange. When the co-operation stretches over a region, the concept of industrial ecosystem is more adequate than EIP. Examples of successful EIPs and industrial ecosystem projects are the industrial ecosystems project in the Rotterdam harbor area (The Netherlands), the Houston Ship Channel co-operation (US), and, most famously, the industrial symbiosis project in Kalundborg (Denmark). In addition to these examples, there are a large number of successful smaller initiatives with only a few companies involved and just one or two environmental aspects taken into consideration. Many EIPs projects, in a wide range of countries, are currently under investigation or being implemented (for a more detailed description of examples and potentials of EIPs, see Lowe *et al.* 1997 and Van Koppen and Mol 2001).

From existing projects and ongoing initiatives, it has become clear that it is possible to establish EIPs and industrial ecosystems. Yet, it is certainly not easy. Quite often potential options of optimizing resource management are not realized because of technological, economic and social reasons (Van Koppen and Mol 2001).

2.2.2.2 *Product life-cycle management*

When the IE perspective is applied to product life-cycle management, the system under investigation is not a park or a region, but a product chain. Product life-cycle management is an important field of environmental innovation, as is illustrated by the increasing application of its most important tools: LCA and eco-design, or design for environment (DfE). LCA is the most elaborate method, to date, for assessing the environmental impacts of a product or service over its entire life cycle. DfE is a set of heuristics for designing products that, over their whole life cycle, have a better environmental performance. LCA, in full or simplified modes, is frequently used as part of DfE. Many companies – and probably a majority of the environmentally aware multinationals, such as Unilever, Shell, Dow, etc. – nowadays apply LCA or similar methods to screen their products. Some of them have integrated LCA into ongoing programs of eco-design. Notwithstanding the achievements of forerunners, such as Novotex (green cotton wear), DaimlerChrysler (development of the "Smart" automobile), and Philips (electrical and electronical equipment), DfE is still in its infancy.

IE, in the context of product life-cycle management, is closely related to the concept of dematerialization. Dematerialization means producing goods and services of a higher total value, while using less energy and materials, and generating less pollution and ecosystem disruption (Bartelmus 2003). When the efficiency of fossil fuel utilization is concerned, the term decarbonization is sometimes used instead of dematerialization.

Evidence from environmentally advanced countries (such Germany, The Netherlands, and others) shows absolute dematerialization trends for some polluting substances, for example SO_2. But for other categories of environmental impacts, such as the emission of CO_2, dematerialization is much harder to achieve, and relative increases in eco-efficiency are counteracted by consumption growth (Bringezu 2003).

2.2.3 Strengths and weaknesses of IE

In summary, IE:

- promotes an integrated and comprehensive view of the industrial system (production and consumption) and its relations with the biosphere;
- puts an emphasis on material flows within and outside the industrial system;
- considers technological trajectories as a crucial element for industrial transition.

Although it is widely acknowledged that such an integrated and systemic view on environmental improvement is of great importance, the actual achievements of IE to date are rather modest. Doubtlessly, one important reason is that the ambitions of IE are high, and therefore difficult to realize. Another reason is that the toolkit of IE is rather skewed: it is strong on flows but weaker on technologies, and it still lacks elaborated tools for analyzing actor involvement. As is argued in this chapter, a more balanced combination of these aspects would help to better understand and conduct eco-industrial design. This argument is in line with recent efforts to strengthen technological and social science approaches within IE (for technological approaches see for instance Bourg and Erkman 2003, and several contributions in this volume and other volumes of the IWA Integrated Environmental Technology Series; for social approaches see for instance Van Koppen and Mol 2001 and Cohen-Rosenthal and Musnikow 2003).

2.3 ANGLES OF IE DESIGN

2.3.1 Flows, technologies, and actors in IE

2.3.1.1 A lesson from P2 projects

To introduce the different angles to environmental design, experiences with P2 projects are a good starting point. From the 1970s many of these projects have been carried out worldwide, under names, such as waste and emission prevention (WEP), toxics use reduction (TUR), P2, and cleaner production (CP). A pioneer example was the pollution prevention pays (P3) program at 3M Company (US), which started in 1975. Encompassing thousands of preventive actions, the 3P program demonstrated that preventive environmental care could be cost-neutral or even cost-saving (Schmidheiny 1992). P2 projects of this type aim at developing and implementing options for preventive environmental care within individual companies. Several methods have been developed to guide such projects (e.g. De Hoo *et al.* 1991; World Bank Group 1999). Although these methods may diverge in some of the details, they have three main elements in common:

- *Analyzing flows.* An important analytical step in P2 is demarcating the relevant parts of the process and analyzing the environmentally relevant flows – of water, energy, and substances – within these parts.
- *Identifying and assessing improvement options.* A crucial synthetic step in a P2 project is the identification of technical and operational options for improvement. This step usually follows on to the flows analysis, but cannot simply be deduced from it. Additional technical expertise and creativity is needed to develop feasible options for improvement.
- *Establishing and motivating a team.* A competent and motivated group of persons committed to the goal of prevention is needed to bring a project to a good end. Experience shows that often the presence of a key actor with sufficient authority, knowledge, and enthusiasm – in other words, a *champion* for the project, is decisive for success or failure.

Developing preventive environmental care, therefore, has three main aspects: flows, technologies, and actors. These aspects, or angles, are not only relevant to P2 within a company, but they also apply to environmental improvement at the scale level of industrial systems, as IE is aiming at.

2.3.1.2 A triangle model

IE design, as is illustrated in Figure 2.1, involves three main aspects:

(1) *Environmental flows and impacts*: The assessment of relevant environmental flows (energy, materials, emissions) and environmental impacts is vital to eco-industrial design. It is needed to identify environmental hotspots; that is, those flows and impacts that are the most urgent and important candidates for improvement. It is also needed to evaluate technological and operational innovations with regard to their impact on all-over environmental performance. An assessment of water flows in an EIP, for instance, may reveal that large amounts of water from private wells are used, causing a fall of the groundwater level, and resulting in damage to natural ecosystems. Reduction of water consumption, then, is an environmental hotspot. An analysis of the energy consumption over a product chain may reveal that the use of the product by the consumer is the major cause of energy consumption over the products life cycle. In this case, the use phase of the product would be the hotspot, and product design for low energy use would contribute most effectively to the over-all environmental performance of the product.

(2) *Technological and operational improvements*: It is equally vital to eco-industrial design that feasible options for improvement are identified and elaborated. Often, some options for improvement can be found in operational procedures (good housekeeping). Major environmental improvements, however, tend to imply technological change. The identification, development, and optimization of technologies for environmental improvement constitute an essential step in eco-industrial design. Reducing industrial water use, to take up this example again, is dependent on technologies for treatment and

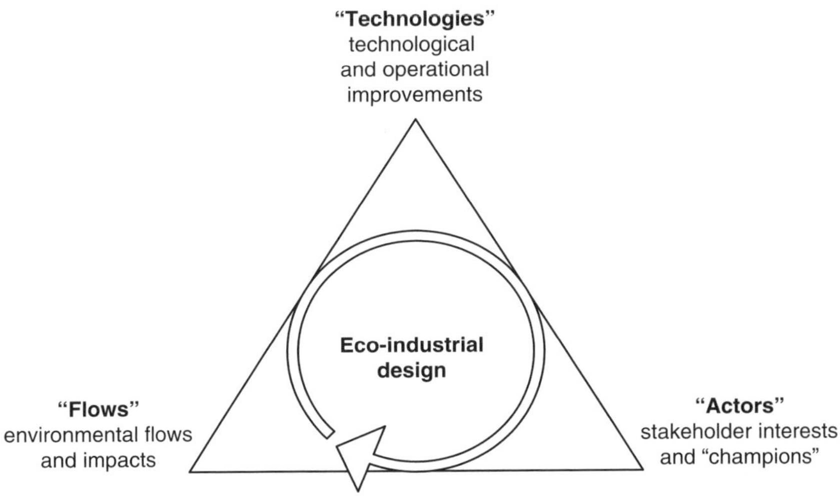

Figure 2.1 A triangle approach to eco-industrial design.

recycling of water, and – in the case of cooling water – on technologies of heat transfer. In the example of reducing the energy consumption of consumer products, designers would need to identify from the wide range of energy-saving technologies those options that would be appropriate to the specific product and to the consumer practices of using it.

(3) *Stakeholder interests and champions*: None of the previous steps would be effective without actors that have an interest in environmental change and actors who want to take the lead. Therefore, it is necessary to identify actual and potential stakeholders and their interests (including financial interests) and to identify potential champions; that is (groups of) actors willing and able to initiate and direct the process of change. Industrial symbiosis in water use is dependent on the willingness of companies in an industrial park to co-operate, which again is dependent on the costs and benefits of water-saving innovations. Developing energy-saving consumer products is dependent on companies' engagement in eco-design, as well as on consumers' willingness to buy these products and to properly use them.

As Figure 2.1 aims to visualize, the combination of angles is decisive for a successful IE design. Moreover, the procedure of design principally involves a cyclical process that can start at any angle. Environmental scientists tend to start from an analysis of flows and impacts. Technological experts, however, often start from a specific set of technologies and look for relevant environmental hotspots as well as interested actors to further develop improvement options. Companies and environmental organizations would usually start as interested parties and from the stakeholder angle look into technological and environmental aspects of IE. Finally, it is important to realize that each angle influences and constrains the other ones. For example, the economic interests of companies – conditioned by the pricing of water and effluent discharge – will set constraints to the feasibility of water recycling options. The range of options in energy-saving equipment, obviously, will be co-determined by consumer attitudes toward the operation of such equipment. Persons involved in eco-industrial design – engineers, marketers, regulators, environmental organizations – usually take into account such conditioning factors, but not always in an explicit way. Making these angles explicit and dealing with them methodically may contribute to clarifying and better directing the process of design. The rest of this chapter sets out to do this in a preliminary and exemplary way. The three angles will be referred to in shorthand as "flows", "technologies", and "actors".

2.4 ASSESSING FLOWS, TECHNOLOGIES, AND ACTORS

In applying the three angles described above, the following sections explore some of the methods that are available for eco-industrial design. While the nature of this exploration is more general, the focus of the concrete examples is on biomass.

2.4.1 Environmental flows and impacts

Several methods for analyzing flows and impacts have been developed by IE researchers and other environmental scientists. The most important methods are material flows analysis (MFA) and substance flows analysis (SFA) on one hand, and LCA and ecological footprint analysis (EFA) on the other hand. These methods will be shortly described here. More extensive reviews, with an emphasis on MFA and SFA, can be found in Van der Voet (1996) and Daniels (2002), among others. An overview of LCA is provided in the ISO Standards 14040 to 14043. For EFA the work of Wackernagel and Rees (1995) is seminal.

2.4.1.1 MFA and SFA

MFA and SFA focus on the quantitative analysis of flows within a geographically demarcated system, such as a region, a country, a city, or an industrial park. The two methods are very similar in many respects, and sometimes confounded, but they may be distinguished by the principle that MFA is concerned with aggregate flows (mass flow, biomass flow, or total nutrients flow), while SFA is focused on single substances (cadmium, nitrogen, PCBs). When MFA and SFA methods are applied on micro levels, such as a company, a farm, or a lake, they are comparable to methods, such as mass balances, process flow diagrams, and input–output analyses. Given the principles of IE, it is obvious that they have a central role to play in eco-industrial design. MFA and SFA methods can be used to identify hotspots of materials use and materials dissipation in the system, and may help assessing options for recycling.

A very concrete practical application of flow analysis is making inventories of rest materials generated by companies, and matching them with company requests for materials needed as an input in the production process. In several countries and regions, databases for rest materials have been established that can be searched by interested companies, to stimulate the off-site reuse of materials.

As their names suggest, the emphasis in MFA and SFA is on mass flow, not on energy. With good reason, IE has been criticized for its lack of attention to energy (e.g. O'Rourke *et al.* 1996). More recently, however, industrial metabolism approaches have been extended as to encompass energy flows (e.g. Haberl 2002).

2.4.1.2 LCA and EFA

LCA and EFA focus on the comparative analysis of impacts, related to a product (LCA) or another set of human activities (EFA). Basically, LCA and EFA aim to assess whether a set of activities, taken as a whole, is better or worse for the environment in comparison to another set of activities. LCA is often used for comparing the overall environmental performance of products. EFA is a popular method for comparing the impact of individual persons, organizations, cities, or countries. Both methods have a broad and overlapping range of potential applications. Generally, however, LCA is regarded as the more scientifically based method (but hard to grasp for non-experts) and EFA as method with a better appeal to the public (but less grounded in environmental science). Both methods can be important for IE because they can help identifying activities with high environmental impact. They can also be used for evaluating the relative performance of different improvement options. LCA is the tool *par excellence* for integrated environmental assessment of production and consumption chains. Aspects of energy use (such as fuel consumption and CO_2 emission) are included in LCA and EFA. However, it should be mentioned that the results of even a complex tool like LCA are highly contingent on implicit and explicit assumptions. One of the assumptions of LCA is that the impact of emissions or extractions is independent of their geographical place. Another assumption is that all environmental impacts can be aggregated into a limited group of impact categories (a group *not* including, among others, biodiversity loss, risks of environmental calamities, soil erosion, occupational health, and animal welfare).

Though LCA and EFA are helpful tools to augment our insight in environmental flows and impacts (as demonstrated by several other contributions in this volume), it would be an error to take the results as a definite, objective ground for policy and management decisions.

2.4.1.3 Biomass flows

When looking at the case of biomass, it is easy to see that it presents an important but complex area of investigation. Because of its omnipresence, and the serious environmental risks it poses through water pollution and global warming, the biomass cycle represents undoubtedly the most important flow in IE. Typically, biomass flows comprise converging and diverging fluxes of different substances. Moreover, both in natural and anthropogenic flows, biomass may convert into energy, as well as the reverse. Until recently it was not unusual to design biomass processing facilities from a partial perspective, thus optimizing only a limited set of parameters. Examples are the optimization of organic waste water treatment installations without taking into account energy use and CO_2 emission; or the development of organic waste processing facilities without taking into account the caloric value of the waste to be disposed. From LCA and LCA-like studies it has become clear that options for environmental improvement should to be evaluated from an integrated material and energy perspective (e.g. Haberl 2002; Vogt *et al.* 2002).

2.4.2 Technological and operational improvements

As insightful as the assessment of flows may be, the success of IE is highly dependent on the development and implementation of technological and operational options for improvement (operational change being defined as any change in processes other than technology change, e.g. changes in direction of flows, changes in process handling, changes in monitoring and control).

2.4.2.1 The P2 hierarchy

Since the range of potential options is almost infinite, we need heuristics for identification and selection of relevant options. Perhaps the most well-known heuristic in environmental science is the waste management hierarchy, which for the purpose of P2 projects in industry was modified into the P2 hierarchy. An example of this hierarchy, by the authors referred to as WEP hierarchy, is presented in Table 2.2.

The options on top of the ranking are considered most preferable and most preventive. The dotted line indicates where, according to the authors of the WEP hierarchy (De Hoo *et al.* 1991), the border is between preventive and non-preventive options (why off-site reuse is not considered preventive will be discussed later). For each of the preventive categories, P2 literature offers lists of frequently found options, which can be used as checklist in the process of identifying feasible technological and operational improvements.

In adapting the WEP hierarchy to an IE approach, particularly the categories 1 and 6 would need further elaboration. Product design – highly ranked, but marginally applied in practical P2 projects – constitutes a key activity of IE, and it has several branches. Product design may start, fundamentally, from the function of the product and seek for alternative ways to fulfill that function, for example replacing the product by a service (for chemicals, see Reiskin *et al.* 2000). In its most comprehensive form, eco-design is concerned with all impacts of the production, consumption, and disposal stages of the product's life cycle. But it can also be applied with a more limited focus, for instance by specifically aiming at remanufacture or recycling of the product (design for disassembly, design for recycling).

Off-site reuse, which was not regarded preventive at all in WEP, by contrast is a major road to environmental improvement in IE, particularly in EIPs. Reuse often will not mean that materials are recycled within the same product chain, but rather that they are part of a cascade from high quality to low quality use. There are many options for cascading, and recently models have been developed for calculating cascading optima, particularly for water. One of them is the so-called Water Pinch

Table 2.2 The WEP hierarchy (De Hoo *et al.* 1991).

1	Changes in products (eco-design)
2	Changes in process inputs (less or environmentally preferable materials, less or renewable energy)
3	Changes in process technology (integrated, cleaner technology)
4	Changes in process operation (good housekeeping)
5	On-site reuse (reuse of raw material or auxiliaries, or useful application for other purposes)
6	Off-site reuse
7	Emission control measures (end-of-pipe technology)
8	Controlled treatment and disposal of waste (incineration of waste with utilization of energy value; secure land filling)

method (see Baetens 2001; suggestions have been made to extend this method to other substances). In general, one can say that the array of potentially interesting technological and operational options is much broadened by IE, but that apart from the rather simple (and disputable) hierarchies presented, only a limited number of models and procedures are available for the structural development of IE technology as yet (for an effort to create more sophisticated categorizations of prevention technology, see Jones and Klassen 2001).

2.4.2.2 Biomass technologies

With regard to biomass flows, two related categories of options seem particularly interesting.

(1) *Technologies for biomass–energy conversion*: Linking biomass and energy flows is vital for finding optimal solutions to both renewable energy needs and waste disposal problems. It seems fruitful, in this context, to depart from hierarchies that prefer reuse to incineration, and to opt for integrated solutions that optimize the total of energy and carbon flows. Many successful examples of technologies for generating energy from biomass exist (e.g. incineration of biomass, anaerobic digestion and methane generation, bio-ethanol production) but there is a vast array of options for innovation, development, and up-scaling.

(2) *Life-cycle design of bio-based materials*: Materials synthesized from non-renewable carbon (or petrochemicals) may be replaced by bio-based materials, either produced in agriculture or recovered from organic waste. Applications of this principle may vary from very conventional options (e.g. use of natural fibers instead of synthetic fibers in textiles) to high-tech innovations, such as the use of lignine (as a by-product from paper industry) in printed wiring boards for computers (Kosbar *et al.* 2001).

What counts, in all these efforts, is an integrated approach to energy and material flows, in order to identify technologies with an optimal environmental efficiency for both energy flows and material flows. It should be noted, in this context, that there is not always sufficient grounding for the often-expressed feeling that "natural" (low-tech, biological) biomass flow options perform better than "chemical" (high-tech, synthetic) ones. It appears to be more sensible to deal with petrochemical and natural carbon flows from an integrated perspective. Not only from a geological history point of view fossil fuel flows are linked to biomass flows, but these flows are also cross-linked in modern technology. Earlier, the example of bio-based alternatives to petrochemical products was mentioned. Reversely, we may think of synthetic products that, through biodegradation, feed back into biological flows after their disposal, for example into compost. From a global warming point of view, the general heuristic in IE improvement must be to replace fossil fuel flows by biomass flows. In addition to decarbonization, therefore, we might also speak of re-carbonization: replacement of non-renewable carbon (i.e. CO_2-generating carbon) with renewable carbon (CO_2-neutral carbon) in industrial flows.

2.4.3 Stakeholder interests and "champions"

Equally vital in eco-industrial development are the actors that bring about industrial change. Two categories can be distinguished: stakeholders and champions. The stakeholders comprise all groups and individuals that, in some way or another, have an interest in and an influence on IE. Stakeholders can be company boards, employees, non-government organizations (NGOs), stockholders and financing organizations, local residents, governmental bodies, scientists, media, customers, clients and consumers, and several other sorts of groups. Their interests vary from financial interest and job opportunities to social responsibility and environmental care. Their influence can be direct and indirect. Together, stakeholders constitute the social and economic setting of any trajectory of change.

A technological innovation or significant operational change also needs a person, or a small group of persons taking initiative and directing the process. In literature, they are sometimes called the "champions" of change. The role of such key actors can be based on formal authority (e.g. when the director of a company or governmental organization initiates an environmental project) or on informal leadership qualities (as often in non-governmental organizations and in bottom-up processes of change). For any eco-industrial improvement to succeed, there must be a support base among stakeholders, plus one or more key actors to set and keep in motion the process of change.

2.4.3.1 Social analysis in IE

Although there is a considerable amount of research into social processes of technological innovation, and several authors have acknowledged the need for integrating social theory into IE (e.g. Van Koppen and Mol 2001, and several contributions to Cohen-Rosenthal and Musnikow 2003), achievements in this area are modest to date. Some industrial ecologists refer to the "systems approach" as the most appropriate tool for analyzing policy making and social change (e.g. Lowe *et al.* 1997). The systems approach, which plays a pivotal role in IE, is very helpful in analyzing flows, and to some extent useful in structured approaches to industrial design. For social analysis, however, it is not adequate, since it lacks the interpretative methodology and the social concepts needed for understanding social dynamics, and it might even falsely suggest that social relationships and natural science systems can be mapped in similar ways. Concepts that are central to the social dimensions of IE – such as ecological rationality, trust, legitimacy, corporate identity, risk perception, environmental quality, and IE itself – are all subject to different interpretations and debates. The analysis of actor involvement in IE, therefore, cannot work around social interpretation and debate and has to move beyond the systems approach. How this can be done is a complex issue. Within the scope of this paper, it can only be dealt with in an exemplary, indicative way.

2.4.3.2 Key actors and stakeholders: some observations

Existing successes in IE appear to be molded upon specific stakeholders–key actors combinations. At the heart of successful EIP initiatives (such as Kalundborg in Denmark, and in the Netherlands for instance industrial ecosystems and Rietvelden-De Vutter), we often find an informal group of active entrepreneurs, with close and often long-lasting ties among each other, as well as good relationships with one or more persons within local government. Together, these entrepreneurs and officials form small working parties that act as drivers and facilitators of change. Environmental improvements are often embedded in a range of other activities, which are directed to promoting the interests of the industries in the park or region (e.g. by improving infrastructure and facilities, enhancing public image, and joining forces in lobbying and procurement). Such interests represent the major stakeholder interests involved. Particularly in the US and Canada, we also see local university groups which act as drivers in the establishment of EIPs. In Canada, for example, Dalhousie University (Halifax) was an important agent of change in the case of Burnside Industrial park (see Côté *et al.* 1994). The moderate success of many of these university efforts, however, also reveals that it is quite hard for external groups to concretely implement innovations in EIPs. One of the key social characteristics of successful implementation of industrial ecosystems is a relationship of trust between the initiators – trust that is usually built over a longer period of co-operation. Another key characteristic is the acknowledgement of common interests among the stakeholders involved – interests that go beyond economic and environmental performance, and include a shared feeling of responsibility for the prosperity and development of a region, a municipality, or an industrial zone.

In product life-cycle management, the key actor is generally a company that combines environmental commitment with concern for its environmental reputation and, in some cases, an interest in marketing green products. The main stakeholders involved are the companies in the product chain and the customers or consumers. A precondition for successful life-cycle management is that the company has sufficient dominance in the chain to effectuate modifications in relevant parts of the product life cycle. Depending on the specific product domain, key actors may be retailers (Body Shop, Migros), manufacturers (Novotex, Xerox), or mining and process industries (Shell, Dupont). In some occasions, when a dominant chain actor is lacking, other types of organizations such as product associations can act as chain directors (in the Netherlands, for instance, product associations have taken initiatives in life-cycle management of agricultural products, such as meat, vegetables, and eggs). These kinds of associations, however, have much more difficulties in developing technological innovations than large companies.

"Vertical" networking of companies over a product chain is much more compatible with modern business practice than "horizontal" networking within a region or industrial park. Vertical networking is a common feature in corporate strategies to date, as is manifest in concepts like co-makership and product stewardship. Logistic control and quality assurance over the product chain have become important factors in corporate management, particularly among large companies operating on international markets. In such cases, environment-oriented product life-cycle management can be an additional dimension of the company's general concern for product chain management. Only companies with the ambition of having a green identity and aiming at green product markets are likely to establish life-cycle management structures that are primarily geared to environmental performance. Key parameters in the environmental strategy of large companies are consumer trust (with branding as a major instrument), legitimacy in the eyes of the public (sometimes referred to as the company's "licence to produce") and corporate reputation on financial markets (since financiers and stockholders increasingly take into account environmental performance as an investment criterion).

2.4.3.3 Social aspects of biomass flows

Social parameters are not as specifically bound up with categories of substances as flows and technologies are. Therefore, it would be difficult to specify social characteristics of biomass flows. One particular feature, though, can be mentioned. With regard to controlling powers, there is a typical difference between oil, gas, and petrochemical flows on one hand, and flows of agricultural products on the other hand. In general, the flows that originate from fossil fuels are dominated by a few large and powerful corporations (whether private or state-owned). In the case of agricultural flows social power is much more distributed: over farm enterprises and agricultural organizations, food manufacturers and retailers, governmental agencies controlling land use and food quality and supply, and last but not least, consumers. This implies, for example, that a petrochemical company planning to partially shift from mineral oil to biomass as carbon input, will have to deal with a supply chain that, as it consists of many individual farmers, requires an organizational arrangement quite different from oil mining. Another general implication is that within agricultural product chains it is much harder to allocate sufficient resources for technological innovation than it is in the fossil fuel chains. These and other social factors may contribute to the fact that cross-linking fossil fuel and biomass flows, despite apparent advantages in terms of IE, has hardly taken off yet.

2.5 FINDING FLOW–TECHNOLOGY–ACTOR COMBINATIONS

The key argument of this chapter is that a successful IE innovation requires what might be called a flow–technology–actor (F–T–A) combination. That is, a feasible combination of a flow that needs modification in order to reduce environmental impacts, a potential technological option for realizing the modification, and an actor with the capacity and stakeholder support to develop and implement the technology. Eco-industrial design, in perspective of this chapter, is a systematic effort to find such F–T–A combinations, by matching potential options and limiting conditions from three angles: analyzing flows and hotspots, scanning and selecting innovative technologies, and assessing stakeholder networks and potential key actors. By way of illustration, three examples of innovative F–T–A combinations in the area of biomass will be described.

2.5.1 Entrepreneurial innovation in solid-waste processing

The first examples concerns VAR, a company based near Apeldoorn, in the East of the Netherlands (VAR 2003). VAR is a medium-sized representative of an emerging group of private companies operating in the area of solid-waste treatment. The input flows involved in the company operations include household and industrial waste. These waste streams are processed into compost, wood, rubble, metals, secondary solid fuels, and sand, all of which are recycled to the market. The company is also charged with the excavation and sanitation of an old, uncontrolled landfill. Non-recyclable materials from the old landfill and from the input flows are deposited in a new, controlled landfill that is operated by the company. Furthermore, the company has developed an installation for wet oxidation of sludge, using a 1200 m deep U-shaped tube to achieve high pressure and temperature circumstances.

In its efforts to process waste flows in commercially viable ways, the company applies conventional technologies as well as innovative ones. In the excavation of the old landfill and in the separation of household waste, conventional mechanical, and even manual techniques are utilized. More innovative technologies developed by VAR include low-emission open air composting methods and the wet oxidation technique. VAR has also been innovative in product development, particularly with regard to high and low caloric secondary solid fuels, which are sold to the German cement industry, among others. The secondary fuels comprise of biomass (wood and paper) but also plastics.

The example of VAR illustrates that environmental entrepreneurship in waste processing can generate new technologies and products, which contribute to off-site recycling of materials and energy. At the heart of the VAR case is the company, which acts as the key driver of change. The stakeholder interests include municipalities and industries who pay for the disposal of their waste, as well as other industries who are willing to pay for the products from recycling, such as cement industries for secondary fuels, and building industries for rubble. To capitalize on these interests, the company is constantly developing new technologies and markets. In comparison to most public agencies, it is less restricted in looking for technologies and markets, as is illustrated by the expansion of its range of processes and products. Because the company realizes that the public will look critically to a private enterprise operating in this traditionally public area, it puts a high value on external communication, as is manifested by the company's investments in on-site excursions for interested groups, consultations with residents and authorities, reporting and certifications (VAR 2003).

2.5.2 Regional biomass networks under government control

In her dissertation on IE in Vietnam, Tran Thi My Dieu (2003) elaborates a model industrial ecosystem for agro-food production. As is illustrated in Figure 2.2, this industrial ecosystem comprises arable and livestock farms, food processing industries (both human food manufacturers and animal fodder producers),

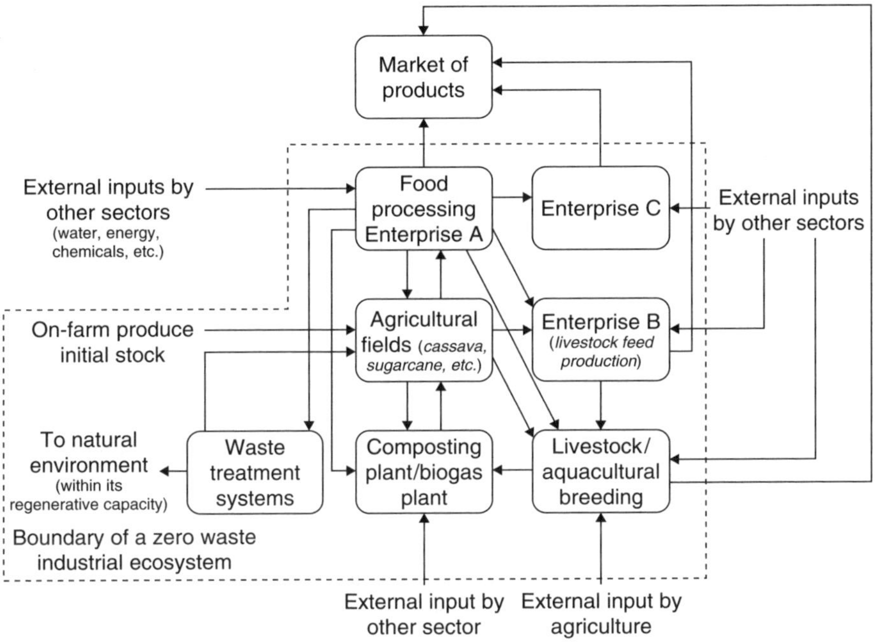

Figure 2.2 Industrial ecosystem for food production (Tran Thi My Dieu 2003).

and waste treatment facilities. The ambition of the full system is to achieve zero waste production by recycling rest materials. The model development starts from an assessment of the relevant flows, which mainly consist of agriculture-related biomass and nutrient streams. Tran Thi My Dieu (2003) elaborates the model in more detail for tapioca-related flows, among others. Most of the technologies are conventional, but more advanced technologies can be added, particular in the case of large and medium-sized companies.

While the relevancy of the flows addressed by the model is clear, and most of the technologies needed are available, the realization of such a model is dependent on key actors and stakeholders. Farms and industries involved may benefit from savings in materials and energy use, and the reduction of waste and effluent charges. It is not likely, however, that there are actor groups among them with sufficient power and authority to initiate the full system. Therefore, the initiative of government, or perhaps of a public–private organization will be crucial to bring about the transformation towards an agro-industrial ecosystem. (Semi)governmental organizations can stimulate the exchange of flows by formulating standards and permit requirements to farms and industries, and by providing infrastructure facilities and subsidies. Building on the research of Dieu, among others, Vietnamese authorities possibly will take such steps in the next years. As far as technological innovation is needed, other organizations such as research institutes and technologically resourceful companies have to play a major role.

2.5.3 Global branding of bio-based materials

The third example concerns a bio-based material, produced by a globally operating company. Together with research institutes the company, Cargill Dow LCC, has developed technologies for producing a biodegradable plastic-like material from plant sugars (field corn) through fermentation, distillation, and polymerization. The product, a polylactide polymer, which is branded NatureWorks™ PLA, can be used for a range of fiber and packaging products. In these products, the polymer can replace oil-based plastics and fibers, thereby saving in fossil fuel and reducing CO_2 emission. According to Cargill Dow, the PLA polymer has excellent properties as packaging material or fiber and is fully compostable. In contrast to petrochemical plastics and fibers, it has also an environmental appeal to consumers (see Figure 2.3). The company, founded in 1997 and based in Minnesota (USA) is focused on developing, manufacturing, and marketing PLA-products. In 2002 the production capacity of PLA amounted to 140,000 MT. Cargill Dow has offices in Japan and Europe to promote the worldwide use of their products (Cargill Dow, 2003).

In this case technologies, rather than flows or actors, are at the base of eco-industrial development. The techniques for producing PLA are central to the company and the product. Obviously, flows and actors are important too. The techniques are molded upon the biomass flows in agricultural products and geared to the aims of utilizing CO_2-neutral and biodegradable materials. And without the support of a company with sufficient production and marketing capacities, they would have little chance of competing against the commercial powers that back up petrochemical flows. As the example illustrates, global branding can be an IE strategy as well.

2.6 CONCLUSIONS

In this chapter, some key considerations for eco-industrial design have been presented. Most cases in eco-industrial development, like the examples given above, are not the result of a single design project, but involve many steps of thinking and practicing over several years in time. Often, the practitioners of IE did not explicitly approach design from the three angles of flows, technologies, and actors, as described in this article. Nonetheless, all three angles have been important to the development outcome. Taking them into account explicitly and methodically, therefore, can help directing the process of eco-industrial design. How this can be embedded in design procedures has to be further elaborated, and will highly depend on the context of design.

It was also argued that particularly the angles of technologies and actors need further elaboration. When technologies are given, IE offers several tools to assess and rank them according to their impacts on environmental flows. There are far less heuristics, however, to identify and prioritize technologies given a particular environmental hotspot. Heuristics for methodically assessing actors are lacking as well. It is positive that increasing efforts are made to fill these gaps, but more, and more focused efforts are needed.

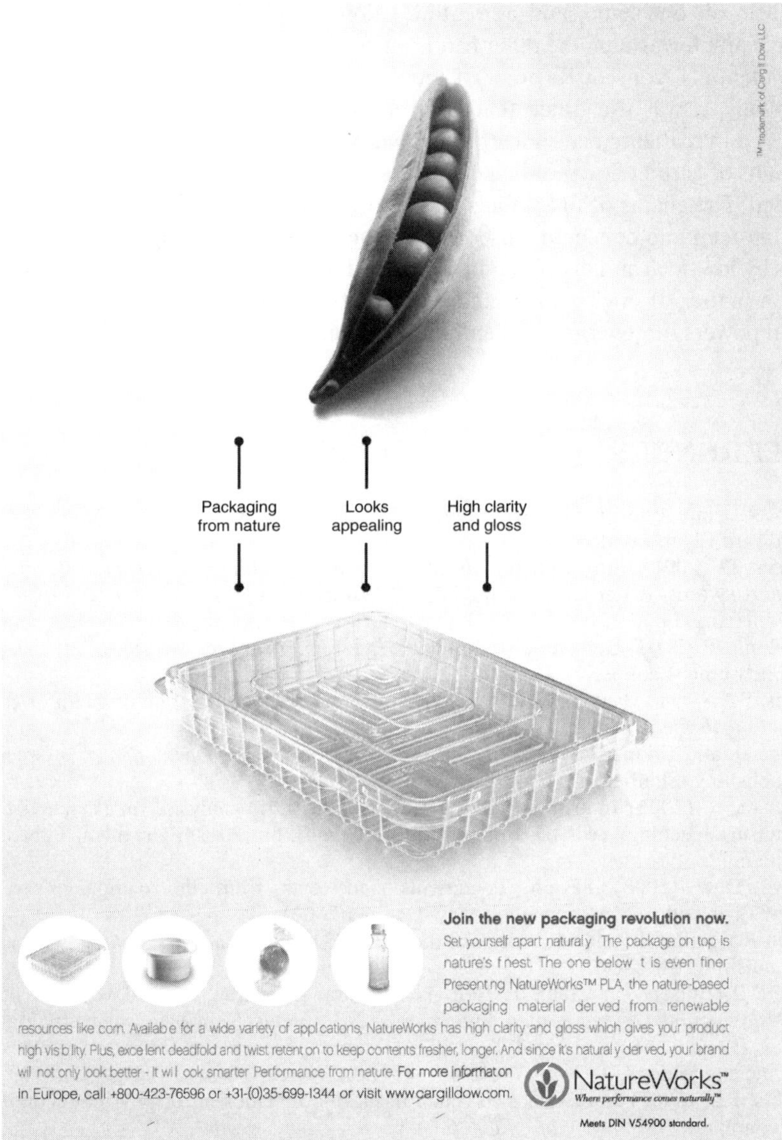

Figure 2.3 Advertisement for NatureWorks™ PLA (reprinted with permission of Dow Cargill LLC).

Overlooking the issue of biomass flows, which ran through this chapter as a case, some additional conclusions can be made. With regard to flows, the need of integrated assessments of energy use, use of materials, and pollution has been highlighted. In analyzing technologies, special attention has been paid to technologies crossing the border line between fossil fuel, energy production, and synthetic products on one hand, and agricultural flows, including agricultural products, wood, and biowaste on the other hand. Such technologies, it seems, offer interesting potentials from an IE point of view. Concerning the social aspects of these potentials, it was illustrated that different actors and different strategies can be involved in realizing eco-industrial innovations. It was also observed that the two domains of carbon flows, indicated above, are not only physically but also socially divided. The chains of fossil fuels and petrochemistry are generally characterized by high-tech and concentrated-power settings, the chains of agriculture and biomass by low-tech and distributed-power settings. Cross-linking these chains is not only a matter of environment and technology; it also implies challenges to the social powers of governmental and industrial actors.

REFERENCES

Ayres, R.U. and Ayres, L. (1996) *Industrial Ecology: Towards Closing the Materials Cycle*. Edward Elgar, London.

Baetens, D. (2001) *Water Pinch analysis*. In: *Water Recycling and Resource Recovery in Industry* (eds. P. Lens, L. Hulshoff Pol, P. Wilderer and T. Asano), pp. 205–228, IWA Publishing, London.

Bartelmus, P. (2003) Dematerialization and capital maintenance: two sides of the sustainability coin. Ecol. Econ. **46**, 61–81.

Boons, F.A.A. and Baas, L.W. (1997) Types of industrial ecology: the problem of coordination. *J. Clean. Prod.* **5**(1–2), 79–86.

Bourg, D. and Erkman, S. (eds.) (2003) *Perspectives on Industrial Ecology*. Greenleaf Publishing, Sheffield.

Bringezu, S. (2003) Industrial ecology and material flow analysis. In: Perspectives on Industrial Ecology (eds. D. Bourg and S. Erkman), pp. 20–34, Greenleaf Publishing, Sheffield.

Cargill Dow (2003) Several documents and texts from the company's website www.cargilldow.com (consulted May–October 2003).

Cohen-Rosenthal, E. and Musnikow, J. (eds.) (2003) *Eco-industrial Strategies*. Greenleaf Publishing, Sheffield.

Côté, R.P., Ellison, R., Grant, J., Hall, J., Klynstra, P., Martin, M. and Wade, P. (1994) *Designing and Operating Industrial Parks as Ecosystems*. Dalhousie University, Halifax.

Côté, R.P. and Cohen-Rosenthal, E. (1998) Designing eco-industrial parks: a synthesis of some experiences. *J. Clean. Prod.* **6**, 181–188.

Daniels, P.L. (2002) Approaches for quantifying the metabolism of physical economies: a comparative survey. *J. Ind. Ecol.* **6**(1), 65–88.

De Hoo, S., Brezet, H., Crul, M. and Dieleman, H. (eds.) (1991) *PREPARE Manual for the Prevention of Waste and Emissions*. Ministry of Economic Affairs, The Hague.

Erkman, S. (1997) Industrial ecology: an historical view. *J. Clean. Prod.* **5**(1–2), 1–10.

Frosch, R.A. and Gallopoulos, N.E. (1989) *Strategies for Manufacturing*. Scientific American 1989 (September), 144–152.

Graedel, T.E. and Allenby, B.R. (1995) *Industrial Ecology*. Prentice Hall, Englewood Cliffs.

Jones, N. and Klassen, R.D. (2001) Management of pollution prevention. Integrating technologies in manufacturing. In: *Greener Manufacturing and Operations. from Design to Delivery and Back* (ed. J. Sarkis), pp. 56–70, Greenleaf Publishing, Sheffield.

Haberl, H. (2002) Accounting for the energetic metabolism of societies: concepts and empirical examples. *J. Ind. Ecol.* **5**(1), 11–34.

Kosbar, L.L., Gelorme, J.D., Japp, R.M. and Fotorny, W.T. (2001) Introducing biobased materials into the electronics industry. *J. Ind. Ecol.* **4**(3), 93–105.
Lowe, E., Moran, S.R. and Holmes, D. (1995) *A Fieldbook for the Development of Eco-industrial Parks.* Prepared for the U.S. Environmental Protection Agency. Indigo Development, Oakland.
Lowe, E.A., Warren, J.L. and Moran, S.R. (1997) *Discovering Industrial Ecology. An Executive Briefing and Sourcebook.* Battelle Press, Columbus.
O'Rourke, D., Connelly, L. and Koshland, C.P. (1996) Industrial ecology: a critical review. *Int. J. Environ. Pollut.* **6**(2–3), 89–112.
Reiskin, E.D., White, A.L., Johnson, J.K. and Votta, T.J. (2000) Servicizing the chemical supply chain. *J. Ind. Ecol.* **3**(2–3), 19–31.
Schmidheiny, S. (1992) *Changing Course. A Global Business Perspective on Development and the Environment.* MIT Press, Cambridge (MA).
Tibbs, H. (1991) *Industrial Ecology, An Environmental Agenda for Industry.* Arthur D. Little, Inc. An updated version of the paper was published in Whole Earth Review, Winter 1992, no. 77, 4–16.
Tran Thi My Dieu (2003) *Greening Food Processing Industry in Vietnam.* Dissertation Wageningen University, Wageningen.
Van der Voet, E. (1996) *Substances from Cradle to Grave.* Dissertation Leiden University, Leiden.
Van Koppen, C.S.A. and Mol, A.P.J. (2002) Ecological modernization of industrial ecosystems. In: *Water Recycling and Resource Recovery in Industry* (eds. P. Lens, L. Hulshoff Pol, P. Wilderer and T. Asano), pp. 132–158, IWA Publishing, London.
VAR (2003) *Milieujaarverslag 2002.* VAR B.V., Twello. See also the VAR website www.var.nl (consulted May–October 2003).
Vogt, R., Knappe, F., Giegrich, J. and Detzel, A. (2002) *Ökobilanz Bioabfallverwertung.* Erich Schmidt Verlag, Berlin.
Wackernagel, M. and Rees, W. (1995) *Our Ecological Footprint. Reducing Human Impact on the Earth.* New Society Publishers, Gabriola Island.
World Bank Group (1999) *The World Bank Pollution Prevention and Abatement Handbook 1998.* World Bank, Washington DC.

3

System analysis of organic-waste management schemes – experiences of the ORWARE model

Jan-Olov Sundqvist

3.1 INTRODUCTION

Waste management in Sweden is rapidly changing. Due to political decisions several actions are taken towards a more sustainable waste management. Producer's responsibility on newsprint, packages, tires, and electric and electronic scrap has been introduced, and the responsibility is connected to recycling targets. From 2000 there is a tax on all waste that is landfilled. From 2001 there is a new national ordinance on landfilling, based on the European (EU) Landfill Directive 1999/31/EC. According to the national Solid Waste Ordinance, landfilling of combustible waste is prohibited from 2001 and landfilling of organic waste is prohibited from 2005. The EU directive on incineration of waste (2000/76/EC) has been implemented into a Waste Incineration Ordinance. All these actions will cause changes in the waste management now and in the future. As an example, at the moment Sweden has

© 2004 IWA Publishing. *Resource Recovery and Reuse in Organic Solid Waste Management.* Edited by Piet Lens, Bert Hamelers, Harry Hoitink, Werner Bidlingmaier. ISBN: 1 84339 054 X. Published by IWA Publishing, London, UK.

26 incineration plants for municipal solid waste and about 20 more are planned. Also biological treatment and material recycling is increasing.

The Swedish energy system is also in the position of many changes. Recently, one nuclear power reactor has been closed down and the government's aim is to close more reactors as renewable energy sources are introduced into the market. The use of fossil fuels is supposed to decline, which demands for other energy sources, of which waste is one.

All this means that the treatment capacity for incineration and biological treatment as well as material recycling has to increase in order to meet the landfill restrictions. This give rise to the question: which treatment options are preferable from an environmental, energy and economic point of view.

A system study has been carried out of how energy, material and plant nutrients in waste are utilised at the best with respect to environment, energy and economy. This chapter describes the part of the study that considers management of biodegradable waste. The full study is published (Sundqvist *et al.* 2002) (in Swedish with an English summary). A shorter review in English has also been presented (Sundqvist 2002).

3.2 LIFE CYCLE ASSESSMENTS

3.2.1 LCA

LCA (life cycle assessment) is a process to assess the potential environmental burdens associated with a product, a process or an activity. Characteristic parts in an LCA are identifying and quantifying energy and material flows, and evaluating the environmental impacts associated with these flows. The assessment should encompass the entire life cycle of the studied system (the studied system can be a product, a process or an activity), including material and energy raw ware acquisition, manufacturing, usage and waste treatment.

3.2.2 The LCA framework

The interest for LCA has increased dramatically since around 1990, resulting in development and increased harmonisation of the methodology. ISO standards have recently been issued (ISO 1997, 1998, 1999, 2000).

The framework outlined here is based on the ISO Standard 14040. According to the standard references a complete LCA consists of the following interrelated components:

(1) Goal definition and scoping
(2) Inventory analysis
(3) Impact assessment
    (3.1) Classification
    (3.2) Characterisation
    (3.3) Valuation or weighting
(4) Interpretation.

In the goal definition and scoping, the purpose of and the range covered by the study should be defined. This includes definition of system boundaries, data requirements, assumptions and limitations.

In the inventory analysis, the inputs and outputs of the system under study are analysed. The system is usually a product through its lifetime. The inputs to the system are, e.g. energy and raw materials. The outputs from the system are, e.g. emissions from processes during raw material acquisition, manufacturing, transportation, usage and waste management. The inventory analysis results in tables of inputs and outputs of the system or systems under study.

Impact assessment is a process to characterise and evaluate the influence of the inputs and outputs identified in the inventory analysis. The impact assessment of an environmental LCA should consider the following major categories:

- Resource depletion
- Impacts on human health
- Ecological impacts.

Each of these major categories is further divided into several impact subcategories.

The impact assessment is divided into three steps: classification, characterisation and valuation. In the classification, the different inputs and outputs are assigned to different impact categories. An analysis and quantification of each impact category is made in the characterisation step. Valuation or weighting is the step in which the data of the different specific impact categories are weighted so that they can be compared.

3.2.3 Some important terms and methods used in LCA

3.2.3.1 Functional unit

The functional unit is the basis for the calculations in a quantitative LCA. It is a product, a material or a service for which the environmental loading are quantified. In an absolute LCA the whole life cycle of a specific product, material or service is studied, and the different parts of the life cycle are compared with each other. In this case the functional unit should be, e.g. one item of the studied product.

On the other hand, in a comparative LCA different products are compared with each other. In that case it is not always relevant to compare the direct products, but more to compare the functions of the products. For example, crushed tomatoes can be bought in either Tetra Brik packages (a composite package based on cardboard covered with aluminium foil and polyethylene foil) with a volume of 0.5 l, or in tin cans with a volume of 0.4 l. It is not relevant to compare one piece of Tetra Brik package with one piece of tin can, instead the comparative LCA should be based on, e.g. 2 l of each – i.e. four pieces of the Tetra Brik and five pieces of the tin can. The functional unit in this example is 2 l of package volume.

The choice of appropriate functional units is of great importance for the LCA. A relevant choice of the functional units is needed for relevant results. For waste management systems it is often preferable to work with several functional units, each one representing an essential utility that is produced from waste. For example, organic degradable waste can be incinerated, anaerobic digested or composted. When incinerated, the product (in Sweden) will be district heating. When anaerobically digested, the products can be district heating, biogas for vehicle or for electricity depending on how the biogas is used. When aerobically digested or composted we also produce a fertiliser. When assessing management of organic waste several functional units must be used:

- treatment of a specific amount of waste,
- production of a specific amount of district heating energy,
- production of a specific amount of electricity,
- production of a specific amount of nitrogen (N) and phosphorus (P) fertiliser (eventually also K fertiliser),
- production of a specific amount of vehicle fuel (e.g. the amount necessary to drive a buss or a car X km).

In all scenarios and cases the same functional units must be used. If the functional unit is not produced from waste, it has to be produced from another source.

3.2.3.2 Emission factor

In an LCA the used data is often presented as emission factors and energy factors. Information in databases is often expressed as emission factors. The emission factor gives the emission for a process or sub-process in the life cycle, in relation to an input or output parameter, e.g. per weight of product, per weight of a certain element in the product, or related to the energy content of the product. For example, emission of HCl from waste incineration may be expressed as kg HCl emitted per kg Cl in the input to the incinerator. Energy consumption for transport can be presented as MJ fuel (or litres of diesel oil) per kg of transported product and per km transport distance.

3.2.3.3 System boundaries

The system boundaries define the system that is studied. An LCA is based on the material flows and energy flows over the system boundaries. It is of absolute necessity to have well-defined system boundaries, in order to obtain unambiguous results. Usually the system boundaries can be:

- Geographical boundaries: e.g. disposal of waste generated within a municipality.
- Time boundaries: e.g. the waste generated during 1 year.
- Functional boundaries: e.g. wastes that can be used for biological treatment.

3.2.3.4 Allocation

A traditional problem in LCA is how to deal with processes or groups of processes with more than one input and/or output. Some examples are processes or productions with co-products of economic value (multi-output processes), and waste treatment where several different waste components are treated in the same process with common consumption of raw material and common formation of emissions (multi-input processes). In LCA allocation can be defined as the act of partitioning in some proportionate shares the responsibility for environmental impacts caused by processes in a life cycle (Lindfors *et al.* 1995). General allocation problems in connection with LCA have been discussed by Lindfors *et al.* (1995). Allocation problems connected to the waste stages in LCA have been discussed in several reports (e.g. Sundqvist *et al.* 1997, 1999).

3.3 THE ORWARE MODEL

The simulation model ORWARE has been used for assessing waste management schemes for biodegradable waste and other wastes. The framework of the model has been developed during the past 10 years in several research projects. Descriptions of ORWARE can be found in several reports and articles (e.g. Carlsson 1997; Dalemo *et al.* 1997, 1998, Soneson *et al.* 1997; Sundqvist *et al.* 1999, 2002; Björklund 1998; Eriksson *et al.* 2002).

ORWARE is a model for the calculation of substance flows, energy flows, environmental impacts, and costs of waste management. It was first developed for systems analysis of organic waste management, hence the acronym ORWARE from ORganic WAste REsearch, but now the model covers inorganic fractions in municipal waste as well.

ORWARE consists of a number of separate sub-models, which may be combined to design a waste management system. Each sub-model describes a process in a real waste management system, e.g. waste collection, waste transport, or a waste treatment facility (e.g. incineration).

3.3.1 Methods and general description of the model

All sub-models in ORWARE calculate the turnover of materials (including the emissions), energy and financial resources in the process. Processes within the waste management system are e.g. waste collection, incineration, anaerobic digestion, composting, material recycling and landfill disposal. Materials turnover is characterised by (1) the supply of waste materials and process chemicals, (2) the output of products and secondary wastes and (3) emissions to air, water and land. Energy turnover is the use of different energy carriers such as electricity, coal, oil or heat, and recovery of, e.g. heat, electricity, hydrogen, or biogas from waste treatment processes. The financial turnover is defined as costs of individual processes. The financial costs are calculated in a life cycle perspective from the cradle to the grave.

The sub-models may be combined to describe the waste management system in a city or municipality (or other system boundary). Such a conceptual ORWARE model of a complete waste management system is shown in Figure 3.1.

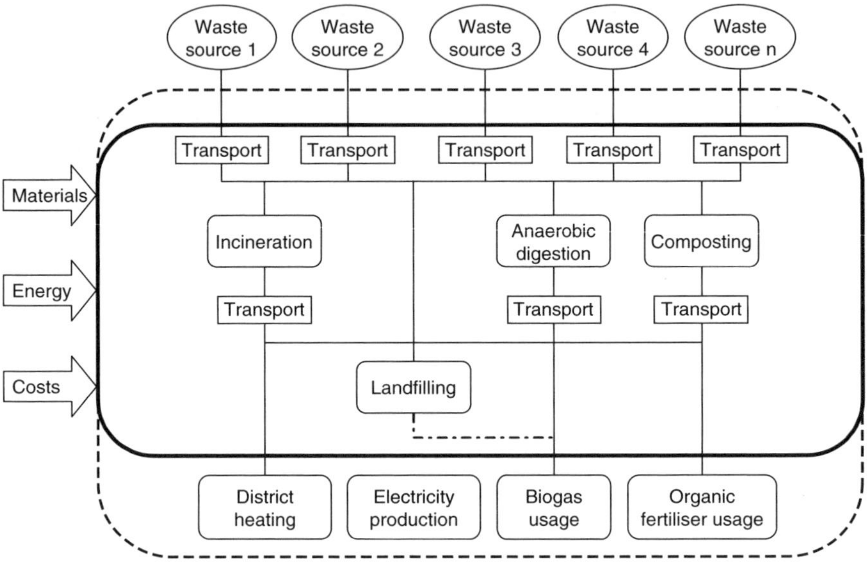

Figure 3.1 A conceptual model of a system for management of organic wastes. The conceptual model is built up from different sub-models.

At the top of the conceptual model in Figure 3.1 there are different waste sources, followed by different transport and treatment processes. The dashed line in the figure defines the boundaries of the waste management core system, where wastes are treated and different products are formed.

Another important topic is the "from-cradle-to-grave" perspective. Both upstream processes ("cradle") and downstream processes ("grave") are considered. Examples of upstream and downstream processes are:

- If electricity is consumed in a process, the environmental impact and resources consumption from the production of electricity is included as an upstream process. If the electricity is produced from coal condense power also the landfilling of coal ash is included as a downstream process.
- If oil is consumed in a process, the environmental impact and resource consumption related to the crude oil extraction, the oil refining and the distribution of oil products are included.

3.3.2 LCA in ORWARE

The substance flow analysis carried out in ORWARE generates data on emissions from the system. The individual emissions are aggregated into different environmental impact categories, according to LCA practice (see, e.g. ISO 14042 Life Cycle Impact Assessment). The following impact categories are assessed: global warming, acidification, eutrophication and photooxidant formation. Also ecotoxicity and human toxicity can be studied, but the models for weighting different toxic emissions are not fully developed so the results have to be considered carefully.

The system boundaries are of three different types: time, space and function. In an analysis of a certain system, the temporal system boundaries vary between different studies (depends on scope of the study) and also between different submodels. Most of the process data used are annual averages, but for the landfill model as well as the arable land model also long-term impacts are included (Björklund 1998; Sundqvist *et al.* 1999).

There is a geographical boundary delimiting the waste management system as shown in Figure 3.1, whereas emissions and resource depletion are included regardless of where they occur. The system boundaries in ORWARE are chosen with an LCA perspective, thus including in principle all processes that are connected to the life cycle of the waste management system. We have included environmental impact and resource consumption for upstream and downstream processes. Construction, demolition and final disposal of capital equipment are not included regarding energy consumption and emissions but are included for economy. This is according to common LCA practice. A literature survey was made about energy input during the construction phase and compared to the energy input and output during the whole life cycle (Sundqvist *et al.* 2002). It was found that the energy consumption and environmental impact from the construction phase could be neglected. However if only processes with a low energy turnover are studied, the energy consumption from construction may be of importance, e.g. a comparison between composting and landfilling (without gas recovery).

The studied system works with several functional units. The major function of the waste management system is to treat the waste. The waste management system can produce different products (functions) from the waste:

- *Energy*: district heating, electricity and biogas (the biogas can be used as vehicle fuel or as an energy source for electricity and district heating).
- *Fertilisers*: a digestion residue or compost containing, e.g. N and P.

Each of these waste products has an alternative virgin raw material source with a production process that has been included in the studied system. All studied systems produce the same amount of energy (district heating, electricity and vehicle fuel) and fertilisers (N and P), see Figure 3.2. This alternative system outside the waste system, which produces the same products as the waste management system, is called the compensatory system. The use of the compensatory system enables a quantitative comparison of environmental, economic and energy parameters between the use of waste as raw material and the use of virgin raw materials.

Compensatory systems also have upstream and downstream processes. For electricity, the upstream process includes production of the fuel used for production of electricity, and the downstream process is disposal of waste from the energy generation. Therefore, each treatment alternative in ORWARE has its own unique design of the core system as well as different compensatory systems.

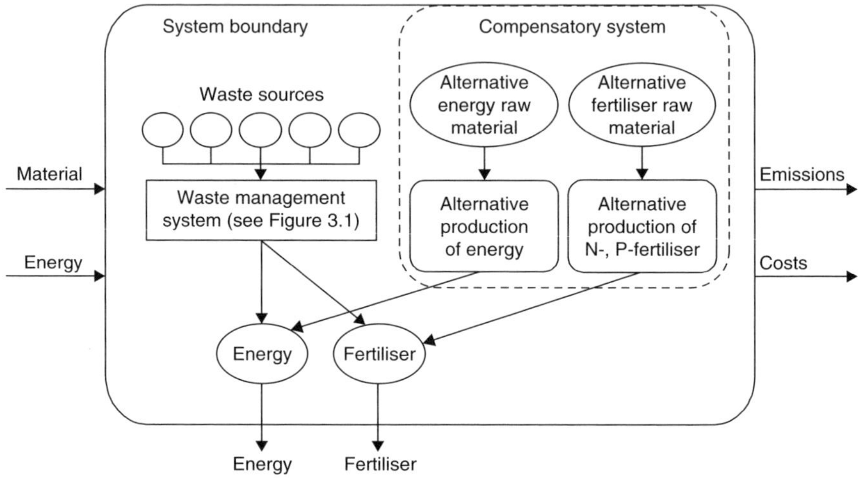

Figure 3.2 The studied system: the waste management system (according to Figure 3.1) and alternative production of energy, material and nutrients/fertiliser in the compensatory system (Sundqvist *et al.* 1999, 2002).

3.3.3 Overview of some sub-models in ORWARE

3.3.3.1 Collection and transports

The transport sub-model has been described in Dalemo *et al.* (1997). The transport sub-model is divided into three blocks: "garbage truck", "ordinary truck", and "truck and trailer". They have the same structure while the parameters differ. From different demographic data the model first calculates the transport distance. For each transport block there is specific fuel consumption (MJ per km). The emissions are calculated from the fuel consumption by emission factors.

3.3.3.2 Incineration

The original incineration model has been described in Dalemo *et al.* (1997) and Björklund (1998). Since then the model has been updated (Sundqvist *et al.* 1999, 2002). The incineration plant that is modelled is a modern plant that with good margins fulfils the requirements in the EU waste incineration directive. The emission data is based on emissions from a real incineration plant. The plant has an advanced flue gas cleaning system, including a flue gas condensation step where energy is recovered by heating pumps. The incineration plant produces district heating.

3.3.3.3 Landfill

A conventional municipal solid-waste landfill has been modelled. The landfill model is described in detail in earlier reports (Björklund 1998; Fliedner 1999).

Some basic theory of how landfilling can be handled in LCA is given in, e.g. Sundqvist (1999). The landfilled waste undergoes different degradation processes, similar to anaerobic digestion. The degradation occurs rather slowly.

The time aspects in our landfill model for organic materials are based on the two time horizons:

- the *surveyable time* period, which is the time until some kind of pseudo-steady-state is obtained. For a conventional municipal solid-waste landfill this corresponds to the end of the methane production phase, which is estimated to be of the magnitude of one century.
- the *hypothetical, infinite time* period, which is the period until the landfilled material is completely released to the environment.

In this chapter, only the emissions during the surveyable time period are presented.

During the surveyable time close to 100% of sugar, starch, hemicellulose, fats and proteins are degraded and about 70% of the cellulose. Plastics are degraded to 3% and humus and lignin are assumed to be undegradable. Landfill gas, consisting of mainly methane, carbon dioxide and water is produced during the degradation. We have assumed that 50% of the theoretically produced landfill gas is recovered (during the surveyable time period) and used as a fuel in a gas engine producing electricity and heat. The rest of the landfill gas will migrate to the ambient air. During the passage through the landfill cover about 15% of the methane gas is oxidised into carbon dioxide. The remaining methane gas is emitted as a greenhouse gas.

The leachate water is treated in a local treatment plant, reducing chemical oxygen demand (COD)-, N- and P-emissions to water.

3.3.3.4 Anaerobic digestion

The original anaerobic digestion model has been described in earlier reports (Dalemo *et al.* 1996, 1997). The model has since then been adapted to a thermofile process.

The incoming domestic food wastes are packed in plastic bags. In a bag separator the bags are cut and emptied and separated from the degradable material. Slaughterhouse wastes are hygienised at 70°C or 130°C before digestion. The water content is adjusted to about 85% and the waste is fed into the digestion reactor. The hydraulic residence time in the reactor is 20 days.

The study addressed two ways to use the biogas:

(1) The gas is upgraded to 97% CH_4 and compressed to 250 bars. The gas is then used as fuel for busses.
(2) The gas is combusted in a gas engine, where both electricity and heat (for district heating) is produced.

The digestion residue or digestate is spread on agricultural land, thus substituting chemical P and N fertiliser. During the spreading a lot of ammonia can be released. A new, improved spreading technology was modelled where the digestate is worked down in the soil simultaneously with the spreading.

3.3.3.5 Composting

In the study, open windrow composting was modelled. About 50–75% of the organic material is degraded into carbon dioxide, water and compost. The degradation rate is different for different organic compounds. During the process ammonia is released to air. In a separate study the emissions from "closed" compost or reactor compost were estimated, where the exhaust gases are treated to separate out ammonia. The compost is spread on agricultural land, thus substituting chemical N- and P-fertiliser.

3.4 EARLIER ORWARE STUDIES

Several ORWARE studies have been published. The real first study was reported several years ago (Dalemo *et al.* 1997; Soneson *et al.* 1997). Since then several case studies has been done. In this chapter the focus is on a larger project during 1998–2001 which was carried out in two stages. In the first stage, case studies were made in three municipalities (Sundqvist *et al.* 1999, 2001; Eriksson *et al.* 2002). The second stage involved some model improvements, and a major sensitivity analysis, and several sub-studies where different assumptions were tested and analysed (Sundqvist *et al.* 2002).

During the first stage case studies were made in three different municipalities with very different characteristics were studied (Sundqvist *et al.* 1999, 2001; Eriksson *et al.* 2002):

- Stockholm, the capital of Sweden, is a big city with an incineration plant and a system for district heating. There is no arable land within the municipality borders (arable land is needed for spreading of the organic fertiliser produced from biological treatment of the organic waste). Stockholm City has approximately 750,000 inhabitants.
- Uppsala is a relatively big municipality, also with an incineration plant and a system for district heating. Arable land can be found close to the city area. Uppsala municipality has approximately 186,000 inhabitants.
- Älvdalen is a municipality with a rather low population living on a large area. Älvdalen has no incineration plant and no system for district heating. There is hardly any agricultural land within the municipality. Älvdalen municipality has 8100 domiciled inhabitants plus a great number of tourists from season to season. In Älvdalen some of the most famous Swedish ski-centres are located which means that during short time periods, tourists produce large amounts of waste with a low degree of source separation. In Älvdalen, the source separated wastes and the rest fraction is collected at a central place in each village. As an average, there are about 135 people in each village.

The results showed that the same conclusions could be drawn in all three municipalities. The ranking between different waste management options was the same. The following conclusions for organic wastes were drawn for all three municipalities:

(1) The study showed that decreased amounts of waste to landfilling, and increased amounts to incineration, anaerobic digestion and composting were positive, from an environmental and energy point of view, as well as from a welfare economic and a life cycle cost point of view. This means that landfilling of energy-rich waste or biodegradable waste should be avoided as much as possible. The negative results of landfilling depends on both environmental impact from the landfill (especially greenhouse gases), and a low recovery of resources, which give rise to more energy consumption, environmental impact and costs in the compensatory system.
(2) Incineration should be a basis in the waste management systems of the three municipalities studied, even if the waste has to be transported to a regional facility. Once the waste is collected, longer regional transports are of little significance from an energetical, environmental and economic point of view, as long as the transports are carried out in an efficient way.
(3) When comparing biological treatment and incineration, no unambiguous conclusions can be drawn. There are benefits and drawbacks associated with all these waste management options:
   - Anaerobic digestion of easy degradable waste gives a higher welfare economic cost than incineration, and has both environmental advantages and disadvantages. Conclusions regarding energy use depends upon how the biogas is used.
   - Composting of easy degradable waste is comparable to anaerobic digestion from a welfare economic aspect, but gives higher energy consumption and environmental impact.

In the study also plastic recycling and cardboard recycling were studied.

3.5 FURTHER ANALYSIS OF ORWARE DATA – OBJECTIVES AND SYSTEM BOUNDARIES

Data presented in this section is from the final report of the latest study (Sundqvist *et al.* 2002).

3.5.1 Objectives

The objectives of the second stage were to study how general the results from the first stage were, and to identify parameters that could shift the results.

3.5.2 System boundaries

Based on the results in the first stage, the second stage was based on a hypothetical municipality, where different parameters were varied more systematically in

Table 3.1 Impact categories that have been studied.

Environmental impact	Global Warming Potential Acidification Potential Eutrophication Potential Photooxidant formation, divided into – VOC – NO_X Ecotoxicity Human toxicity
Energy consumption	Total consumption of primary energy carriers Consumption of non-renewable primary energy carriers
Economy	Financial life cycle costs Environmental costs (by three different methods for valuation of emissions and energy resources) Welfare economy (which is the sum of the two above)

sensitivity analyses. The chosen hypothetical municipality was based on Uppsala, which has 186,000 inhabitants and consist of a city area, a suburban area and a rural area. The municipality has a waste incineration facility and a district heating system. All transport distances and similar demographic data were taken from the Uppsala case study, but were varied in the sensitivity analysis.

The waste studied is domestic waste and similar waste from business and industry. The total amount of waste is approximately 69,000 tons/year. In all scenarios newsprint paper (75%), glass packages (70%) and metal packages (50%) are assumed to be sorted out and recycled "outside" the studied system (is not included in the 69,000 tons of waste). The remaining paper, glass and metal are present in the waste studied.

The time frame studied was 1 year. The waste amount corresponds to 1 year's generation of waste in the studied municipality.

Important assumptions are the choices of upstream and compensatory energy sources. In this study, the electricity has been assumed to be marginal production, produced by natural gas combi power stations. This assumption has a high implication on the results. The design of the district heating systems is special for each municipality. There are about 100 separate district heating systems in Sweden. The fuel that competes with waste can be peat, wood chips, oil or coal, depending on the heat demand, when the demand is raised, the prices for different fuels, existing combustion facilities, etc. Biofuel (wood chips) was chosen to be the compensatory heat, because biofuel seems to be the most common fuel for district heating today. In the sensitivity analysis also oil was studied as compensatory heat.

The studied parameters are according to Table 3.1 (see also Tables 3.2 and 3.3 for weighting factors).

The weighting factors used for characterisation and valuation is presented in Tables 3.2 and 3.3. The impact category "photooxidant formation" is divided into two sub-classes: NO_X and VOC, which are obtained from the inventory result.

Table 3.2 Environmental impact categories studied and weighting factors used.

	Global warming kg CO_2-equiv./kg	Acidification kg SO_2-equiv./kg	Eutrophication kg O_2-demand/kg	Photooxidant formation	
				NO_X kg NO_X/kg	VOC kg ethene-eq./kg
CO_2 (fossile)[a]	1				
N-NO_X[a]		0.7	6	1	
N-N_2O[a]	310				
S-SO_2[a]		1			
CH_4[a]	21				0.006
N-NH_3[a]		1.88	16		
HCl[a]		0.88			
N-NH_4[b]			15		
N-NO_3[b]			4.4		
COD[b]			1		
P[b]			140		
VOC[a]					0.416
CO[a]					0.03

[a] Emission to air.
[b] Emission to water.

3.5.3 Scenarios

The wastes included in the study are shown in Table 3.4.

The formulated scenarios are given in Table 3.5. They were chosen to illustrate the alternatives that are discussed in several Swedish municipalities today. The landfilling scenario was chosen to illustrate the need for decreased landfilling – the earlier studies had already shown that conventional landfilling was a bad option (Dalemo *et al.* 1997; Sonesson *et al.* 1997; Björklund 1998).

Some of the specific data used in the study are:

- The thermal efficiency (defined as produced amount of district heating, divided by the lower heating value in the waste) of the waste incinerator is 91%, exclusive the flue gas condensation. The contribution from the flue gas condensations depends on moisture content in the waste, 80% of the condensation heat is recovered. The NO_X-emission from the incinerator is 75 mg/MJ.
- The fuel used in compensatory district heating is biofuel with a total thermal efficiency of 109%, inclusive heat recovered from flue gas condensation.
- The fuel used in compensatory electricity production is natural gas in a combi cycle.
- Distances to all treatment plants are 7 km for waste from flats and apartments and, 10 km for waste from one-family houses and 15 km for waste from rural areas.
- The average distance from anaerobic digestion plant or composting plant to arable land is 8 km.

All these assumptions were varied in a sensitivity analysis.

Table 3.3 Economic weighting factors used in the study.

	Economic weighing		
	ORWARE SEK/kg*	EPS 2000 SEK/kg*	ECOTAX'99 SEK/kg*
CO_2 (fossilt)[a]	0.4	0.92	0.40
Particles (dust)[a]		306	31.5
N-NO_X[a]	54	18	34.50
N-N_2O[a]	124	326	88,000
S-SO_2[a]	34	27.80	53.30
CH_4[a]	8.4	23	3.40
N-NH_3[a]			46.80
HCl[a]	68		
N-NH_4[b]	47	−3.60	54.50
N-NO_3[b]			15.8
COD[b]	3	0.01	3.80
P[b]	439	0.50	
VOC[a]	1.49	18	121
CO[a]	0.11	2.80	0.60
Pb[a]	310,000	24,735	7,800,000
Pb[b]	310,000		96,400
Pb[c]	310,000		3700
Cd[a]	1,123,000	87	3,730,000
Cd[b]	1,123,000		617,000
Cd[c]	1,123,000	46	30,000
Hg[a]	232,000	522	3,910,000
Hg[b]	232,000		20
Hg[c]	232,000	1649	1300
Cu[a]	0	0	3,910,000
Cu[b]	0	0	20
Cu[c]	0	0	1300
Cr[a]	0	170	599,000
Cr[b]	0	0	570
Cr[c]	0	0	
Ni[a]	0	0	551,000
Ni[b]	0	0	3380
Ni[c]	0	0	3400
Zn[a]	0	360	120,000
Zn[b]	0	0	6.70
Zn[c]	0	0	3100
Consumption of biomass	0	0.34	0
Consumption of crude oil	0	4.30	11.70
Consumption of coal	0	0.42	0.16
Consumption of natural gas	0	9.35	8.83

[a] Emission to air.
[b] Emission to water.
[c] Emission to land.

Table 3.4 Amount of and composition of waste (tons/year).

	Detached houses	Flats	Rural houses	*Domestic waste, sum*	Waste from business*	*Total amount of waste*
Degradable waste	5642	9490	2655	*17,787*	5645	*23,432*
Non-combustible residue	549	924	258	*1732*	1408	*3140*
Combustible residue	1930	3246	908	*6085*	7370	*13,455*
Diapers	831	1398	391	*2621*		*2621*
Rubber, textiles, etc.	401	674	189	*1264*		*1264*
Dry paper	2762	4645	1300	*8706*	3678	*12,384*
Cardboard	787	1324	370	*2481*	1096	*3577*
Plastic sheets and bags	327	549	154	*1030*	488	*1518*
Plastic containers	223	375	105	*702*	340	*1042*
Laminate	163	275	77	*515*		*515*
Glass	950	1598	447	*2996*	340	*3335*
Metals	282	474	133	*889*	1592	2481
Sum	*14,848*	*24,972*	*6988*	*46,809*	*21,957*	*68,765*

* Exclusive construction and demolition wastes.

Table 3.5 Description of scenarios.

- Incineration of all waste
- Landfilling of all waste
- Anaerobic digestion of biodegradable waste.* The biogas is used as fuel for buses. The rest of the waste is incinerated
- Anaerobic digestion of biodegradable waste.* The biogas is used for production of district heat and electricity. The rest of the waste is incinerated
- Composting of biodegradable waste* in open windrows. The rest of the waste is incinerated
- Composting of biodegradable waste* in a closed reactor. The rest of the waste is incinerated

* 70% of the biodegradable waste is sorted out at source and collected for biological treatment.

3.5.4 Sensitivity analyses

Several important choices are made in the goal definition and scoping stage, e.g. when choosing system boundaries and when using process data, see Section 3.3 above. The importance of different "municipality specific" and "site specific" parameters were studied in a sensitivity analysis. The results were thoroughly analysed to identify parameters that could have a potential influence on the results. The following parameters were found to be of interest in the sensitivity analysis:

- Different methods of electricity production in the compensatory system: natural gas combi, coal condense or "average" Swedish electricity. The Swedish average electricity production is based on mainly hydropower and nuclear power, which gives very small environmental impact with the impact categories that have been studied.
- Different fuels for district heating in the compensatory system: biofuel or oil.

- A variation of performance characteristics on waste incineration and alternative district heating production. Especially thermal efficiency and NO_X-emissions are of interest.
- Combined heat and power production from waste or heat station only.
- Different distances to waste treatment plants (from the centres of population).
- Different distances to arable land for compost and anaerobic digestion residue.
- Different economic weighting for the environmental costs.

3.6 FURTHER ANALYSIS OF ORWARE DATA – RESULTS

3.6.1 Environmental impacts

In the diagrams the waste system is displayed as one data category (the black part in the staples). This category comprises all processes in the waste management system as collection, transports, and the type of treatment used in the specific scenario including downstream processes as landfilling of residues and spreading of organic fertiliser.

3.6.1.1 Emissions of greenhouse gases

Figure 3.3 presents the results for global warming (emissions of greenhouse gases). Greenhouse gases are especially fossil carbon dioxide (CO_2) from combustion of fossil fuels and plastic, and methane gas (CH_4) from landfilling of organic waste. Landfilling gives the worst impact due to methane emissions from the landfill. It should be observed that the landfill has a landfill gas recovery system and produces electricity from the landfill gas. However, the efficiency of the gas recovering system is poor – only 50% of the generated methane gas is recovered (Björklund 1998; Sundqvist 1999). Anaerobic digestion has lower emissions of greenhouse gases than incineration, because fossil fuels are replaced when utilising the biogas.

3.6.1.2 Emissions of acidifying substances

Emissions of acidifying substances are given in Figure 3.4. Acidifying substances are mainly gases such as SO_2 (e.g. from fossil fuels), HCl (from waste incineration), NO_X (from all combustion processes: incineration, district heating production, engines, etc.), and ammonia from the composting process and the spreading of compost and anaerobic digestion residue. Landfilling gives the highest emissions of acidifying gases, due to NO_X-emissions from the landfill gas combustion, and from district heat production in the compensatory system. Windrow composting gives a high emission due to ammonia releases from the compost process. As expected the emissions from the reactor compost is lower than from the windrow compost. Anaerobic digestion with production of heat and electricity gives high NO_X-emissions from the combustion engine.

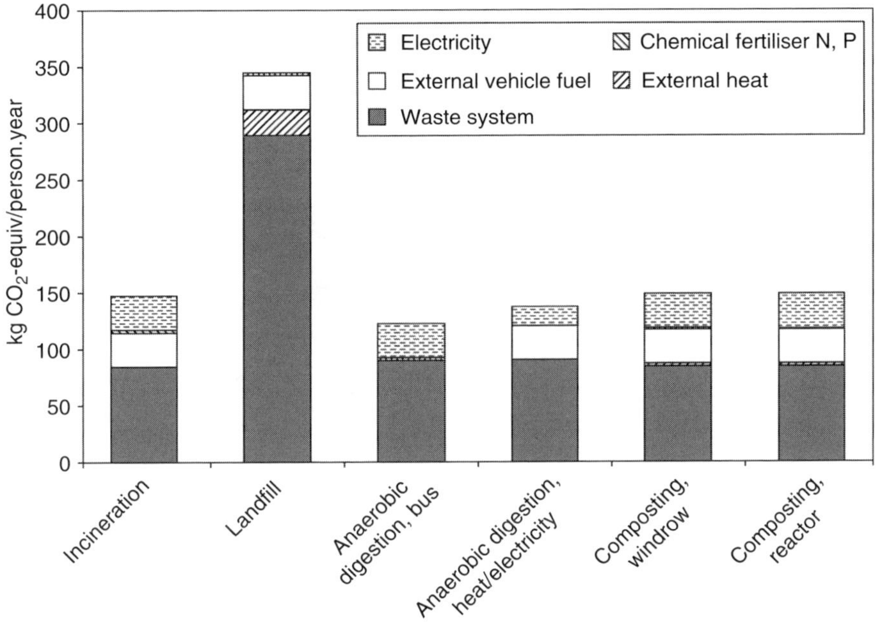

Figure 3.3 Emissions of greenhouse gases.

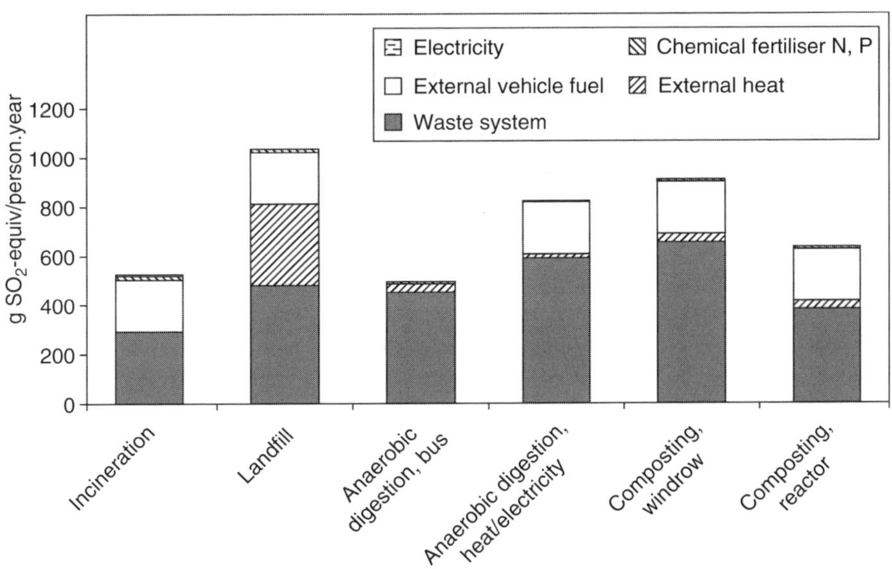

Figure 3.4 Emission of acidifying substances.

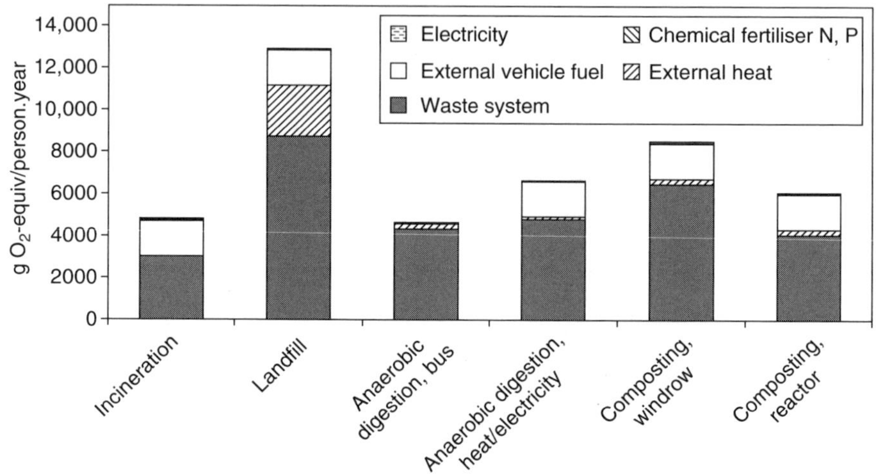

Figure 3.5 Emissions of eutrophicating substances.

3.6.1.3 *Emissions of eutrophicating substances*

The emissions of eutrophicating substances are given in Figure 3.5. Eutrophicating substances are N- and P-compounds in water, COD in water, NO_X in combustion gases, and ammonia (NH_3) releases from spreading of anaerobic digestion residue and compost. Landfilling gives the highest eutrophication impact depending on N- and P-compounds in the leachate water. Anaerobic digestion and composting causes emissions from spreading of the digestion residue respectively compost. The spreading model is based on new spreading techniques where the material (digestion residue or compost) is cultivated into the soil and immediately covered with soil to decrease the release of ammonia. In the first case studies (Sundqvist *et al.* 1999) the digestion residue and compost were assumed to just be spread on the soil, which caused a lot more releases of ammonia, so the spreading of compost and digestion residue caused higher eutrophicating emissions than spreading of chemical fertiliser.

3.6.1.4 *Emissions of photooxidant formers*

Photooxidant formers have been divided into VOC (volatile organic compounds) and NO_X. The VOC and NO_X emissions are shown in Figures 3.6 and 3.7. Methane is included in the VOC. Landfilling gives the highest emissions due to the methane emissions. Anaerobic digestion gives higher emissions than incineration, depending on emissions from the biogas use. Landfilling gives the highest NO_X-emissions, depending on emissions both from the landfill gas combustion, and from the district heat production in the compensatory system. The two anaerobic digestion alternatives give different results. Using the biogas as bus fuel

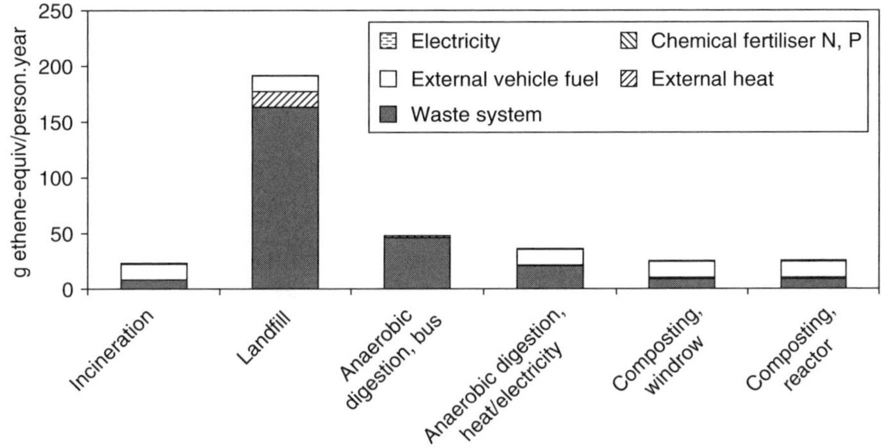

Figure 3.6 Emissions of photooxidants – VOC.

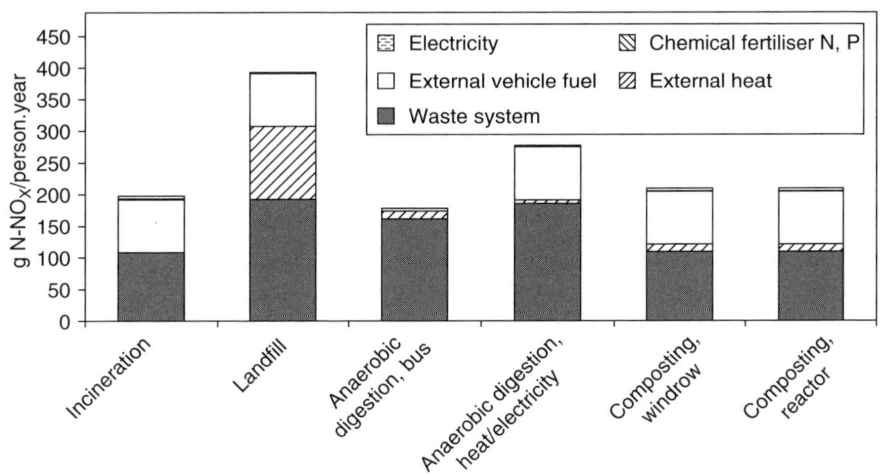

Figure 3.7 Emissions of photooxidants – NO_X.

gives lower emissions of NO_X than using the biogas for electricity and heat production in a gas engine.

3.6.2 Energy – consumption of primary energy carriers

In Figures 3.8 and 3.9, the energy consumption, measured as consumption of energy primary carriers, for the different scenarios is given. There is a net consumption of

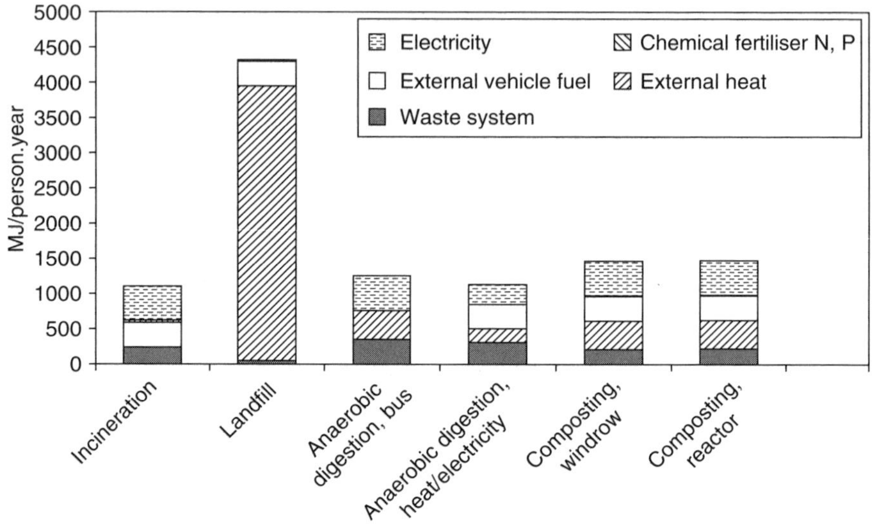

Figure 3.8 Consumption of primary energy carriers – total consumption for different processes.

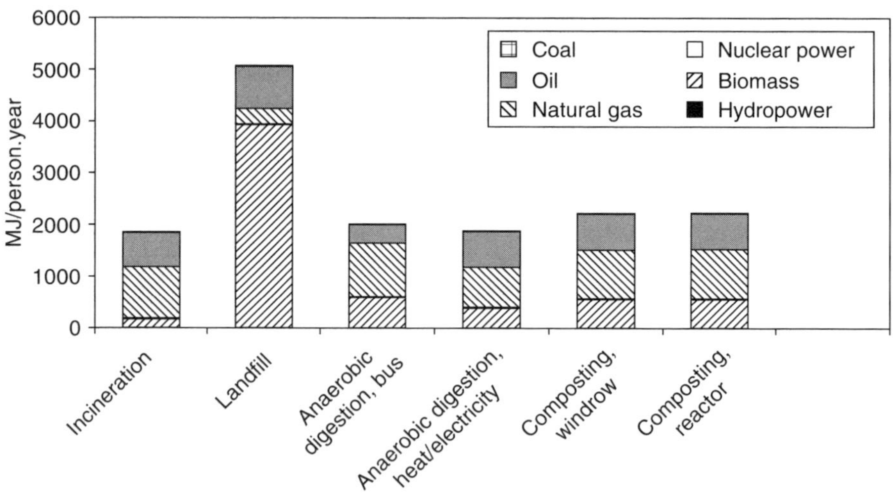

Figure 3.9 Consumption of primary energy carriers – consumption of different primary energy carriers.

energy for the whole system (including the compensatory system), because of the choices of functional units. In general the differences in energy consumption between the scenarios are small, except for the landfill scenario in which consumption of energy resources is much higher. This is because of the production of district heating, fuels and fertilisers in the compensatory system.

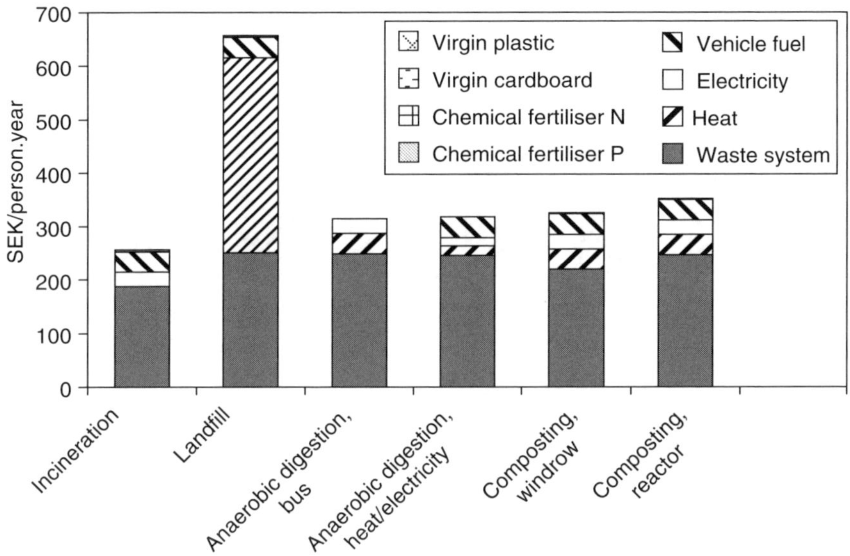

Figure 3.10 Financial life cycle costs (total system).

Another result from the study, not shown in the diagrams, is that the energy consumption for collection and transport of waste is small compared to the energy consumption of the other processes in the studied system.

3.6.3 Economy

3.6.3.1 Financial life cycle costs

The financial life cycle costs are shown in Figure 3.10. Biological treatment (anaerobic digestion and composting) is more expensive than incineration. Landfilling is the most expensive waste treatment due to the landfill tax, and because of the costs for producing new district heating and vehicle fuel in the compensatory system.

3.6.3.2 Environmental costs

The environmental costs have been calculated with three different methods. The result from the "ORWARE method" is shown in Figure 3.11. The other methods give some different results. The costs are lower in "ORWARE" than in EPS 2000, which in turn is lower than ECOTAX. ORWARE has no valuation of the natural sources, which EPS 2000 and ECOTAX 99 have. ORWARE shows that biological treatment gives higher environmental costs than incineration. EPS 2000 shows

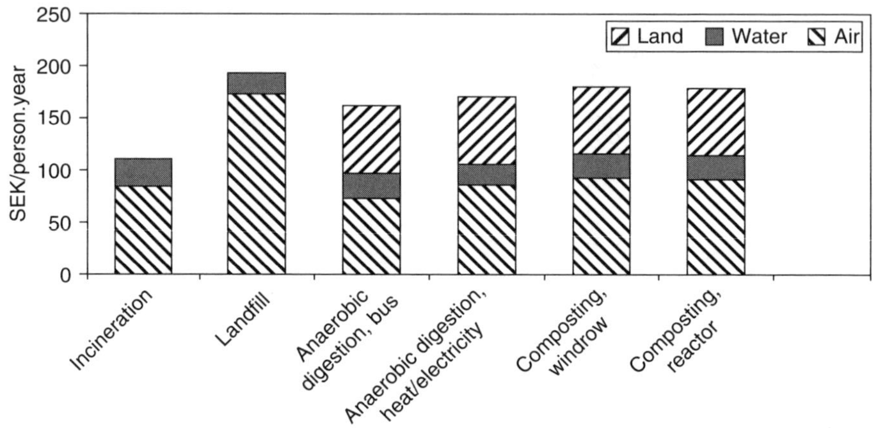

Figure 3.11 Environmental costs – weighting factors according to ORWARE.

that anaerobic digestion has lower environmental costs than incineration, but composting has higher costs. ECOTAX 99 shows that the environmental cost for anaerobic digestion is approximately the same (or slightly higher) than for incineration, and that the environmental costs for composting is higher. All three methods show that landfilling has the highest environmental costs.

3.6.3.3 Welfare economy (financial plus environmental costs)

The total welfare costs have been calculated as the financial life cycle costs, excluding environmental taxes and fees, plus the environmental costs (with three different methods, see above). All three environmental valuation methods show that biological treatment (both anaerobic digestion and composting) is more expensive than incineration. Landfilling is the absolutely most expensive alternative in all three methods.

3.6.4 Sensitivity analysis

The results from the sensitivity are summarised in Table 3.6.

3.7 THE RELEVANCE FOR OTHER COUNTRIES

The study was based on typical Swedish conditions. The most specific parameter is probably the performance of the incineration plant. The energy in the waste is

Table 3.6 Summary of sensitivity analysis.

Scenario	Changes compared to base scenario
Production of electricity	
Coal condense	Anaerobic digestion with production of electricity and heat gives slightly lower total energy consumption than incineration. Ranking is not changed
Swedish average electricity*	No change in ranking
District heating production	
Oil	The two anaerobic digestion alternatives gives higher emissions of greenhouse gases than incineration
Thermal efficiency	No change in ranking
NO_X-emissions (high NO_X-emission for waste incineration, low NO_X-emission for biofuel or oil)	No change in ranking
NO_X-emissions (low NO_X-emission for waste incineration, high NO_X-emission for alternative biofuel or oil)	Incineration gives lower NO_X-emission, lower emissions of acidifying substances, lower emissions of eutrophicating substances than anaerobic digestion
Power production	
Combined power and heat production	No changes in ranking
Transport distance	
500 km distance to incineration	No change in ranking for environmental impacts categories and energy consumption. But one of the three environmental cost valuation methods anaerobic digestion gives lower welfare costs than incineration
150 km distance to incineration	No change in ranking
Spreading of anaerobic digestion residue and compost	
Distance to arable land 50 km	No change in ranking
The compost and the digestion residue can not be spread but have to be incinerated	The welfare costs for composting becomes lower than for anaerobic digestion, but still higher than incineration
Valuation of resources	
Doubled price for energy and doubled environmental costs for greenhouse gases.	No change in ranking, but all financial life cycle costs, and welfare costs increases
Energy price and valuation of greenhouse gases increases by a factor = 5	No change in ranking, but life cycle costs and welfare costs for all scenarios increases
The phosphorus price increases by a factor = 10	No change in ranking. The phosphorus price has to increase by a factor 100 to make the welfare costs for anaerobic digestion and composting equal to the costs for incineration
Landfill as a carbon sink	The difference between landfilling (of all waste) and incineration (of all waste) decreases, but landfilling is still the most favourable

* The Swedish average electricity production is based on mainly hydropower and nuclear power, which gives very small environmental impact with the impact categories that has been studied.

used for district heating production. The incineration plant has a flue gas condensation system, which gives a very efficient flue gas cleaning, but also a high-energy recovery rate. The net energy recovery is about 100% (relative the lower heating value), and the energy is used for district heating which can be sold during almost the whole year. The high-energy recovery rate results in relative favourable results for incineration. In several other countries the opportunities for selling or using energy are more limited.

The modelled municipality has a city area, a suburban area and a rural area. The majority of the waste is collected from the city and suburban region. The structures of the city and suburban area are estimated to not differ from other EU municipalities and cities. The rural part in the study was typical Swedish, but it is of lower relative importance.

3.8 CONCLUSIONS

The most obvious conclusion is that landfilling should be avoided. Wastes that can be incinerated (combusted), anaerobically digested or composted should not be landfilled. This is valid even if landfill gas is recovered, and the leachate is collected and treated. This is due to that the resources in the waste are inefficiently utilised when landfilled, making it necessary to produce materials, fuels and nutrients from virgin resources.

It is impossible to draw unambiguous conclusions of which of the other treatment options is "most preferable". There are advantages and disadvantages with all treatment options. In a systems perspective there are small differences between incineration and aerobic digestion of easy degradable organic waste. Anaerobic digestion and incineration should not be seen as competing options, but as complementary options. Since it is impossible to obtain 100% biological treatment, there will always be some combustible waste that has to be incinerated. There should always be possibilities for incineration, even if the waste has to be transported to a regional incineration plant.

Composting of degradable waste has hardly any advantages with respect to environment and energy turnover when being compared to incineration or anaerobic digestion. Composting gives a higher welfare cost than anaerobic digestion and incineration.

Transports of waste, when the waste has been collected, is of very low importance considering the consumption of energy resources, environmental impact and costs, if the transport is performed in an efficient manner. The type of collection system has very low influence on the total consumption of energy resources, the total environmental impact and the total welfare costs.

Home composting was not analysed in the study. Judging from the study it seems that home composting has no, or only a few, advantages compared to incineration: the saved collection and transport work is of minor importance, and open composting gives lower energy utilisation and more emissions of acidifying and eutrophicating substances than incineration. However, that conclusion is valid if the compost only substitutes chemical fertiliser in the garden. Most home gardeners use peat or bark for soil improvement, which can be substituted by the own compost. Home composting may be motivated in that perspective.

REFERENCES

Björklund, A. (1998) *Environmental Systems Analysis of Waste Management with Emphasis on Substance Flows and Environmental Impact*, Licentiate thesis. The Royal Institute of Technology, Department of Chemical Engineering and Technology, Sweden (ISSN 1402-7615, TRITA_KET_IM 1998:16, AFR-Report 211).

Carlsson, M. (1997) *Economics in ORWARE – A Welfare Analysis of Organic Waste Management.* Report 117. Department of Economy, Swedish University of Agricultural Sciences.

Dalemo, M., Sonesson, U., Björklund, A., Mingarinai, K., Frostell, B., Jönsson, H., Nybrant, T., Sundqvist, J.-O. and Thyselius, L. (1997) ORWARE – a simulation model for organic waste handling systems, Part 1: Model description. *Resour. Conserv. Recycl.* **21**, 17–37.

Dalemo, M., Sonesson, U., Jönsson, H. and och Björklund, A. (1998) Effects of including nitrogen emissions from soil in environmental systems analysis of waste management strategies. *Resour. Conserv. Recycl.* **24**, 363–381.

Eriksson, O., Frostell, B., Björklund, A., Assefa, G., Sundqvist, J.-O., Granath, J., Carlsson, M., Baky, A. and Thyselius, L. (2000) ORWARE – a simulation tool for waste management. *Resour. Conserv. Recycl.* **36/4**, 287–307.

Eriksson, O., Carlsson-Reich, M., Frostell, B., Björklund, A., Assefa, G., Sundqvist, J.-O., Granath, J., Baky, A. and Thyselius, L. (2002) Municipal solid waste management from a systems perspective (accepted by J. Clean. Product.).

ISO (1997) *Environmental Management – Life Cycle Assessment – Principles and Framework*. ISO 14040.

ISO (1998) *Goal and Scope Definition and Inventory Analysis*. ISO 14041.

ISO (1999) *Life Cycle Impact Assessment*. ISO 14042.

ISO (2000) *Life Cycle Interpretation*. ISO 14043.

Lindfors, L.-G., Christiansen, K., Hoffman, L., Virtanen, Y., Juntilla, V., Hanssen, O.-J., Rønning, A., Ekvall, T. and Finnveden, G. (1995) LCA-Nordic Technical Reports No 1–9. TemaNord 1995:502. Nordic Council of Ministers, Copenhagen, Denmark.

Sonesson, U., Dalemo, M., Mingarinai, K. and Jönsson, H. (1997) ORWARE – a simulation model for organic waste handling systems, Part 2: Case study and simulation results. *Resour. Conserv. Recycl.* **21**, 39–54.

Sundqvist, J.-O. (1999) *Life Cycle Assessments of Solid Waste*. AFR Report 279 (Swedish Environmental Protection Board 1999, http://www.naturvardsverket.se).

Sundqvist, J.-O. (2001) Material and nutrient recycling and energy recovery from solid waste: a systems perspective. In: *Water Recycling and Resource Recovery in Industries: Analysis, Technologies and Implementation* (eds. Lens P *et al.*), 2002 IWA Publishing.

Sundqvist, J.-O. (2002) LCA for treatment and disposal of urban waste. In: *Proceedings from International Conference on Life Cycle Analysis, Efficiency and Cost-Benefit* to be held next 7th November 2002 in Madrid.

Sundqvist, J.-O., Finnveden, G., Albertsson, A.-C., Karlsson, S., Berendson, J., Höglund, L.O. and Stripple, H. (1997) *Life Cycle Assessment and Solid Waste – Stage 2*. AFR-Report 173; AFR, Stockholm, Sweden.

Sundqvist, J.-O., Baky, A., Björklund, A., Carlsson, M., Eriksson, O., Frostell, B., Granath, J., Thyselius, L. (2000) *Systemanalys av energiutnyttjande från avfall – utvärdering av energi, miljö och ekonomi*. Översiktsrapport. (*System Analysis of Energy from*

Waste – Assessment of Energy, Ecology and Economy.) Summary Report Stockholm December 1999. IVL Report B 1379 (In Swedish, with English summary), http://www.ivl.se/reports

Sundqvist, J.-O., Baky, A., Carlsson, M., Eriksson, O., Granath, J. (2002) *Hur skall hushållsavfallet tas omhand – Utvärdering av olika behandlingsmetoder.* (*How Shall Municipal Solid Waste be Disposed – Assessment of Different Treatment Methods.*) Stockholm February 2002. IVL Report B 1462 (In Swedish, with English summary), http://www.ivl.se/reports

4

Biowaste management from an ecological perspective

F. Knappe, R. Vogt and B. Franke

4.1 INTRODUCTION

The German "Commercial and Industrial Waste Management Act" (Kreislaufwirtschafts- und Abfallgesetz – KrW-/AbfG) promotes the treatment of waste materials. With respect to the recycling of waste, two ecological aspects need to be considered; while priority has to be given to recycling systems that are "more compatible with the environment", those that are of "high grade" should be used as well. Both have to be taken into account by waste generators and waste proprietors as far as they are technically feasible and economically reasonable.

Consequently, the selection of a particular waste management option requires the proof that it is more environmentally friendly than other existing alternatives. An environmental ranking of the complete systems of waste management is needed, the simple comparison between the techniques for composting and fermentation is not sufficient. A research project with the support by the German Environmental Foundation (Deutsche Bundesstiftung Umwelt) resulted in a novel approach

 Resource Recovery and Reuse in Organic Solid Waste Management. Edited by Piet Lens, Bert Hamelers, Harry Hoitink, Werner Bidlingmaier. ISBN: 1 84339 054 X. Published by IWA Publishing, London, UK.

(Giegrich 1999) and was finalized in 2002 (Vogt *et al.* 2002). It allows the comparison of recycling systems for biological waste in order to determine which of various options is most compatible with the environment. The method is based on the life cycle assessment approach as standardized in the DIN EN ISO 14040ff. Hence the environmental analysis is not limited to the performance of recycling plants, but extends to the environmental impacts of the entire recycling system of specific biological wastes from the source of generation to its application as a secondary material (Figure 4.1).

According to the life cycle assessment approach, different systems are only comparable when they comply with the principle of the equivalent benefit. Therefore, not only the waste treatment itself has to be analyzed but also the benefit derived from the use of the resulting products. The system boundaries of the life cycle assessment include the environmental impacts that would have been associated with the production of substituted primary products. The environmental impacts from these "equivalent processes" are quantified and credited to the analyzed recycling route.

Usually, the environmental effects from the generation of the waste itself are not considered in the life cycle assessment of wastes, because wastes are not produced on purpose. Therefore, the environmental impacts from the equivalent processes have often a very high influence on the results of the ecological valuation, especially when high quality products are substituted. For example, despite higher energy consumption and plant-specific emissions, a waste management system that with a high quality secondary product can be the better option from an ecological point of view due to the saving of harmful effects that would have been associated with the production of primary product that is being replaced.

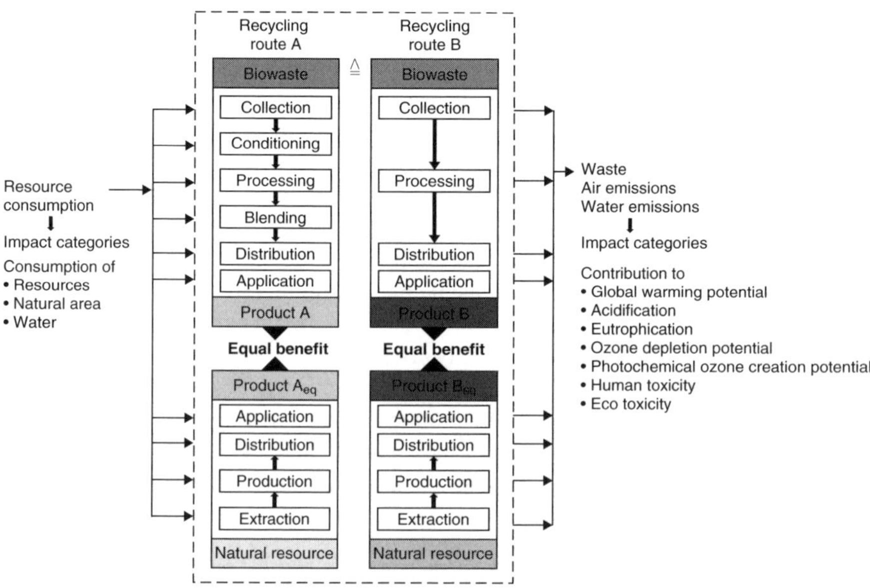

Figure 4.1 Method for comparative evaluation of recycling systems for biogenic waste.

Figure 4.1 shows two different recycling systems for a certain biogenic waste that result in two different secondary products (products A and B). Each secondary product substitutes a certain primary product (products A_{eq} and B_{eq}). The environmental impacts from the production of these primary products may be significantly different. It is therefore possible that recycling route A has the least environmental impacts and costs, although recycling route A is more expensive, as it may cause the less environmental effects and hence less indirect costs due to averted impacts from manufacturing the primary product.

4.2 MATERIALS AND METHODS

4.2.1 Types of organic waste management

The first step in the project (Vogt *et al.* 2002) was the definition of biogenic waste and the selection of treatment possibilities. The following rationale was used: the focus should be on biogenic waste that requires pretreatment on the basis of current or future regulations, for example mulch or fertilizer in agriculture or garden areas. For these waste types, a range of treatment and utilization options should exist, including the fermentation of liquids, the composting of solids and the combustion of wastes with high energy content.

Aside from the standard waste treatment options, the ecological impact of the use in agriculture or goods production has to be analyzed. The list of biogenic wastes along with waste management options that are analyzed in the project is shown in Table 4.1. For each treatment option, various technologies were selected for the investigation, for example high-tech as well as low-tech composting

Table 4.1. Selection of biogenic waste and an array of types of waste management.

Commercial biogenic waste	Types of waste management
Apple juice production residues	animal feeding; agricultural utilization; composting; pectin production
Biowaste from households (source separated)	fermentation; co-fermentation with slurry; co-fermentation with sewage sludge; composting
Crude wood	composting; energetic utilization; mulching
Distillery slop	co-fermentation with slurry, with sewage sludge and with biowaste from households; agricultural utilization
Fatty waste	co-fermentation with slurry, with sewage sludge and with biowaste from households; feedstuffs production; processing of fats
Kitchen wastes from canteens	co-fermentation with slurry, with sewage sludge and with biowaste from households; animal feeding
Yard wastes from public play grounds and graveyards	mulching; composting

systems. The range of fermentation technology is represented by wet and dry fermentation processes.

The discussed options are described below, briefly focusing on the aspect of the marketing of compost products or other goods deriving from waste management. The product of the recycling route composting is either premature or mature compost. The fermentation products are either directly used in agriculture as fertilizer or as mature compost. The product of co-fermentation with slurry is not dewatered but used completely and directly in the agriculture on nearby farmland. The product of co-fermentation with sewage sludge is dewatered sludge and then used in various applications (22% incineration, 52% agriculture and 26% recultivation).

The necessary environmental data to model the selected recycling routes were mainly derived from literature (e.g. Bidlingmaier and Müsken 1994; Eurich-Menden *et al.* 1996; Gronauer *et al.* 1997; Körner *et al.* 1997; Bidlingmaier and Schmelz 1998; KTBL 1998) and from questionnaires sent to process operators. Since information gathered this way usually refers to technical rather than environmental aspects, further assumptions were necessary and were validated by a sensitivity analysis. For example, most empirical studies on co-fermentation processes report a synergistic effect resulting in a higher biodegradation rate of organic matter. The reason for this synergetic effect is not known and cannot be properly quantified. In our study a 60% higher degradation rate for co-fermentation processes in relation to standard fermentation processes was assumed (Bidlingmaier and Schmelz 1998). A sensitivity analysis showed that this assumption is not a decisive one with respect to the final results (see e.g. Table 4.2 and Figure 4.13).

4.2.2 System boundaries

The system boundaries for each modeled recycling route are shown using representative examples: the recycling of biowaste from households with wet fermentation (Figure 4.2), the recycling and local combustion of wood chips (Figure 4.3) and the processing of fatty waste to biofuel (Figure 4.4).

An alternative to the fermentation, the recycling of fatty waste, could be the transesterification with methanol and subsequent processing to cooling lubricants or biodiesel as a fuel (Bock 2001). Using biodiesel in district heating and power plants are common, whereas the use as biodiesel in vehicles is rather uncommon in Germany. A substitution of conventional diesel according to the heat value is the intention of this recycling process. The use as a cooling lubricant also requires several steps of processing the fatty waste. Separation of unsaturated fatty acid and transesterification with recycling of methanol is the intention of using the fatty waste as cooling lubricants.

There are several waste management options for uncontaminated crude wood: aside from using it for the production of compost and mulch with fertilizing properties substituting bark products, it can be processed into wood chips. Wood chips are common in local heating plants, substituting especially heating oil. In addition, ashes can be used as a fertilizer especially in gardens.

Table 4.2 Description of the analyzed scenarios for the utilization of biowaste from household, fatty waste and crude wood.

Scenario	Description
Composting	
PrivatComp.	Private composting in the backyard; result: premature compost but used like mature compost
CompClosed_1	Composting in a close treatment system over the hole composting process; result: premature compost
CompClosed_2	Composting in a close treatment system over the hole composting process; result: mature compost
CompOpen_1	Composting in a just roofed system; result: premature compost
CompOpen_2	Composting in a just roofed system; result: mature compost
Fermentation	
FermDry_1	Dry fermentation, one-stage, thermophilic; result: biogas and premature compost
FermDry_2	Dry fermentation, one-stage, thermophilic; result: biogas and mature compost
FermWet_1	Wet fermentation, one-stage, despoil; result: biogas and premature compost
FermWet_2	Dry fermentation, one-stage, mesophilic; result: biogas and mature compost
Co-FermSlurry	Co-fermentation with slurry; result: biogas and utilization of slurry in the farmland nearby
Co-FermSludge	Co-fermentation with sewage sludge; result: biogas and disposing of dewatered sludge
Others	
Mulch	Comminution of crude wood and utilization as mulch
Heating	Comminution of crude wood into wood chips and utilization as heating material in households; using air pollution control in a low standard
Coolant	Processing of fatty waste and utilization as coolant
Biodiesel	Processing of fatty waste to fatty acid methyl ester and utilization as biodiesel
Incinerator	Disposal of biowaste with refuse bin and incineration

As previously mentioned, the substitution effects of secondary products can have a significant influence on the results: the higher the utilization value, the higher the substitution effects in most cases. The study considered usage scenarios for the residues that are representative for the situation in Germany.

In principal, different secondary materials have a large range of potential applications that depend on the technology. In order to estimate the potential for secondary products to replace primary products, the market situation in Germany was examined, considering public acceptance issues as well. For example, mature compost derived from fermentation with subsequent composting has a low content of salt and nutrients matching the properties of peat. It could thus be used as a substitute for peat in humus products of high quality. Until today compost products already replace peat in commercial garden mulch, amounting to 15% of the

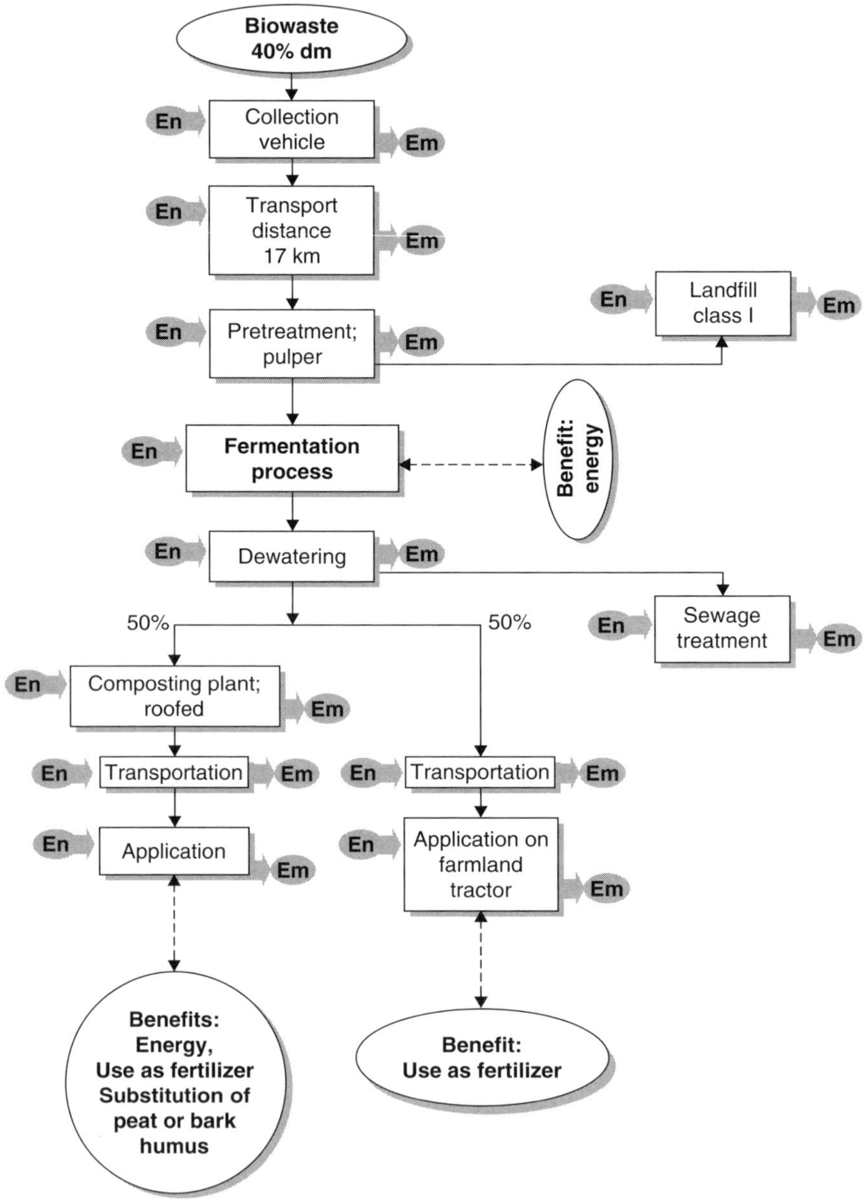

Figure 4.2 Waste management system for biowaste from households with fermentation (En = energy and feedstock; Em = emissions).

total substitution potential in Germany. Accordingly, the environmental credit for using secondary products can be assessed depending on the marketing potential. Table 4.3 shows the average utilization of mature compost and mature composted fermentation residue in Germany.

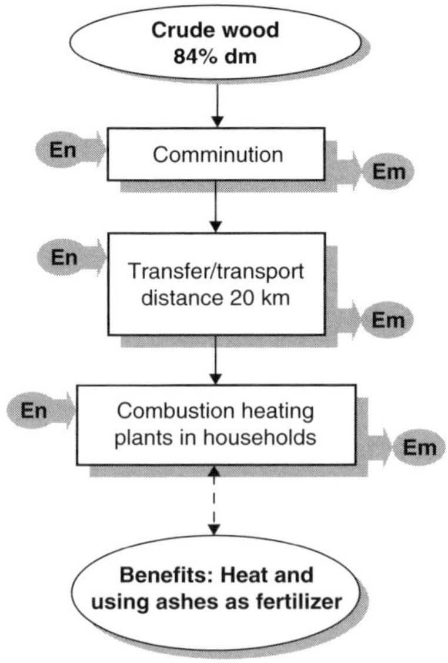

Figure 4.3 Waste management system for wood with local combustion (En = energy and feedstock; Em = emissions).

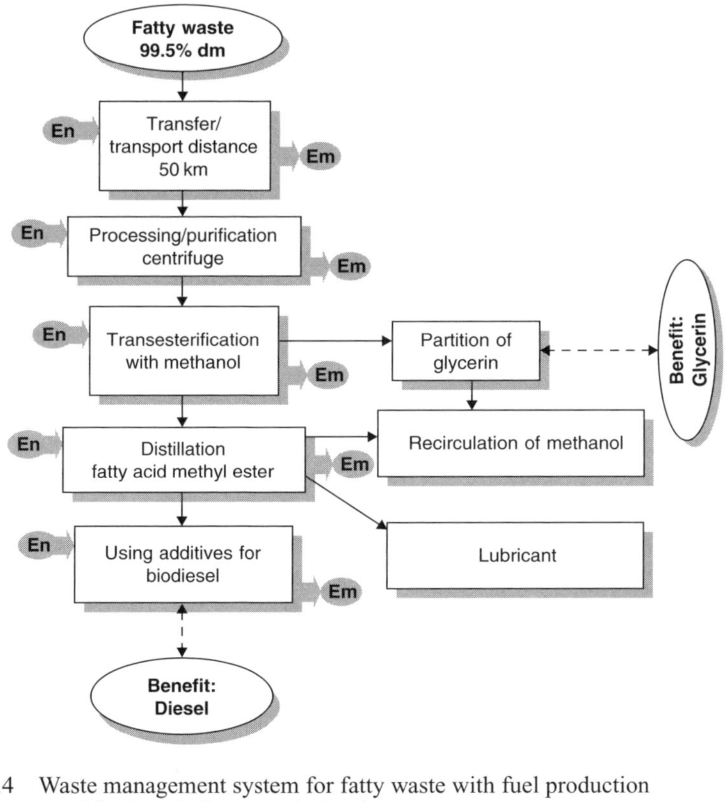

Figure 4.4 Waste management system for fatty waste with fuel production (En = energy and feedstock; Em = emissions).

Table 4.3 Use of the secondary products compost and composted fermentation residue in Germany.

Application	Average utilization of mature compost (%)	Average utilization of composted fermentation residue (%)	Substitute
Recultivation, others	39	37	Earth fill
Private gardening	21	3	50% peat, 50% bark humus; fertilizer
Municipalities	6	1	50% peat, 50% bark humus; fertilizer
Agriculture	6	46	Fertilizer
Fruit farming, Horticulture	13	3	Fertilizer
Market Gardening	8	2	100% peat
Earth industry	7	8	100% peat (substrates)

4.3 RESULTS

4.3.1 Biowaste from households

The results for some impact categories are shown in Figures 4.5–4.8. Since the utilization and the corresponding substitution benefits are of primary importance, the results for the global warming potential depend largely on the substitution effects. The waste management systems with fermentation processes as main treatment show evident net benefits in the impact category *global warming potential* (Figure 4.5). The substitution effects are marginally higher than those resulting from the treatment processes alone. The advantages of waste management systems with co-fermentation are evident when compared to other types of digestion. The results are based on the presumption of higher biogas efficiency in the co-fermentation process for which about 80% of the substitution effects result from electricity generation. Fermentation processes with mature compost production are marginally less beneficial since due to the positive effects of substitution of peat.

The substitution of peat is the main reason for the high rank of several composting processes that produce mature compost. The resulting benefits can outweigh the direct effects that are dominated by the emissions of N_2O and CH_4, accounting for nearly 30% on the direct effects. Due to the lack of viable options to reduce N_2O emissions by biofilters, there is no difference between the enclosed or open composting systems. The negative effect of the systems with fermentation is mainly due to the N_2O emissions from land spreading. As there is no loss of nitrogen during the treatment process itself, the emissions resulting from the application of slurry are the highest. The scenario "composting in the backyard" is extraordinary beneficial due to very high substitution effects. While composting in the backyard results just in premature compost, this kind of compost in the backyard is usually applied like mature compost and thus competes with bark

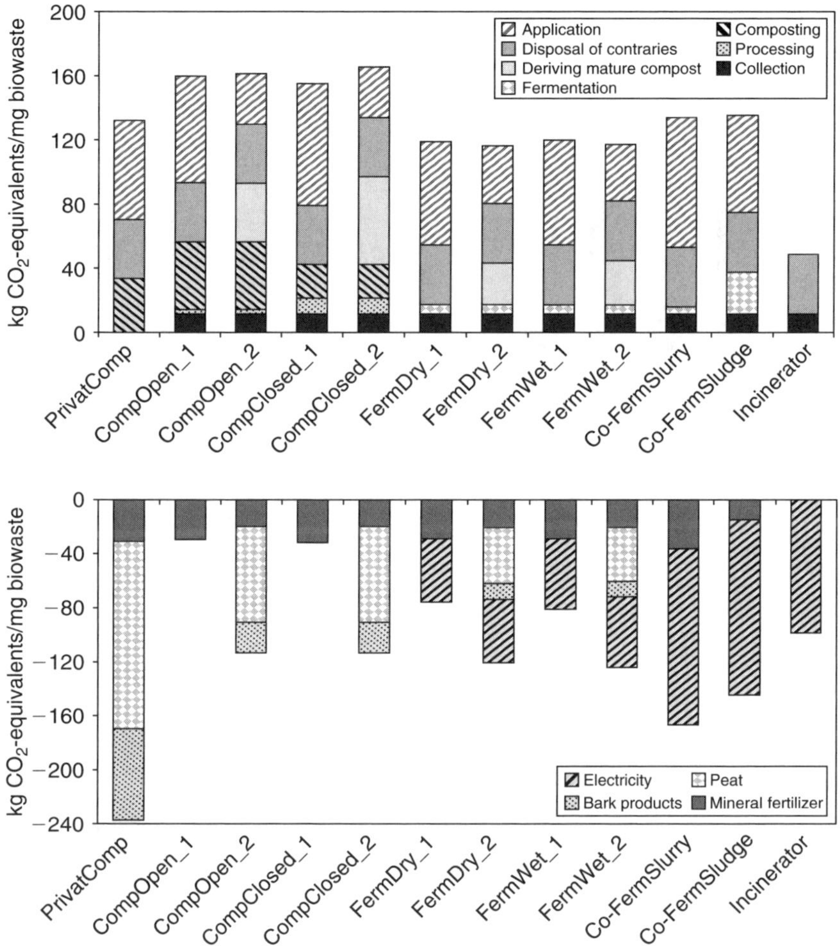

Figure 4.5 Results of the ecological valuation for the global warming potential (top: direct effects; bottom: substitution effects).

products and peat. Due to the larger quantity of premature compost in comparison to mature compost the substitution effects are the highest.

The ranking of systems differs for the terrestrial eutrophication potential (Figure 4.6). The substitution effects are very small compared with the direct effects. The latter are dominated by ammonia emissions from the application of residues or from open composting. Biofilter have a high efficiency (~96%) to reduce ammonia emissions. Waste management systems that produce the largest amount of mature compost show the largest values. While composting in open process is combined with high emissions, composting in a closed system denotes the other extreme. There are negligible differences between those management systems based on biowaste digestion. Due to the residual content of nitrogen,

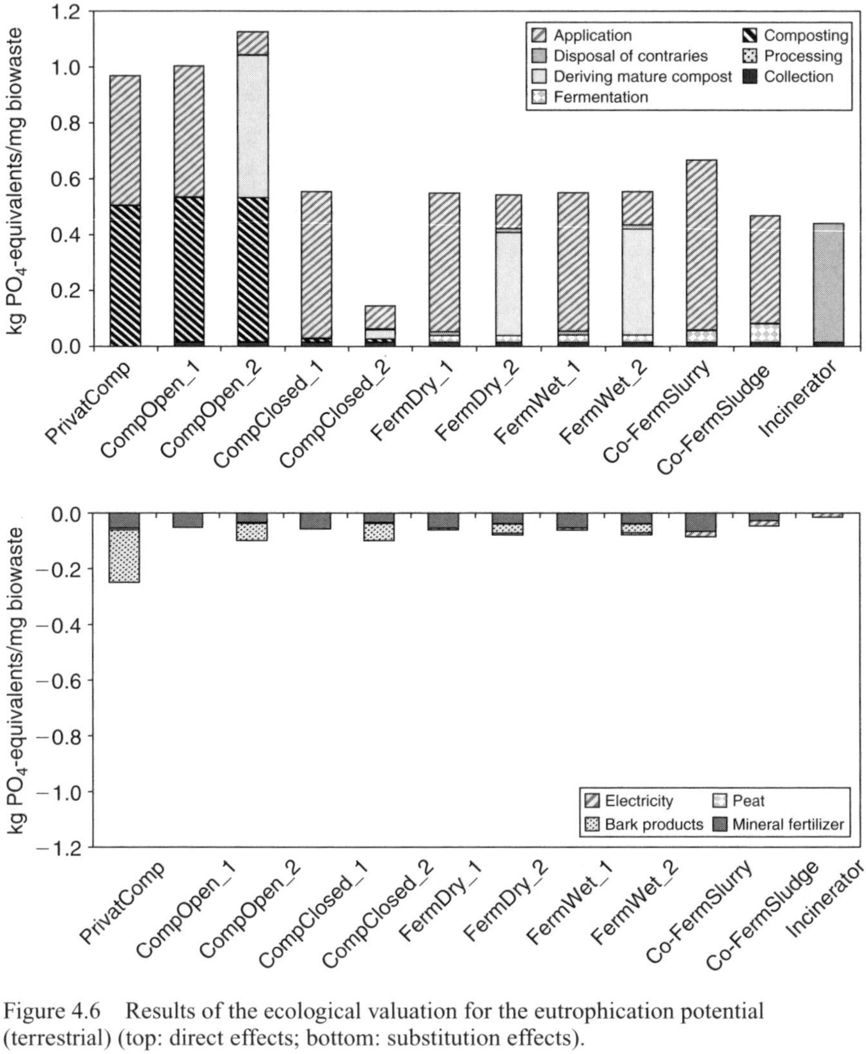

Figure 4.6 Results of the ecological valuation for the eutrophication potential (terrestrial) (top: direct effects; bottom: substitution effects).

ammonia emissions occur either in the composting process itself or after application to soil. Processing of mature compost in a closed system would lead to results comparable to the management system CompClosed_2. The assumption that 22% of the residues from co-fermentation with sewage sludge are incinerated is the reason for slightly better results for that waste management system.

Due to the importance of ammonia emissions, the results for the *PM10 risk* are similar to that for eutrophication potential (Figure 4.7 compared to Figure 4.6). The negative effects to 85% (for private composting 100%) depend on the emissions of ammonia as a precursor for secondary particulates. Waste management systems with a significant portion of combustion processes are different. For

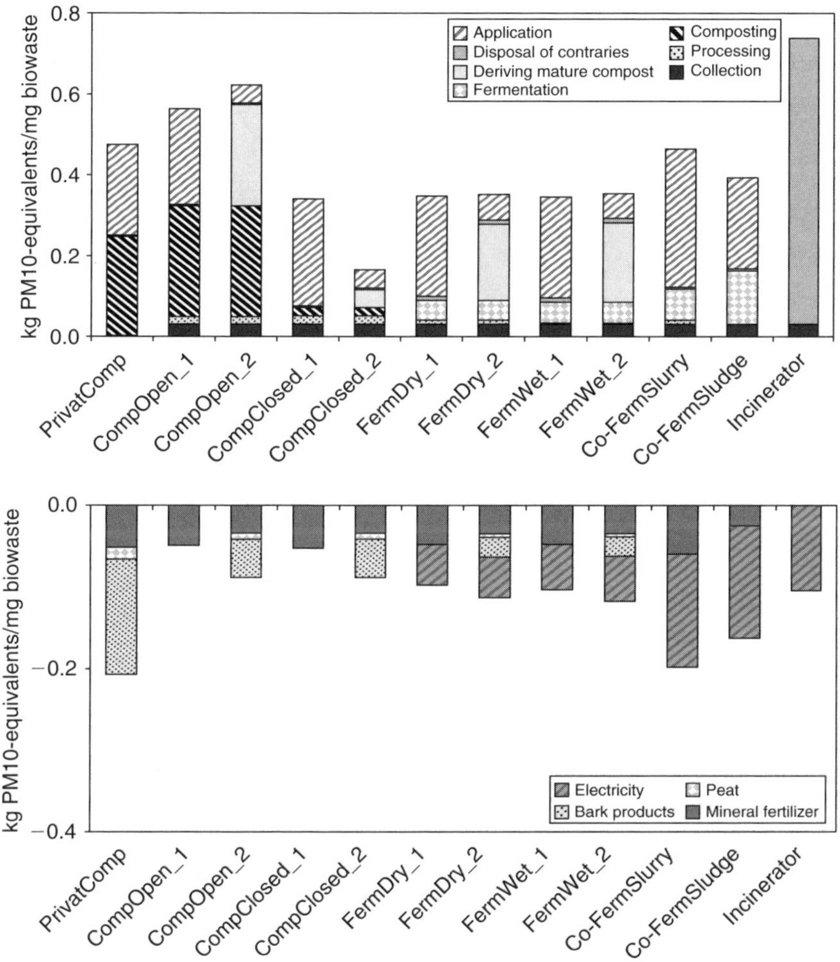

Figure 4.7 Results of the ecological valuation for PM10 risk (Humantox) (top: direct effects; bottom: substitution effects).

co-fermentation with sewage sludge, the emissions of NO_x resulting from transportation and partial incineration of remaining sewage sludge are more important (see Figure 4.7 "application"). The latter type of emissions account for the results of the remaining biowaste in refuse bin combined with incineration.

Management systems with the digestion of biowaste lead to substitution effects that depend on the rate of biogas production or the calorific value of biowaste. The substitution of bark products is successful as well. The processing of bark products needs the application of nitrogen combined with the emission of ammonia as well.

The application of the residues on farmland, horticulture or for recultivation measures is associated with an *entry of heavy metals* and other harmful substances *into soils* (Figure 4.8). In most of the cases the net results for that

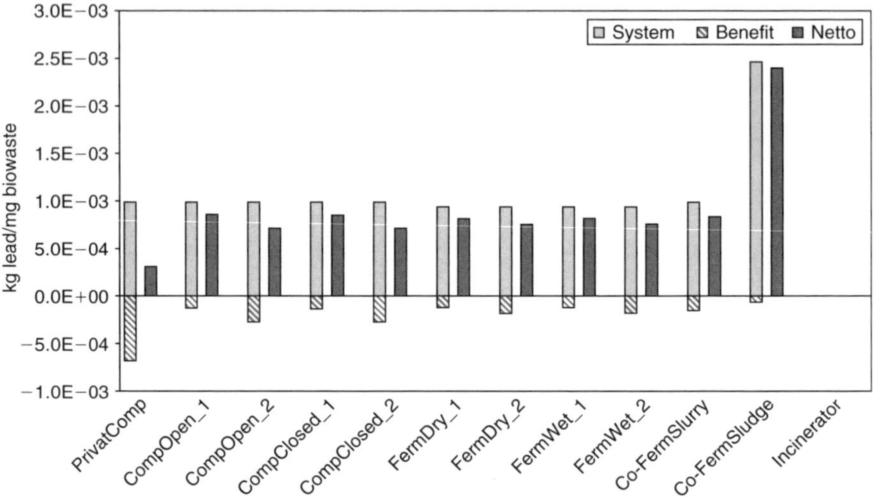

Figure 4.8 Results of the ecological valuation for entry of heavy metals into soil (lead).

impact category depend on the content of the relevant substances in the biogenic waste itself. The less the loss during the process of waste treatment and the more residues are applied in open field utilizations, the higher is the negative impact in this impact category.

In most cases, these negative effects exceed the substitution potentials as shown in case of lead (Figure 4.8). The negative results for the management system based on co-fermentation together with sewage sludge requires explanation. About 20% of the used auxiliary agents to dewater sewage sludge in waste-water treatment plants in Germany are ferrous salts. However, these ferrous salts are contaminated with heavy metals that accumulate in the dewatered sewage sludge and increase the load of especially lead when applied to farmlands or somewhere else.

One exception is noteworthy, however, artificial fertilizers (nitrogen and especially phosphate) contain significant amounts of cadmium and the content can be as high as that of the several biowaste treatment residues when normalized to the fertilizing effect. This is also true for the compost that is substituting bark humus. Hence, the substitution effect for cadmium in using the residues from composting processes leads to the highest benefits. To a certain extent this is also true for composted fermentation residues that partially substitute fertilizer and bark humus: the substitution effects yield better net results than in the case of other options with fermentation processes. Co-fermentation with sewage sludge is associated with the lowest substitution effects. Only 78% of the residues from this recycling route substitute fertilizer, the rest is incinerated and not combined with any substitution effects in this impact category.

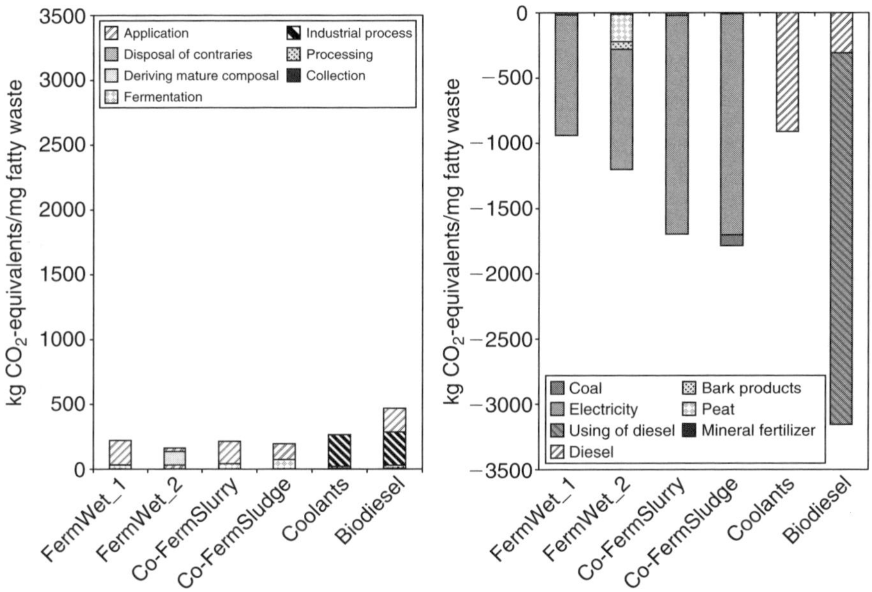

Figure 4.9 Results of the ecological valuation for global warming potential.

4.3.2 Fatty waste

Due to the high calorific value the results for the management systems of fatty waste are very different from those for biowaste from households. The substitution effects dominate the results for the impact category *global warming potential* whereas the direct emissions of N_2O and CH_4 in utilization processes are negligible (Figure 4.9). For all systems in which biogas produced, the positive effects result from substitution of fossil fuels. Co-fermentation of biowaste is modeled based on the increased rate of biogas production (see above). The processing of fatty waste to fatty acid methyl ester and utilization as biodiesel leads to the highest substitution effects.

In contrast to the results for biowaste from households, the negative effects resulting from utilization of fatty waste itself dominate the impacts from *eutrophication*, mainly due to the emissions of ammonia and nitrogen oxides (Figure 4.10). The specific amount of NO_x emissions from biodiesel use are hardly the same as from fossil diesel use. The substitution effects outweigh the negative effects from the utilization process of deriving and using of biodiesel.

The results for *PM10 risk* are very similar to the results for eutrophication potential (Figure 4.11). In both cases, the emissions of nitrogen oxides contribute to 70% within the utilization process itself and of 55% within the equivalent processes. Emissions of soot are also important. As shown in Figure 4.11, all management systems for fatty waste have favorable balances.

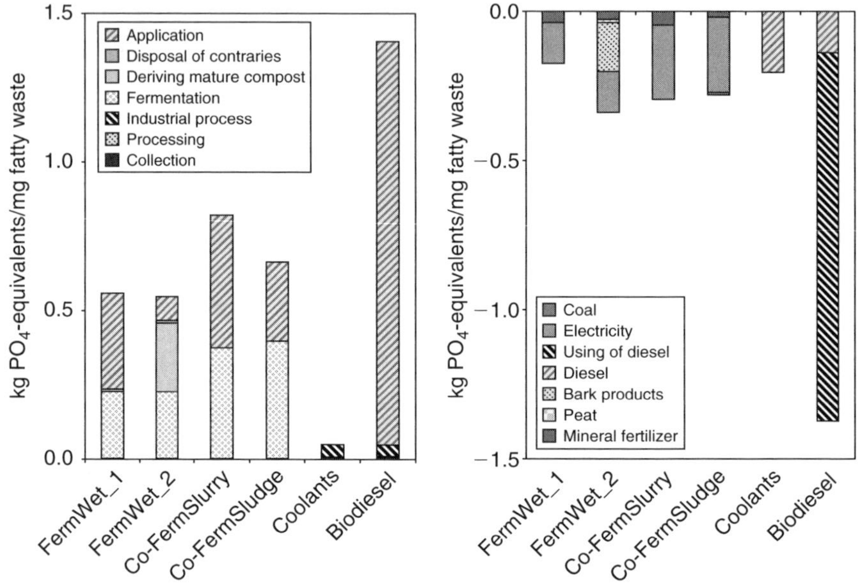

Figure 4.10 Results of the ecological valuation for eutrophication (terrestrial).

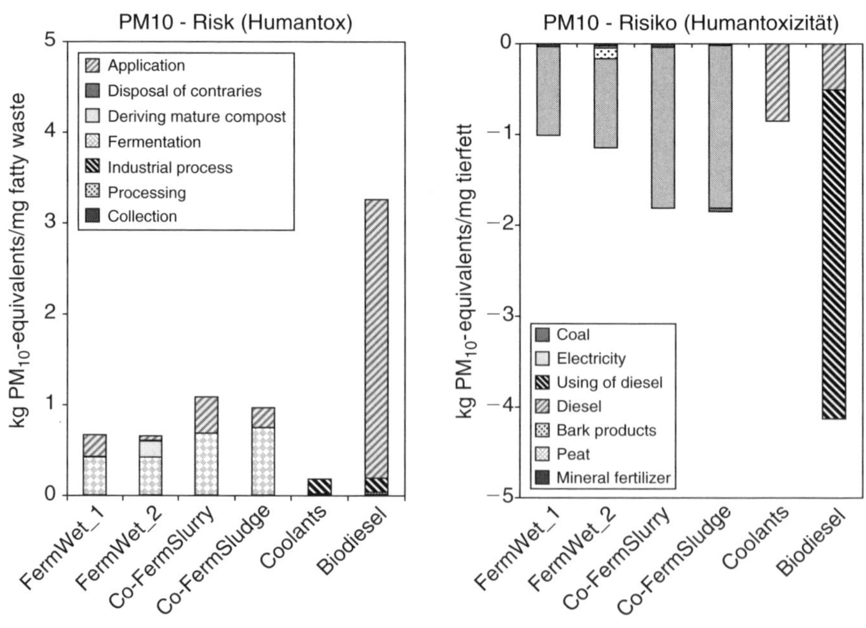

Figure 4.11 Results of the ecological valuation for PM10 risk.

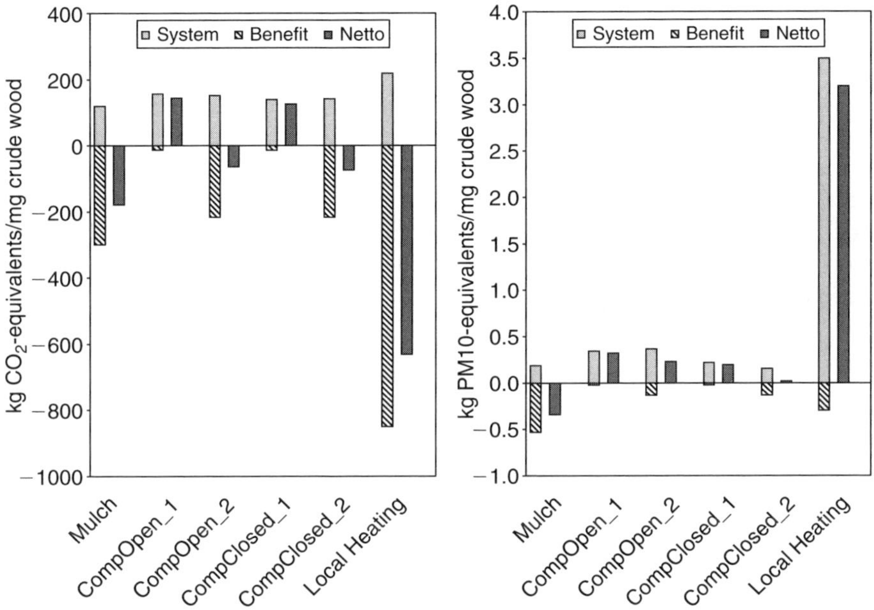

Figure 4.12 Results of the ecological valuation for global warming potential and PM10 risk (Humantox).

4.3.3 Crude wood

The environmental impacts combined with the utilization of crude wood are discussed using the impacts categories global warming potential and PM10 risk as examples (Figure 4.12). Aside from several systems for compost production, comminution of crude wood into wood chips and subsequent utilization as heating material in households are viable waste management options. The use as heating material leads to large substitution effects with net savings regarding the global warming potential. The promotion of wood chip use in Germany in recent years was mainly due to these benefits.

The application of crude wood in local heating systems is linked with negative effects if there is no emission control to reduce primary particulates. The specific emissions are about 10 times higher than the emissions from the use of heating oil. There is no reliable data available for emission rates of local heat plants. Since data for cogeneration plants was used instead, the results are uncertain.

4.4 DISCUSSION

The analyzed impact categories show advantages and disadvantages for several management systems. Which waste management system should be preferred since there is no uniformity in the results? The answer to this question requires a weighting process between the pros and cons based on the relative importance of a given ecological criterion and the second about the significance of the absolute value for each criteria.

In contrast to arithmetic procedures such as eco-indicators and eco-points, the authors prefer a verbal argumentative weighting process. This procedure is also in accordance with ISO 14040 that does not allow a mathematical comparison between different impact categories, because it is impossible to compare CO_2-equivalents to SO_2-equivalents. The verbal argumentative method is also based on work carried out for the German Federal Environmental Agency (UBA 1999); the main criteria can be summarized as follows:

- *Specific contribution of environmental impacts*: To measure the importance of the differences between the results of alternative waste management systems, the contribution can be related to the total emissions in the same region (e.g. Germany). The "specific contribution" of a recycling option is thus calculated as the average value per inhabitant (EDW).
- *Ecological risk*: This criterion addresses the different importance of the several impact categories according to the scientific knowledge, depending for example on the matter of reversibility and spatial dimension.
- *Distance to target*: This criterion expresses the distance between the actual situation on environmental effects and the political professed environmental goals.

The "ecological risk" and the "distance to target" are combined to the "ecological importance" of impact categories. The ecological importance of the impact categories is described by four classes (A)–(D). The impact categories global warming potential and carcinogenic risk potential both belong to class (A) and are of very high ecological importance. The impact categories eutrophication potential (air), acidification potential, eco-toxicity represented by the parameters copper and zinc (water) as well as the input of cadmium in soils belong to class (B) and are of high ecological importance, while the photochemical ozone creation potential, the eutrophication potential (water), the input of lead in soils and the demand on fossil resources belong to class (C) and are of medium ecological importance. The demand on the mineral resource phosphate ore is categorized in class (D), or of small importance.

4.4.1 Biowaste from households

The results in Figure 4.13 are given as the specific contribution of each recycling route and represent the relative difference of each option compared to the self-composting in the backyard. The differences are shown in values of 0.1 EDW (i.e. the statistical average for one person in Germany). The results are to read as in the following example: for the impact category entry of lead into soil, the recycling option based on co-fermentation with sewage sludge is worse than composting in the backyard; the difference represents nearly 0.2 EDW. In other words: compared to self-composting, the co-fermentation of 1 mg of biowaste from households together with sewage sludge is associated with an additional entry of lead into soil that is equivalent to that of 0.2 average persons in Germany per year within 1 year. In contrast, the

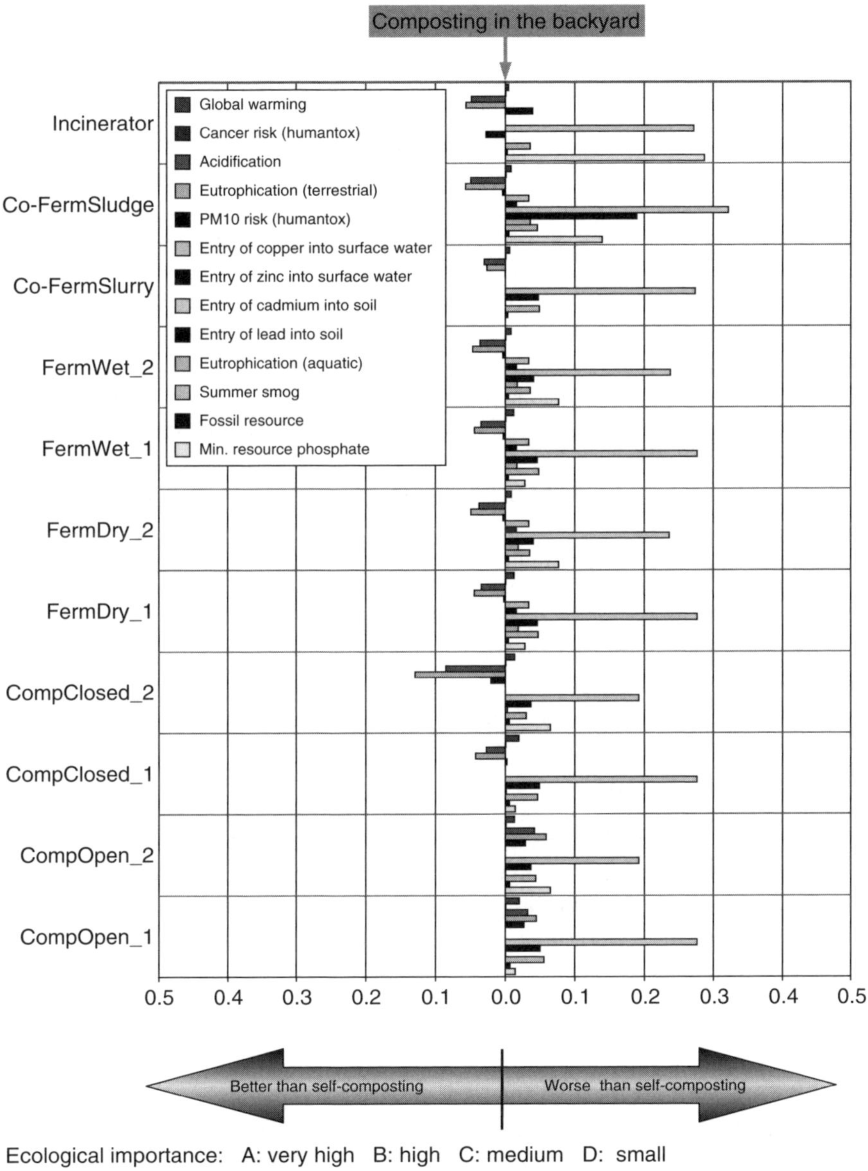

Figure 4.13 Results for 1 mg of biowaste from households – pros and cons of the different management systems in relation to the option self-composting.

incineration of biowaste together with other refuse has an advantage that represents about 0.05 EDW. At first sight, there seems to be only a very small difference between the options, but this is only due to the fact that the shown results represent the treatment of only 1 mg of biowaste (functional unit). While the results can be scaled to mass, the presented environmental ranking between the alternatives will remain.

The results presented in Figure 4.13 show the importance of the verbal argumentative interpretation. The results differ by impact categories for a given waste management system. Within the impact categories of a very high ecological importance (A), the differences between the alternatives are less important. There is a small advantage for composting in the backyard (global warming potential) and a disadvantage for the composting alternatives. Within the impact categories of high importance (B), there are advantages to the fermentation of biowaste from households and especially composting with mature compost as main output within a completely closed process. This alternative of waste management shows clear advantages compared to self-composting regarding acidification and eutrophication. The opposite result regarding the entry of heavy metals into soil is significant; it is caused by the presumption of using the premature compost as mature compost in the backyard (see above) in combination with the marginal heavy metal content of biowaste itself. There are no relevant differences within the alternative options. The disadvantage of the composting alternative is mainly due to the high nitrogen emissions. The recycling route composting in the backyard shows definitive disadvantages as far as the emissions of ammonia and N_2O are concerned. The differences are greatest for the recycling route CompClosed_2.

The results illustrate the advantage of recycling options with emission reduction (especially of ammonia) and with the utilization goal to derive in mature compost to be used in horticulture replacing peat and bark products. Due to the low content of nutrients (especially nitrogen), the main profit in the reuse of biowaste is the content of organic matter and the potential for humus regeneration. Recycling routes that combine this benefit with biogas generation and use are equally beneficial.

4.4.2 Crude wood

The summary presented in Figure 4.14 shows different results for the certain waste management systems. Compared to the local heating all alternatives show disadvantages in most of the impact categories. There are few but prominent exceptions concerning PM10 risk and some categories that are combined with ammonia emissions.

With nearly 0.4 EDW (average value per person per year) the prominent advantage of all recycling alternatives is a lower PM10 risk. Combustion with pollution control (fabric filters) leads to a significant reduction of the primarily particulates emissions in small and very local heating plants.

Advantages in acidification and eutrophication potential can be traced back to the avoided emissions of ammonia linked with the substituting bark products. However, the use of crude wood as mulch material is unlikely to compete with bark products or peat, the aspect of disposal may outweigh in practice. Without this substitution effect, the recycling option mulching is much worse than all of the discussed alternatives.

4.4.3 Fatty waste

The use of fatty waste as a co-substrate in digestion tanks of sewage treatment plants is common in Germany. The results presented in Figure 4.15 show disadvantages in

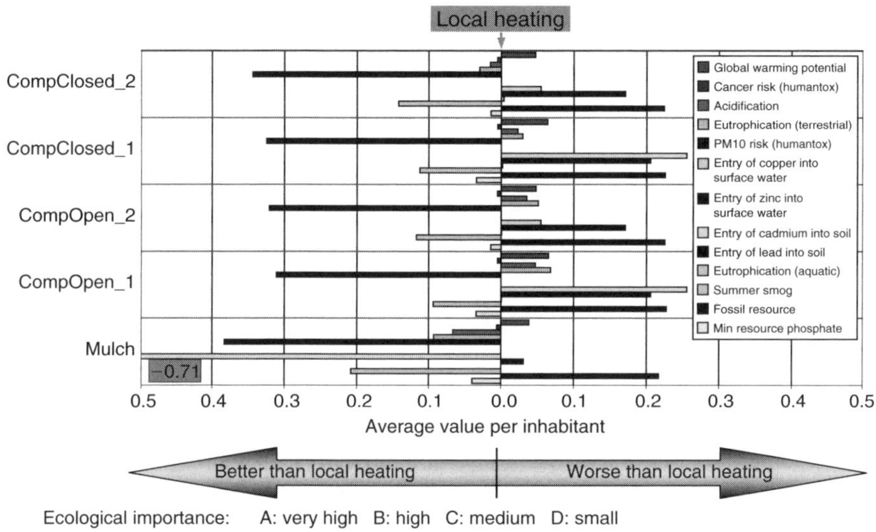

Figure 4.14 Results for crude wood – pros and cons of the different management systems in relation to the option local heating.

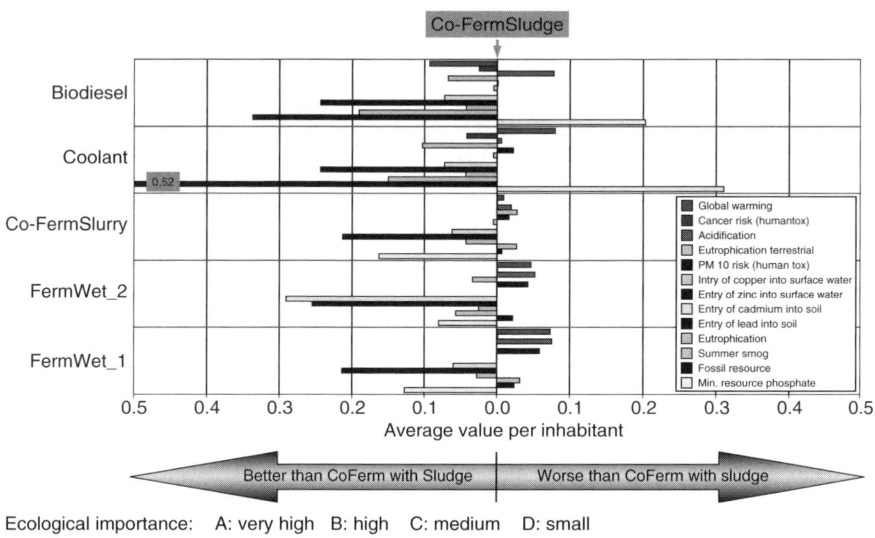

Figure 4.15 Results for fatty waste – pros and cons of the different management systems in relation to the option co-fermentation with sewage sludge.

environmental impacts for other recycling of fatty waste relative to the co-fermentation with sewage sludge. The processing to fatty acid methyl ester and utilization as biodiesel is beneficial.

The better results for the impact categories global warming, terrestrial eutrophication and PM10 risk are mainly based on the assumption that a higher specific rate of biogas production is obtained than in fermentation processes with biowaste from households. Since this may be debated, a positive interference between the substrates should be unattached from the kind of material. There is no data available for a specific biogas-rate in co-fermentation with biowaste from households.

4.5 CONCLUSION

There are a lot of options to utilize biowaste. From environmental perspective they have different assets and drawbacks. Priority has to be given to recycling systems that are more compatible with the environment. The utilization and the corresponding substitution benefits are of primary importance. But emissions during the waste treatment are not negligible. In many cases they have a good share in overall results. The diminution of emissions especially of laughing gas and ammonia from composting processes demands attention.

REFERENCES

Bidlingmaier, W. and Müsken, J. (1994) *Handbuch Abfall – Vergärung und Kompostierung von Bioabfällen, Methodenvergleich*, Landesanstalt für Umweltschutz Baden-Württemberg.

Bidlingmaier, W. and Schmelz, K.-G. (1998) Nutzung freier Faulraumkapazitäten zur gemeinsamen anaeroben Behandlung von Klärschlamm und Bioabfällen. In: *Anaerobe biologische Abfallbehandlung* (eds. Bilitewski *et al.* Beiträge zur Abfallwirtschaft), Band 7, pp. 65–79, TU Dresden.

Bock, R. (2001) Kühlschmierstoffe aus tierischen Fetten. In: *BSE und die Auswirkungen auf die Entsorgungswirtschaft*, Heft 41, pp. 171–181, Schriftenreihe des ANS Braunschweig.

Eurich-Menden, B., Wegener, H.-R. and Hackenberg, S. (1996) *Humuswirtschaft in Deutschland – Literaturstudie zur vergleichenden Darstellung des Kreislaufs organischer Substanz im Naturhaushalt unter Berücksichtigung des Umfelds in der Industrie-Gesellschaft Deutschlands*, Justus-Liebig-Universität Giessen, Institut für Bodenkunde und Bodenerhaltung, Fachbereich 17: Agrarwissenschaften und Umweltsicherung, Gießen.

Giegrich, J. (1995) Bilanzbewertung in produktbezogenen Ökobilanzen, UBA-Texte 23/95, Berlin.

Gronauer, A. *et al.* (1997) Bioabfallkompostierung. Verfahren und Verwertung, im Auftrag des Bayerisches Landesamt für Umweltschutz, LfU Schriftenreihe Heft 139, München.

Körner, I., Brilsky, H., Jensen, U., Ritzkowsky, M. and Stegmann, R. (1997) Possibilities for the regulation of the composting process to optimize the nutrient composition of compost. In: *Orbit 97 Conference*, Harrogate, UK, 3–5 September 1997.

KTBL – Kuratorium für Technik und Bauwesen in der Landwirtschaft (1998) *Kofermentation* KTBL-Arbeitspapier 249, Darmstadt.

Narodoslawsky, M. and Obernberger, I. (1996) From waste to raw material – the route from biomass to wood ash for cadmium and other heavy metals. *J. Hazard. Mater.* **50**, 157–168.

UBA – Umweltbundesamt (1999) Bewertung in Ökobilanzen. Methode des Umweltbundesamtes zur Normierung von Wirkungsindikatoren, Ordnung (Rangbildung) von Wirkungskategorien und zur Auswertung nach ISO 14042 und 14043, *UBA Texte* 92, Berlin.

Vogt, R., Knappe, F., Giegrich, J. and Detzel, A. (2002) Ökobilanz Bioabfallverwertung. Untersuchung zur Umweltverträglichkeit von Systemen zur Verwertung von biologisch-organischen Abfällen. *Initiativen zum Umweltschutz* 52, Berlin.

Zöttl, H.-W. (1994) Rindenmulch und Rindenhumus. In: *Bodennutzung und Bodenfruchtbarkeit*, Bd. 6, *Recycling kommunaler und industrieller Abfälle in der Landwirtschaft*, S. 57–65, Münster.

PART TWO: Socio-economic aspects of solid-waste management

5

Costs and benefits of bioprocesses in waste management

D. Hogg

5.1 SYSTEM ANALYSIS OF SOLID-WASTE MANAGEMENT

There is an increasingly wide-ranging array of possible approaches to seeking to inform decision-making regarding "optimal" waste management solutions in a given context. Since the implementation of European Union (EU) Directive 85/337/EWG, an environmental impact assessment (EIA) has been an obligatory element of licensing procedures for certain public or commercial projects. The EIA is required to identify and describe relevant impacts of the project regarding the environmental media and compartments. There are no further EU regulations providing technical guidance on how the EIA is to be carried out. The decisions concerning environmental compatibility or incompatibility are usually made with regard to legal environmental quality standards (EQS). Therefore, the EIA normally

Edited by Piet Lens, Bert Hamelers, Harry Hoitink, Werner Bidlingmaier. ISBN: 1 84339 054 X. Published by IWA Publishing, London, UK.

uses exposure models to infer whether or not the associated emissions lead to the exceeding or otherwise of existing standards. As such, it does not necessarily determine impacts, *per se*, but frequently leads to statements as to whether or not existing limit values are likely to be breached as a consequence of the project's operation. The EIA method is therefore suitable for assessing site-specific issues in the context of existing legislation, but it does not always assess the damages associated with "below limit value" emissions.

Risk analysis methodologies resemble EIAs in a number of respects. Often an EIA includes elements of a risk analysis when the direct impacts of a technique on the surrounding environmental media are estimated. In appropriate cases, specific toxicological aspects are analysed in detail. On the other hand, risk analysis is not standardised as a tool, and the user is free to choose the scope and processing methods. Risk analysis also tends to relate to the risk of specific consequences arising. It is also frequently the case that risk analysis, by its very nature, seeks to quantify things which are not readily quantifiable. This means that frequently, "any number" is deemed better than none, leading to situations in which the boundaries of probabilistic risk and radical uncertainty are crossed in the attempt to arrive at "a number" as the answer to the problem in hand.

Two techniques used with increasing regularity, though apparently by different communities, are life-cycle approaches and economic valuation. Both have some strengths and weaknesses.

5.1.1 Life-cycle approaches

The most standardised environmental assessment methodology is life-cycle assessment (LCA). The ISO standards 14040 to 14043 define the basic steps of this instrument. Several subsequent standards illuminate or instruct further technical aspects. During the last 10 years, LCA methods have become established as one of the most commonly applied approaches to value systems of production (ecological product balance from cradle to grave) or waste management systems. The methodological approach (defining general system boundaries and balancing the material and energetic inputs and outputs) allows a comparison of systems with varying levels of complexity. In terms of system analysis the LCA approach can be regarded as superior to both the risk analysis and EIA approaches, not least since it does not limit itself to specific limit values, or the risk of certain problems occurring.

Despite these advantages, the following drawbacks are worthy of mention:

- *Weightings in impact assessment categories*: The weightings attributed to specific pollutants in specific impact categories (e.g. human toxicity) are not universally agreed. These may be said to be subject to continuous review, which is as it should be.
- "*Data dumping*": One frequently finds that life-cycle modelling makes use of data from an array of different sources of differing vintage. As one seeks greater "completeness" in the data gathering exercise, the temptation inevitably arises

to include whatever data is available for even the most obscure auxiliary input. Partly for this reason, one frequently finds that life-cycle studies are carried out using inventory data spanning periods of a decade or so. This is a similar problem to that one finds in issues of risk analysis. It barely needs pointing out that this leads to serious questions concerning the validity of the data, and one has to state quite bluntly that the whole enterprise will always suffer from this type of problem since one is necessarily "running to stand still" to keep ahead of the "data gathering" game. Indeed, the environmental optimist would rather hope that the data *is* out-of-date (since if it is not, this implies nothing has improved in the intervening years). Although it is sometimes claimed that this should not be a major problem, inspection of the weightings applied in specific impact categories used in LCA suggests otherwise. Where specific inventory items are weighted heavily in the impact assessment categories, the inclusion or absence of these can have significant impacts on the results. Hence, unless inventory data for processes which are the subject of the comparison are all measuring the same range of emissions, the use of datasets of differing completeness can lead to bias.

- *Insensitivity to location*: The analysis of impacts in LCA does not (usually) account for location-specific impacts. To do so would require location-dependent weightings in the impact assessment phase. As such, arguably, it is most useful for assessing global impacts or other impacts unrelated to the populations exposed.
- *Insensitivity to time*: Life cycle analyses are frequently insensitive to the issue of time. This means that not only are "future impacts" weighted equally with current ones, but slow releases of pollutants over the life cycle are potentially treated in the same way as rapid releases of the same pollutant. This neglects the fact that pollutant concentrations may be critical in determining impacts, especially where threshold effects are likely to occur.
- *Inability to account for disamenity and other "impacts" not associated with "flow of pollutants"/materials*: The approach is not sensitive to impacts which are not associated with changes in flows of pollutants. This means that, for example, visual impacts and impacts on biodiversity (through ecological disturbance) are not accounted for.
- *Weighting across impact assessment categories*: Whilst LCA may be able to assess impacts within categories (however imperfectly) through weighting the effect of specific pollutants in those impact categories, it is relatively weak on assessing how the outcomes in the different impact assessment categories should be traded off against one another. How much "global warming" is to be equated with "how many units of human toxicity"?

5.1.2 Economic valuation (benefits assessment)

Economic valuation approaches usually lean on some of the other approaches, but add to them valuation of effects which are less easily quantified through other approaches, for example, the impact of loss of amenity, as elicited through questionnaire surveys. One apparent benefit of valuation is that all impacts are reduced to a common metric – that of money – so that the last of the problems discussed above in the context of LCA is, supposedly, overcome. But the flip-side is that not all impacts can be readily quantified. Even where they can be, it may be costly to do so, because the methodologies used may demand local surveys to elicit the preferences of those located near a particular activity. In addition, rather like the attempts to "aggregate" contributions to a specific impact index under LCA, the attempts to do this through economic valuation are far from universally accepted. Some studies have begun to use LCA impact indices as a basis for what economists call "benefits transfer" (e.g. RDC-Environment and Pira International 2001). This is a questionable exercise. Problems with the approach, therefore, are (for a discussion see Hogg *et al.* 2000, 2002a):

- *An inability to account for all pollutants*: At least in life-cycle studies, a wide range of pollutants can be accounted for, albeit imperfectly. Under valuation studies, the impact of such a range is difficult to assess since credible valuation studies covering such a range of pollutants are not available.
- *The issue of location specificity*: Different studies are carried out either using "bottom-up" approaches, which can account for location and the exposure of populations, or "top-down" approaches, which make use of unit values for estimated damages from other studies. Bottom-up studies are usually relatively expensive to carry out and involve mapping changes in pollutant concentrations to effects, and then monetising those effects. In top-down studies, the transfer of the valuation figures from one study to another is problematic given the fact that populations exposed are different in different locations. Furthermore, effects such as disamenity are likely to vary in different locations, so they demand location-specific analysis, or at worst, some form of meta-analysis.
- *The issue of time*: The issue of time is dealt within economic valuation studies through use of a process known as discounting. Future impacts are attributed to reduced significance through use of a discount rate, supposedly reflecting preference of society for benefits today rather than in the future (a social discount rate). This is also used effectively to determine how financial resources should be allocated to address problems which have impacts beyond the present. Some LCA practitioners have criticised this process as being inconsistent with basic tenets of sustainability (Helweg 2002).
- *Uncertainty in valuation figures*: There are two sources of uncertainty in the valuation figures used in cost–benefit analyses. The first is the nature of the effect upon which a value is being placed. Most environmental impacts are shrouded in a degree of uncertainty, and in some cases, the extent of knowledge is "radically uncertain". This is a point which is not always appreciated in such studies, which frequently posit "unique values" for unit damages associated with a given pollutant (and the same is true, of course, for the way emissions are attributed in life-cycle inventories). The other source of uncertainty is the valuation process itself. There are several different approaches used to value environmental effects. None are completely free of epistemological or

moral problems. Given these two sources of uncertainty, the benefits or costs attributed to a given environmental effect typically cover a wide range, usually varying by a factor of 10 or so. By way of example, one can readily appreciate that valuing the environmental damages associated with the emission of a ton of carbon dioxide are unlikely to be knowable with any great certainty.

5.1.3 Selection of system analysis methodology

From the above discussion, it seems clear that in order to evaluate the environmental effects of waste management options, one is faced with a variety of approaches, each of which suffers from shortcomings. Perhaps the most important issue in seeking to carry out such analysis is, to state what seems to be obvious, to do the analysis well. Typically, this requires clear statement of limitations of the study, and almost certainly, modesty in arriving at conclusions associated with the study. Sensitivity analysis is clearly to be encouraged. A problem here is that in more complex analyses, sensitivity analysis is time consuming and the results can be laborious to present.

In a recent study, an attempt was made to understand the nature of the costs and benefits of separate collection of compostable materials and their subsequent treatment through composting and anaerobic digestion (AD) (Hogg *et al.* 2002a). The study looked also at options other than separate collection and treatment, and at the fees paid for the typical waste treatment options. This chapter, however, is principally concerned with the *external* costs and benefits of composting and AD. The full study can be found on the European Commission's website.

5.2 EMISSIONS FROM THE COMPOSTING PROCESS

Full life-cycle analyses of composting processes are difficult to carry out. One of the difficulties associated with the approach to life-cycle analysis regarding composting is the variety of processes available and the fact that different processes make use of different feedstocks. Ideally, as with all treatments, one seeks to relate specific emissions and outputs to the input materials. This has only been attempted seriously in one of the models that were surveyed, the ORganic WAste REsearch (ORWARE) model, developed in Sweden (see Chapter 3 in this volume). More recent work undertaken in the US appears to have made a similar attempt (Komilis and Ham 2001).

5.2.1 Gaseous emissions

Tables 5.1 and 5.2 below show estimates of the gaseous emissions from good practice composting facilities and AD facilities. The methane estimates from compost plants have been estimated as non-zero. Some would argue for a zero value at best practice facilities. The pollutants mentioned are those to which an economic value has been attributed.

Table 5.1 Gaseous emissions data from composting process (all figures are g/ton of waste composted).

Gas	g/ton of waste composted
CO_2	350,000
CH_4	983
N_2O	11
Dioxins/furans	0
Ammonia	371
Volatile organic compounds (VOCs)	24

Table 5.2 Gaseous emissions data from AD process (all figures are g/ton of waste digested).

Gaseous compound	g emitted/ ton digested
CO_2	440,000
CH_4	0
NO_x	10
N_2O	0
SO_x	2.5
HCl	0.011
HF	0.0021
H_2S	0.033
HC	0.0023
Halogenated HC and polychlorinated biphenyls (PCBs)	0.00073
Dioxins/furans (TEQ)	1E-08
Ammonia	No data
Cadmium	9.4E-07
Chromium	1.1E-07
Lead	8.5E-07
Mercury	6.9E-07
Zinc	1.3E-05

The energy used at the plant also contributes to the emissions associated with the composting process. Windrow composting systems seem to use less electricity and more liquid fuel whilst reactor systems use more electricity and less fuel. Energy use at compost systems might amount to approximately 50 kWh/ton and 1 l of diesel per ton of waste input. For AD systems, there is a utilisation of energy which is more than offset by the generation of energy from the system itself (see Table 5.3).

For the calculation of external costs associated with the process itself, the air emissions are the key ones to which costs can readily be attached. The results of the analysis of the external costs of the associated air emissions are shown in Table 5.4 and Table 5.5, shown for "high" and "low" values attached to the damage associated with different pollutants. Note that the unit damage costs used for the global warming externalities vary with the discount rate chosen for the economic analysis since the different gases have extended residence times in the atmosphere and

Table 5.3 Assumptions concerning energy production from AD.

Parameter	Low value	High value
Biogas yield	70 m^3/ton waste	140 m^3/ton waste
Percentage methane	55%	60%
Calorific value of biogas	385 kWh/ton waste	840 kWh/ton waste
Electricity generated (30% efficiency)	116 kWh/ton waste	252 kWh/ton waste
Electricity for export (net of energy use)	81 kWh/ton waste	176 kWh/ton waste
Heat recovered for CHP option (70%)	189 kWh/ton waste	412 kWh/ton waste
Heat exported for CHP option (net of energy use)	151 kWh/ton waste	329 kWh/ton waste

Table 5.4 External costs of gaseous emissions from composting process including energy use (€/ton) in Italy.

	Low value	High value
Greenhouse gases		
Carbon dioxide (process)	−8.56	−9.01
Methane	−0.44	−0.48
Nitrous oxide	−0.09	−0.16
Other air emissions	−0.02	−0.07
Energy use	−0.85	−1.53
Fuel emissions	−0.36	−0.80
Total external costs	**−10.33**	**−12.04**

Table 5.5 External costs of gaseous emissions from AD process including energy generation (€/ton) in Italy.

	Low value	High value
Greenhouse gases		
Carbon dioxide (process)	−10.76	−11.33
Methane	0.00	0.00
Nitrous oxide	0.00	−0.71
Other air emissions	−0.08	−0.13
Avoided external costs from energy generation		
Electricity	1.03	4.50
CHP	1.48	6.45
Total external costs, no displaced burdens	**−7.72**	**−12.16**
Total external costs, displaced burdens from electricity	**−9.81**	**−7.66**
Total external costs, displaced energy from CHP	**−9.37**	**−6.70**

their contribution to global warming occurs over an extended period of time. The damage costs associated with energy generation were taken from work undertaken as part of the European Commission's ExternE programme and are specific to the country being examined in respect of energy generation (European

Commission 1999). Note that the results for only one country, Italy, have been presented. The externalities associated with energy generation in Italy are broadly representative of an average figure for the EU.

Note that the key contributing costs are those from carbon dioxide, energy use, and other air emissions, specifically particulate matter. These deserve some comment. Some studies do not attribute damages to carbon dioxide from compost since it is argued that these are emissions which would have occurred anyway had the material been left to degrade in the natural environment. The argument that this would be part of some natural cycling of carbon is, however, questionable since the materials which are degrading are frequently produced under techniques which require substantial human intervention, for example, paper from monoculture forests, or crops which are heavily dependent on significant quantities of external inputs. Therefore, the emissions are shown explicitly. This question is addressed later in considering the contribution made by compost to soil organic matter content (see Section 5.3.3).

5.2.2 Other emissions from biowaste treatment processes

Other emissions from the composting process which are of potential concern include leachate, odours, microbial pathogens and dust. These are discussed in the following sections.

5.2.2.1 Leachate

Leachate is not problematic at well-controlled plants, especially where these are covered. Digestion systems may need to have wastewater treated and this impose additional costs. It is possible, especially with dry digestion systems, to reduce this requirement if the output digestate is composted along with other (dry) feedstocks.

5.2.2.2 Odours

Odours are more likely to be a problem at open-air windrows and at plants which accept highly fermentable materials such as food wastes. Licensing laws in a number of countries require that plants be located away from dwellings, usually with the aim of reducing odour-related disamenity. Olfactometry tests are increasingly common. Lastly, it is important to note that control of odours has been an active area of research in recent years and problems associated with odours are likely to be significantly reduced relative to previous years. Obviously, the increasing shift to in-vessel technologies quickens this process.

5.2.2.3 Microbial pathogens and dust

In some cases, siting requirements, especially of open-air windrow systems, reflect concerns regarding the potential for bioaerosols to cause harm to immuno-suppressed citizens. The debate concerning the impact of bioaerosols is still live, though countries with lengthy experience with composting report few, and in some cases, no health-related effects from their compost systems. Generally, it is felt that problems are more likely to affect plant operators than the general public.

Exposure to any of these problems, to the extent that they are not controlled, will obviously be affected by the proximity of citizens to the plant (so that plant location affects the degree to which any of these issues is likely to lead to problems).

5.2.3 Disamenity

Since, to the best of our knowledge, no study has looked at the disamenity associated with compost or digestion facilities, there is no basis for quantification. Certainly, these would be expected to deviate more significantly from zero where one or more of the following are true:

(1) The composting process is poorly managed (and odours are prevalent).
(2) The compost plant accepts inappropriate materials; that is, those which are likely to give rise to problems in the context of the process technology being used.
(3) The plant is of a significant scale, so that visual intrusion becomes an issue, as do transport movements. This may imply some double counting where transport externalities are considered.
(4) The composting process occurs in close proximity to housing.

All of these increase the potential for significant disamenity, though any attempt to estimate disamenity may have to be undertaken with care to avoid double counting of the effects of odour.

The key difference between composting and digestion facilities relates to the reduced potential for odour-related problems from the process, and to reduced release of bioaerosols, at least relative to open-air facilities. Some of the problems mentioned above may still be of potential concern for digestion systems, not least where aerobic composting follows the digestion process. This may be good practice as regards preparation of a stable product for land application.

5.3 BENEFITS FROM COMPOST UTILISATION

Table 5.6 provides a summary of some of the advantages and disadvantages of using compost compared to mineral fertilisers in agriculture. Similar approaches have been adopted for assessing the benefits from compost and from digestate following a post-digestion aerobic treatment.

The addition of compost to soil results in a reduction in bulk density, an increase in soil porosity and an increased water retention. All these factors have a positive effect on plant growth and subsequent crop yields. They may also act to reduce the potential and/or frequency of flooding in periods of high rainfall, and for soil erosion.

Table 5.6 Advantages and disadvantages of using compost compared to mineral fertilisers.

Material	Advantages	Disadvantages
Compost	Improves soil structure Controls erosion Supplies wide range of nutrients Method of waste disposal Increases activity (and presence) of soil micro-organisms	Dilute source of nutrients Even application can be difficult High C : N ratios may rob soil nitrogen
Mineral fertilisers	Convenient Lower transport and handling costs Quick crop response	Easily leached Overuse may lead to breakdown of soil structure Supply only major nutrients

The issue of soil erosion is especially relevant to countries where the soils have a low organic matter content, as is common in Mediterranean countries. Montanarella (1999) estimated that over 150 million hectares of European soils are suffering from erosion, the problem being more acute in southern countries. Soil erosion is a major socio-economic and environmental problem throughout Europe. It reduces the productivity of the land and degrades the performance and effectiveness of ecosystems.

5.3.1 Displacement of alternative nutrient sources

When compost is applied to the soil, it may displace nutrients which are otherwise applied through other means. Typically, it seems reasonable to assume that the product being displaced is synthetic fertiliser. This will not always be true.

The assumption made is that 10 tons of dry matter is applied per hectare, which is equivalent to approximately 16.7 tons of compost, derived from approximately 47.7 tons of waste material. This is based on assumptions that 1 ton of waste material, consisting both kitchen and garden waste leads to production of 350 kg compost with dry matter content of 60%. The modelling further assumes that the compost has the following composition in terms of nutrients:

(1) Nitrogen: 1.5% dry matter
(2) Phosphorous (as phosphorous pentoxide, P_2O_5): 1.0% dry matter
(3) Potassium (as K_2O): 1.2% dry matter

The mineralisation rate of the nutrients is assumed to be 30% for all nutrients. This determines the time profile of the displacement effect, which in turn effects the external benefits associated with displacement via the discounting mechanism.

For synthetic fertilisers, a loss rate of 23% is assumed for nitrogenous fertilisers (based on Hydro Agri Europe 1995). At the same time, the loss rate from compost is assumed to be negligible given the small proportion of total nitrogen which is non-organic nitrogen in composts. This means that more nutrient has to be applied

Table 5.7 Evolution in nitrogen displacement associated with 10 tons dry matter of compost applied to farmland.

Year	Displacement of nitrogen (kg)	Cumulative displacement
1	58.4	58.4
2	40.9	99.4
3	28.6	128.0
4	20.0	148.0
5	14.0	162.1
6	9.8	171.9
7	6.9	178.8
8	4.8	183.6
9	3.4	186.9
10	2.4	189.3

in a given year in the synthetic form than would be available in mineralised form from the composted materials. For an application of 10 tons dry matter per annum in 1 year, the nitrogen displacement would follow the evolution set out in Table 5.7. Equivalent projections for phosphorus and K displacement have been assumed.

5.3.1.1 *Emissions associated with fertiliser manufacture*

The use of compost as a replacement for fertiliser will displace the pollution and other impacts associated with fertiliser production. These are discussed in this section to derive an externality cost per ton of fertiliser produced and therefore a subsequent benefit per ton of compost/digestate applied. This study used data for the production of sodium, phosphorus and potassium (NPK) fertiliser because this is the most widespread in terms of use, and the most readily available from a data perspective. No quality data for extractive processes was obtained for use in the analysis. Mining phosphate rock is an energy intensive activity and approximately 3.3 tons of phosphate rock are required to produce 1 ton of P_2O_5 (100%) (Bocoum and Labys 1993). Energy use for producing phosphate rock has been estimated at 73.5 kWh/ton (UNEP and UNIDO 1998). Additional energy consumption for phosphate fertiliser is attributed on this basis. As no information concerning emissions from potash production were available, the emissions data for K_2O are likely to grossly understate the environmental benefits of displacement effects. It is important to note that these externalities are associated with best available technologies for both new and existing plants. Therefore the costs in terms of gaseous emissions and energy requirements are likely to be underestimated as not all plants will be using best available technologies.

5.3.1.2 *Phosphate fertiliser – phosphogypsum and process wastewater disposal*

One study sites a major review, undertaken by the US Environmental Protection Agency (EPA), of the production and environmental impacts of phosphoric acid production at 21 locations in the US (Soulsby *et al.* 2000). Significant impacts of process wastewater and phosphogypsum disposal have been identified. The study reports that no economic valuation studies are available in the literature that can provide support in estimating the monetary value of the externalities associated with the processes. However, based on US data, the same study looked at the costs to industry of the environmental regulation involved.

As an admittedly crude estimate of the externalities associated with P_2O_5 production which may be avoided when compost is applied, the estimated external costs were taken as equivalent to £59.91 (€95.86) per ton P_2O_5, or €0.096 per kg P_2O_5.

5.3.1.3 Greenhouse gases from nitrogenous fertilisers

Nitrous oxide emissions from soil are complex since the gas is simultaneously produced and consumed in soils through processes of denitrification, nitrification, nitrate dissimilation and nitrate assimilation. The rates at which these processes occur are affected by temperature, moisture, the presence of plants and the soil composition, as well as the (related) activity of bacteria in the soil column.

It is generally accepted that nitrogenous fertilisers increase fluxes of N_2O. Different fertilisers appear to be more or less susceptible to the loss of nitrogen as nitrous oxide. Ammonia products appear most susceptible, with anhydrous ammonia and aqua ammonia losing between 1% and 5% of nitrogen as nitrous oxide. Other products such as sodium nitrate appear to lose much less nitrogen in this way (Lashof and Tirpak 1990; Erlich 1990). The emissions depend upon temperature, soil moisture, fertiliser type, fertiliser amount, the timing and mode of application, and the type of soil and crop cultivated (McTaggart *et al.* 1998).

In the analysis, it was assumed that in the "low" case, 0.5% of nitrogen applied as fertiliser is lost as N_2O. In the "high" case, it was assumed that 2% would be lost. These figures were combined with the nitrogen replacement figures for the compost as outlined in earlier sections. For the 3% discount rate case, the benefits range from €0.21–€1.46, a range which arises from combining low emissions with low unit damage costs, and high emissions with high unit damage costs.

5.3.1.4 Effects on nitrate leaching from soil

Nitrogen from mineral fertiliser is the major source of nitrogen input in the EU. Nitrogen in commercial fertilisers is readily soluble to facilitate uptake by crops, which in conjunction with excessive application can pose a threat to the environment and in some cases affect the fertility of the soil itself. Losses to the environment can be minimised if sustainable agricultural practices are followed and reasoned fertilisation is used. This must account for weather conditions to reduce the incidence of runoff and must involve applying at the appropriate stage of crop growth, using appropriate doses, etc.

Nitrogen when applied as uncomposted animal manures or inorganic nitrogen fertilisers also have the potential to volatise and lose more than 50% of their nitrogen to the atmosphere within the first few days following application to land. When animal manures are spread nitrogen is lost to the atmosphere through volatilisation as ammonia or as the greenhouse gas N_2O.

Nitrogen supplied from compost is not immediately available. Approximately 40% is available in the 1st year following application, 20% in the 2nd year and 10% in the 3rd, slowly decreasing every subsequent year (the rates depend on climate, etc.). Therefore, composting when managed correctly is a form of nitrogen conservation. As most of the nitrogen in compost is not in a form that is immediately available to the soil, there is less risk of nitrogen volatilisation and nitrogen leaching. This is especially relevant in nitrate sensitive areas.

Nutrients that are not taken up by plants may be metabolised by micro-organisms in the soil which will improve soil fertility. However, this is a slow process and there is a risk that soluble nutrients such as nitrate will run off into surface water or percolate into groundwater reservoirs. Combined, excessive amounts of nitrogen and phosphorous can result in eutrophication in lakes, rivers and coastal areas, resulting in the proliferation of toxic blue-green algae. Soils can also be at risk of eutrophication, where excess nutrients deplete the soil of oxygen, resulting in a reduction of natural micro-organisms and subsequent reduction in soil fertility.

The effects of displacing the equivalent quantity of nitrate fertiliser from the soil are not only that one avoids burdens associated with their manufacture. In addition, there is a reduction in the leaching of nitrate into groundwater. Work on economic valuation of nitrate pollution of groundwater is relatively scarce. There are a number of difficulties associated with this, not the least of these being the fact that such leaching as occurs today may only affect people a generation or more in the future. The cost of removing nitrate from groundwater is not insignificant. A recent study in the UK estimates the cost of nitrate removal at £18.8 million/year (approximately €28 million) in capital expenditure in the years 1992–1997 plus £1.7 million operating expenditure (approximately €2.6 million) (Pretty *et al.* 2000). The study estimated that 80% of nitrate originated from agriculture, giving a cost of removal of £16.4 million/annum (approximately €25 million/annum).

Current estimates do not provide a reliable basis upon which to value the "displaced leaching" associated with the nitrogen displacement. However, the effect should not be ignored.

5.3.2 Disease suppression

It has been shown that compost can help to control plant diseases and subsequently reduce crop losses in both agriculture and horticulture (see, e.g. Hoitink and Fahy 1986; Ringer *et al.* 1997; DeCeuster and Hoitink 1999). Disease control in compost has been attributed to four main mechanisms. These are:

- Successful competition for nutrients by beneficial micro-organisms.
- Antibiotic production by beneficial micro-organisms.

- Successful predation against pathogens by beneficial micro-organisms.
- Activation of disease resistant genes in plants by composts.

Compost quality is critical to the disease suppressive characteristics of compost. Composts that are allowed to mature have greater suppressive qualities than immature composts.

The beneficial effects of compost use can help growers to save money and reduce their reliance on pesticides, subsequently conserving natural resources. Disease suppressive soils are a well-known phenomenon, especially in organic agriculture. Supressiveness in soils is related to changes in microbial populations, which the addition of compost enhances. There is also evidence to suggest that the application of compost can help to control numbers of parasitic nematodes by providing nutrients to the soil that encourage the growth of fungi and bacteria which compete with, or destroy the nematodes.

A number of attempts have been made to estimate in money terms the environmental costs of using pesticides (Pimentel *et al.* 1993; James 1995; Steiner *et al.* 1995; Foster *et al.* 1998; Pretty *et al.* 2000). These studies suggest that the external costs associated with pesticide use are not trivial. They lie between 25% and 125% of the total private costs of pesticide use. Here it is assumed that external costs of pesticide use lie between €13 and €19/kg of active ingredient used. On a per hectare basis, in 1993–1995, purchases varied between 1.2 and 16.3 kg of active ingredient used per hectare of arable and horticultural land across Europe (including set aside). It was assumed that pesticide use could be reduced by an equivalent of 20% of their use on arable land. Clearly this would vary according to the country concerned, the crops being grown on land where compost was applied and, therefore, the pesticides currently in use. On this basis, high and low levels of the external benefits that might be derived from compost use through the avoided external costs of pesticide use were calculated. On this basis, the figure would be equivalent to a benefit of €0.40–€0.60 per ton of waste composted for the Italian situation.

These estimates above are subject to considerable uncertainty. However, they represent a first attempt to capture the external benefits associated with reducing pesticide use through the application of compost. Of course, to the extent that other soil management practices change, these will have implications for pesticide use also.

It should be noted that the external benefits from the displacement, whilst they are uncertain, are based on estimates of pesticide externalities that are generally thought to be conservative. No account has been taken of the avoided external costs of the production of pesticides which might otherwise have been used. This is the exact opposite of the situation in respect of fertilisers where the attempt was made to capture the external benefits from avoiding production, but the avoided external costs of, for example, nitrate leaching following application were not estimated. It should also be noted that applications of pesticide vary across crops. Horticultural applications are typically associated with higher applications of pesticides. To the extent that composts are used in such situations, the benefits may again be an underestimate, notwithstanding the somewhat arbitrary nature of the assumption regarding the proportion of pesticides displaced.

5.3.3 Greenhouse gas emissions from compost in soil

The use of compost in agriculture can have a positive effect on soil carbon levels and subsequently act as a carbon reservoir. Compost does not result in the permanent and irreversible locking up of all carbon in soil. What compost can do is reverse the decline in soil organic matter which has occurred in relatively recent decades through contributing to the stable organic fraction in soils (effectively locking up carbon). It is also important to realise that whilst the debate concerning "sequestration" has emerged as a topical one in the wake of the debate on climate change, the role played by soil organic carbon is far more complex, and potentially far more important, than the single role played in terms of carbon sequestration. It is clear, for example, that the effects of soil organic matter on soil biota are at the heart of the disease suppressing effects of compost. The interrelationship between carbon and nitrogen largely determines the magnitude of soil microbial populations. Utilisation of carbon and nitrogen by microbes is also responsible for the turnover between organic and mineral forms of nitrogen. Hence, the biomass production potential of soil is largely dependent upon the ability of a soil to support microbes such as bacteria and fungi.

Three pools of organic carbon are available for microbial utilisation:

(1) *The active soil fraction*: turnover of around 2 years, and representing short-term sequestration of carbon – provides source of energy for microbes, and soil carbon and nitrogen supply necessary components for amino acid synthesis.
(2) *The slow or decomposable soil fraction*: turnover time 2–3 years – of great importance to developing good soil structure – disturbed by cultivation and other disturbances – provides a source of carbon for biological digestion by microbes, so linking to the active pool – it can be viewed as mature compost.
(3) *The passive soil organic fraction*: turnover time of the order 1000 years – resistant to oxidation processes – acts as a "cement" that binds particles.

Only the first two of these pools contain carbon in readily available forms for microbial utilisation. The last pool contains carbon in a highly stable form. Some microbes can utilise this pool so depletion does occur. It can also be replenished from active and slowly decomposable fractions. It is the fact that this passive pool of carbon can be maintained or increased that leads to the idea that the passive pool can act to "sequester" carbon in the soil. Clearly, this long turnover time does appear to imply that, for all intents and purposes (certainly for any economic analysis deploying a non-zero discount rate), this carbon is not released into the atmosphere.

The dynamics of soil organic carbon where it is applied in composted form were modelled (Hogg *et al.* 2002a). The pathways modelled are outlined in Figure 5.1. The application of compost is assumed to lead to the readily available carbon being mineralised at Y% whilst X% of the readily available organic carbon is converted to stable organic matter. Of this stable organic matter, some carbon is mineralised, but at a much lower rate than that at which the readily available matter is converted to stable organic matter.

This means that as compost is applied to the soil, there is a build up of organic matter in the soil since the rate at which the material is mineralised is, overall, much

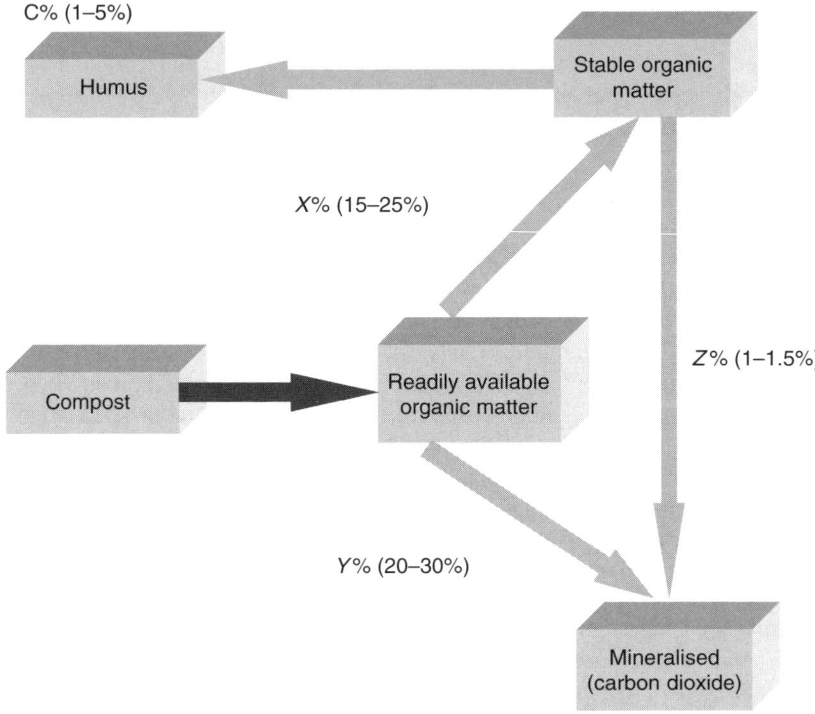

Figure 5.1 Basic description of modelling of fate of carbon in compost/soil.

slower than the rate at which the material is applied. Eventually, the rate of loss of the remaining organic matter equals the rate of application so that the application of compost at different rates each year tends to take the soil organic matter, asymptotically, towards an equilibrium concentration of organic matter. In practice, this equilibrium level will depend upon other factors such as soil cultivation practices, and specifically, the attitude towards tillage of the soil. Already, in North America, one company (TransAlta) has looked at the possibility of securing carbon activities against the resort by farmers to no-till farming practices on the basis that this reduces the rate at which soil organic carbon is mineralised into the atmosphere.

Using the figures $X = 25\%$, $Y = 20\%$, $Z = 1\%$, and with an initial organic matter concentration of 2%, one can plot the effects of different rates of compost application. This is shown in Figure 5.2.

The production of compost and incorporation in topsoil has the potential to act as a reservoir for carbon. When combined with responsible agricultural practices it could have a positive impact on reducing the rate of global warming. However, as compost is applied, the processes of mineralisation do lead to releases of carbon dioxide. The progressive loss of soil organic matter over time following 1-year's application of compost suggests that after 50 years, some 13% of the original carbon applied still remains in the soil. This implies a potential to "buy time" whilst other measures in respect of climate change mitigation measures are allowed to take effect.

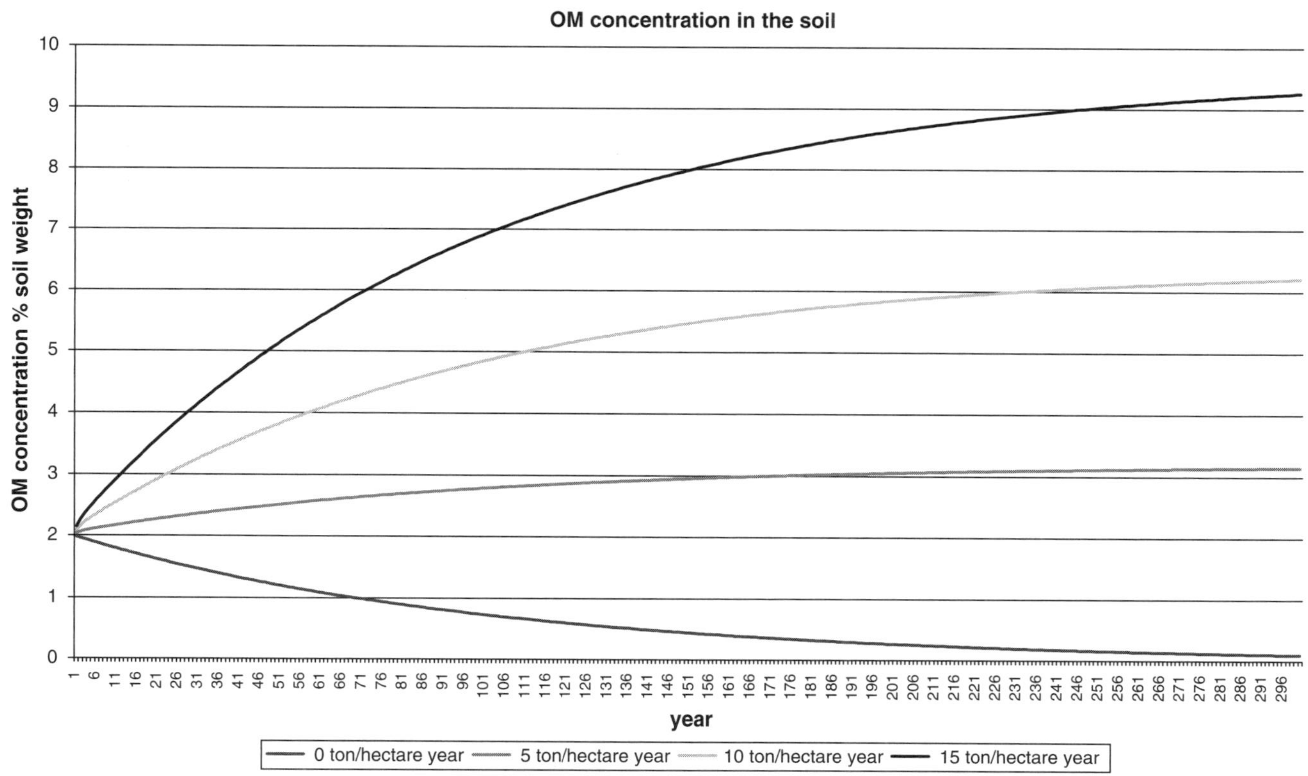

Figure 5.2 Effect of different rates of compost application on soil organic matter levels.

Note that the combustion of waste to generate energy, whilst it may appear superficially attractive, is a poor idea where organic materials are concerned. These have low calorific value, and there is a suggestion that, for example, if these were combusted alone, their high moisture content/low calorific value would imply that limited energy would be recovered using thermal technologies.

5.3.4 Compost as an alternative to peat

Compost may, in some countries, substitute for peat in pot cultivation (and home gardening in some countries). The main use of peat in amateur gardening is a soil improver. Compost can be used as a substitute for peat as a soil improver. It performs well and is likely to be accepted by the public as an alternative to peat. However there are several barriers to overcome before this is the case. Retailers and consumers must be assured that compost is of an acceptable quality (fit for purpose) and that there is uniformity between batches.

The other main use of peat is as the main constituent in growing media. Growers need to be assured that compost is free from contaminants and that it will not have any phyto-toxic effects. These assurances need to be given based on substantiated research and adequate quality and user guidelines.

5.3.4.1 Environmental costs of peat extraction

The environmental costs of peat extraction are difficult to quantify in monetary terms. The primary losses to the environment through peat extraction are:

- loss of biodiversity;
- loss of landscape and recreational value;
- loss of palaeoecological and archaeological value;
- Increased carbon emissions (loss of carbon reservoirs).

Loss of biodiversity Many rare and protected species thrive in Europe's peatlands and bogs. The bog moss *Sphagnum imbricatum* is entirely restricted to bogs and is the principal peat forming species in oceanic peatlands. It is becoming increasingly rare as more sites are being developed. There is also the loss of rare and unique plants which have potential medicinal properties. These benefits are difficult to value although studies concerning biodiversity loss have reported high values reflecting the willingness of citizens to pay for conservation.

Loss of landscape and recreational value In Europe areas of peatlands and bogs have a cultural importance as some of the last true remaining wilderness areas. They attract visitors for this reason. Travel-cost and contingent valuation studies capture consumer surplus associated with, and preferences for, respectively, the continued existence of these landscapes. One study in the UK translates values for the Somerset levels into a value of £7245 per hectare (Ingham 1996, citing Willis *et al.* 1993), equivalent to approximately €12,200 per hectare in current terms. Another UK study estimated a preservation value of £68.4 million, or £4.1 million/annum using a 6% discount rate (Hanley and Craig 1991), equivalent to approximately €116 million in total, or €7 million/annum.

Loss of palaeoecological and archaeological value Peatlands and bogs contain a rich archive of information about our history. Examination of peatlands provides an insight into past climates, culture and economy. These non-use benefits of peat are lost once they are developed for exploitation.

Greenhouse gas emissions (loss of carbon reservoirs) The greenhouse gas emissions associated with peatbogs are extremely complex, and they change once the process of development (for extraction) occurs. In northern peatlands, the anaerobic conditions and cold temperature result in increased sequestration of carbon (relative to other wetlands). Wetlands store carbon in short- and long-term reservoirs. Storage occurs when primary production is high and exceeds the rate of decomposition, or when the rate of decomposition is slowed by a process known as anoxia, and cold temperatures (leading to accumulation of undecomposed organic matter). Unperturbed peatbogs, whilst they may act as a sink for carbon, may also emit methane. However, as long as they are unperturbed, they most likely retain a balance between methane emissions and carbon sequestration.

Drainage and degradation of peatlands increases carbon dioxide emissions. It also increases nitrous oxide emissions significantly (Freeman *et al.* 1993; Roulet *et al.* 1993; Regina *et al.* 1998). Wetlands store carbon in short- and long-term reservoirs. Storage occurs when primary production is high and exceeds the rate of decomposition, or when the rate of decomposition is slowed by a process known as anoxia, and cold temperatures (leading to accumulation of undecomposed organic matter). It has been estimated that peatlands contain between 329 and 528 billion tons of carbon (equivalent to 1200–1900 billion tons of carbon dioxide). Unless the bogs are disturbed by extraction, drainage or other human intervention, much of the carbon will remain *in situ* for near geological timescales.

Drainage of peatlands and other wetlands acting as carbon reservoirs will result in the oxidation of the organic matter, releasing it to the atmosphere as carbon dioxide, methane and other greenhouse gases. Conversely, restoration or creation of new wetlands may provide additional carbon sinks (Environment Canada 1998).

5.3.4.2 Quantification

The use of peat would seem to incur considerable environmental costs. Although these can be captured in various ways, it is very difficult to impute an environmental cost per ton of peat extracted. Data was taken from the Finnish life-cycle study used in the ExternE National Implementation study (European Commission 1999). This was adapted since that study, which covers emissions from various aspects of the peat fuel life-cycle, includes emissions from power production and from restoration. The former are subtracted from the data for obvious reasons. The latter are subtracted because at non-zero discount rates, the effects of the restoration "savings" are likely to be relatively small.

These gaseous emissions are used as the basis for the external cost savings from compost use where it displaces peat. Peat is replaced by compost more on the basis of volume than on weight. The density of peat is low, estimated here at 2000 kg/m^3. The density of compost, on the other hand, is of the order 500 kg/m^3 for a compost with dry matter content 60%. This implies that to replace 1 ton of peat would require compost resulting from 7.14 tons of waste material.

The external cost savings would appear to be a significant underestimate of the avoided external costs to the extent that pressure to develop new peatbogs is reduced. This is due to the fact that the non-use values of peatbogs appear to be significant.

5.3.5 Other potential benefits

There are a number of other potential benefits from the application of compost. Some are discussed below, but no valuation data were available for inclusion in the study.

5.3.5.1 Reduced requirement for liming

Since compost acts as a buffer against falling pH, there is likely to be a reduced requirement for liming, which might be required to offset the increased acidity of the soil. The extraction of lime implies quarrying for the material, though sometimes, industrial by-products are applied in this context.

5.3.5.2 Reduced susceptibility to soil erosion

The condition of a soil surface determines whether rainfall infiltrates the soil or simply runs off. Soil therefore regulates and partitions water. When water runs off land, it tends to carry soil particles. This results in costs to farms in terms of lost productivity and off-farm impacts ranging from damage to commercial and recreational fishing, increased pressure on water treatment facilities, increased flood damages and requirement for repairs from redredging damaged waterways.

It is increasingly recognised that off-farm costs of soil erosion are probably greater than on-farm ones. The off-farm costs associated with soil erosion in the US due to waterways alone were estimated at \$2–\$17 billion (National Research Council 1989; Ribaudo 1989; Pimentel *et al.* 1995). In the UK, a 1996 study estimated soil erosion impacts between £23.8 and £50.9 million (1991 prices) with off-farm losses responsible for as much as 80% of this figure (Evans 1996). In future, severe storms may generate the bulk of soil erosion losses, and this may be a possible "positive feedback" associated with global warming in the future. Air-borne soil particles may also have impacts on human health, and their presence could be reduced through greater use of organic matter to bind soil into stable aggregates.

Management factors play a role in reducing erosion, but so also does the soil texture and organic matter content. It is difficult (for obvious reasons) to estimate the incremental reduction in soil erosion associated with applications of compost.

However, the benefits associated with reduced soil erosion are potentially significant. No quantification was attempted.

5.3.5.3 Benefits from improved infiltration

Benefits from improved infiltration of water arise through reduced risk of flooding (and soil erosion – see above) and reduced requirement for irrigation water. No quantification has been carried out.

5.3.5.4 Reduced irrigation requirement

Water holding capacity can be increased by as much as 3–5% through application of soil organic matter. The avoided environmental burden is difficult to assess at the margin, though it is possible to place values upon water in specific contexts. Indeed, many argue for the use of tradable permits as an allocation mechanism to ensure efficient use of water. Other benefits relate to the increased survival rates of unmanaged young trees in dry periods.

It is difficult to quantify these savings, either in terms of environmental benefits or financial savings, because of the varying nature of the demand for water for agriculture across countries. The financial savings to be realised depend very much on the charging regime for water. In many countries, water for agriculture is still made available on a flat fee or per hectare basis. Consequently, there is no marginal benefit to be gained from reduced consumption. Such savings may become more important in the future, however, as it is likely that more and more countries will move towards marginal cost pricing for water resources.

5.3.5.5 Improved tilth

The fact that compost improves soil structure means that it is actually easier to work the soil with agricultural machinery. There are likely to be savings in fuel use resulting from this change in soil quality. No attempt has been made to quantify these.

5.4 CONCLUDING OBSERVATIONS

Putting the picture together suggests that the external benefits of compost application (and energy recovery in the case of digestion) can offset, though not completely, the external costs of composting and digestion (see Tables 5.8 and 5.9).

It is an open question as to whether a "complete" analysis, if it were possible, would show positive numbers, or whether the numbers would remain negative. It is the view of this author, however, that, when compared with other waste management options, the unquantifiable external *costs* are likely to be less significant than those unquantifiable external costs for the other principal waste management

Table 5.8 Net costs and benefits of compost treatment and application (in €/ton of waste) in Italy.

	Low value	High value
Greenhouse gases		
Carbon dioxide (process)	−8.56	−9.01
Carbon dioxide (post-application)	−3.13	−3.30
Methane	−0.44	−0.48
Nitrous oxide	−0.09	−0.16
Other air emissions	−0.02	−0.07
Energy use (kWh)	−0.85	−1.53
Fuel emissions (l)	−0.36	−0.80
Total external costs	**−13.46**	**−15.34**
External benefits from nutrient displacement	0.13	1.74
External benefits from pesticide reduction	0.40	0.60
External benefits from avoided nitrous oxide emissions	0.04	0.65
External benefits from avoided process wastewater disposal	0.01	0.03
External benefits from avoided peat extraction	0.36	0.52
Net externality	**−12.52**	**−11.80**

Table 5.9 Net costs and benefits of AD treatment and compost application (in €/ton of waste) in Italy.

	Low value	High value
Greenhouse gases		
Carbon dioxide (process)	−10.76	−11.33
Carbon dioxide (post-application)	−3.13	−3.30
Methane	0.00	0.00
Nitrous oxide	0.00	−0.71
Other air emissions	−0.08	−0.18
Avoided external costs from energy generation		
Electricity	1.38	5.38
CHP	1.97	7.72
Total external costs, no displaced burdens	**−13.97**	**−15.51**
Total external costs, displaced burdens from electricity	**−12.59**	**−10.13**
Total external costs, displaced energy from CHP	**−12.00**	**−7.79**
External benefits from nutrient displacement	0.12	1.67
External benefits from pesticide reduction	0.40	0.60
External benefits from avoided nitrous oxide emissions	0.04	0.65
External benefits from avoided process wastewater disposal	0.01	0.02
External benefits from avoided peat extraction	0.36	0.52
Total external costs, no displaced burdens	−13.05	−12.05
Total external costs, displaced burdens from electricity	−11.67	−6.67
Total external costs, displaced energy from CHP	−11.08	−4.33

options. At the same time, the *benefits* of energy recovery techniques (energy itself) are relatively well characterised. The *benefits* of compost utilisation are less well understood and less amenable to ready quantification. This chapter, and the work on which it is based (Hogg *et al.* 2002a) should be seen, therefore, as an

Table 5.10 Private savings associated with avoided pesticide and fertiliser use.

Memorandum items	Compost		AD	
	Low value	High value	Low value	High value
Private savings from avoided fertiliser use	0.77	1.93	0.58	1.45
Private savings from avoided pesticide use	1.00	1.00	1.00	1.00

early attempt to point the way forward to more complete assessments of costs and benefits of composting and digestion processes.

There is an interesting question as to whether the private savings which arise from the use of compost should be considered in an analysis of "external costs and benefits" of the use of compost. Theoretically, and in many cases, in practice also, such savings should not be included. People who buy compost would be expected to "internalise" such savings in their decision as to whether or not to purchase the product. In this case, these benefits would be internalised in the market price for compost.

However, there may be effects which arise from the use of compost which might not be obviously attributable to the use of compost itself. A good example would be the disease suppressing effect of compost. Reduced outlay on pesticides might not necessarily be linked to the application of compost, not least since it is not easy to know what the counterfactual scenario would have been. Arguably, the more people understand the benefits associated with compost, the smaller is the justification for considering the private savings as "an external benefit". In the world of perfect information, the effect is internalised in the decision-making process (and arguably, the market for compost would improve where such benefits were understood and realised).

For completeness, we estimate the potential financial savings here (see Table 5.10). If one were to include estimates of savings in terms of the "fertilisers and pesticides avoided", the picture would appear more positive.

5.5 CONCLUSIONS

5.5.1 Composting and digestion versus other waste treatments

Note that these externalities, though negative, are less negative than those of, for example, incineration (see Hogg *et al.* 2002a). The approach taken in this chapter is not directly comparable with figures in other studies since most other studies ignore carbon dioxide emissions from biogenic wastes. The approach used here can account for the effects of the difference in the time profile of carbon dioxide emissions between, for example, the combustion of biowastes, and the composting of biowastes, followed by the application of compost to land.

Though the estimation of external costs is an incomplete and uncertain science where waste treatments are concerned, there are good reasons to suppose that composting and digestion would remain superior in an "ideal" analysis. This relates to the fact that the majority of emissions from compost appear to be relatively benign. Those from incinerators, on the other hand, are not, and indeed, whilst the effects of some are subject to scientific debate (as with some emissions from compost), the effects of others – particulates, dioxins, sulphurous oxides, oxides of nitrogen – on human health are relatively well established.

Three key observations appear to come from this study:

(1) First, the potential benefits from compost are wide-ranging and include benefits which are only relatively poorly understood at present. As a material for use in sustainable, knowledge intensive (as oppose to "synthetic external input intensive" agriculture), compost may have an important role to play.
(2) Second, composting as a waste (resource?) treatment method for dealing with biodegradable materials, implies some negative externalities, but these are rarely "large ones". Other non-landfill waste treatment technologies have been shown to generate negative externalities of a potentially more significant magnitude (as well as, in the case of incinerators, hazardous waste materials). Judged by this criterion, the process is relatively benign. This supposition is strengthened to the extent that plants for composting and digestion may be located in rural areas, whereas incinerators rarely are. This means that the size of the populations exposed to the pollutants emitted are frequently very different, a fact that is routinely ignored in life-cycle comparisons of the treatments.
(3) Of the two treatments, composting and AD, the environmental performance of AD is superior. The issue then arises as to what the cost differentials might be, and whether the costs of AD, which appear to be higher in most cases than the costs of composting, are justified by the additional benefits. Our impression is that as process management improves at AD plants, and as regulatory pressures increase upon compost plants (not always, it would seem, with strong justification), the costs of composting food wastes, and the costs of AD, are becoming more comparable. It seems likely that it will be possible to treat garden wastes more cheaply through windrow techniques, and it may lead to strategies in which the collection of food and garden wastes is kept separate where possible, so as to optimise the balance of costs and benefits, private and external, of the treatment of biowastes.

5.5.2 Separate collection of biowastes

On the basis of the above considerations, the rationale for encouraging effective source separation of biodegradable wastes seems relatively strong. This is particularly true in Southern Europe where the need for organic material in soil is of paramount importance. However, temperate zones are also losing organic matter from the soil. Maintaining soil health and quality in the context of a simultaneous movement towards these twin goals suggests the wisdom of collecting source-separated organic wastes for the creation of quality composts.

Further support for a strategy focused on source separation is to be found in the fact that the separate collection of compostable materials does not necessarily incur significant additional costs over standard refuse collection. This depends upon the specifics of the collection system, but moving to less frequent refuse collection enables the costs of systems collecting compostables separately to be constrained at levels similar to those where no separate collection occurs.

Hence, the private costs of separate collection and composting/digestion can be quite favourable, depending upon the prevailing regime governing the costs of other waste treatment options. In countries such as Austria, Netherlands, Belgium (Flanders), and Germany, separate collection is well advanced. It should come as no surprise to learn that it is in these countries where the private costs of incineration are relatively high (because of the regulatory demands on flue gas emissions, the costs of treating ash residues, and the more limited support for "renewable energy" from incineration) and where there are constraints on landfilling of untreated waste (either through bans on landfilling in Flanders and Netherlands, or requiring pre-treatment through mechanical biological treatment in Germany and Austria). Hence, there is an argument to be advanced that if the (regulatory and economic) regime for determining the costs of other waste treatment options were more harmonised across the EU, separate collection of compostables would flow naturally as a consequence of the fact that the alternatives to separate collection and composting would be a higher cost combination of refuse collection and more expensive residual waste treatment.

Note that this chapter has not considered the significance of the issue of standards, though this has been addressed elsewhere (Hogg *et al.* 2002b; Chapter 11). It is clear that standards, to the extent that they draw distinctions between wastes and products, encourage the use of biowaste treatment for the treatment of wastes which are separated at source for the development of quality products. Quality assurance schemes can help to support such standards in developing the market for quality composts derived from source-separated biowastes.

REFERENCES

Bocoum, B. and Labys, W.C. (1993) Modelling the economic impacts of further mineral processing: the case of Zambia and Morocco. *Resour. Policy* **19**(4), 247–263.

DeCeuster, T.J.J. and Hoitink, H.A.J. (1999) Prospects for compost and biocontrol agents as substitutes for methyl bromide in biological control of plant diseases, Compost. Sci. Util. **7**, 6–15.

Ehrlich, A. (1990) Agricultural Contributions to Global Warming. In: *Global Warming: The Greenpeace Report* (eds. J. Leggett), Oxford University Press, Oxford.

Environment Canada (1998) *Wetland Conservation: An Excerpt from the Sinks Table Options Paper*, November.

European Commission (1999) *ExternE: Externalities of Energy: Vol. 10, National Implementation*, Office for Official Publications for the European Communities, Luxembourg.

Evans, R. (1996) *Soil Erosion and its Impact in England and Wales*, Friends of the Earth Trust, London.

Foster, V., Mourato, S., Tinch, R., Ozdemiroglu, E. and Pearce, D. (1998) Incorporating external impacts in pest management choices. In: *Bugs in the System – Redesigning the Pesticide Industry for Sustainable Agriculture* (eds. Vorley, W and D. Keeney), Earthscan Publications Ltd, London.

Freeman, C., Lock, M. and Reynolds, B. (1993) Fluxes of CO_2, CH_4, and N_2O from a Welsh peatland following simulation of water-table draw-down: potential feedback to climate change. *Biogeochemistry* **19**, 51–60.

Hanley, N. and Craig, S. (1991) The economic value of wilderness areas. In: *Environmental Policy and the Economy* (eds. F. Dietz, F. Van der Ploeg and J. van Straaten), Amsterdam, North Holland.

Hogg, D., Broome, E. and Vaze, P. (2000) *Beyond the Bin: The Economics of Recycling.* Final Report to Waste Watch, www.wastewatch.org

Hogg, D., Thompson, J., Wood, K., Penschke, A., Economides, D. and Papageorgiou, S. (2002a) *Economic Assessment of Options for Dealing with Biodegradable Waste*. Report to DG Environment, European Commission, by Eunomia, Scuola Agraria del Parco di Monza, HDRA Consultants, ZREU and LDK.

Hogg, D., Barth, J., Favoino, E., Centemero, M., Caimi, V., Amlinger, F., Devliegher, W., Brinton, W. and Antler, S. (2002b) *Comparison of Compost Standards Within the EU, North America and Australasia*, Final Report to WRAP, Banbury, Oxon.

Hoitink, H.A.J. and Fahy, P.C. (1986) Basis for the control of soil borne plant pathogens with composts. *Annu. Rev. Phytopathol.* **24**, 93.

Hydro Agri Europe (1995) *Important Questions on Fertilizer and the Environment*, Hydro Agri Europe, Brussels.

Ingham, A. (1996) *The Use of Economic Instruments to Protect Raised Lowland Peatbogs*, Report to Royal Society for Protection of Birds, Department of Economics, University of Southampton.

James, Paul C. (1995) Internalising externalities: granular carbofuran used in rapeseed in Canada. *Ecol. Econ.* **13**(3), 181–184.

Komilis, D. and Ham, R.K. (2001) Emissions during solid waste composting based on life cycle analysis principles. *Proceedings of the 2nd ISWA Specialised Conference on Biological Treatment of Waste: A Solution for Tomorrow*, 91–114.

Lashof, D. and Tirpak, D. (eds.) (1990) *Policy Options for Stabilising Global Climate*, Hemisphere, London.

McTaggart, I.P., Ball, B.C. and Watson, C.A. (1998) Influence on land use in Scotland on soil N_2O emissions and atmospheric CH_4 uptake. *Paper for the British Society of Soil Science Annual Meeting*, Queen's University, Belfast, September 1998.

Montanarella, L. (1999) *Soil at the Interface Between Agriculture and Environment*. European Commission.

National Research Council (1989) Problems in US agriculture. In: *Alternative Agriculture*, National Academy Press, Washington DC.

Pimentel, D., Acquay, H., Biltonen, M., Rice, P., Silva, M., Nelson, J., Lipner, V., Giordani, S., Horowitz, A. and D'Amore, M. (1993) Assessment of environmental and economic impacts of pesticide use. In: *The Pesticide Question: Environment, Economics and Ethics* (eds. D. Pimentel, and H. Lehmann), 47–84, Chapman and Hall, London.

Pimentel, D., Harvey, C., Resosudarmo, P., Sinclair, K., Kurz, D., McNair, M., Crist, S., Shpritz, L., Fitton, L., Saffouri, R. and Blair, R. (1995) Environmental and economic costs of soil erosion and conservation benefits. *Science* **267**(5201), 1117–1123.

Pretty, J.N., Brett, C., Gee, D., Hine, R.E., Mason, C.F., Morrison, J.I.L., Raven, H., Rayment, M.D. and van der Bijl, G. (2000) An assessment of the total external costs of UK agriculture, *Agr. Syst.* **65**, 113–136.

RDC-Environment and Pira International (2001) Evaluation of costs and benefits for the achievement of reuse and recycling targets for the different packaging materials in the frame of the Packaging and Packaging Waste Directive 94/62/EC. *Draft Final Report to DG Environment, European Commission*, RDC-Environment and Pira International, May 2001.

Regina, K., Nykaken, H., Maljanen, M., Silvola, J. and Martikainen, P. (1998) Emissions of N_2O and NO and net nitrogen mineralization in a boreal forested peatland treated with different nitrogen compounds. *Can. J. Forest Res.* **28**, 132–140.

Ribaudo, M. (1989) *Water Quality Benefits from the Conservation Reserve Programme*. Agricultural Economic Report No. 606, USDA Economic Research Service, Washington DC.

Ringer, C.E., Millner, P.D., Teerlinck, L.M. and Lyman, B.W. (1997) Suppression of seedling damping-off disease in potting mix containing animal manure composts. *Compost Sci. Util.* **5**, 6–14.

Roulet, N., Ash, R., Quinton, W. and Moore, T. (1993) Methane flux from drained northern peatlands: effect of a persistent water table lowering on flux. *Global Biogeochem. Cycle.* **7**(4), 749–769.

Soulsby, P.G., Hickman, G.A.W. and McMahon, P.L. (2000) The environmental and agricultural economics of recovering value from recycling treated biosolids to land. Paper for the *Chartered Institution of Water and Environmental Management/Aqua Enviro Technology Transfer Conference "Wastewater Treatment: Standards and Technologies to Meet the Challenges of the 21st Century"*, Leeds, 4–7 April.

Steiner, R., McLaughlin, L., Faeth, P. and Janke, R. (1995) Incorporating externality costs into productivity measures: a case study using US agriculture. In: *Agricultural Sustainability: Environmental and Statistical Considerations* (eds. V. Barbett, R. Payne and R. Steiner), John Wiley and Sons, New York.

UNEP and UNIDO (United Nations Environment Programme and United Nations Industrial Development Organisation) (1998) Mineral fertilizer production and the environment – Part 1: The fertilizer industry's manufacturing processes and environmental issues. *Report prepared in collaboration with the International Fertilizer Industry Association. Technical Report No. 26 – Part 1*. UNEP and UNIDO, Paris and Vienna.

Willis, K., Garrod, G. and Saunders, C. (1993) *Valuation of the South Downs and Somerset Levels and Moors Environmentally Sensitive Landscapes by the General Public*. Report to MAFF by the Centre for Rural Economy, Newcastle-upon-Tyne.

6

Modelling the municipal solid-waste management problem in an applied spatial general equilibrium framework

H. Bartelings, R.B. Dellink and E.C. van Ierland

6.1 INTRODUCTION

One of the fundamental questions regarding waste management concerns the "optimal" quantity of waste a society should generate. Most environmental scientists argue that waste generation should be minimized. From an economic point of view, it may, however, not be desirable to fully close the material cycle. It is often forgotten that both recycling and reuse of materials have financial and environmental impacts, which makes it undesirable to completely eliminate the waste generation (Pearce and Turner 1993). Both environmental scientists and environmental economists, however, agree that too much waste is currently being generated (*cf.* Graig 2001). The question remains how much waste should be generated and how

Edited by Piet Lens, Bert Hamelers, Harry Hoitink, Werner Bidlingmaier. ISBN: 1 84339 054 X. Published by IWA Publishing, London, UK.

the waste market can be regulated to produce the "optimal" quantity of waste. Fricker (2003) argues that the only sustainable way of reducing waste generation is by reducing the consumption. This view is, however, not supported by economic theory. In this section, a number of policy instruments that can be used to control waste generation, will be discussed.

Waste management is in most countries still dominated by inefficient pricing, institutional barriers and legal structures. The primary virtue of the pricing mechanism, i.e. a well-functioning market, is that it gives consumers an idea of the costs of producing a particular product and offers producers insight into how consumers value a product (Löfgren 1995). Naturally, the pricing mechanism only supplies the correct information if the market is undistorted. Solid-waste management pricing is mostly based on a flat fee system. Households pay a fixed charge, the so-called flat fee, for the collection of waste. As the amount of the fee is independent of the quantity of waste that is actually generated, consumers have no price incentive to reduce the generation of waste, and thus larger quantities of waste are disposed of than is socially desirable.

Figure 6.1 illustrates a possible demand curve for waste collection services, i.e. the amount of waste that is generated (and thus has to be collected). As the price of these services declines, the demand for these services increases. In the case of household waste disposal, the price of disposing one extra unit of waste equals zero, as the price is independent of the quantity of waste disposed off.

If the price, i.e. the marginal costs of waste disposal, is equal to zero then Q^0 will be the optimal quantity of waste disposal. If, however, the marginal costs of waste disposal are positive, then the demand for solid waste services is clearly higher than optimal. For example, assume that the social costs of waste disposal are equal to P^*, then the optimal demand for solid waste services will be equal to Q^*. Society faces a net total cost equal to the triangle *abc* caused by the inefficiently high demand for waste disposal services. Only when the disposal fee is equal to the exact marginal costs of waste disposal, the demand for waste disposal services will equal the optimal quantity of waste disposal (Jenkins 1993).

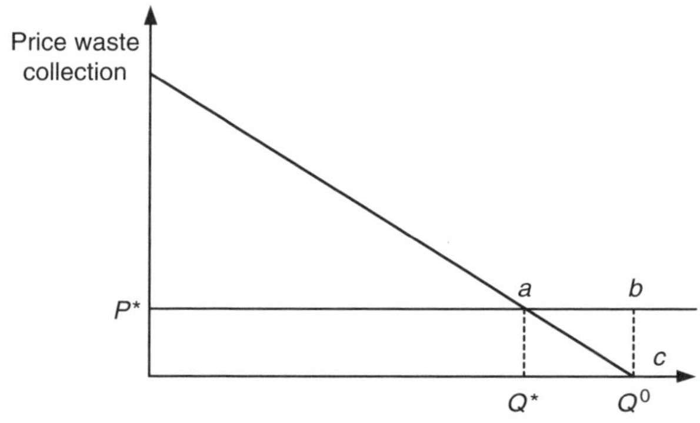

Figure 6.1 The demand curve for waste collection services (Jenkins 1993).

Finding the optimal disposal fee, however, is not as simple as it sounds. The optimal disposal fee should cover both the marginal financial and the marginal environmental costs of municipal solid waste disposal and treatment. It is, therefore, important to quantify all external effects caused by waste treatment (see Chapter 5). However, as Figure 6.1 clearly demonstrates, the flat fee pricing scheme will always lead to a non-optimal quantity of waste generation since the marginal costs of waste disposal are most assuredly positive.

By replacing the flat fee pricing system with a unit-based pricing system, it is in theory possible to alleviate the market distortion. In a unit-based pricing system, households pay a variable fee to the municipalities for the collection of municipal solid waste; the fee charged will in some way depend on the actual quantity of waste generated. Several differentiating pricing systems are possible: for example a *weight-based pricing system*, which bases its price of collection on the total weight of waste collected; a *frequency-based pricing system*, which bases the price of collection on the frequency it is collected or a *volume-based pricing system*, which bases its price on the volume of waste collected.

Empirical studies, like Hong *et al.* (1993), Miranda *et al.* (1994) and Sterner and Bartelings (1999) conclude that waste generation is sensitive to user fees. The introduction of user fees can lead to a substantial reduction in municipal solid waste generation, especially if they are combined with programs that increase the public awareness about the municipal solid waste problem. The imprudent construction of waste collection fees, however, might not have the desired effect and can encourage illegal dumping, burning or other improper kinds of disposal (see e.g. Fullerton and Kinnaman 1995; Palmer and Walls 1997).

Thus far studies analysing unit-based pricing have focussed only on the effect unit-based pricing will have on the *quantity* of waste generated. Unit-based pricing can, however, also influence the *quality* of waste generated. Policy makers expect households to separate organic waste and rest waste. If households start to pollute organic waste by dumping rest waste in it, it may have grave consequences on the quality of organic waste and thereby on the quality of compost produced from this organic waste. Especially if policy makers introduce unit-based pricing for the collection of rest waste but not for the collection of organic waste, households will have an incentive to pollute organic waste to reduce their waste bill.

In this chapter, a model to simulate how the introduction of a unit-based pricing scheme will influence both the quality and quantity of municipal solid waste is presented. Depending on the quality and quantity of waste collected, municipalities can choose to send the waste either to a small, a medium or a large waste treatment centre. Due to economies of scale, a large waste treatment centre will ask a lower price to treat waste. They will also be able to treat a lower quality of waste than a small waste treatment centre.

This model is applied to a stylised setting of the waste market in a region of The Netherlands. This example illustrates how the quality and quantity of waste generated significantly influences the choice of waste treatment methods, their size and location. Furthermore, it demonstrates that the choice of optimal waste treatment methods, size and location can significantly influence the cost-effectiveness of a policy change. A unit-based price for the collection of rest waste is simulated and shows that although such a policy change reduces waste generation, in line with recent studies such as Fullerton and Wu (1998) and Choe and Fraser (1999), it also decreases the quality of organic waste, thus increasing the composting and transport costs of waste. The model presented in this chapter is applied to the municipal solid waste market in The Netherlands, but is written in general terms and can easily be adapted to the municipal solid waste market in any other industrialized country.

6.2 MODELLING THE MUNICIPAL SOLID-WASTE PROBLEM TAKING INTO ACCOUNT THE SPATIAL ASPECTS

6.2.1 General introduction to the model structure

In the model only one producer who makes a composite consumption good is distinguished. The private households and the government consume the consumption good. Four municipalities are distinguished in the model (note that the municipalities and the government are separate entities; the government can be interpreted as a national government and the municipalities as producers of the service waste collection. Modelling the links between the municipalities and the government is beyond the scope of our analysis and therefore abstracted from). In each municipality, two types of consumers are modelled: a "traditional" consumer and a "green" consumer. The "traditional" consumer has little preference for the environment. The "green" consumer has some preference for a clean environment. Each municipality differs in the number of consumers and the share of "traditional" and "green" consumers.

Consumption of the "produced good" results in waste. Consumers can choose to generate three types of waste, namely "rest waste", "low quality organic waste" and "high quality organic waste". The private households must invest labour in the separation of organic from rest waste. Generating high quality organic waste requires more labour than generating low-quality organic waste.

Consumers either pay a flat fee for the collection of waste or a unit-based price. The flat fee pricing scheme will be implemented in the model using a subsidy-*cum*-tax scheme. This means that consumers pay the equilibrium price for collection of waste to the municipalities. They are reimbursed with the same amount of money in the form of a subsidy from the government. Thus the perceived price of collection for the consumers equals zero. The consumers pay a direct tax to the government to reimburse the costs of the subsidy. In such a way, the consumers do pay for waste collection, but the amount they pay is not coupled with the quantity of waste they generate, in line with the definition of a flat fee.

The municipality, who is a “producer” of collection services, collects rest waste and sends it to an incinerator or a landfill. Low and high quality organic wastes are both collected together and sent to a composting unit. The costs of composting depend on the quality of the organic waste stream. If the overall quality of organic waste is low (i.e. there is a large share of low quality organic waste in the entire organic waste stream), it will be difficult to compost the waste. Part of the waste will then be rejected by the composting unit and sent to an incinerator.

6.2.1.1 Spatial aspects in the model: transport costs and economies of scale

In the model, the spatial aspects of the waste management problem are integrated in a general equilibrium framework. This means including both transport costs and economies of scale. In the model, municipalities collect waste and send it to a waste treatment unit. The municipality has the choice between sending waste to a small, a medium or a large sized waste treatment unit. The small unit is located close to the municipality; the medium and large waste treatment units are located further away. Thus the larger waste treatment units are more efficient in waste treatment, but they are more expensive with regard to transport costs. The key characteristics of the spatial aspects are depicted in Figure 6.2.

In accordance with the standard specification of general equilibrium models, perfect competition between production sectors was assumed. This implies that there is also perfect competition between waste treatment units. In reality this may not be the case. For example, contracts between municipalities and incinerators exist, which specify the quantity of waste municipalities must deliver to the incinerator. It goes beyond the scope of this chapter to include such imperfect

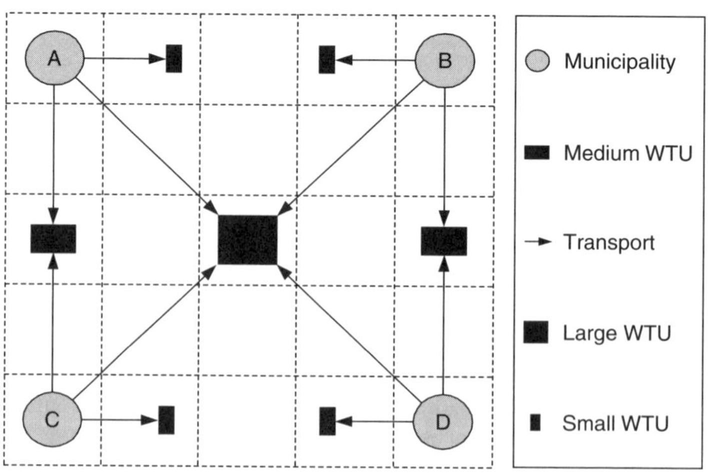

Figure 6.2 Spatial aspects in the model for four municipalities (WTU stands for waste treatment unit).

competition in the model. However, in future research it would be interesting to see how monopolistic behaviour of waste treatment units influence the results.

The three waste treatment units, namely the landfill unit, the incineration unit and the composting unit, offer waste treatment services to the municipalities. These units use capital and labour as inputs to the production process. The composting unit, besides using capital and labour, also uses incineration services to get rid of both the residue and the rejected low quality organic waste.

Due to economies of scale in the production process, the larger waste treatment units offer waste treatment services at a lower price. In the context of our study, several locations are exogenously specified for each waste treatment unit. Given transport costs and waste treatment costs of each unit, municipalities will choose the least expensive option. Except for the differentiated marginal costs, the services provided by the different sizes of a waste treatment unit are identical.

As mentioned before low quality organic waste must be cleaned before it can be composted. It is therefore more expensive to compost low quality organic waste than high quality organic waste. The costs of composting low quality organic waste will also be differentiated according to the size of the composting unit. Smaller composting units tend to charge a higher price to compost low quality organic waste than larger units.

6.2.2 The model represented in equations

6.2.2.1 Welfare, production, consumption and waste generation

The model is built as a standard general equilibrium model in the Negishi format. In the Negishi format, the equilibrium solution is found with the help of an iterative process. Given initial values for the Negishi weights based on the income of a consumer, the model is solved and prices for each commodity are calculated as shadow prices. Subsequently, the budget constraint for each consumer is checked. If one or more consumers in the model spend more or less than their income, the Negishi weight for that consumer is adjusted. The model is then solved again with the adjusted Negishi weights. The process continues until the budget constraints of all consumers hold. For more information about the Negishi format and why the iterative process of the Negishi format results in a general equilibrium solution, consult Negishi (1972) or Ginsburgh and Keyzer (1997).

Total welfare (TWF) is defined as the sum of the utilities (u_i) of each consumer i. The utility of each consumer is weighted by a factor α_i, the so-called Negishi weight. Each consumer derives utility from the consumption of the consumption good (x_i^{g}).

$$\mathrm{TWF} = \sum_i \alpha_i \ln[u_i(x_i^{\mathrm{g}})] \tag{6.1}$$

The production function of the consumption good is given by well-known constant elasticity of substitution (CES) function. The producers of the consumption good use only capital (k), labour (l) as inputs for production:

$$q_{\mathrm{g}} \leq \mathrm{CES}(k_{\mathrm{g}}, l_{\mathrm{g}}; \sigma_{\mathrm{g}}^{k,l}) \tag{6.2}$$

Note that the notation $z = \text{CES}(x, y; \sigma)$ reflects the following function:

$$z = \left(x^{\left(\frac{\sigma-1}{\sigma}\right)} + y^{\left(\frac{\sigma-1}{\sigma}\right)} \right)^{\frac{\sigma}{\sigma-1}}$$

Consumption of the consumption good results in waste generation. Waste generation has been specified as a percentage β of consumption x_i^{g}. It was assumed that only the private households generate waste; the government does not generate waste:

$$W_c = \beta^{\mathrm{g}} x_c^{\mathrm{g}} \tag{6.3}$$

where c is a subset of i including only the private households.

The private households deal with the waste they generate by demanding the so-called waste collection services provided by the municipality. They can substitute the demand for collection services of rest waste (x^{r}) for the demand for collection services of organic waste (x^{o}). Private households may also choose to generate *low* ($x^{\mathrm{o,l}}$) or *high* quality organic wastes ($x^{\mathrm{o,h}}$) as specified in the following CES function, which simulates limited substitution possibilities between the different types of waste:

$$W_c = \text{CES}(x_c^{\mathrm{r}}, \text{CES}(x_c^{\mathrm{o,l}}, x_c^{\mathrm{o,h}}; \sigma_c^{\mathrm{l,h}}); \sigma_c^{\mathrm{r,o}}) \tag{6.4}$$

If the private households decide to generate organic waste, they will have to spend labour (lw) on separating organic waste from rest waste. Generating high quality organic waste involves more labour than generating low quality organic waste. The production function for organic waste of quality f is defined as: (f = *low*, *high*):

$$x_c^{\mathrm{o},f} \leqslant \mu^f \, \mathrm{lw}_c^f \tag{6.5}$$

where μ reflects the units of labour necessary to produce a unit of organic waste of quality f.

The amounts of labour spent on generating organic waste affects effective labour supply. The labour supply L of each consumer i is calculated as the exogenous labour endowment, $\bar{L}$, minus the total amount of labour used for generating both types of organic waste:

$$L_c = \bar{L}_c - \sum_f \mathrm{lw}_c^f \tag{6.6}$$

6.2.2.2 *Transport costs*

The municipality collects the waste generated by private households and sends it to a waste treatment unit. The municipality must pay transport costs for transporting waste to a waste treatment unit. Transport costs are modelled as if the municipality demands transport services of a transport company. Waste treatment units will also charge a price to treat the waste. Each different waste treatment method (index m = *incineration, landfilling and composting*) will charge a different price, according to the marginal waste treatment costs.

The production function for waste collection services is given by a nested Leontief–CES function. It depends on the inputs of capital, labour, transport services (ts) and waste treatment services (wts). Municipalities can choose between different waste treatment methods, i.e. incineration, landfilling, composting and different sizes of waste treatment units. Each waste treatment unit comes in three different sizes: small, medium and large (index s = *small*, *medium*, *large*). The production of waste collection services of municipality j is given as a nested CES/Leontief function that combines primary inputs capital and labour (k,l) with the different potential waste treatment services of type m and size s ($\text{wts}^{m,s}$) and transport services (ts):

$$q_{\text{cs}} \leqslant \min(\text{CES}(k_{\text{cs},}\ l_{\text{cs}};\sigma_{\text{cs}}^{k,l}), \text{CES}(\text{wts}^{m,s};\sigma_{\text{cs}}^{m,s}), \text{ts}_{\text{cs}}) \tag{6.7}$$

Municipalities cannot make a completely free choice between the three different waste treatment options. Organic waste will have to be transported to a composting unit. Rest waste can either be brought to an incinerator or a landfill.

Transport costs depend on the quantity of waste transported and the distance travelled. The possible locations for each waste treatment unit are exogenously determined. The total distance from the municipality to all waste treatment units is given in a transport matrix T. Thus, the demand for ts by municipality j is calculated as the sum of the transport costs to each waste treatment unit m:

$$\text{ts} \leqslant \sum_{m} T^{m}\,\text{wts}^{m} \tag{6.8}$$

The transport services are produced with the input of capital and labour. The production function for transport services is defined as follows:

$$q_{\text{ts}} \leqslant \text{CES}(k_{\text{ts}}, l_{\text{ts}}; \sigma_{\text{ts}}^{k,l}) \tag{6.9}$$

6.2.2.3 *Economies of scale*

As mentioned above, the marginal costs of waste treatment services produced by a larger unit are lower than the marginal costs of waste treatment services produced by a smaller unit, due to economies of scale. This is modelled as if the larger unit uses a more advanced technology A, which means that fewer input factors are necessary to produce the same quantity of services. In the small waste treatment unit, A is equal to unity. In the medium and large waste treatment unit A is larger than unity, thus less production factors need to be used in the production process. Economies of scale are introduced exogenously in the model. This means that the used technology does not depend on the quantity of waste treated in the waste treatment unit.

The production functions for the different waste treatment units of type s and size m are defined as follows:

$$q_{m,s} \leq A_{m,s}\ \mathrm{CES}(k_{m,s}, l_{m,s}; \sigma^{k,l,e}_{m,s}) \tag{6.10}$$

Finally, as in any general equilibrium model, balance equations have been included to determine that demand equals supply for each commodity in the model. The equations involved are straightforward and not reproduced here.

To recapitulate, the model maximizes utility of the consumers and simultaneously determines the optimal method and location of waste treatment, the composition of municipal solid waste generated, and finally the social costs of waste treatment, which consist of financial and environmental costs.

6.3 MODEL APPLICATION AND NUMERICAL ANALYSIS

The model discussed above is applied in a stylised example with numerical data from the Randstad, a large region in The Netherlands. This example focuses on the waste market. The economic data used in the numerical example are based on national accounts for The Netherlands in 2000 (Statistics Netherlands 2002). These data are supplemented with detailed data on the waste sectors (collection, incineration, landfilling and composting) and data necessary for the subsidy-*cum*-tax scheme (fee and subsidy), based on WMC (2000, 2001, 2003a).

6.3.1 Calibration of the model

6.3.1.1 Waste generation in the benchmark case

In total, consumers generate 1733 thousand tons of rest waste and 413 thousand tons of organic waste. All organic waste collected is assumed to be sent to a composting unit, consistent with the laws concerning organic waste in The Netherlands (WMC 2002). Rest waste can either be incinerated or landfilled. In the benchmark case, a large quantity of waste is landfilled. Landfilling is less expensive than incineration. A large tax on landfilling, however, is included which raises the price including the tax of landfilling to a level equal to the price of incineration, in line with empirical observations. Given that the price of landfilling is equal to the price of incineration, municipalities have no preference based on price differences for either landfilling or incineration. Since 2000, such a landfilling tax has actually been introduced in The Netherlands.

Consumers pay a fixed amount of money for the collection of rest waste, the so-called flat fee. Consumers pay a total amount of €510 million for the collection of waste. On an average, the fee paid by the consumers covers only 95% of the real

Table 6.1 Transport distances.

Size of waste treatment unit	Distance from municipality (in km)
Small	35
Medium	75
Large	150

Table 6.2 Value technology parameter waste treatment.

	Small	Medium	Large
Composting	1.0	1.1	1.3
Landfilling	1.0	1.1	1.3
Incineration	1.0	1.3	1.7

costs (WMC 2001). This means that the real costs of waste collection and thus the amount spent on the subsidy on waste collection equals roughly €540 million.

6.3.1.2 Transport costs, economies of scale and differentiated composting costs

The costs of transporting a tonne of waste to a waste treatment unit depends on the distance travelled. In Table 6.1, the transport distances are presented. These distances are based on the average distances from a municipality to a small, medium or large facility in The Netherlands (WMC 2003b).

A larger waste treatment facility will be able to offer waste treatment services against a lower price due to economies of scale. In this model, economies of scale are introduced for the three sizes of waste treatment units by changing the technology parameter in the production function. In Table 6.2, the values of the technology parameters are shown.

A lower quality of organic waste will cost more to compost due to two reasons: firstly the composting process itself is more expensive because more residue is produced, which has to be incinerated and secondly organic waste either needs to be cleaned or is rejected and sent to an incinerator before composting, which also increases the costs. It will be more expensive to treat low quality organic waste in small composting units. In Table 6.3, the marginal costs of composting 1000 tons of high and low quality organic wastes, respectively, are given.

6.3.1.3 Municipalities

The region was assumed to consist of four representative types of municipalities:

A: a large municipality with a relatively high percentage of traditional consumers,

Table 6.3 The marginal costs of composting low and high quality organic waste (Euro per tonne of waste).

Costs	Small	Medium	Large
Composting low quality	258	203	101
Composting high quality	38	34	29

Table 6.4 Differences between municipalities.

Municipalities	Percentage of waste generated (%)	Share of consumers (%)	
		Traditional	Green
A	40	72	28
B	40	68	32
C	10	65	35
D	10	50	50

B: a large municipality with a relatively high percentage of green consumers,
C: a small municipality with a relatively high percentage of traditional consumers,
D: a small municipality with a relatively high percentage of green consumers.

Each of these four types of municipalities collect waste from private households living in that municipality.

It is presumed that each large municipality generates 40% of the total quantity of waste in the economy. Each smaller municipality generates 10% of the total quantity of waste. As explained in Section 6.2, two types of consumers are distinguished: "traditional" and "green" consumers. Based on the MOSAIC system described by Beker (2002), consumers in The Netherlands may be divided into 32.5% "green" consumers and 67.5% "traditional" consumers.

Table 6.4 shows the percentages of traditional and green consumers in each municipality.

6.3.1.4 Generation of organic waste

Private households can generate three types of wastes: rest waste, low and high quality organic wastes. To generate organic waste, either high or low quality, consumers have to incur costs. There are no costs involved in the generation of rest waste. Low and high quality organic wastes are defined as follows: high quality organic waste is organic waste without any pollution and thus 100% pure. It was assumed that low quality organic waste is somewhat polluted by organic waste and only 70% pure. Composting high quality organic waste will not result in any residue, whereas composting of 100 tons of low quality organic waste will result in 30 tons of residue that has to be incinerated.

According to anecdotal evidence obtained from the Waste Management Council as well as the Ministry of Spatial Planning, Housing and the Environment, no recent research has been done to determine the overall quality of organic waste collected by municipalities. Composting units check the quality of organic waste beforehand and decide whether the waste delivered by the municipality can be composted or not. Their records, however, have not been released to the public. Therefore, it was impossible to determine how much high and low quality organic wastes is actually generated in the benchmark case. As an indication of the quality of organic waste, the amount of residue that is produced during the composting process was used. On an average, composting of 100 tons of organic waste of a mixed quality results in 7 tons of rest waste (Beker 2002). Given that composting of high quality organic waste results in no residue and that composting of low quality organic waste results in 30% residue, it can be calculated that the overall mixture of organic waste consists of 23.3% low quality organic waste and 76.7% high quality organic waste.

In line with the way in which "traditional" and "green" consumers are defined, "traditional" consumers generate more low quality organic waste than "green" consumers. Based on overall quality of the organic waste stream and the percentage of "green" consumers and "traditional" consumers, it can be calculated that if the "green" consumer generates 90% high quality organic waste and 10% low quality organic waste then the "traditional" consumer generates 70% high quality organic waste and 30% low quality organic waste.

The calculations on the quantity of labour necessary to generate organic waste are based on the estimation of social costs of waste handling as found in Bruvoll (1998). Bruvoll estimated the social costs of waste handling at 145 dollar per tonne of waste. The amount of time spent on waste handling per week was estimated at 30 min a week. This is comparable to other empirical studies, such as Sterner and Bartelings (1999). About 54% of all recyclable and organic waste collected consists of organic waste (WMC 2003c). If one supposes that the costs are directly related to the amount of waste collected, the average costs of handling organic waste are equal to 78 dollars per tonne. It can thus be expected that generation of low quality organic waste is less costly than generation of high quality organic waste, though the generation of both high and low quality organic wastes will involve many of the same costs. For example, both high and low quality organic wastes must be separated from rest waste and the organic waste bin will, in both cases, have to be cleaned. Therefore, generating high quality organic waste is expected to be 10% more expensive than the average costs and generating low quality organic waste will be 10% less expensive than average costs, reflecting the differences in associated opportunity costs.

Consumers are obviously not able to create high quality organic waste from just any type of rest waste, but can only put more effort in preventing organic waste to enter the rest waste stream. An upper limit on the quantity of high quality organic waste consumers can separate from rest waste has therefore been introduced. About 32% of the rest waste stream consists of organic waste (Beker

Table 6.5 Substitution elasticities for production factors and waste categories.

	CS rest	Traditional consumer	Green consumer
Substitution elasticity incineration and landfilling ($\sigma^{i,l}$)	6	–	–
Substitution elasticity organic waste and rest waste ($\sigma^{o,r}$)	–	0.6	0.3
Substitution elasticity high and low quality compost ($\sigma^{h,l}$)	–	0.9	0.1

Note: "Good" stands for the produced good; "CS rest" stands for the collection of rest waste and "CS organic" stands for the collection of organic waste.

2002). Therefore, consumers will be able to increase their generation of high quality organic waste to a maximum of 32% of their original production of rest waste.

6.3.1.5 Substitution elasticities

Several studies have estimated the own price elasticity of the generation of rest waste. In this chapter, the substitution elasticities between rest waste and high and low quality organic wastes have been based on the price elasticity estimated by Fullerton and Kinnaman (1996). They established an own price elasticity of −0.058. Comparable estimates of price elasticities can be found in Wertz (1976), Hong *et al.* (1993), Jenkins (1993) and Linderhof *et al.* (2001). Since no information is available about the quality of organic waste, it is difficult to calibrate the substitution elasticity between low and high quality organic wastes. It was asserted that the demand is inelastic and thus the substitution elasticity is smaller than unity. By definition, the substitution elasticity will be larger for the "traditional" consumer than for the "green" consumer. Substitution possibilities between different types of waste are given in Table 6.5.

The municipalities can choose between treating a waste in a landfill unit or an incinerator. According to current laws in The Netherlands landfilling of combustible waste is prohibited. Since this chapter wants to show, however, how the choice of the preferred waste treatment option is influenced by the transport costs, quality and quantity of waste generated, it was assumed that municipalities can choose to landfill all waste. The choice between landfilling and incineration will only depend on the prices of the waste treatment options. Furthermore, to ensure that municipalities can easily substitute incinerating for landfilling their waste, a large substitution elasticity between the two options has been introduced.

6.3.2 Scenarios

6.3.2.1 Scenario 1 benchmark: flat fee pricing system

In scenario 1, consumers pay a flat fee for the collection of organic waste and rest waste. Thus they have no price incentive to lower the quantity of waste they generate. Municipalities collect the waste and choose the location of the facility to which they transport it. The benchmark case reflects the historical situation. Over the past few decades, landfilling has been the most popular waste treatment option. Until the mid-1990s, most of the waste was landfilled. Incineration only became popular after 1996 because the government stimulated incineration by both prohibiting and taxing the landfilling of combustible waste.

6.3.2.2 Scenario 2 benchmark: selective unit-based pricing system

In scenario 2, a unit-based price for the collection of rest waste is introduced. This means that consumers pay the equilibrium price for the collection of rest waste, which reflects the marginal costs of producing these services. The consumers still pay a fixed amount of money for the collection of organic waste, so the marginal costs of producing organic waste remain equal to zero. It may be expected that this policy change will provide consumers with an incentive to reduce generation of rest waste in favour of generation of organic waste. Consumers will start to separate more organic waste and rest waste. To separate more waste, they will need to incur more costs in terms of labour use.

6.3.2.3 Scenario 3 benchmark: full unit-based pricing system

In scenario 3, a unit-based price is introduced for the collection of both organic waste and rest waste. In this scenario, consumers pay the marginal costs of generating waste. Since composting is slightly less expensive than incineration or landfilling, consumers will pay somewhat less for the collection of organic waste than for the collection of rest waste. Consumers, however, do not only pay a fee for the collection of organic waste, but they also incur costs generating organic waste, as they have to invest labour in waste separation. Thus, the total costs of producing low and high quality organic waste combined with the collection costs are higher than the collection costs of rest waste. Therefore, only a minimal change in the generation of organic waste and rest waste may be expected. Introducing a unit-based price on the collection of rest waste and organic waste can give the consumer an incentive to minimize their total waste generation by consuming less.

6.3.3 Results: comparing the different pricing mechanisms

The results show that the total quantity of waste generated is hardly affected by the introduction of either a selective unit-based pricing system or a full unit-based pricing system. The composition of the waste stream, however, does differ between the three different pricing systems. This is illustrated in Figure 6.3.

A selective unit-based pricing system causes the most significant substitution of organic waste for rest waste. A full unit-based pricing system has hardly any effect on the quantity of rest waste and organic waste generated. In the selective

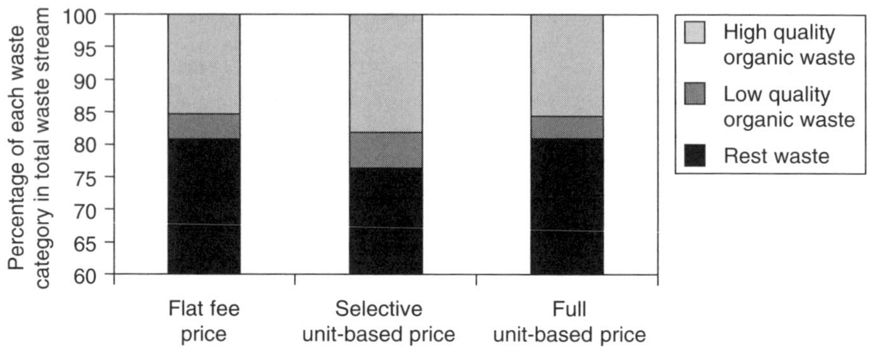

Figure 6.3 The composition of the waste stream for the three different pricing mechanisms.

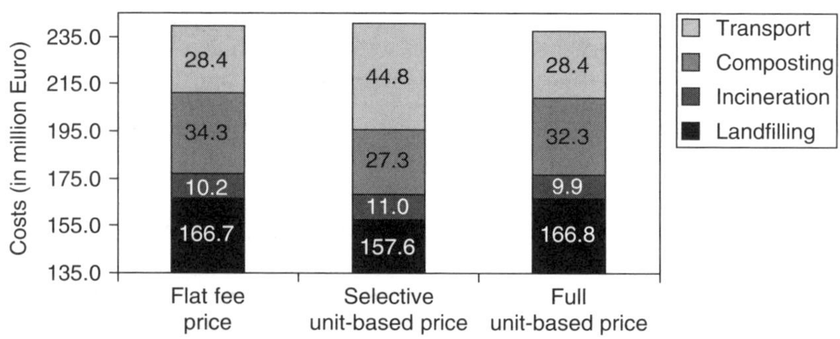

Figure 6.4 The total costs of waste treatment for the three different pricing mechanisms.

unit-based pricing system, however, there is a substantial increase in the quantity of low quality organic waste. Since consumers do not need to fear any penalties, they start polluting the organic waste stream with rest waste. This waste leakage effect greatly increases the cost of composting organic waste. In a full unit-based pricing system, there is barely any incentive for the consumers to start polluting waste; waste leakage is, therefore, not such a big problem.

The cost of waste treatment differed significantly between the different pricing systems. This is shown in Figure 6.4. In a flat fee pricing system, all organic waste is treated in a small composting unit.

Households generate mostly high quality organic waste, but also some low quality organic waste. Composting of low quality organic waste is expensive, especially if it is treated in a small composting unit. Although it is quite expensive for the municipalities to compost low quality organic waste in a small composting unit, the increased costs of composting are not so high that they offset the costs of transporting the waste to a large composting unit. If a selective unit-based pricing system is introduced the situation is quite different. In the larger municipalities, the quality of the organic waste stream declines so much that these municipalities

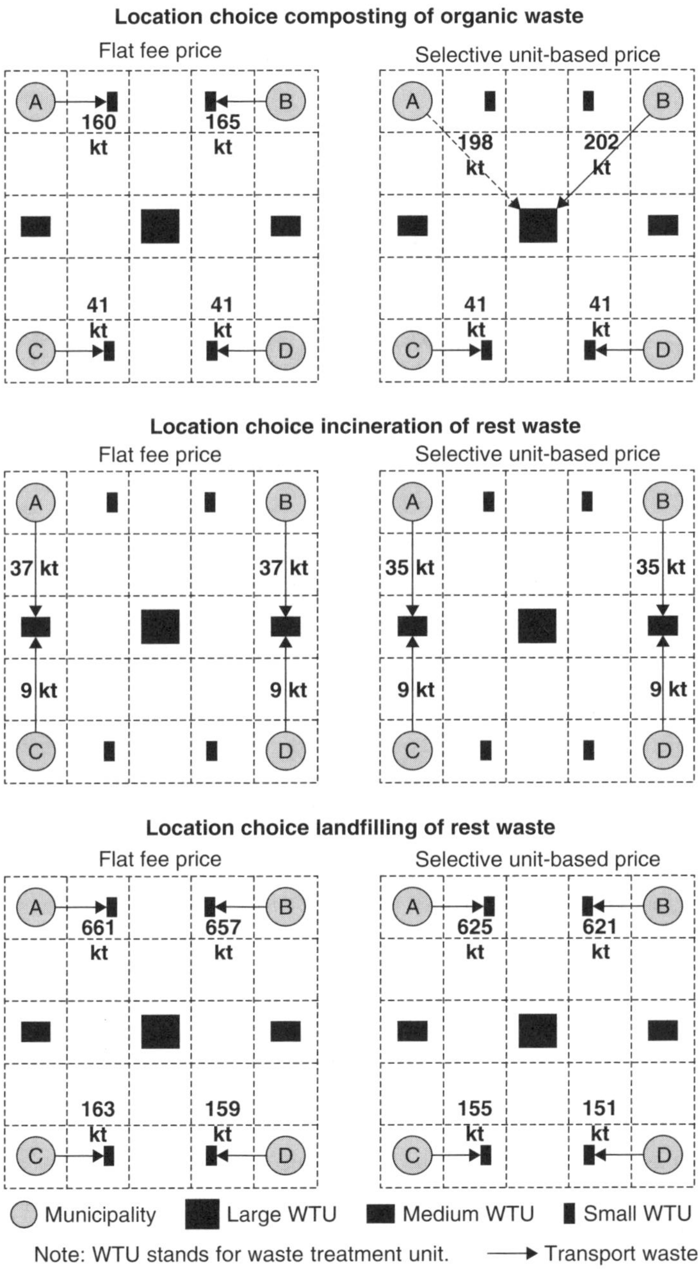

Figure 6.5 Transport of organic and rest waste for each municipality.

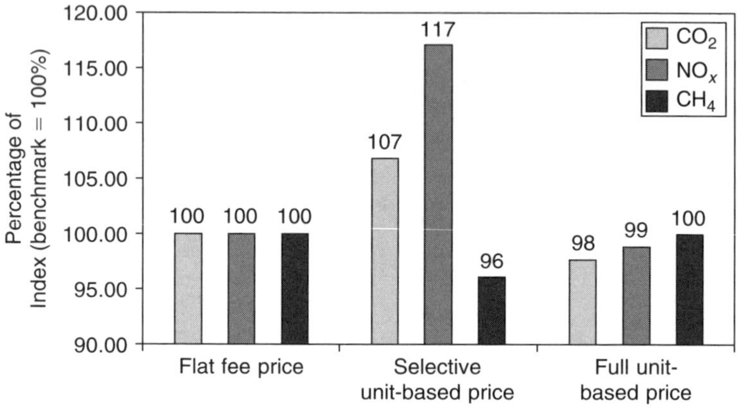

Figure 6.6 Normalized CO_2, NO_x, and CH_4 emissions caused by waste treatment and transport as compared to flat fee scenario (flat fee = 100%).

choose to transport the waste to a large composting unit. This results in sharply increasing transport costs and decreasing composting costs, as shown in Figure 6.4.

An illustration of the transport flows of waste in a flat fee pricing system and a selective unit-based pricing system is given in Figure 6.5. The distribution of costs in the full unit-based pricing system is quite similar to the distribution of costs in the flat fee pricing system. The overall costs spent on waste treatment are highest in the selective unit-based pricing system, due to pollution of the organic waste stream, which can also be called waste leakage.

Finally, it is also interesting to compare the emissions of CO_2, NO_x and CH_4 in the different pricing systems. These emissions are shown in Figure 6.6.

In the selective unit-based pricing system, both CO_2 emissions and NO_x emissions increase. This is caused by two factors. Firstly, the transport of waste increases sharply in the selective unit-based pricing system. As transport lead to both CO_2 and NO_x emissions, these emissions increase. Secondly, in the selective unit-based pricing system, more low quality organic waste is generated. Composting low quality organic waste creates a residue, which is incinerated. Incineration of this residue results in extra CO_2 emissions and NO_x emissions. A positive effect of the selective unit-based pricing system is that it reduces the quantity of waste that is landfilled, thus generating substantially fewer CH_4 emissions.

The emissions in the full unit-based pricing system are slightly lower than the emissions in the flat fee pricing system. The full unit-based pricing system is slightly better for the environment than the flat fee pricing scheme. Based on the results presented in Figure 6.6, one may jump to the conclusion that a selective unit-based pricing scheme is worse for the environment than either of the two other pricing schemes. However, this would depend on an economic valuation of the damage caused by CO_2, NO_x and CH_4 emissions. It is, however, clear that in our model the selective unit-based pricing scheme aggravates the problem of acidification.

6.4 DISCUSSION AND CONCLUSIONS

In this chapter, it has been demonstrated how spatial aspects of the waste management problem can be included in a general equilibrium framework. A general equilibrium model is a suitable tool for modelling the interactions between the choices that the consumers make about the quality and quantity of waste they generate, the choices that the municipalities make about how and where the waste should be treated and the costs of waste treatment.

In a stylised example with numerical data based on the Randstad in 2000, it was shown how the optimal waste treatment choice and the choice for the optimal location of the waste treatment unit can be influenced by a policy change for the consumers. The results show that if a unit-based price for the collection of rest waste is implemented, consumers will start to generate less rest waste and more organic waste. More waste will be composted, which will result in lower waste treatment costs and less environmental damage. There is, however, also a negative effect. The quality of organic waste may be seriously affected. The "traditional" consumers in particular are likely to start disposing rest waste in the organic waste bin, thus polluting the organic waste stream. Since the quality of the organic waste stream greatly declines, it consequently becomes far more expensive to compost the waste. If the quality of organic waste is too poor, part of the waste cannot be composted, but must be incinerated instead. This means that the environmental gains will be much smaller than can be expected based on the reduction in waste generation without considering waste leakage. Although less rest waste is incinerated or landfilled, more organic waste has to be composted, thus leading to higher CO_2 emissions.

The results of the numerical example show that in the larger municipalities the quality of the organic waste stream declines so much that waste will have to be treated in a large composting facility instead of a small one, thus incurring more transport costs. In this example, it would not be cost-effective to introduce a unit-based price for the collection of rest waste in larger municipalities since the increased costs of composting due to the decrease in the quality of organic waste are not offset by the benefits of lower generation of rest waste. Only in small municipalities, with a relatively large number of green consumers, will unit-based pricing for the collection of rest waste be interesting.

The introduction of unit-based pricing on the collection of rest waste as well as on the collection of organic waste creates fewer problems with regard to waste leakage. This policy option, however, does not stimulate the consumer to separate its waste. Therefore, there are hardly any benefits in introducing such a pricing scheme, given the restrictions of the model. It is not advisable for municipalities to introduce this policy option, considering that introducing such a pricing scheme will probably involve high transaction costs. It is important to bear in mind that only the option of substituting rest waste for organic waste has been analysed. Particularly in smaller municipalities, consumers will have the option of reducing their organic waste stream by composting it at home. The full unit-based pricing scheme may well provide an incentive for more waste to be composted at home, thus reducing the organic and rest waste streams. Municipalities with a relatively large share of consumers that are not able to engage in home composting, however, do not need to consider this kind of pricing scheme.

Naturally, there are a number of caveats and uncertainties in this analysis. Several extensions, such as, for example, home composting, recycling, detailed modelling of prevention or more categories of waste could be added to the model to make it more realistic. The current analysis, however, demonstrates that the interactions between the quality of waste, the method of waste treatment, transport costs, the presence of economies of scale and environmental damages can have significant implications for the success of a policy change and therefore should be considered when deciding on which policy should be implemented in order to minimize waste generation.

REFERENCES

Beker, D. (2002) *The Composition of Residual Household Waste in the Netherlands, 2000 and 2001. RIVM report 776221006*. RIVM, Bilthoven.

Bruvoll, A. (1998) Taxing virgin materials: an approach to waste problems. *Resour. Conserv. Recycl.* **22**, 15–29.

Choe, C. and Fraser, I. (1999) An economic analysis of household waste management. *J. Environ. Econ. Manag.* **38**, 234–246.

Fricker, A. (2003) Waste reduction in focus. Futures. *J. Forecast. Plann. Policy*. **35**(5), 505–519.

Fullerton, D. and Kinnaman, T.C. (1995) Garbage, recycling and illicit burning or dumping. *J. Environ. Econ. Manag.* **29**, 78–91.

Fullerton, D. and Kinnaman, T.C. (1996) Household responses to pricing by the bag. *Am. Econ. Rev.* **86**(4), 971–984.

Fullerton, D. and Wu, W. (1998) Policies for green design. *J. Environ. Econ. Manag.* **36**, 131–148.

Ginsburgh V. and Keyzer, M. (1997) *The Structure of Applied General Equilibrium Models*. The MIT Press, London.

Graig, P.P. (2001) Energy limits on recycling. *Ecolog. Econ.* **36**(3), 373–384.

Hong, S.R., Adams, M. and Love, H.A. (1993) An economic analysis of household recycling of solid waste: the case of Portland, Oregon. *J. Environ. Econ. Manag.* **25**, 136–146.

Jenkins, R.R. (1993) *The Economics of Solid Waste Reduction, the Impact of Users Fees.* Edward Elgar, Aldershot.

Linderhof, V., Kooreman, P., Allers, M. and Wiersma, D. (2001) Weight-based pricing in the collection of household waste; the Oostzaan case. *Resour. Energ. Econ.* **23**, 359–371.

Löfgren, K.-G. (1995) Markets and externalities. In: *Principles of Environmental and Resource Economics* (eds. H. Folmer, H.L. Gabel, and H. Opschoor) Edward Elgar, Aldershot.

Miranda, M.L., Everett, J.W., Blume, D. and Barbeau Jr A.R. (1994) Market-based incentives and residential municipal solid waste. *J. Policy Anal. Manag.* **13**(4), 681–698.

Negishi, T. (1972) *General Equilibrium Theory and International Trade.* North-Holland publishing company, Amsterdam.

Palmer, K. and Walls, M. (1997) Optimal policies for solid waste disposal taxes, subsidies, and standards, *J. Public Econ.* **65**, 193–205.

Pearce, D.W. and Turner, R.K. (1993) Market-based approaches to solid waste management. *Resour. Conserv. Recycl.* **8**, 63–90.

Statistics Netherlands (2002) *National Accounts of The Netherlands.* Voorburg: Statistics, Netherlands.

Sterner, T. and Bartelings, H. (1999) Household waste management in a Swedish municipality: determinants of waste disposal, recycling, and composting. *Environ. Resour. Econ.* **13**, 473–491.

Wertz, K.L. (1976) Economic factors influencing household production of refuse. *J. Environ. Econ. Manag.* **2**, 263–272.

WMC (2000) *Municipal Solid Waste Management Taxes in 2000*, Report no. WMC 2000-03, Waste Management Council, Utrecht (in Dutch).

WMC (2001) *Waste Treatment in the Netherlands: Data 2000*, Report no. WMC 2001-04, Waste Management Council, Utrecht (in Dutch).

WMC (2002) *Municipal Solid Waste Management Taxes in 2002,* Report no. WMC 2002-05, Waste Management Council, Utrecht.

WMC (2003a) *Waste Treatment in the Netherlands: Data 2002*, Report no. WMC 2003-18, Waste Management Council, Utrecht.

WMC (2003b) *National Waste Management Plan 2002–2012.* Waste Management Council, Utrecht (in Dutch).

7

Impacts of European legislation on biotreatment and recovery of organic solid waste

R. Braun and R. Kirchmayr

7.1 INTRODUCTION

Anaerobic digestion (AD), together with composting, represents a sustainable, natural route of treatment and recycling of wastes of biological origin and a wide range of useful industrial organic by-products. AD can provide a method accessing the energy and nutrients contained in organic material. Therefore, the utilisation of AD for organic waste management permits a significant movement up the waste hierarchy over current management methods.

For decades AD has been used for industrial wastewater treatment, stabilisation and volume reduction of sewage sludge, animal manure and digestion of organic wastes. These years of experience have shown that, if correct processing methods are applied, the spreading of digested material on land of whatever type has not caused any health problems. AD also provides a method to increase the measurability and

 Resource Recovery and Reuse in Organic Solid Waste Management. Edited by Piet Lens, Bert Hamelers, Harry Hoitink, Werner Bidlingmaier. ISBN: 1 84339 054 X. Published by IWA Publishing, London, UK.

plant availability of the nutrients contained in organic material. With correct storage and application methods of the digested material, best agricultural practice for the application methods and restrictions on the season of application, the risk of volatilisation and runoff of nutrients can be greatly reduced compared to storage and application of untreated organic waste and manure.

Therefore AD makes nutrient management much more accurate and reduces the risks of excess application of artificial nutrients on top of organic based nutrients. This will greatly help to achieve the targets of the EC Water Framework Directive 2000/60/EC. It is important to utilise management methods that enable all nutrients to be recycled beneficially if the Water Framework Directive has to be achieved. AD will help to achieve this goal. More recently AD has received increasing attention as a mainstream energy conversion process, based on renewable agricultural biomass (energy crops), as well as on co-digestion of various industrial by-products, including animal by-products (ABP) and wastes.

7.2 IMPLICATION OF EUROPEAN LEGISLATION ON COLLECTION, TREATMENT AND RECOVERY OF WASTES – AN OVERVIEW

Induced by a steadily increasing intensity of collection, treatment and recovery of biowastes, numerous EC Regulations and guidelines have been issued in this area or are currently under development. Most of these regulations profoundly influence the technological developments and practical applications of AD.

Among the most important EC Regulations are:

- 75/442/EEC (Waste Directive)
  Council Directive 75/442/EEC of 15 July 1975 on waste *Official Journal L* 194, 25/07/1975 P. 0039–0041.
- 1986/278/EEC (Sewage Sludge Directive)
  Council Directive 86/278/EEC of 12 June 1986 on the protection of the environment, and in particular of the soil, when sewage sludge is used in agriculture *Official Journal L* 181, 04/07/1986 P. 0006–0012.
- 2000/60/EC(Water Framework Directive)
  Directive 2000/60/EC of the European Parliament and of the Council of 23 October 2000 establishing a framework for Community action in the field of water policy *Official Journal L* 327, 22/12/2000 P. 0001–0073.
- 1999/31/EC (Landfill Directive)
  Council Directive 1999/31/EC of 26 April 1999 on the landfill of waste *Official Journal L* 182, 16/07/1999 P. 0001–0019.
- COM/2002/179 final (Soil Protection Strategy)
  Communication from the Commission to the Council, the European Parliament, the Economic and Social Committee and the Committee of the Regions – Towards a Thematic Strategy for Soil Protection.

- 2001/77 EC (Electricity from Renewable Energy Sources Directive)
  Directive 2001/77/EC of the European Parliament and of the Council of 27 September 2001 on the promotion of electricity produced from renewable energy sources in the internal electricity market
  Official Journal L 283, 27/10/2001 P. 0033–0040.
- Biological Treatment of Biowaste, Working Document 2nd draft (2001).
- (EC) No. 1774/2002 (ABP Regulation)
  Regulation (EC) No. 1774/2002 of the European Parliament and of the Council of 3 October 2002 laying down health rules concerning ABP not intended for human consumption
  Official Journal L 273, 10/10/2002 P. 0001–0095.

7.2.1 Council directive 75/442/EEC of 15 July 1975 on waste

Directive 75/442/EEC contains the definition of wastes, together with guidelines for waste classification as well as the exclusion of specific wastes (e.g. radioactive materials, animal carcasses, waste waters).

Member States shall take the necessary measures to ensure the prevention of endangering human health and harming the environment through the disposal of wastes.

In article 3 Member States are requested to take appropriate steps to encourage the prevention, recycling and processing of wastes. Any process for the re-use of wastes and the extraction of raw materials and energy from wastes shall be forced.

7.2.2 The sewage sludge directive 1986/278/EEC

The directive 1986/278/EEC "Protection of Environment and Soil at the Utilisation of Sewage Sludge in Agriculture" defines limiting values for heavy metals and organic trace compounds. Furthermore hygienic requirements for handling and application of sewage sludge on agricultural soils are defined.

In addition, the Regulation on Organic Farming 2092/91/EWG defines heavy metal limiting values for compost deriving from source sorted collection of municipal biowaste.

7.2.3 The water framework directive 2000/60/EC

The water framework directive affects water industry, agriculture, development and construction industry and all businesses having discharge consents, trade effluent licences or abstraction licences.

The aim of this Directive containing 72 pages is to establish a framework for the protection of waters. As its name suggests, the Directive sets out a framework for action rather than imposing a set of rules.

7.2.4 Council directive 1999/31/EC on the landfill of waste

In the EC directive on the landfill of waste the goals of organic waste reduction in landfills (based on the year 1975) are defined as follows:

- reduction to 75% by the year 2006,
- reduction to 50% by the year 2009,
- reduction to 35% by the year 2016.

7.2.5 Towards a thematic strategy for soil protection: COM(2002) 179 final

The Commission will present a thematic strategy on soil protection in 2004. The strategy is one of the seven "thematic strategies" foreseen under the EU's 6th Environment Action Programme. It will consist of legislation on a Community information and monitoring system on soil, as well as of a set of detailed recommendations for future measures and actions. The monitoring system will be built on existing information systems and databases and ensure a harmonised way of establishing the prevailing soil conditions across Europe.

By the end of 2004 a directive on compost and other biowastes will be prepared with the aim to control potential soil contamination and to encourage the use of certified compost.

7.2.6 Directive 2001/77/EC on the promotion of electricity produced from renewable energy sources in the internal electricity market

The document states that at the moment the exploitation of renewable energy sources is underused in the Community. For this reason the directive aims to promote an increase in the contribution of renewable energy sources to electricity production in the internal market for electricity and to create a basis for a future Community framework thereof. To ensure increased penetration of electricity produced from renewable resources, the Member States are requested to set appropriate national indicative targets. The EC "White Paper's" indicative target of 12% by the year 2010 provides a useful guidance.

Biogas is one of the possible renewable alternatives and its broader penetration as an energy source should therefore benefit from these efforts.

7.3 BIOLOGICAL TREATMENT OF BIOWASTE, WORKING DOCUMENT

Currently the second draft version of the forthcoming regulation "Biological Treatment of Biowaste" is under discussion.

The current version has to be harmonised with the recently published ABP Regulation (EC) No. 1774/2002 (Section 7.4). Furthermore it has to be adopted to the requirements laid down in the "Thematic Strategy for Soil Protection" – COM (2002) 179 final (Section 7.2.5).

A revised third version of "Biological Treatment of Biowaste" has been announced for the year 2004.

The forthcoming regulation will contain:

- a list of allowable wastes for biotreatment,
- directives for the collection, handling and treatment of wastes,
- approval of treatment plants and allowable processing emissions,
- quality classes for compost and residues of biotreatment,
- control and analysis of end products and,
- application standards for the end products.

In the current version, the hygienisation of biowaste has to be guaranteed by a minimum temperature of 55°C for at least 24 h, at an average hydraulic dwell time in the reactor of at least 20 days.

Otherwise the following treatment is required:

- pre-treatment at 70°C for 1 h or,
- post-treatment of the solid digestate at 70°C for 1 h or,
- composting of the solid digestate.

The quality standard defines two categories of solid end product (compost) and one category of stabilised waste (Table 7.1). According to the specific category the utilisation possibilities and quantities for soil application will be regulated.

Table 7.1 Limiting values for heavy metals, trace organic compounds and impurities according to the forthcoming EC Regulation on biological treatment of biowaste.

Parameters (mg/kg total solids)	Compost/solid end products		Stabilised waste
	Category 1	Category 2	
Cadmium	0.7	1.5	5
Chromium	100	150	600
Copper	100	150	600
Mercury	0.5	1	5
Nickel	50	75	150
Lead	100	150	500
Zinc	200	400	1500
PCB	–	–	0.4
Polycyclic aromatic hydrocarbons (PAH)	–	–	3
Impurities > 2 mm	<0.5%	<0.5%	<3%
Gravel/stones > 5 mm	<5.0%	<5.0%	

7.4 ABP REGULATION (EC) NO. 1774/2002

The Regulation (EC) No. 1774/2002 of the European Parliament and the Council from October 3, 2002 laying down health rules concerning ABP not intended for human consumption (ABP Regulation) has to be applied in all Member States since May, 1st 2003. The Commission has already amended the original Regulation with Regulations (Regulations (EC) No. 808/2003 to 813/2003: *Official Journal L* 117, 13.05.2003) and proposed transitional measures in several Member States through Decisions of the Commission (2003/320/EC to 2003/329/EC: *Official Journal L* 117, 13.05.2003 and 2003/334/EC: *Official Journal L* 118, 14.05.2003). Other amendments are still under discussion.

7.4.1 Goals

ABP are defined as "*all animals or parts of animals not intended for human consumption*". This includes dead farm animals, animal manure and catering waste. "Catering waste" contains "*all waste food including used cooking oil originating in restaurants, catering facilities and kitchens, including central kitchens and household kitchens*".

It has been estimated that only 68% of a chicken, 62% of a pig, 54% of a bovine animal and 52% of a sheep or goat are directly consumed. In the EU more than 14.3 million tons (1998) of ABP deriving from meat of healthy animals not intended for human consumption are processed per year. Subsequently, these materials will be transformed into a variety of products used in pet and animal feeding, as well as in cosmetics, pharmaceuticals or other technical use. In 1998, 16.1 million tons of ABP (therefrom 14.1 million tons deriving from healthy animals) were processed into 3 million tons of meat and bone meal and 1.5 million tons of fat (COM(2000)574).

Since inappropriate processing standards and the use of rendered products and catering waste are supposed to be the main reason for major pandemic outbreaks of bovine spongiform encephalopathy (BSE) and foot and mouth disease (FMD), rigorous measures had to be taken. The new regulation will require major changes in processing procedures by both waste producers and waste managers. In Regulation 3 "risk-categories" are classified and new rules for the collection, treatment and disposal of ABP, including animal manure and catering waste (kitchen waste, restaurant waste etc.) are set up.

7.4.2 Classification of ABP in three categories

Based on their potential risk to the public, to animals or to the environment, the Regulation (EC) No. 1774/2002 (ABP Regulation) classifies all ABP and their processed products and wastes into three categories and defines the appropriate treatment and utilisation possibilities.

In the following, a short survey of the amended regulation (EC) No. 1774/2002 published on October 10, 2002 in the *Official Journal L* 273 (95 pages including 11 attachments) will be given.

Category 1 Category 1 concentrates on ABP representing the highest risk to environment, animals or humans. This category contains, among others, the following materials:

- Animals or materials that are infected by transmissible spongiform encephalopathies (TSEs: BSE, mink spongiform encephalopathies (MSE), feline spongiform encephalopathies (FSE), scrapie etc.) or suspected of being infected.
- Specified risk materials (SRM) representing the material of (healthy) animals having the highest potential of containing the TSE agent such as the scull of 12 months old sheep and cattle or the intestines of cattle of all ages.
- Animals or parts of them with exceeding residues of environmental contaminants (e.g. dioxins, polychlorinated biphenyl (PCBs)).
- Catering waste originating from international means of transport.
- Animal waste collected in the wastewater stream of Category 1 processing plants with a particle size >6 mm. Category 1 processing facilities have to apply a wastewater pre-treatment system removing all animal material with a particle size of more than >6 mm.

Category 2 This category includes the following materials:

- Animals, or parts of them, bearing a risk of being contaminated or transmitting any animal diseases (e.g. animals dying on farms or being killed in the context of disease control measures).
- ABP with exceeding values of veterinary drugs.
- Animal waste collected in the wastewater stream of slaughtering facilities (Category 2 processing plants), with a particle size >6 mm (fat scraper contents, sand trap contents, oil- and sludge residues). Category 2 processing facilities have to apply a wastewater pre-treatment system removing all animal material with a particle size of >6 mm.
- Animal manure, contents of gut, stomach and intestines (separated from the intestines), milk and colostrum from animals not suspected to spread any diseases.

Category 3 Category 3 contains all animal materials deriving from healthy animals and slaughtered for human consumption, but not intended for human consumption because of being rejected as unfit for human consumption or simply because of commercial reasons.

Catering waste (with the exception of wastes from international means of transport which is classified as Category 1) is classified as material of Category 3.

7.4.3 Compulsory ABP treatment and recovery processes

The ABP Regulation (EC) No. 1774/2002 assigns to each ABP a compulsory treatment procedure and the corresponding utilisation possibilities according to the previously described three categories. Detailed requirements for all steps of the collection, transport and handling of the ABP are laid down. Furthermore requirements of processing and control measures for treatment plants processing ABP are defined.

In the following, a short survey of the principal treatment requirements of the respective categories will be given.

7.4.3.1 Materials of Category 1

Materials of Category 1 have to be collected without undue delay and marked (if possible with smell) or sterilised (50 mm, 133°C, 3 bar, 20 min) and marked, followed by incineration in approved incineration plants.

If national landfill-legislation does not ban ABP's, materials of Category 1 (with the exception of TSE contaminated- or suspected materials) may also be sterilised (50 mm, 133°C, 3 bar, 20 min), marked and buried in approved landfills.

Catering wastes from international means of transport may, after sterilisation, be buried in approved landfills.

Category 1 material may also be processed with other processes which are to be approved by the scientific committee.

7.4.3.2 Materials of Category 2

Category 2 materials may be (co-)incinerated directly, sterilised and incinerated, or sterilised and processed for other uses than for animal feeding, e.g. in an oleochemical plant or in a biogas or composting plant. The end product can be used as fertiliser, soil conditioner or for technical products (but not for medical products).

If there is no risk of transmitting any infectious disease, manure (unprocessed or transformed in a biogas plant), rumen, gut and intestine contents, milk and colostrum may be brought out on land or used in an approved pet food plant. Manure can be used as unprocessed raw material in an approved biogas or composting plant.

Category 2 materials may also be processed with other processes which are to be approved by the scientific committee.

7.4.3.3 Materials of Category 3

Processing and utilisation possibilities laid down for material of Categories 1 and 2 may also be applied for materials of Category 3.

Alternatively, following appropriate treatment in approved processing plants Category 3 materials may be processed to pet food, pharmaceutical and cosmetic products. Category 3 materials may further be transformed in approved biogas and composting plants.

Catering wastes (with the exception of catering wastes from international means of transport, Category 1) may be processed in an approved biogas or composting plant according to national legislation.

Table 7.2 ABP processed in biogas plants, according to ABP Regulation (EC) No. 1774/2002.

Category	ABP	Requirements
1	Not envisaged	
2	Manure, stomach- and gut contents, milk, colostrum, all without any pre-treatment	Absence of infectious diseases; Biogas plant approved according to national legislation
	All other Category 2 materials	Sterilisation (133°C, 3 bars, 20 min, steam pressure) and marking; Biogas plant approved according to (EC) No. 1774/2002, article 15
3	All Category 3 materials	Biogas plant approved according to (EC) No. 1774/2002, article 15
	Catering waste except Category 1 catering wastes	Biogas plant approved according to national legislation (according to (EC) No. 1774/2002)

7.4.4 ABP for biotreatment

The ABP Regulation (EC) No. 1774/2002 suggests biogas recovery or composting for a variety of ABP's (Table 7.2). Whether biotreatment will be applied or not is determined by the demands of pre-treatment, process equipment requirements and the permitted utilisation of the end product (digestate or compost). The resulting treatment costs will decide the appropriate process selection.

Biotreatment is not foreseen for materials of Category 1. Only incineration or in some cases burial in an approved landfill are permitted.

After sterilisation with steam pressure (133°C, 3 bar, 20 min) all materials of Category 2 are admissible for biotreatment in approved (according to article 15, (EC) No. 1774/2002) biogas or composting plants.

Manure, stomach- and gut contents, milk and colostrum (Category 2) are exempted from the above requirements, if the absence of infectious diseases can be guaranteed. The respective biogas plant has to be approved according to national legislation.

Biogas or composting plants approved according to article 15 of the ABP Regulation may transform all ABP of Category 3.

Category 3 catering wastes may be transformed in biogas plants according to national legislation based on the requirements of the ABP Regulation.

7.4.5 Practical approach of the AD of materials of Category 2 and Category 3

Manure or manure with other biogenic wastes or raw materials (energy crops, silage) not covered by the ABP Regulation, may be processed in biogas plants without any pre-treatment. The fermentation end product is classified as "manure" and may be used and applied on farm or pasture land like unprocessed manure without any requirements from this regulation.

To avoid an unwanted uptake by ruminants, with the exception of manure (or digestate deriving from manure), the application of other organic fertilisers (deriving from ABP) on pasture land is prohibited (article 22, (EC) No. 1774/2002).

A forthcoming new European Regulation, (SANCO/2380/2003) will lay down the pre-treatment requirements for the application of manure and organic fertilisers (deriving from ABP) on farm or pasture land. The trade of manure (digestate) between Member States is restricted.

Additional hygienic requirements for manure products placed on the market have been laid down. A heat treatment of 60 min at 70°C or an equivalent treatment according to article 33 (2) (EC) No. 1774/2002 is obligatory. End products have to be free from *Salmonellae* (absence in 25 g of end product) and *Enterobacteriaceae* (less than 1000 colony forming units per gram end product).

As indicated earlier, catering waste is classified as material of Category 3. Catering wastes are defined as "*all waste food including used cooking oil originating in restaurants, catering facilities and kitchens, including central kitchens and household kitchens*" (ABP Regulation amended by (EC) No. 808/2003). Catering wastes from international means of transport are classified as material of Category 1 and have to be disposed of. Until the Commission decides to lay down other regulations, catering waste (Category 3) or mixtures with manure may be processed in biogas or composting plants approved in accordance with national legislation. In this case the national authority may derogate from the requirements laid down in the Regulation (EC) No. 1774/2002 as long as the process guarantees an equal reduction of pathogens. The use of catering waste as swill for pig feeding is prohibited (article 22, (EC) No. 1774/2002). Only Germany and Austria may derogate therefrom under very strict treatment and control measures until October 2006.

Therefore the AD of catering wastes may increase considerably in future.

7.4.6 Requirements for the approval of biogas plants according to article 15, Regulation (EC) No. 1774/2002

Biogas or composting plants processing and converting ABP have to be approved in accordance with article 15 of the regulation (EC) No. 1774/2002.

Only manure, stomach and intestine contents (separated from stomach and intestines), milk, colostrum (Category 2) or catering waste and substrates not covered by the ABP Regulation treating biogas and composting plants may be approved partly diverging from the requirements laid down in article 15 of the ABP Regulation.

For the approval of biogas or composting plants article 15 demands five major conditions to be fulfilled. They have to:

(1) meet the specific requirements for the approval of biogas or composting plants (Annex VI, Chapter II, Part A; Section 7.4.6.1);
(2) handle and transform ABP in accordance with the hygiene requirements and processing standards
(Annex VI, Chapter II, Parts B and C; Section 7.4.6.2);
(3) be checked by the competent authority (in accordance with article 26; Section 7.4.6.3);

(4) have established and implemented methods of monitoring and checking the critical control points (Section 7.4.6.4);
(5) ensure that the digestion and compost residues comply with the microbial standards (Annex VI, Chapter II, Part D; Section 7.4.6.5).

7.4.6.1 Specific requirements for the approval of biogas and composting plants

Specific requirements for the approval of biogas and composting plants are defined in Regulation (EC) No. 1774/2002 Annex VI, Chapter II, Part A.

Premises Biogas plants must be equipped with a pasteurisation/hygienisation unit which cannot be bypassed. Composting plants have to be equipped with a closed composting reactor without bypass. These units have to be equipped with:

- installation for monitoring temperature against time;
- recording devices to record the results of these measurements continuously;
- an adequate safety system to prevent from insufficient heating;
- adequate facilities for cleaning and disinfecting vehicles and containers transporting untreated ABP on leaving the plant.

However, a pasteurisation/hygienisation unit is not mandatory for biogas plants transforming only ABP that have undergone steam pressure sterilisation at 3 bars, 133°C for 20 min.

Each biogas or composting plant has to have a laboratory on its disposition. This may be an own certified laboratory or an external laboratory. The laboratory has to be equipped with facilities to carry out the necessary analysis and further to be approved by the competent authority.

7.4.6.2 Handling and transformation of ABP

Hygiene requirements Specific hygienic requirements are defined in ABP Regulation (EC) No. 1774/2002 Annex VI, Chapter II, Part B and C:

(1) Only the following ABP may be transformed in a biogas or composting plant:
    - Category 2 material, when using processing method 1 (steam sterilisation: 50 mm, 133°C, 3 bar, 20 min) in a Category 2 processing plant;
    - Manure and digestive tract content;
    - Category 3 material.
(2) These ABP must be transformed as soon as possible after arrival. They have to be stored properly and treated.
(3) Containers, receptacles and vehicles used for the transport of untreated material must be cleaned in a designated area. This area must be situated or designed to avoid the risk of contamination of treated products.
(4) Preventive measures against birds, rodents, insects or other vermin must be taken systematically. A documented pest control programme must be used for that purpose.

(5) Cleaning procedures must be documented and established for all parts of the premises. Suitable equipment and cleaning agents must be provided for cleaning.
(6) Hygiene control must include regular inspections of the environment and equipment. Inspection schedules and results have to be documented.
(7) Installations and equipment must be kept in good state of repair and measuring equipment must be calibrated at regular intervals.
(8) Digestion residues must be handled and stored at the plant in a way that recontamination is precluded.

Processing standards Category 3 material used as raw material in a composting plant or a biogas plant equipped with a pasteurisation/hygienisation unit has to meet the following minimum requirements:

- maximum particle size before entering the unit: 12 mm;
- minimum temperature in all material in the unit: 70°C;
- minimum time in the unit without interruption at the minimum temperature: 60 min.

However, pending the adoption of rules in accordance with article 6(2)(g), the competent authority may authorise the use of specific requirements other than those laid down in this chapter, if catering waste is the only ABP used as raw material in the respective biogas or composting plant, it has to be guaranteed that an equivalent effect regarding the reduction of pathogens is achieved. These specific requirements may also be applied to catering waste or the mixture with manure, digestive tract content (separated from the digestive tract), milk and colostrum considering the resulting material as deriving from catering waste.

If manure, digestive tract content (separated from the digestive tract), milk and colostrum are the only ABP's being treated in the respective biogas or composting plant, the competent authority may authorise the use of specific requirements other than the ones specified in this chapter, declaring that:

- the material is not considered to bear a risk of spreading any serious transmissible disease;
- the residues or compost are considered to be untreated material.

7.4.6.3 Check by the competent authority

(In accordance with ABP Regulation, article 26, Official control and lists of approved plants.)

Article 26 defines the frequency of inspections and supervisions of approved plants by the competent authority. Furthermore, a list of approved plants with unique identification numbers assigned to each plant has to be established by the national administration.

7.4.6.4 Establishment and implementation of monitoring and checking methods of the critical control points

The ABP Regulation underlines the principle of self responsibility of the plant owners and requires the establishment of a self control system according to the principle of Hazard Analysis and Critical Control Point (HACCP – concept). The potential hygienic risks must be minimised by applying strict criteria for ABP acceptance or refusal, pasteurisation and hygienic operation (e.g. definition of clean and unclean divisions). An example of critical control point concepts for biogas plants has been established by Baggesen (2002).

7.4.6.5 Microbial standards for the digestion residue and compost

Microbial standards for digestion residues and compost are defined in the ABP Regulation (EC) No. 1774/202, Annex VI, Chapter II, Part D.

Samples of digestion residues or compost taken during or on withdrawal from storage at the biogas or composting plant must comply with the following standards:

- *Salmonella*:
  No Salmonella must be present in five samples of 25 g each.
- *Enterobacteriaceae*:
  Out of five samples of 1 g each:
  Three samples with <10 cfu/g (colony forming units per gram) of *Enteriobacteriaceae*.
  Two samples with a value between 10 cfu/g and 300 cfu/g of *Enteriobacteriaceae*.

7.5 CONCLUSIONS

Recently issued and forthcoming EC waste- and veterinary regulations will greatly influence the development and application of "biotreatment" processes (AD or composting), as well as the reuse of treated end products (compost, digestate) or untreated organic by-products on arable and pasture land.

Existing and upcoming veterinary and waste regulations need to be further harmonised and their practicability has to be proved.

Since the upcoming EU landfill ban for organic wastes will continue to push source separate collection of wastes, huge amounts of organic wastes will result in most EU countries, waiting for an alternative treatment or for recovery.

If the safety and security requirements laid down in the respective regulations are strictly followed, biotreatment and especially AD can become a mainstream treatment and recovery process, both for biowastes and selected ABP.

In many areas of application, AD is not concerned at all by the strict requirements of the new ABP Regulation (EC) No. 1774/2002.

Manure, stomach, gut and digestive tract contents, milk and colostrum (ABP, Category 2 materials) may be used without pre-treatment for biogas or compost production. Only the trade of manure between Member States is restricted. The utilisation of co-substrates for the AD of manure is not restricted by the ABP Regulation. National regulations concerning substrates not abrogated by European legislation have to be considered.

Category 3 catering wastes (excluding those from international means of transport) can be processed as a single substrate or in mixtures (e.g. manure) in biogas or composting plants approved in accordance with national legislation.

The treatment of ABP (Category 3 material with pasteurisation and Category 2 material with sterilisation) requires an approved biogas plant in accordance with article 15 of the Regulation (EC) No. 1774/2002. These laid down, highly demanding requirements will lead to an increase of the technical state of the art of biotreatment plants. Therefore, the practicability of biological waste treatment in an agricultural context will become much more difficult.

REFERENCES

Baggesen, D. (2002) Animal waste in co-digestion plants in Denmark – new challenges and opportunities. In: *Impacts of Waste Management Legislation on Biogas Technology*. (ed. R. Braun), Proceedings Workshop IFA Tulln, 12.–14.9.2002, A-3430 Tulln, Konrad Lorenzstrasse 20.

Kirchmayr, R., Scherzer, R. and Braun R. (2003) Chancen und Auflagen durch die EU-Verordnung zur Verwendung von nicht für den menschlichen Verzehr bestimmten tierischen Nebenprodukten. Presented at: 12. Jahrestagung Fachverband Biogas e.V. Borken bei Kassel, Germany, February 4–6.

75/442/EEC: Council Directive 75/442/EEC of 15 July 1975 on waste (*Official Journal L* 194, 25/07/1975 P. 0039–0041).

1986/278/EEC: Council Directive 86/278/EEC of 12 June 1986 on the protection of the environment, and in particular of the soil, when sewage sludge is used in agriculture (*Official Journal L* 181, 04/07/1986 P. 0006–0012).

1999/31/EC: Council Directive 1999/31/EC of 26 April 1999 on the landfill of waste (*Official Journal L* 182, 16/07/1999 P. 0001–0019).

2000/60/EC: Directive 2000/60/EC of the European Parliament and of the Council of 23 October 2000 establishing a framework for Community action in the field of water policy (*Official Journal L* 327, 22/12/2000 P. 0001–0073).

COM (2000) 574 final Proposal for a regulation of the European Parliament and the Council laying down health rules concerning animal by-products not intended for human consumption. European Commission, Brusseles.

2001/77 EC: Directive 2001/77/EC of the European Parliament and of the Council of 27 September 2001 on the promotion of electricity produced from renewable energy sources in the internal electricity market (*Official Journal L* 283, 27/10/2001 P. 0033–0040).

Biological Treatment of Biowaste, Working Document 2nd draft (2001) http://europa.eu.int/comm/environment/waste/facts_en.htm

COM/2002/179 final: Communication from the Commission to the Council, the European Parliament, the Economic and Social Committee and the Committee of the Regions – Towards a Thematic Strategy for Soil Protection.

EC No. 1774/2002: Regulation (EC) No. 1774/2002 of the European Parliament and of the Council of 3 October 2002 laying down health rules concerning animal by-products not intended for human consumption (*Official Journal L* 273, 10/10/2002 P. 0001–0095).

SANCO/2380/2003: Commission Regulation implementing and amending Regulation EC No. 1774/2002 as regards the organic fertilisers and soil improvers (1st Draft).

8

Products from waste: quality assurance systems

J. Barth

8.1 INTRODUCTION

What makes a product a product? This question develops to the main focus of the recycling industry after it has learned to collect and to treat a waste stream successfully. The close connection to the waste with its often very suspicious images and potentially harmful ingredients makes it obvious that it cannot be the recycling industry alone which defines the product property and thus makes a "product" out of treated waste. Possible harmful effects on the environment and humans require precaution and an effective monitoring of the waste.

8.1.1 Pro "waste property" arguments

- Complete governmental control of treatment and reuse.
- Environmental safety is guaranteed.

 Resource Recovery and Reuse in Organic Solid Waste Management. Edited by Piet Lens, Bert Hamelers, Harry Hoitink, Werner Bidlingmaier. ISBN: 1 84339 054 X. Published by IWA Publishing, London, UK.

The market shows a suspicious attitude towards recycled goods too and expects guaranteed product qualities and properties from product which can be marketed free of any waste restrictions and supervisions.

8.1.2 Pro "product property" arguments

- Free trade, even over borders.
- Flexible production in the plants.
- Market can be fully developed.

So there are good arguments for both approaches (Figure 8.1).

8.1.3 The role of standardisation and certification

At this point standardisation and certification comes into the play. A range of different levels and types of standards can meet the needs of the monitoring governmental waste bodies and in addition guarantee the requirements of the industry and the market.

It is obvious that only the product property of solid organic waste will be accepted by the market and thus leads to the sustainable solution. Nevertheless with the exception of Austria, no European government dares to loosen the grip on the recycled compost and grants quality compost the product property whished by the industry and expected by the market.

In this contradictory situation most of the European countries develop expectations that the coming European Biowaste Ordinance solves the problem, the countries cannot solve internally. Especially the industry relies on the European legislation that quality compost will get the necessary product status to allow free trade over boarders.

Figure 8.1 No reason to keep high-quality recycling products under the waster regime.

8.2 STANDARDISATION AS A PRECONDITION FOR THE PRODUCT PROPERTY

The standardisation levels can be summarised in a way that the statutory standards act as a kind of platform, just fulfilling the precautionary requirements resulting from the waste origin and giving the product a legal framework. On that platform the industry can act and fulfil voluntarily the expected standards of the consumers and the market (Figure 8.2).

An interesting aspect is to detect the relation between voluntary and statutory standard (Table 8.1). What should be voluntary in the eyes of the industry and what not?

In principle, the more voluntary the system is, the more easily it can react on the market needs. The standards in the different European countries do not reflect this philosophy. We have on the one hand Austria and Germany with a compost ordinance which regulates nearly everything from allowed raw materials to product

Figure 8.2 From collected material to marketed products – the range of compost standardisation (Hoog *et al.* 2002).

Table 8.1 Recommended extent of statutory and voluntary standards.

Type of standard	Statutory	Voluntary
Precautionary standards (PTEs)	Yes	No
Product quality	No	Yes
Production standards (except hygiene)	No	Yes
Product specifications	No	Yes
Application restrictions and recommendations	Yes	Partly

and production requirements up to application restrictions. We have on the other hand Belgium with only few heavy metal limit values and Sweden with no regulations in force at all.

Nevertheless the industry in Germany and Austria can live quite well with their ordinances which give a huge framework on statutory requirements for the compost production. In both countries the development of the ordinances is based on the experiences and standards of voluntary quality assurance systems (QASs). So it was quite well known for years what the industry can fulfil and what not. This ordinances supports the industry's effort as they define compost as a product in Austria and likely as a product in Germany.

8.3 LEGAL POINT OF VIEW OF A PRODUCT STATUS

8.3.1 Product properties and legislation

Discussions about the legal status of composts suitable to specifications as "products" or as "wastes for recycling" are still running in many countries – especially in Germany and Austria through their legal framework of compost ordinances.

Biowastes (separate collected organic waste from households and gardens) which are treated in a utilisation process could actually loose their waste properties and the result of the treatment can be seen as a product. The German Government has been asked by an inquiry of the European Commission dated 1999/08/13 to give an opinion on this behalf and answered that in their opinion the waste property of compost ends *when it is actually used*, which means: not after it is put into circulation, but with its real application on the land.

The German Environmental Ministry substantiates this with the following pros and cons (Kehres 2001).

8.3.1.1 Waste property until an environmentally compatible utilisation

According to article 1 letter a) of the *EC Directive 75/442/EWG* all material or objects are defined to be "waste" which are falling under the groups listed in annex I and which the owner wants to dispose off. Doubtless to say that the biowastes which arise at the "primary producer" (compare article 1 letter b) of the Directive 75/442/EWG are fulfilling this waste definition. The Directive, however, does not regulate the length of time of the waste property but mentions in article 4 that human beings and environment have to be circumstantially protected. So the waste property of material continues until it is environmentally compatible applied (Figure 8.3).

8.3.1.2 Waste until the very end of the utilisation procedure

A precondition for the end of the waste property is the end of the utilisation process by *a terminating utilisation procedure.* The definition of the utilisation

Figure 8.3 The legal point of view – waste property ends on the soil.

process must be realised in accordance to article 1 letter f) of the Directive 75/442/EWG with the utilisation processes named in annex II B. The end of *a preparatory utilisation process* does not yet finalise the legal waste utilisation.

8.3.1.3 Compost is not listed in the European waste catalogue

A signal for a terminating utilisation process and thus the loss of waste property could be that the produced compost is not distinctly listed in the European waste catalogue (EWC). The EWC is not a complete index. From this fact it cannot be deduced that compost cannot be a waste.

8.3.1.4 Market value of compost

The fact that compost has an *objective market value* and/or is item of trade agreements or acceptance contracts can be another argument for the end of the waste property. This would point out the ability for utilisation as "secondary raw material" which is definitely named in article 3, para 1 letter b) sub letter i) of the Directive 75/442/EWG as the target of material utilisation (Figure 8.4).

8.3.1.5 Definition as secondary raw material comparable to primary goods

The utilisation could be terminated with the production of compost if it could be defined as a "secondary raw material" [compare article 3 para 1 letter b) sub letter i)

Figure 8.4 Quality gives value and markets to compost.

of the Directive 75/442/EWG]. This means that the compost has properties, which can be compared with the substituting primary raw materials. It would be the case if corresponding product standards would be kept and compost can be used like a corresponding product that is produced from primary raw materials.

8.3.1.6 Controlled application with no risk potential

It is decisive whether the produced compost still possesses a waste-specific dangerous potential. Composting alone cannot yet remove a waste-specific dangerous potential. Biowastes are sanitised; however still contain harmful matter and impurities. As a rule harmful matter in biowastes cannot be separated completely. Following from this, compost must only be utilised in a *controlled* manner.

8.3.1.7 Summary of the legal requirements

The German Environmental Ministry concludes that on account of the typical waste properties of composts coming from treated biowastes, dangers and impacts may arise (especially through excessive application and thus increased fractions of harmful matter) through inexpertly application which must be prevented. In the frame regulated by the German Biowaste Ordinance compost is the result of a treatment process R3 of annex IIB of the Directive 75/442/EWG thus not being a product but waste according to article 1 letter a) of this Directive. The utilisation is terminated not before it is applied on the land (procedure R10 of the annex IIB of the Directive 75/442/EWG).

Figure 8.5 Quality labels show continuous and independent quality assurance QAS in European treatment plants and is a precondition for a product status.

The German Environmental Minister Conference (UMK) from October 1999 also shares this point of view and stipulates in general the question of the termination of waste properties of waste material. The UMK is of the opinion that waste properties in treated wastes are terminated "when the utilisation of the treated wastes takes place". The UMK attaches, however: "Are these wastes fulfilling the properties of a product that was produced from primary raw materials for the same purpose and no waste-specific impacts do arise for the public welfare, the waste property terminates already after the treatment is finished".

Thus quality assured compost – like it is the result of the German RAL Compost Quality Assurance System (RAL = German Institute for Certification and Standardisation) or similar systems in other European countries – could be also looked upon as a product (Figure 8.5).

8.3.2 How to make compost a product? The quality assurance concepts

It is a matter of fact for the around 550 large European manufacturer of quality assured composts and digestion residuals (=digestate) that the quality assurance process only labels high-quality products from which no waste-specific impacts for the public welfare are proceeding and the quality and effectiveness can be compared with products from primary raw materials.

This quality assurance processes comprise these elements (Hoog *et al.* 2002):

- Raw material/feedstock type and quality.
- Limits for harmful substances/potential toxic elements (PTEs).
- Hygiene requirements.
- Quality criteria for the valuables (e.g. organic matter).
- External monitoring of the product and the production.

- In-house control at the plant for all batches (temperature, pH, salt).
- Quality label or a certificate for the product.
- Annual quality certificate for the plant and its successful operations.
- Product specifications for different application areas.
- Recommendations for use and application information.

Following the target to establish compost as a product it is the main obligation of the industry to build up such QAS and to transform them to an independent body which must guarantee:

- standardised products,
- defined high qualities,
- specifications for use,
- monitoring by an independent organisation.

Compost quality assurance has gradually evolved differently in various parts of the world as political and industrial developments have taken place (Table 8.2). There is a number of areas seemingly close agreement across national boundaries. Less surprising are the areas of difference.

In addition, most countries, in the context of the development of systems for composting, have in place QASs, which either stand freely, or are supportive of the existing statutory standards. These systems have a variety of objectives, but they aim to ensure production of quality products to specific standards, and to facilitate marketing through use of quality symbols, and through ensuring that products are tailored to specific end users (through discussions with potential end users).

Quality assured composts/digestates fulfil by far the legal product requirements. The reason why will be explained in the following with the German System, which is the most developed system in the world but which is comparable to the Dutch, the Belgium (Flanders), the Swedish and the Austrian approaches.

8.3.2.1 General attitude of compost as a product

The general attitude of compost as a product is based on the RAL procedure for the introduction of the QAS for the product group "compost". RAL, the German Institute for Quality Assurance and Certification as a national federation and holding company is carrier of all the quality signs in Germany.

When a RAL quality sign is created for a product group under the guidance of the RAL the trade associations, business circles, experts and public authorities are taking part in the decisions. If agreements between the parties concerned are achieved and a quality assurance organisation has been founded is the corresponding quality sign acknowledged by the RAL and officially announced in the Federal Gazette by the Ministry of Economics. The specifications of the quality standards agreed upon are stipulated in the Quality Guidelines and are continuously improved.

The RAL introduction procedure is a procedure for an objective definition of composts suitable to specifications, which correspond to a high-class product according to the general attitude. Composts which are subject to the RAL Quality

Table 8.2 Compost QASs in place (Hoog *et al.* 2002).

Country	Status of quality assurance/certification activities and organisation
Austria	Fully established QAS (in redesign phase because of influences of the new statutory compost ordinance) – carried by the Austrian compost quality association KGVÖ together with the Austrian ÖNORM Standardisation Institute Agricultural Composting Associations (ARGE Kompost; BKAL) have established a QAS in five provinces. Establishment of common rules for acknowledgement is in progress. No quality symbol.
Belgium	Fully established QAS in the Flanders region (Flemish compost promotion organisation VLACO)
Denmark	Recently started with QAS for compost (criteria, standardised product definition, analysing methods) (DAKOFA – Association of waste processors – compost division)
Finland	No official efforts until now
France	Proposal for quality criteria, research program for a quality management system (Research carried out by the French EPA – ADEME)
Germany	Fully establish QAS for compost and anaerobic digestion products (German Compost Quality Assurance Organisation BGK together with RAL German Institute for Certification and Standardisation)
Greece	No efforts until now
Ireland	A first draft of a proposal for quality assurance is presented (managed by the Irish Composting Association CRE)
Italy	Proposal of a QAS, the former standards form the beginning of the 1990s are in revision (Italian Compost Consortium CIC)
Luxembourg	Statutory system similar to German QAS exists as part of the licensing procedure for composting plants
The Netherlands	Fully established quality assurance and certification system (Association of waste processors VVAV together with the Dutch certification organisation KIWA). A second system with less requirements for green waste exists (carried by BVOR).
Portugal	Proposal for quality assurance exists
Spain	Draft statutory standard for Spain and Catalonia (Division at the Environmental Ministry)
Sweden	Just started with QAS for compost and digestion products (Swedish public cleansing organisation RVF together with the Swedish Standardisation Institute SP)
UK	Quality standard and QAS run by the Composting Association is in effect, accompanied by the national standard PAS 100
US	Test guidelines and four product standards of the US Composting Council, and some internal State standards. Quality approval in the US is a state-by-state matter, and also a private marketing matter (an example of the latter are use categories by the organic publishing company Rodale Inc.). Some states, which have large volumes of yard-waste compost, have advanced compost guidelines, such as Washington and California, whilst other states, such as Pennsylvania, where agriculture is very

(Continued)

Table 8.2 *(Continued)*

Country	Status of quality assurance/certification activities and organisation
	conventional, have no compost standards at all. Compost in these states is only regulated if it falls under other laws such as solid waste, or voluntary rules, such as by the Pennsylvania Composting Association. Approximately 38 states have separate guidelines (US Composting Council, State Departments for Transportation)
Canada	National BNQ Standard for "organic soil conditioners – Compost" (The Composting Council of Canada has renewed discussions recently to focus on end market needs, information and enhanced product performance analysis)
Australia	Australian Standard for composts, soil conditioners and mulches (Standards Australia)
New Zealand	Strictly market oriented "standard for production and labelling of compost"

Assurance and which correspond to the Quality Guidelines are certified as products. Composts without new specifications remain wastes in the sense of the EWC (EWC 19 05 01).

8.3.2.2 Objective market value of compost products

The production of composts from separate collection of biowastes has increased enormously during recent years. Today in Germany nearly 5 million tons of compost raw material in more than 440 compost and digestion plants, which are subject to RAL Quality Assurance, are processed to different compost products.

Markets for compost products exist in a sufficient number and can be developed further on. Thus the price is dependent mostly on the product quality, the application range, the marketing measures and the market conservation and lies in the range between 1.5 and 16 Euro/ton according to a survey of the German Federal Humus and Soil Association BHE in 2000 (Kehres 2001).

8.3.2.3 Equality with other fertilisers

According to the Fertiliser Law (DüMG) composts are secondary raw material fertilisers. The admission of compost as a fertiliser is given through the Fertiliser Ordinance (DüMV) since 1997. There new specific secondary raw material fertilisers were defined and standards stipulated for their labelling as a fertiliser fitting to legal standards. As rule composts correspond to the admitted fertiliser type "Organic NPK-fertiliser". Composts which on account of low contents of plant nutrients do not correspond to an admitted type of fertiliser can be put in circulation as a means of soil additive.

Composts are usually applied for fertilising and for soil improvement in an interval of several years. This corresponds at usual application quantities of 20–30 tons of dry matter per hectare in 3 years to a fertilisation of e.g. 300 kg N, 150 kg P_2O_5 and 250 kg K_2O and 1000 kg CaO per hectare. This is a common fertilisation method made in practical operation like with other fertilisers which were produced from primary raw materials.

In order to compare secondary raw material fertilisers like farm manure, digestion residues and biocomposts the German Quality Assurance Organisation introduced in 2001 a new quality parameter (Kehres 2002). Besides the examination parameters the compost or digestion product is evaluated according to its efficiency (precaution–use ratio). Hereby the ratio between the precautious equivalence which is made by means of the percentage exhaustion of the limit value of the relevant heavy metals and the efficiency equivalence which includes the valuable ingredients (nutrients, organic substances and alkaline effective components) is used. This new parameter will help to establish compost and digestates as products on the market because it shows comparable benefits to other fertilisers.

8.3.2.4 Conformity to statutory standards

The RAL Quality Assurance takes into account further legal regulations and standards besides the fertiliser and legal waste rules. The following laws and ordinances are involved:

- Closed Loop Waste Management Act (KrW-/AbfG).
- Biowaste Ordinance (BioAbfV).
- Fertilising Law (DüMG).
- Fertiliser Ordinance (DüMG).
- Federal Soil Protection Act (BBodSchG).
- German DIN Norms and Standards of Landscaping Association.

This proves that the RAL Quality Assurance fulfils a multiple variety of most different requirements and gives proof for a corresponding conformity of the relevant standards. This will be also valid for future European regulations. Thus the ability for utilisation of quality assured composts as “secondary raw materials” in the sense of article 3, para 1 letter b) sub letter i) of the Directive 75/442/EWG is given.

8.3.2.5 Guaranty of application safety and user protection

The German Compost Quality Organisation BGK issues annually an external monitoring attestation for each product which is subject to the RAL Quality Assurance combined with recommendations for application. The external monitoring attestation serves as a quality evidence for the customers and as evidence of a regular quality monitoring for the authorities in charge. Dangers or impacts through inappropriate utilisation of the compost in the sense of article 4 of the Directive 75/442/EWG are effectively prevented on account of the said quotations of declaration.

8.3.3 Conclusion for the product status

During the last years the RAL Quality Assurance for Compost has established itself in Germany as the product certification and as a standard for compost products. The legal arguments for waste properties of composts mentioned in the first part are not relevant for certified products from the RAL Quality Assurance of Compost. Similar activities are at a starting point in most of the European countries.

A suitable instrument for the qualification of composts as a product is a product certification like it has been successfully established during recent years through the German RAL Quality Assurance. The compost producers wish and need the chance to make their secondary raw material fertilisers a real alternative and to establish a profitable market for their products. In the long run the marketing of wastes will not run because of the lack of markets.

8.4 CONCLUSION

Waste or product? Recycled goods under control of authorities or free tradable on the market? The biological waste management industry in some European countries shows already solutions for this decisive question when it comes to acceptance of recycled material by users. Other countries are on a good way. Quality assurance and product specifications are preconditions to show and do demonstrate that the quality of recycled goods fulfil precautionary consumer protection requirements and product standardisation needs of the market. This gives the European legislation a chance to grant product status for example to the end products of composting and anaerobic digestion processes of separately collected organic waste.

REFERENCES

Hoog, D., Hogg, D., Barth, J., Favoino, E., Centemero, M., Caimi, V., Amlinger, F., Devliegher, W., Brinton, W. and Antler, S. (2002) *Independent Comparison of Compost Standards within the EU, North America and Australasia*. Edited by WARP Waste Resource and Action Programme. In *www.wrap.org.uk/reports.asp*

Kehres, B. (2001) Vom Abfall zum Produkt – RAL-Gütesicherung als Instrument der Produktqualifikation von Sekundärrohstoffdüngern und Bodenverbesserungsmitteln: *Humuswirtschaft & Kompost* 03/2001, 233–240 (summarised and modified).

Kehres, B. (2002) Nutzwertindex: Eine Methode zur einheitlichen Bewertung von organischen Bodenverbesserungs- und Düngemitteln (Kompost, Gärprodukte, Gülle, Stallmist, Klärschlamm) *Humuswirtschaft & Kompost* 01/2002.

PART THREE: Bioprocesses in organic solid-waste management

9

Solid-state NMR investigations of organic matter conversions during solid-waste processing

H. Knicker

9.1 INTRODUCTION

The invention of nuclear magnetic resonance (NMR) spectroscopy about 60 years ago started the development of one of the most powerful experimental methods for the elucidation of molecular level structure. As a non-invasive technique, it can be used for the chemical characterization of individual compounds and bulk chemical composition, as well as for the examination of reaction kinetics and physicochemical properties of certain molecular domains (Sanders 1993).

The fast advances in NMR spectroscopy during the 1970s resulted in the development of the high resolution cross-polarization magic angle spinning (CPMAS) technique (Hartmann and Hahn 1962; Andrew 1971; Pines *et al.* 1973; Schaefer and Stejskal 1976) which opened a door for the qualitative and quantitative examination of solid samples. Its introduction into environmental sciences

 Resource Recovery and Reuse in Organic Solid Waste Management. Edited by Piet Lens, Bert Hamelers, Harry Hoitink, Werner Bidlingmaier. ISBN: 1 84339 054 X. Published by IWA Publishing, London, UK

started only 20 years ago (Mikinis *et al.* 1979; Hatcher *et al.* 1980; Newman *et al.* 1980). Since then, solid-state NMR, in particular solid-state ^{13}C, ^{15}N and ^{31}P NMR adapted to a well recognized and broadly used analytical tool in soil, geochemical and environmental science. This is mostly due to the fact that it is a non-degradative method and allows the analysis of heterogeneous and insoluble samples as a whole. Those advantages are certainly responsible for the increasing number of publications, in which solid-state NMR spectroscopy was applied for elucidating the conversion of organic matter in solid waste (Fründ *et al.* 1987; Chefetz *et al.* 1996, 1998a, 2000; Ayuso *et al.* 1997; Benke *et al.* 1998; González-Vila *et al.* 1999; Pichler *et al.* 2000, 2001; Rowell *et al.* 2001) and the dissolved organic fraction formed during composting of municipal solid waste (MSW) (Chefetz *et al.* 1998b; Hartlieb *et al.* 2001). However, despite its increasing popularity, solid-state NMR spectroscopy is often viewed as a mysterious technique, mostly because understanding introducing textbooks on solid-state NMR often requires a deeper knowledge of thermodynamics, physical chemistry and quantum mechanics (Mehring 1983; Stejskal 1994). Thus, the purpose of this chapter is to provide an introduction into basic NMR theory and background for the uninitiated. It is further intended to supply the reader with some experimental details, important for acquiring quantitative NMR data of solid waste and to give some examples how specialized solid-state NMR techniques may help in data interpretation. It is hoped that this chapter will provide enough information to those unfamiliar with NMR and its terminology for them to allow a successful application of this technique, and to encourage the interested readers to try this approach also for their research.

9.2 BASIC NMR THEORY

9.2.1 The resonance condition

NMR spectroscopy takes advantage of the magnetic properties of the observed nucleus that are influenced not only by its chemical environment but also by physical interactions with its environment.

Nuclei with a nuclear spin I have a magnetic moment μ. Experimentally, it has been found that this situation exists for nuclei where either the atomic mass number or the atomic number is an odd value. The nuclei ^{12}C and ^{16}O show no magnetic moment μ and therefore will not allow NMR spectroscopy. Consequently, if the structure of C compounds are investigated the less sensitive ^{13}C (1.1 atm%) ($I = \frac{1}{2}$) can only be observed.

Posting a nucleus with μ into a static magnetic field B_0 causes the nucleus to precess around the z-axis of the B_0-field with the Larmor frequency, ω_0. This is similar to a spinning top precessing about the Earth's gravitation field. The magnitude of ω_0 is the product of B_0 and the gyromagnetic ratio γ and is specific for each nucleus. The latter is a nuclear constant which is unique to each elemental nucleus and encounters nuclear charge number and mass.

The basic principles of quantum mechanics state that the energy of a spin in a magnetic field must be in discrete states described by the magnetic spin quantum number m_I with $2I + 1$ values for m_I: It follows that $2I + 1$ different energy levels of the spins are possible (nuclear Zeeman splitting). In the case of a nuclei assembly with $I = ½$, the nuclei with $m_I = +½$ will align themselves so that their magnetic moment vectors are parallel to the B_0-field and those with $m_I = -½$ are antiparallel to B_0. The value of the corresponding energy level is determined by the strength of B_0, the Planck's constant h divided by 2ω ($\hbar$) and γ:

$$E = -\gamma \hbar m_I B_0 \tag{9.1}$$

At thermal equilibrium the population distribution of the nuclei spins with $I = ½$ among the nuclear Zeeman energy levels is described by the Boltzmann distribution law:

$$N_\beta / N_\alpha = \exp(-\Delta E / kT) = \exp(-\gamma \hbar B_0 / kT) \tag{9.2}$$

N_α represents the number of spins at the lower energy level and N_β the number of spins in the upper energy level, k is the Boltzmann constant, and T is the temperature in degrees Kelvin. Taking all μ of a nuclei assembly together, a net magnetization or polarization along the z-axis of the B_0-field is yielded.

According to quantum mechanics, transitions of the spins between the two energy levels can be induced if the supplied energy corresponds to the energy difference ΔE between the two levels:

$$\Delta E = E_\beta - E_\alpha = -\gamma \hbar \Delta m_I B_0 \tag{9.3}$$

Here, Δm_I is the difference between the ground state or the lowest energy state and the excited state. With the Bohr's frequency law this corresponds to $\Delta E = h\nu = \hbar\omega$ where ν is the resonance frequency in Hertz and ω the resonance frequency in rad/s. From this the basic essence of NMR follows with the consideration of the quantum mechanical rule that only transitions with $\Delta m_I = 1$ are allowed:

$$\omega_0 = -\gamma B_0 \tag{9.4}$$

Equation (9.4) represents the resonance condition and states that for a spin transition from the lower to the higher energy level in a magnetic field B_0 to occur, an energy quantum of ω_0 is necessary. The energy needed is supplied by a pulse of a second oscillating electromagnetic radio frequency field B_1 orientated perpendicular to B_0 and with the frequency ν, and fulfilling the resonance condition. This forces the net magnetization along the z-axis to flip in the direction of the y-axis of the laboratory frame at an angle determined by the intensity and duration of the pulse. Since the transversal magnetization is still preceessing with ω_0 around the z-axis, the magnitude of its y-component is periodically altering. This alteration is detected as a voltage signal oscillating with ω_0 by a receiver, located in y-direction.

After the pulse, the excited spin system begins to relax to its thermal equilibrium and emits energy. The transversal magnetization decreases with the spin–lattice relaxation time constant, T_1, while the net magnetization along the z-axis increases. At the same time, as an entropic process, energy is redistributed among the spins. The coherent phase of the excited spin system is randomized in the x–y-plane at a rate governed by the spin–spin relaxation time constant, T_2. Both relaxation mechanisms result in a reduction of the amplitude of the detected oscillating signal. The relaxation decay pattern is called the free induction decay (FID). Transformation from the frequency domain into the time domain yields the spectrum with a signal at the specific Larmor frequency of the observed nucleus. The intensity of this signal is proportional to the number of excited spins. Thus, the signal intensity is a quantitative means for the determination of the concentration of certain nuclei in a sample.

Equation (9.4) shows that the greater the field strength, B_0, the higher the resonance frequency, ω_0. Equation (9.2) demonstrate further that the application of higher magnetic fields increases the energy difference ΔE and the population difference among the two energy levels. In practice, this results in higher sensitivity of the spin system for an NMR experiment.

9.2.2 The chemical shift

So far we have seen that the detected NMR signal is dependent upon the field strength of the static magnetic field B_0 and the gyromagnetic ratio γ. However, one has to realize that nuclei in a compound are surrounded by electrons. Electrons are moving charges and thus build up a magnetic field which shields the nucleus of interest from the static magnetic field B_0. The nucleus thus, experiences a reduced effective magnetic field, B_{eff}. The strength of this chemical shielding σ is dependent upon the specific electron density around the observed nucleus. Chemically non-equivalent nuclei have different electron densities and therefore are exposed to different magnetic fields. Consequently, their resonance frequency varies, leading to resonance lines which are separated in the spectrum. To determine their location within a spectrum independently from the strength of B_0, the resonance frequency of nucleus is given as chemical shift δ (ppm) with respect to some arbitrary standard according to equation (9.5). Here, the frequency of the examined substance is ν_s and the frequency of a reference substance is ν_{ref}:

$$\delta = ((\nu_s - \nu_{ref})/\nu_{ref}) \times 10^6 \qquad (9.5)$$

Usually tetramethylsilane (TMS) is used as reference for ^1H and ^{13}C nuclei. Unfortunately for nuclei less routinely analyzed by NMR spectroscopy, such as ^{15}N, ^{31}P and some others, no common standards exists. In these cases, a comparison of spectra is still possible by using conversion factors.

In molecules, the electron distribution is anisotropic. This leads to the fact that the actual value of σ is dependent upon the specific orientation of the molecules relative to the B_0-field. For molecules in rapid motion, such as those in liquids, the anisotropy averages on a temporal base to zero. In solids, on the other hand, the molecular motion is slow, and the time dependent chemical shielding of each nucleus is not averaged to its isotropic value. Consequently, each nucleus in a different orientation relative to B_0 is effected by a different B_{eff} and therefore resonates with a slightly different frequency. For a powder, where all possible orientations of a molecule occur, the different δ cannot be resolved by the instrumentation software and a broad resonance line embracing the total chemical shift range of a nuclei is observed in the spectrum. Such resonance lines can have the magnitude of several kHz, which aggravates the interpretation of solid-state spectra.

The fact that nuclei in different chemical environments are excited at different resonance frequencies also needs the generation of the resonance conditions for each individual chemical group. Nowadays, this is achieved by using a very short but intense B_1-field with the frequency ν_1. A frequency band symmetrically to ν_1 is created allowing the simultaneous excitation of all nuclei of one kind but in different chemical environments. After termination of the pulse the overlapping FIDs of the individual nuclei are transformed from the time domain into the frequency domain by the mathematical operation of Fourier transformation (FT-NMR).

9.2.3 The scalar coupling

In molecules, the nuclear spins are coupled to each other by interactions of their surrounding electrons. This is known as scalar, spin–spin, or indirect coupling and can occur on a heteronuclear and homonuclear base. The respective modification of the local magnetic field B_{loc} of the interacting nuclei causes further splitting of the nucleus Zeeman energy levels and thus the splitting of the resonance signal into multiplets.

The frequency difference between the multiplet peaks indicates the coupling constant J, which can also be used as an important parameter for structural analysis. At natural abundance homonuclear coupling between ^{13}C or ^{15}N is of low importance, but the heteronuclear spin–spin coupling between protons and the observed nucleus is a common phenomenon. For large organic molecules it can produce complicated spectra. To avoid those, the ^1H-spin system is decoupled by irradiation with an additional continuously applied B_2-field close to the resonance frequency of the ^1H-spin system. Fast transitions of the spins between their energy levels are induced. The observed nuclei only feel an average interaction and only one resonance line at the resonance frequency of the observed nucleus is detected. However, this decoupling causes a change in the spin population of the relevant energy levels and obscures the intensity of the observed signal. This effect, known as the nuclear overhauser enhancement (NOE) is not equal for nuclei in different physical and chemical environments, and it is obvious, that such decoupled spectra cannot be easily quantified. To circumvent this problem, the inverse gated decoupling technique is used in which the decoupler is only turned on during data acquisition. With this approach, the time of ^1H-saturation is long enough to avoid coupling but too short to detect changes in the spin population among the different energy levels.

9.2.4 Dipole–dipole interactions

At the molecular level the magnetic dipole fields of neighboring nuclei affect each other resulting in a change of B_{loc} of both nuclei. The extent of the change in the strength of B_{loc} is dependent upon μ of both nuclei, the distance r between them, and their orientation relative to B_0. In liquids, the Brownian molecular motion averages the orientation dependent interactions and narrow signals can be obtained. In solids, the rigid molecules stay at a fixed angle in every possible orientation. Comparable to the effect of the chemical shift anisotropy, this orientation dependency leads to broad resonance lines. As in the case of spin–spin coupling, this interaction can be suppressed by decoupling the unobserved nuclei, which is mostly ^1H.

9.2.5 Quadrupole interactions

Up to now, we only considered nuclei with $I = ½$. Some nuclei interesting for organic chemistry, e.g. ^{14}N or ^{17}O however exhibit $I > ½$. Such nuclei have a non-spherical charge distribution and thus a quadrupolar moment. The field gradients of the electron cloud around the nucleus interact with this quadrupole moment which leads to broad resonance lines. This line broadening can be so large that even in the liquid state, ^{14}N line widths of 100–1000 Hz are encountered. Therefore, for the examination of N-containing organic material the much less abundant and, therefore, less sensitive ^{15}N-isotope (natural abundance 0.37%) is commonly used.

9.2.6 Relaxation

After termination of the excitation pulse, the spin systems relax to their thermal equilibrium by releasing energy to the environment. This becomes possible due to the fact that the Brownian molecular motion causes fluctuation of B_{loc} with frequency contributions at magnitudes of the resonance frequency. As well as indirect and direct coupling, chemical shift anisotropy, rotation of the nuclei and their electrons around the binding axis, but also interactions with unpaired electrons contribute to the creation of such fluctuating fields. The latter offer extremely efficient relaxation pathways, which greatly reduces T_1.

T_1 and T_2 relaxation play an important role in practical NMR spectroscopy. To increase the signal-to-noise (S/N) ratio of a spectrum, several single spectra are accumulated and added. In order to obtain quantitative data, the spin system has to return completely to its thermal equilibrium before a new rf-pulse is applied. If there is not enough time between B_1-pulses, some of the nuclei will still be in the excited state when the new pulse occurs. The system will be saturated and these previously excited nuclei will not be detected. As a general rule the delay time between two rf-pulses should be $5 \times T_1$. T_1-values can be as short as a few milliseconds to as large as days and are dependent upon the mobility of the nucleus and therefore upon the type of molecular domain the nucleus resides in. In solids where the molecular motion can be very slow the efficiency of energy donation to the environment by the spin system during relaxation is very low. This can result in very long T_1s.

9.2.7 Magic angle spinning

The orientation-dependent interactions, leading to the line broadening in solid-state NMR are proportional to the term $3\cos^2\beta - 1$. If β is made equal to the magic angle of 54°44′, this term vanishes. Spinning the sample with sufficient speed at this magic angle imitates molecular motion and averages the orientation-dependent interactions. For their complete averaging, the magnitude of the spinning frequency must be at least as high as the line broadening caused by the interaction itself. In the case of the chemical shift anisotropy, this line broadening is dependent upon the applied static field and can have values of up to 500 ppm. This requires spinning speeds of several kHz. As the chemical shift anisotropy is proportional to B_0, it is obvious that for higher magnetic fields increased spinning speeds must be used. In the case of insufficient spinning speed, spinning side bands appear in the spectrum. They are observed on each side of the isotropic signal at a frequency distance equal to the spinning frequency (1 order spinning side bands) or even at their multitude *n* (2 to *n* order spinning side bands). They contain signal intensity and can overlap with other signals. This can aggravate quantification of solid-state NMR spectra (Benke *et al.* 1998). Thus, even though higher magnetic fields increase the sensitivity of an NMR experiment, their application for solid-state NMR spectroscopy is not always an advantage. However, examining nuclei with low sensitivity such as ^{15}N, useful spectra can only be acquired with higher magnetic fields, thus the occurrence of spinning side bands has to be encountered. Elegant experiments have been designed to remove the sidebands such as total side band suppression (TOSS) (Dixon 1982). Applying this pulse sequence, it must be kept in mind that intensity losses may occur.

For the complete removal of the heteronuclear dipolar interaction, unrealistic spinning speeds of approximately 60 kHz have to be achieved. Therefore, high power ^1H-decoupling is employed by using ^1H-amplifiers with a frequency range of several kHz and a power of several hundred watts. To avoid the NOE, the decoupler is only turned on during data acquisition.

9.2.8 The cross-polarization

In a solid-state NMR experiment, the long T_1s of nuclei with low natural abundance (e.g. ^{13}C or ^{15}N) can result in unfeasible long measurement time for achieving spectra with tolerable S/N ratios. This can be avoided by applying the cross-polarization (CP) technique, in which energy or magnetization is transferred from a spin system with a high relative abundance and relatively short T_1 (e.g. ^1H) to a nuclear spin system X (e.g. ^{13}C or ^{15}N) with low natural abundance and long T_1. The X-nuclei are then observed after the magnetization transfer.

However, according to quantum mechanics, any polarization transfer between two spin systems can only occur if ΔE_H between the energy levels of the ^1H-spin system is of equal magnitude than the one of the X-spin system (ΔE_X). With equation (9.4), this is the case if the Hartmann–Hahn condition is fulfilled (Hartmann and Hahn 1962):

$$\Delta E_H = \gamma_H B_{1H} = \gamma_X B_{1X} = \Delta E_X \qquad (9.6)$$

Here, γ_H is the gyromagnetic ratio of ^1H and B_{1H} the ^{1}H-rf-field. B_{1X} and γ_X are the respective parameters for the X-spin system. The Hartmann–Hahn condition can be achieved by simultaneous irradiation of the ^{1}H- and X-spin systems with variable magnetic fields during a contact time t_c. Hereby the fields B_{1H} and B_{1X} have to be adjusted in such a way that the difference between the energy levels of both systems are equal. The spins of both systems are then forced to precess with equal frequencies around their respective rf-fields (spin lock). The CP dynamics can be described by the simplified equation (9.7) (e.g. see Wilson 1987):

$$I(t_c) = I_0 \times 1/\alpha\,(1 - \exp(-\alpha t_c)) \times \exp(-t_c/T_{1\rho H}) \tag{9.7}$$

with $\alpha = 1 + (T_{XH}/T_{1\rho X}) - (T_{XH}/T_{1\rho H})$. The full equation is given in Mehring (1983).

The polarization transfer time T_{XH} expresses the time needed for the CP and is determined by molecular motion and the distance of the X-nucleus from the protons. In general, T_{XH} is shorter for XH_2 and XH than for XH_3, which in turn, is shorter than for non-protonated X-nuclei.

$T_{1\rho H}$ represents the spin–lattice relaxation of the protons during the spin lock state (^{1}H-spin–lattice relaxation time in the rotating frame) and leads to an exponential decrease of the observed signal intensity $I(t_c)$. $T_{1\rho X}$ is the respective relaxation time for the X-spin system. If $T_{1\rho X} \gg T_{XH}$, α simplifies to $1 - (T_{XH}/T_{1\rho H})$. Since $T_{1\rho H}$ starts already at the beginning of the CP process, it is obvious that the maximal signal intensity I_0 of the X nuclei can never be achieved and a signal is only observable if the polarization transfer is complete before $T_{1\rho H}$ becomes affective. This is true if $T_{1\rho H} \gg T_{XH}$. If $T_{1\rho H} \ll T_{XH}$, the signal is completely suppressed.

After t_c the spin systems relax to their thermal equilibriums. Since polarization of the X-spin system is moderated by the polarization of the ^{1}H-spin system and the relaxation behavior of the X-spin system is irrelevant for an efficient CP, only the ^{1}H-spin–lattice relaxation T_{1H} has to be considered for the pulse delay between a new introduction of the Hartmann–Hahn contact. In general, T_{1H} is faster than the one of less abundant X-nuclei such as ^{13}C and ^{15}N. Thus, applying the CP technique allows the accumulation of much more single scans within the same measurement time than an NMR experiment in which the X-spins are directly excited. A further advantage of the CP technique lies in the fact that a signal enhancement can be achieved. This enhancement is dependent upon the ratios of the gyromagnetic ratios of the participating spin systems.

Applying CP at high spinning speeds, which is often necessary to avoid spinning side bands, if high magnetic fields are used, one has to consider that dipolar fluctuations can no longer compensate for slight Hartmann–Hahn mismatches. This leads to a decrease of signal intensity of the effected X-nuclei and is most pronounced for X-nuclei with weak dipolar interactions (i.e. methyl-C, carboxyl-C, C-substituted carbon). The relative contribution of these groups will be underestimated relative to the groups with strong dipolar coupling. In order to overcome this problem a variable amplitude (VA) sequence with a ramped ^{1}H pulse decreasing the power from 100% to 50% is used during t. By doing so, varying Hartmann–Hahn conditions are created (Peersen *et al.* 1993) that eliminates differences in actual spin-lock precession frequencies of the ^{1}H- and X-spins. A constant X-spin magnetization will then result over a wide range of energy mismatches.

9.3 APPLICATION OF SOLID-STATE ^{13}C AND ^{15}N NMR TO SOLID WASTE

9.3.1 Chemical shift assignment

One emphasis of NMR spectroscopy is to determine, via chemical shift assignments, the gross chemical structure of a material, and to quantitatively correlate between the different signal intensities and the chemical composition. Due to the heterogeneity and complexity of waste samples, the broadness of the signals and the fact that signal overlapping can rarely be avoided, detailed information on the presence of specific compounds are difficult to obtain with a common solid-state CPMAS NMR spectrum. However, dividing a solid-state NMR spectrum into chemical shift regions assigned to the most tentative chemical groups contributing to the signals in this region and integrating those regions gives a good overview about the bulk chemical composition of the sample under study. Structural information on organic matter during solid-waste treatment is commonly yielded by solid-state ^{13}C NMR spectroscopy. However, the obtained spectra rarely provide more detailed information on the fate of organic N groups. Here, solid-state ^{15}N NMR has to be applied. Due to the low natural abundance of ^{15}N, the sensitivity of this nucleus is about 50 times lower than that of ^{13}C. Thus, successful application of this technique will only be possible for samples having an N content >1–2% and after accumulating several hundred thousand to million single scans.

Solid-state ^{13}C NMR spectra are often divided into four chemical shift regions (Figure 9.1a). Here, the region between 220 and 160 ppm is assigned to carbonyl/carboxyl/amide C, that between 160 and 110 to sp^2 C (aromatic C or olefine C). The O/N-alkyl region between 110 and 60 ppm includes signals of carbohydrates, alcohols and ether (110–60 ppm) but also N-alkyl C or methoxyl C (60–45 ppm). The region between 45 and 0 ppm is attributed to alkyl C as they occur in fatty acids, amino acids or paraffinic structures. However, if the resolution of a solid-state NMR spectrum allows a more detailed analysis, those regions are further subdivided for improving the signal assignment.

The spectrum of MSW obtained from a landfill in Germany, in Figure 9.1b represents such a well-resolved spectrum. Its dominating signal at 72 ppm is splitted and is attributed to C2, C3 and C5 in carbohydrates. The weaker signals at 63, 89 and 105 ppm derive from their C6, C4 and C1, respectively. The narrow peak at 32 ppm is best explained with methylene groups while its upfield shoulder (26–0 ppm) is typical for terminal methyl C groups. The small resonance at 128 ppm lies in the chemical shift region of sp^2 C and overlaps with the chemical shift of the ring C of polystyrene. The C1 of polystyrene may contribute together with phenolic C from lignin to the intensity in the region between 160 and 140 ppm. Its methylene groups (40 and 43 ppm) can add to the weak shoulder between 40 and 50 ppm.

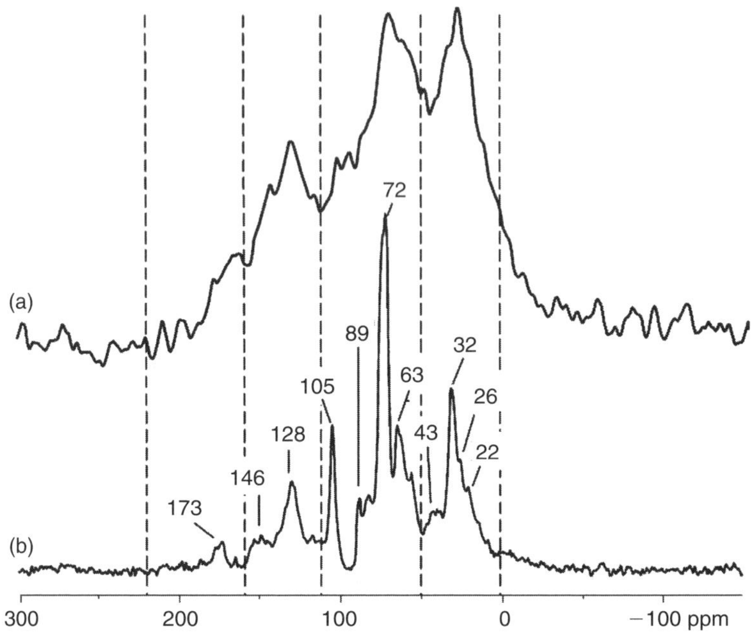

Figure 9.1 Solid-state CPMAS ^{13}C NMR spectra of MSW obtained from a landfill in Germany: (a) before treatment with HF and (b) after treatment with HF (Lindl 2001).

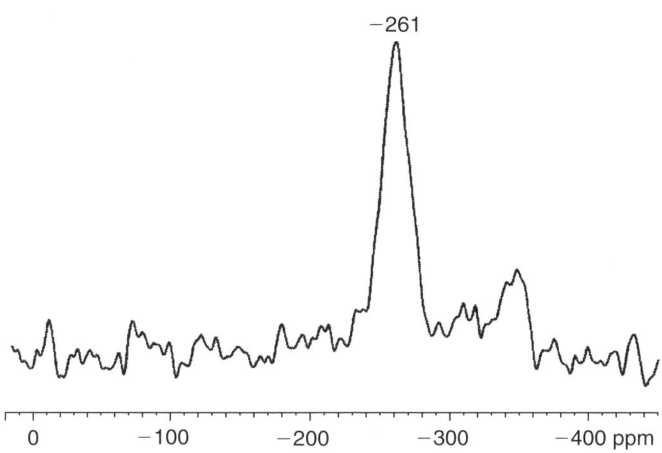

Figure 9.2 Solid-state CPMAS ^{15}N NMR spectrum of HF-treated MSW (Pichler *et al.* 2000).

The solid-state ^{15}N NMR spectrum of MSW offers less signals (Figure 9.2). It is dominated by a broad resonance line between −230 and −280 ppm, which is typically assigned to amides, either deriving from polyamides (plastic) and/or peptide structures of biogenic origin. The latter assignment is supported by the finding that 78% of the N content of the sample was identified as peptides by a chemolytic approach (Pichler *et al.* 2000). The smaller signal at -348 ppm is indicative of free amino groups, occurring as terminal groups of peptides, amino sugars or lysine. Signals attributed to pyrrole-type structures (-170 to -240 ppm) or pyridine-type compounds or imines (-30 to -90 ppm) are not discernable from the noise. From this it can be concluded that if they are present their content will not exceed 10% of the organic N. In this context it should be noted that due to the relaxation and CP behavior of inorganic N compounds, such as nitrate or ammonium, their signals will not quantitatively reflect their concentration in the sample (Knicker 2000).

9.3.2 Considerations for obtaining quantitative solid-state NMR data

One important advantage of solid-state NMR spectroscopy is that it is a non-invasive technique. In contrast to common thermolytic or chemolytic methods during which the macromolecule is degraded into small fragments that are analyzed, solid-state NMR allows the characterization of the sample as a whole. Secondary reactions, such as rearrangement, cracking, hydrogenation and polymerization are excluded. Consequently, one can assume that the obtained NMR spectrum reflects the real chemical composition of the sample, provided the correct acquisition parameters were used and interactions which may obscure the relative signal intensity distribution are avoided. In the following possible parameters and interactions which can lead to non-quantitative solid-state NMR data are introduced and discussed.

9.3.2.1 Paramagnetic compounds

During a CPMAS NMR experiment, interactions of the nuclei with paramagnetic components such as they occur in iron, copper or manganese oxides, can lead to considerable line broadening and shortening of $T_{1\rho H}$ to an extent that the condition $T_{XH} << T_{1\rho H}$ is not any longer fulfilled. A loss of signal intensity is the result. If spin diffusion is efficient, this means if the communication between the spins works well, all signals will be affected to the same extent. The S/N of the respective spectra becomes worse but the intensity distribution stays the same. If, on the other hand, the molecular domains are not in close contact, the impact of paramagnetic material will be selective for the signal of the nuclei in their close vicinity (Smernik and Oades 1999). The intensity distribution will be distorted and the spectrum cannot be quantified. To avoid such disturbance, one can treat the sample with hydrofluoric acid (HF) or HCl (Preston *et al.* 1989; Goncalves *et al.* 2003; Skjemstad *et al.* 1994) to decrease the concentration of paramagnetic constituents without destroying the organic matter. An additional advantage of the technique is that the mineral phase is dissolved and thus the organic matter content is increased. This is of particular interest if the N fraction is intended to be studied, since in many cases solid-state ^{15}N NMR spectroscopy can only be successfully applied if the organic N content of the sample was efficiently enhanced by such a treatment (Figure 9.2).

Table 9.1 Intensity distribution in the solid-state ^{13}C NMR spectra obtained from a MSW sample before and after treatment with 10% HF. The sample was obtained from a landfill site in southern Germany.

	Carbonyl C	sp^2 C	O/N-alkyl C	alkyl C
Before HF	8	23	42	26
After HF	4	17	53	26

However, a loss of some C and N has to be encountered due to removal of water soluble organic compounds and inorganic N with the HF-solution. Preston *et al.* (1989) observed a preferential loss of carbohydrates and some amino acids. Other studies comparing the solid-state ^{13}C NMR spectra of soil samples before and after HF-treatment indicated a non-selective loss of all carbon groups (Goncalves *et al.* 2003). The same finding is reported for sewage sludge samples, where between 10% and 30% were lost (Smernik *et al.* 2003a). Those losses are in the range observed for the samples discussed here. To have a crude estimation if those losses will affect the composition of the organic material to a higher extent, it may be helpful to compare the carbon/nitrogen (C/N) ratio of the sample before and after the treatment.

The impact of the HF-treatment on the quality of a solid-state CPMAS ^{13}C NMR experiment is demonstrated in Figure 9.1. The spectrum obtained before the HF-treatment (Figure 9.1a) shows a low S/N ratio and low resolution. After HF-treatment the spectral quality considerably increased (Figure 9.1b), allowing a more detailed signal assignment. However, looking at the intensity distribution in the spectrum of the HF-treated sample given in Table 9.1 a much higher O/N-alkyl C content and a lower sp^2 C and carbonyl C intensity is detected with respect to the untreated sample.

It is unlikely that O-alkyl C was formed during the HF treatment. A better explanation for the differing intensity distribution in Table 9.1 is that in the untreated sample the paramagnetic compounds interacted preferentially with O-alkyl C groups and decreased their detectable ^{13}C intensity to a higher extent than that of the other functional groups. Consequently, the intensity distribution in the spectrum of the untreated sample will not reflect the real chemical composition of its organic matter. Here, it should be noted that such obscuring of NMR data is not necessarily a fact for all waste samples. However, if the iron content of a sample is high, this means if $C/Fe < 1$, one has to expect it and an HF-treatment is recommended to produce reliable data.

9.3.2.2 Proton-spin–lattice relaxation (T_{1H})

For obtaining quantitative CPMAS NMR spectra saturation has to be avoided. For a solid-state CPMAS NMR spectrum the parameter to be considered is T_{1H}. In a heterogeneous mixture, different molecular domains in various chemical and physical environments may be present. To avoid saturation the longest T_{1H} thus has to determine the pulse delay *d*. T_{1H} can be measured by an inversion recovery experiment. Here, a series of spectra is accumulated in which the magnetization of the ^{1}H-spin system is inverted by application of a 180°-^{1}H-pulse prior to the conventional CP sequence. A variable pulse delay τ after the 180°-^{1}H-pulse allows the inverted ^{1}H-magnetization to relax and the remaining ^{1}H-magnetization is then cross-polarized. The signal intensities $I(\tau)$, of the ^{13}C-signals change as function of τ. T_{1H} is then calculated using the following equation:

$$I(\tau) = I(0)\exp(-\tau/T_{1H}) \qquad (9.8)$$

When evaluating a solid-state NMR spectrum of a heterogeneous mixture, one peak can cover the signals of different groups which vary in their chemical and physical environment. T_{1H}-values determined for such peaks therefore represent the weight average value for all nuclei contributing to the observed signal.

Analyzing T_{1H}-values for typical components of MSW with a Bruker DSX 200, the longest T_{1H} of almost 3 s were measured for cellulose in shredded and milled office paper. For polystyrene a value of 1.6 s and for polyethylene a $T_{1H} = 0.9$ s were determined (Pichler *et al.* 2000). Thus, for the analysis of municipal waste a pulse delay of 15 s should be sufficient to avoid saturation effects.

For solid-state CPMAS ^{15}N NMR spectroscopy on waste, up to now, no T_{1H} values are reported. Since such spectra of samples with natural ^{15}N abundance require the accumulation of at least 400,000 scans, relaxation time measurements are very time consuming. However, comparing the solid-state CPMAS ^{15}N NMR spectrum in Figure 9.2 obtained at a delay time $d = 200$ ms with one acquired at $d = 500$ ms did not reveal any major saturation effects (unpublished data). According to this result it was assumed that a delay time around 200 ms should be sufficient to obtain reliable spectra.

9.3.2.3 CP and single pulse excitation

Different molecular domains in solid waste will also exhibit differing CP dynamics. After performance of a variable contact time experiment and using equation (9.7) the various T_{CH} and $T_{1\rho H}$ values for the peaks at 32, 72 and 128 ppm for the solid-state ^{13}C NMR spectrum of an MSW sample depicted in Figure 9.3 were calculated (Table 9.2).

Those data indicate that the condition $T_{CH} \ll T_{1\rho}H$ is fulfilled and efficient CP can be achieved. For determination of the contact time t_c, which will be applied, one has to bear in mind that according to theoretical considerations t_c should be at least three times larger than T_{CH}. In soil science a $t_c = 1$ ms is often used. The intensity losses calculated from the difference of the theoretically reachable intensity I_0 and the measured $I(t)$ at $t_c = 1$ ms is included in Table 9.3 and comprises between 4% and 10%. If the experimental error is considered, it can be assumed that spectra obtained with this contact time reflect a fairly quantitative intensity distribution. For solid-state CPMAS ^{15}N NMR spectra, an optimal contact time between 0.8 and 1 ms was recently determined using humifying plant residues (Knicker and Lüdemann 1995).

Table 9.2 CP parameters as determined by a variable contact time ^{13}C NMR measurement of a fresh MSW sample (Pichler *et al.* 2000).

	128 ppm	72 ppm	32 ppm
T_{CH} (ms)	0.03	0.2	0.2
$T_{1\rho H}$ (ms)	12.7	12.1	9.5
Intensity loss (%) at t_c = 1 ms	6	9	6

In evaluating the possibility of quantification of a solid-state CPMAS ^{13}C NMR spectrum, one must consider that carbon groups of which the nearest ^{1}H is more than three bonds apart may not be efficiently polarized and thus are not observed, quantitatively (Alemany *et al.* 1983). Such carbons may be constituents of soot or heavily charred residues. One common approach to test if such components are present is the single pulse excitation (SPE) experiment (also called Bloch decay). Here, the ^{13}C is directly observed without prior CP. For avoiding saturation effects, the spin–lattice relaxation time of ^{13}C (T_{1C}) has to be considered for the adjustment of the pulse delay *d*. However, although SPE ^{13}C NMR is often applied in soil (Smernik and Oades 2000) and geoscience, not much effort was conducted to determine this important parameter. On the other hand, in recent studies, for cellulose in office paper and for polystyrene T_{1C} of 53 s and 136 s were measured, respectively (Pichler *et al.* 2000). For the crystalline domains in cotton a T_{1C} = 266 s is reported (Teeäär and Lippmaa 1984). Polyethylene reached even a T_{1C} = 449 s (Pichler *et al.* 2000). Thus, if the latter is present a d = 2250 s (=37.5 min) is needed to avoid saturation. This delay times is considerably longer than those commonly applied and consume infeasible spectrometer times in the order of weeks. The long T_{1C} and T_{1H} of some organic constituents may also be responsible for the recently reported discrepancy between CPMAS and SPE ^{13}C NMR spectra of some sewage sludge obtained at a pulse delay of 1 s (CPMAS) and 90 s (SPE) (Smernik *et al.* 2003a). Consequently, if one intends to test the reliability of solid-state ^{13}C NMR spectra of solid waste by means of the SPE approach, one has to be careful.

9.3.3 Characterization of solid waste by standard CPMAS ^{13}C NMR

9.3.3.1 Alteration of MSW during composting

Figure 9.3 depicts the solid-state CPMAS ^{13}C NMR spectra of HF-treated MSW that was subjected to an aerobic biological treatment. This involved 9 weeks of intensive windrow composting with pressure aeration, followed by 46 weeks of non-aerated windrow composting (Pichler *et al.* 2000). Table 9.3 gives the respective intensity distribution (Pichler *et al.* 2001). After 9 weeks treatment time, the concentration of O-alkyl C in the solid-state CPMAS ^{13}C NMR spectrum declines from 62% to 50% of the total organic C (TOC). At the end of the treatment this

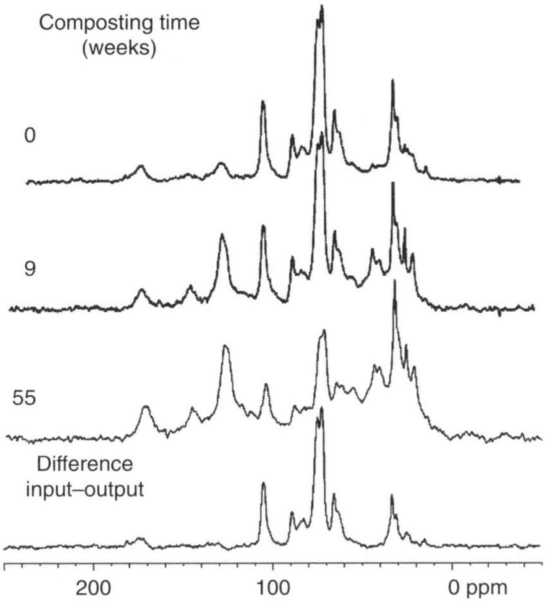

Figure 9.3 Solid-state CPMAS ^{13}C NMR spectra of MSW subjected 9 weeks of windrow composting with pressure aeration, followed by 46 weeks of non-aerated (after Pichler *et al.*, 2001).

Table 9.3 Intensity distribution in the solid-state CPMAS ^{13}C NMR spectra of HF-treated MSW that was subjected to an aerobic biological treatment followed by 46 weeks of non-aerated windrow composting (Pichler *et al.*, 2001).

Composting time	Carboxyl C	sp^2 C	O/N-alkyl C	Alkyl C	CL*
0 weeks	5	10	62	23	4.6
9 weeks	4	19	49	28	7
55 weeks	6	24	34	36	6

*CL: alkyl C/carboxyl C ratio.

value reaches 34%. Simultaneously the alkyl C and sp^2 C content increases. The carbonyl C concentration shows only minor alterations.

If the C loss (TOC loss) during biodegradation is known, the mass loss (ML) of each functional group during composting can be calculated from the intensity in the respective chemical shift region before ($I(0)$) and after ($I(t_{com})$) according to the following equation:

$$\%ML = 100 - ((100 - \%TOC\ loss) \times (I(t_{com})/I(0)) \quad (9.9)$$

Those calculations show that O-alkyl C is the main contributor to the ML with 82% TOC loss after 55 weeks. The lowest TOC loss is observed for the sp^2 C region (22% after 55 weeks), which is best explained with the slow degradation of lignin and aromatic plastic contents. The assumption that plastic material is selectively enriched is supported by the fact that the signals typical for polystyrene at 128, 147, 40 and 43 ppm become more prominent with increasing treatment time. Also the signals at 26 and 23 ppm, most likely assigned to polypropylene or methylene groups in polyamides increase. Calculating the alkyl C/carboxyl C ratio as a parameter for the average chain lengths of alkyl structures further reveals an increase of the chain length from 5 in the control sample to 7 after 9 weeks treatment time. This may be taken as an additional indicator for the accumulation of plastic compounds.

The preferential degradation of carbohydrates and the selective enrichment of alkyl C was also observed during the NMR analysis of MSW subjected to other aerobic and anaerobic treatments (González-Vila *et al.* 1999; Pichler *et al.* 2001). Thus, O-alkyl C represent the most relevant fraction of MSW with regard to landfill emissions. This was confirmed by statistical studies in which different MSW treatments were elucidated by NMR spectroscopy and the respiration rate within 4 days (Pichler *et al.* 2000).

9.3.3.2 Characterization of MSW from a landfill

Figure 9.4 depicts the solid-state CPMAS ^{13}C NMR spectra of HF-treated MSW obtained from a landfill in Germany as a function of landfill depths. The filling of the site occurred during 1979 and 1984. After closing, it was covered for protection (Lindl 2001). The spectrum from the top layer (0–2 m) shows a higher O-alkyl C content than the untreated MSW sample in Figure 9.5, although newspapers found during sampling indicated an age of 10–14 years.

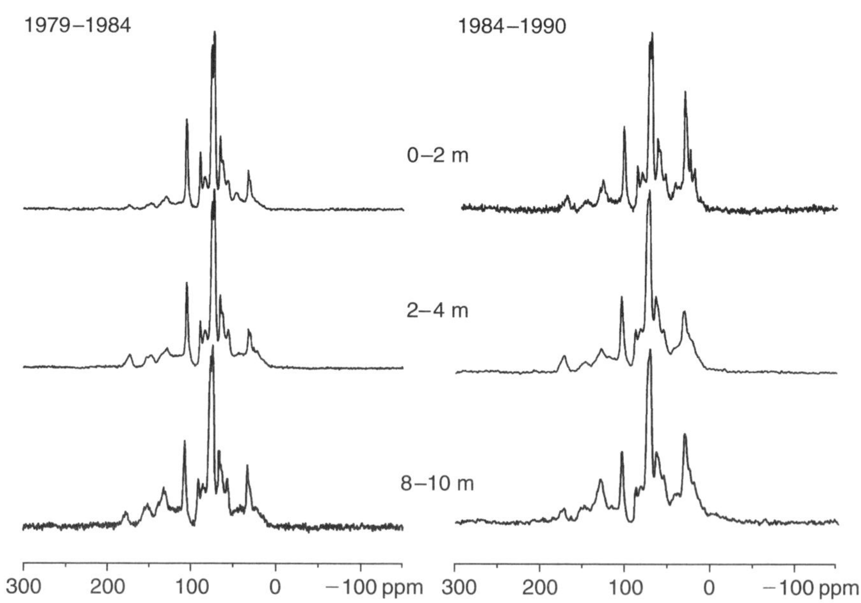

Figure 9.4 Solid-state CPMAS ^{13}C NMR spectra of MSW obtained from a landfill in Germany (after Lindl, 2001).

The low degradation state of this material may be explained with the low water content of 6% allowing the conclusion that this sample was dry stabilized after closing and covering the fill. With increasing depths the pattern expected for biodegradation of organic material, a decrease in O-alkyl concomitantly with the increase of the intensity in all other regions is revealed.

However, examining the degradation processes in a landfill site by determination of the chemical composition is not always as easy. The behavior of the consumer is permanently changing the nature of the waste. This can be observed in the spectra of the waste from the same landfill but from a site that was filled during 1984 and 1990. The sample of the top layer comprise considerably less O-alkyl C and more alkyl C than the respective one of the filling between 1979 and 1984, although it is obviously younger. Visual inspection of the top layer (0–2 m) revealed a content of plastic of 40–50%, which is in accordance to the pronounced and narrow signals in the alkyl C region and the chain length indicator of 7.3. The reason for this high plastic content may be the fact that during this time separating paper for recycling started. For the layer 2–4 m the visually determined plastic content decreased to approximately 10% which is in accordance to the reducing of the alkyl C intensity. In the deepest layer (8–10 m), however, about 50% plastic were visually identified which agrees with the high alkyl C content of the sample.

9.3.4 Advanced NMR techniques

So far it was shown that already standard CPMAS ^{13}C NMR spectroscopy can deliver helpful information for a better understanding of processes occurring during degradation of solid waste as well as for the evaluation of the efficiency of different solid waste treatments. However, modern solid-state NMR spectroscopy offers advanced techniques that offer some more additional information about solid-waste composition.

9.3.4.1 Dipolar dephasing

Dipolar dephasing (DD) solid-state CPMAS NMR spectroscopy is often used to obtain a more detailed assignment of signals to functional groups by allowing a first estimation about the contribution of X-nuclei with weak and strong proton dipolar coupling (Wilson 1987, Rowell *et al.* 2001; Mao *et al.* 2003). Here, directly after t_c but before data acquisition the high power proton-decoupling is interrupted for a certain time delay (interruption delay = t_{dd}). In the obtained spectrum the X-nuclei that are effected by strong proton dipolar coupling exhibit a loss of intensity with increasing t_{dd} due to line broadening. In general their signals disappear after a $t_{dd} > 40\ \mu s$. X-nuclei without directly attached ^1H and ^{13}C with high molecular motion experience weak ^1H dipolar coupling. There signals will still be visible in a spectrum obtained with $t_{dd} > 40\ \mu s$. In this context it should be mentioned that the signal obtained in such an experiment can be modulated by other interactions than coupling between X and ^1H leading to the occurrence of rotational echoes at an integral multiples of the spinner speed. The first echo will be visible at a $t_{dd} = 1/v$, where v is the spinning speed in kHz. Thus, if quantitative information on the degree of protonation of various carbons is needed, the t_{dd} has to be kept below that where rotational echoes interfere. Applying this technique to the fresh and composted MSW confirmed the presence of long methylene as they occur, e.g. in amorphous polyethylene (Pichler *et al.* 2000).

9.3.4.2 Proton spin relaxation editing

Relaxation of an excited nucleus is dependent upon the environment and the mobility of the respective nuclei. Very small molecules with long, and bigger ones with short correlation times will exhibit long T_{1H}. Due to this dependency, T_{1H} values can give valuable information about the mobility of different molecular domains. If those domains are bigger than 30 nm or perhaps even 100 nm (Newman and Condron 1995), a technique called proton spin relaxation editing (PSRE) can be used to create subspectra showing only signals of molecular domains with a certain T_{1H}. For the PSRE approach a linear combination of a standard CPMAS NMR spectrum and a spectrum obtained with the inversion recovery technique is used. To give an example how this technique is used and which information it offers, data previously published by Pichler *et al.* (2000) shall be discussed. This technique was also recently applied to characterize sewage slugde (Smernik *et al.* 2003b).

The subspectrum of the fresh MSW calculated for $T_{1H} = 1.5$ s in Figure 9.5, depicts the typical signal of crystalline cellulose (72 and 74 ppm). The signals at 128, 145 and 32 ppm derive from the aromatic C of polystyrene and polyethylene exhibiting long T_{1H}s of 1.6 and 1 s, respectively. The signals in the subspectrum at 0.3 ms derive from amorphous structures of such as starch, hemicellulose or microbial polysaccharides, as well as from lipids or proteins. Note that after

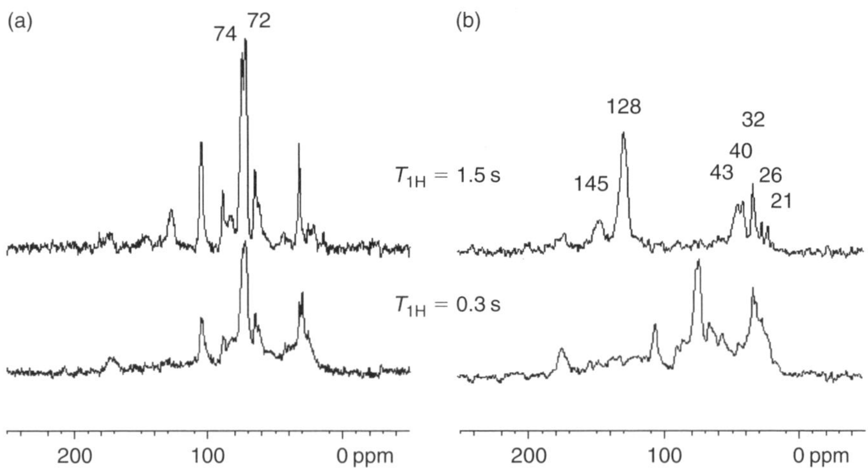

Figure 9.5 PSRE ^{13}C NMR subspectra of (a) fresh and (b) composted MSW (55 weeks).

55 weeks of composting the signals of crystalline cellulose are missing in the spectrum with long $T_{1H} = 1.5$ s, indicating that those constituents were efficiently removed during composting. The signals of plastic material in the aromatic and alkyl region become more prominent, confirming the assumption obtained already by means of routine CPMAS technique that they are selectively enriched. The spectrum of the composted MSW at $T_{1H} = 0.3$ ms on the other hand shows the typical feature of partly degraded biogenic residues or microbial biomass. This material may have accumulated during humification by selective enrichment or synthesis of new microbial biomass. Quantification of the subspectra can be performed by subtracting them from the standard CPMAS ^{13}C NMR spectrum. With this calculation a plastic content of the composted MSW of 37% of the total organic matter was yielded, leaving approximately 63% to amorphous compounds most probably of biological origin.

9.4 CONCLUSIONS AND FUTURE PROSPECTIVES

With this chapter, it was intended to provide some insights into the basic theory of solid-state NMR spectroscopy, mainly of the most often applied solid-state ^{13}C NMR technique. Some examples were given which information can be obtained by this technique, if changes during composting or humification of solid waste is needed. Another example, in which solid-state ^{13}C NMR was used to study the transformation of organic matter during composting of solid waste is that reported by Chefetz *et al.* (2002). They examined the alteration of spent mushroom substrate in a pile and found fast degradation of the organic material at the top of the pile subjected to weathering. The decomposition at the bottom of the pile was significantly slower, probably due to lack of oxygen. As it was shown here, and in many other reports (i.e. Fründ *et al.* 1987; Benke *et al.* 1998; González-Vila *et al.* 1999; Smernik *et al.* 2003a) standard solid-state CPMAS ^{13}C NMR can certainly provide helpful information if the quality, composition and properties of solid waste is the subject of research. Ayuso *et al.* (1997), e.g. compared humic acids extracted from urban waste with those obtained from peats to elucidate the potential of the first as liquid fertilizer. Although they observed structural differences, all humic acids showed an improvement of the plant growth with respect to a control sample. Chefetz *et al.* (1998a) investigated the transformation of humic acids of MSW obtained with (core-HA) and without (HA) prior solvent and sulfuric acid extractions during composting. While the HA showed an increase in relative intensity in the aromatic and phenol region, only minor changes were recorded for the core-HA. From this observation, they concluded that during decomposition the material coating the core-HA (polysaccharides, peptides and lipids) is preferentially degraded, which results in HA structures of higher aromaticity. Analyzing dissolved organic carbon fractions formed during composting of MSW, Chefetz *et al.* (1998b) detected an increase of the fractions with a higher concentration of functional groups with ongoing composting. They suggested that those groups may be able to chelate metals and thus contribute to their improved uptake by plants. Hartlieb *et al.* (2001) determined the structure of dissolved organic matter in an attempt to characterize their mobilization potential for hydrophobic contaminants. In order to predict the potential of composts from biosolids as efficient fertilizers, Rowell *et al.* (2001) related the substrate quality of wheat straw, paper fine and Douglas-fire needle litter to decomposition rates and N release during incubation.

However, during the last years, a considerable amount of new and advanced NMR techniques were developed which were successfully applied to reveal the structure of purified biomolecules. Their application to heterogeneous mixtures such as solid waste will certainly encounter problems concerning spectral resolution and sensitivity. On the other hand, the example of the successful application of the DD and the PSRE technique, given here, show that it may not be infeasible to apply some of those in waste management research. One of those advanced technique which can be helpful in elucidation chemical reactions occurring during biodegradation of waste may be double cross ^{15}N ^{13}C polarization. This technique, recently applied to identify the binding mechanism of the explosive 2,4,6-trinitrotoluene to soil organic matter (Knicker 2003), may also be advantageous if one attempts to obtain information how N-containing pollutants are binding to the organic residue of solid waste. Double CP is based on a magnetization transfer from ^{1}H to ^{15}N and subsequently to ^{13}C. The latter is finally detected. The key of this approach is that the resulting spectrum shows only signals of ^{13}C that couple with ^{15}N. Expanding this experiment to a two-dimensional (2D) version, interactions, indicating covalent bonds, are revealed by cross peaks correlating the signal intensity of the ^{13}C NMR spectrum with that of the ^{15}N NMR spectrum. The application of this technique, however, encounters severe sensitivity problems and is only feasible, if substances are used that are highly enriched in ^{13}C and ^{15}N (Knicker 2002). This, on the other hand, turns into an advantage if one uses isotopically highly enriched reactants. Labeling one of the reactant with ^{13}C and the other with ^{15}N, the respective DCPMAS ^{15}N ^{13}C NMR spectrum will show only signals of the corresponding interacting functional groups, an information that is hardly obtained with standard NMR techniques.

REFERENCES

Alemany, L.B., Grant, D.M., Pugmire, R.J., Alger, T.D. and Zilm, K.W. (1983) Cross polarization and magic angle sample spinning NMR spectra of model organic compounds. 2. Molecules of low or remote protonation. *J. Am. Chem. Soc.* **105**, 2142–2147.

Andrew, E.R. (1971) The narrowing of NMR spectra of solids by high speed specimen rotation and the resolution of solids. *Prog. Nuc. Mag. Res. Spectr.* **8**, 1–39.

Ayuso, M., Moreno, J.L., Hernández, T. and García, C. (1997) Characterisation and evaluation of humic acids extracted from urban waste as liquid fertilizers. *J. Sci. Food Agric.* **75**, 481–488.

Benke, M.B., Mermut, A.R. and Chatson, B. (1998) Carbon-13 CP/MAS NMR and DR-FTIR spectroscopic studies of sugarcane distillery waste. *Can. J. Soil Sci.* **78**, 227–236.

Chefetz, B., Hatcher, P.G., Hadar, Y. and Chen, Y. (1996) Chemical and biological characterization of organic matter during composting of municipal solid waste. *J. Environ. Qual.* **25**, 776–785.

Chefetz, B., Adani, F., Genevini, P., Tambone, F., Hadar, Y. and Chen, Y. (1998a) Humic-acid transformation during composting of municipal solid waste. *J. Environ. Qual.* **27**, 794–800.

Chefetz, B., Hadar, Y. and Chen, Y. (1998b) Dissolved organic carbon fractions formed during composting of municipal solid waste: properties and significance. *Acta Hydrochim. Hydrobiol.* **26**, 172–179.

Chefetz, B., van Heemst, J.D.H., Romaine, C.P., Chrover, J., Rosario, R., Mingxin, G. and Hatcher, P.G. (2000) Organic matter transformations during the weathering process of spent mushroom substrate. *J. Environ. Qual.* **29**, 592–602.

Dixon, W.T. (1982) Spinning-sideband-free and spinning-sideband-only NMR spectra in spinning samples. *J. Chem. Phys.* **77**, 1800–1809.

Fründ, R., González-Vila, F.J., Lüdemann, H.-D. and Martín, F. (1987) Comparison of the solid state CPMAS and solution carbon-13-NMR spectra of humic acids extracted from composted municipal refuse. *Zeitschr. Naturforsch.* **42c**, 205–208.

Goncalves, C.N., Dalmolin, R.S.D., Dick, D.P., Knicker, H., Klamt, E. and Kögel-Knabner, I. (2003) The effect of 10% HF treatment on the resolution of ^{13}C CP/MAS NMR spectra and on the quality of organic matter in Ferralsols. *Geoderma* **116**, 373–392.

González-Vila, F.J., Almendros, G. and Madrid, F. (1999) Molecular alterations of organic fractions from urban waste in the course of composting and their further transformation in amended soil. *Sci. Total Environ.* **236**, 215–229.

Hartlieb, N., Marschner, B. and Klein, W. (2001) Transformation of dissolved organic matter (DOM) and ^{14}C-labelled organic contaminants during composting of municipal biowaste. *Sci. Total Environ.* **278**, 1–10.

Hartmann, S.R. and Hahn, E.L. (1962) Nuclear double resonance in the rotating frame. *Phys. Rev.* **128**, 2042–2053.

Hatcher, P.G., Breger, I.A. and Mattingly, M.A. (1980) Structural characteristics of fulvic acids from continental shelf sediments. *Nature* **285**, 560–562.

Knicker, H. (2000) Biogenic nitrogen in soils as revealed by solid-state carbon-13 and nitrogen-15 nuclear magnetic resonance spectroscopy. *J. Environ. Qual.* **29**, 715–723.

Knicker, H. (2002) The feasibility of using DCPMAS ^{15}N ^{13}C NMR spectroscopy for a better characterization of immobilized ^{15}N during incubation of ^{13}C- and ^{15}N-enriched plant material. *Org. Geochem.* **33**, 237–246.

Knicker, H. (2003) Incorporation of ^{15}N-TNT transformation products into humifying plant organic matter as revealed by one- and two-dimensional solid state NMR spectroscopy. *Sci. Total Environ.* **308**, 211–220.

Knicker, H. and Lüdemann, H.-D. (1995) N-15 and C-13 CPMAS and solution NMR studies of N-15 enriched plant material during 600 days of microbial degradation. *Org. Geochem.* **23**, 329–341.

Lindl, S. (2001) ^{13}C-NMR-festkörperspektroskopische Analyse der organischen Substanz einer Siedlungsabfalldeponie. Diploma thesis, Lehrstuhl für Bodenkunde, Technische Universität München, Freising-Weihenstephan, Germany.

Mao, J.-D., Hundal, L.S., Schmid-Rohr, K. and Thompson, M.L. (2003) Nuclear magnetic resonance and diffuse-reflectance infrared fourier transform spectroscopy of biosolids-derived biocolloidal organic matter. *Environ. Sci. Technol.* **37**, 1751–1757.

Mehring, M. (1983) *High Resolution NMR in Solids*. Springer-Verlag, Berlin.

Mikinis, M.A., Bartuska, V.J. and Maciel, G.E. (1979) Cross-polarization magic angle spinning ^{13}C NMR spectra of oil shales. *Org. Geochem.* **1**, 169–176.

Newman, R.H. and Condron, L.M. (1995) Separating subspectra from cross-polarization magic-angle spinning nuclear magnetic resonance spectra by proton spin relaxation editing. *Solid State Nucl. Magn. Res.* **4**, 259–266.

Newman, R.H., Tate, K.R., Barron, P.F. and Wilson, M.A. (1980) Towards a direct, non-destructive method of characterising humic substances using ^{13}C n.m.r. *J. Soil Sci.* **31**, 623–631.

Peersen, O.B., Wu, X., Kustanovich, I. and Smith, S.O. (1993) Variable-amplitude cross-polarization MAS NMR. *J. Magn. Res.* **104**, 334–339.

Pichler, M., Knicker, H. and Kögel-Knabner, I. (2000) Changes in the chemical structure of municipal solid waste during composting as studied by solid-state dipolar dephasing and PSRE ^{13}C NMR and solid-state ^{15}N NMR spectroscopy. *Environ. Sci. Technol.* **34**, 4034–4038.

Pichler, M., Knicker, H. and Kögel-Knabner, I. (2001) Solid state ^{13}C NMR spectroscopic, chemolytic and biologial assessment of pretreated municipal solid waste. *J. Indust. Microbiol. Biotechnol.* **26**, 83–89.

Pines, A., Gibby, G.G. and Waugh, J.S. (1973) Proton-enhanced NMR of dilute spins in solids. *J. Chem. Phys.* **59**, 569–590.

Preston, C.M., Schnitzer, M. and Ripmeester, J.A. (1989) A spectroscopic and chemical investigation on the deashing of a humin. *Soil Sci. Soc. Am. J.* **53**, 1142–1147.

Rowell, D.M., Prescott, C.E. and Preston, C. (2001) Decomposition and nitrogen mineralization from biosolids and other organic materials: relationship with initial chemistry. *J. Environ. Qual.* **30**, 1401–1410.

Sanders, J.K.M. and Hunter, B.K. (1993) *Modern NMR Spectroscopy. A Guide for Chemists*. Oxford University Press, New York.

Schaefer, J. and Stejskal, E.O. (1976) Carbon-13 nuclear magnetic resonance of polymers spinning at magic angle. *J. Am. Chem. Soc.* **98**, 1031–1032.

Skjemstad, J.O., Clarke, P., Taylor, J.A., Oades, J.M. and Newman, R.H. (1994) The removal of magnetic materials from surface soils. A solid state ^{13}C CP/MAS n.m.r. study. *Aust. J. Soil Res.* **32**, 1215–1229.

Smernik, R.J. and Oades, J.M. (1999) Effects of added paramagnetic ions on the ^{13}C CP/MAS NMR spectrum of a de-ashed soil. *Geoderma* **89**, 219–248.

Smernik, R.J. and Oades, J.M. (2000) The use of spin counting for determining quantitation in solid state ^{13}C NMR spectra of natural organic matter. 2. HF-treated soil fractions. *Geoderma* **96**, 101–129.

Smernik, R.J., Oliver, I.W. and Merrington, G. (2003a) Characterization of sewage sludge organic matter using solid-state carbon-13 nuclear magnetic resonance spectroscopy. *J. Environ. Qual.* **32**, 1516–1522.

Smernik, R.J., Oliver, I.W. and Merrington G. (2003b) Advanced solid-state carbon-13 nuclear magnetic resonance spectroscopic studies of sewage sludge organic matter: detection of organic “domains”. *J. Environ. Qual.* **32**, 1523–1533.

Stejskal, E.O. (1994) *High Resolution NMR in Solid State: Fundamentals of CP/MAS*. Oxford University Press, New York.

Teeäär, R. and Lippmaa, E. (1984) Solid state ^{13}C NMR of cellulose. A relaxation study. *Polym. Bull.* **12**, 315–318.

Wilson, M.A. (1987) *NMR Techniques and Applications in Geochemistry and Soil Chemistry*. Pergamon Press, Oxford.

10

Microbial ecology of compost

M.N. Marshall, A.P. Reddy and J.S. VanderGheynst

10.1 INTRODUCTION

Composting is the decomposition of organic residues controlled by the management of a number of factors including feedstock composition, moisture, oxygen and temperature. This management favours particular consortia of microorganisms that rapidly colonize and decompose the residues. As decomposition proceeds, significant changes occur in the physiochemical properties of the compost. The associated microbial community adapts in response to these changes and continues the decomposition process. This coupling between the microbial community and process dynamics combined with the inherent heterogeneity of compost makes composting microbial ecology a challenging, yet fascinating, area of study.

Given the critical role microorganisms play in decomposition, relatively little information is available on the microbial community in compost compared to the amount of information on the composting process itself. Much of this is due to the limited number of reliable tools available for monitoring microbial communities in

 Resource Recovery and Reuse in Organic Solid Waste Management. Edited by Piet Lens, Bert Hamelers, Harry Hoitink, Werner Bidlingmaier. ISBN: 1 84339 054 X. Published by IWA Publishing, London, UK

heterogeneous, mixed culture systems such as compost. During the past 10 years significant advances have been made on techniques for examining the microbial ecology of environmental systems, which have, to a limited extent, been used to monitor microorganisms in compost. These tools in combination with traditional microbiological methods have allowed investigators to gain insight into the types of microbial communities present during composting and their influence on the decomposition process.

Our first objective in this chapter is to provide a brief overview of the methods that have been utilized for monitoring microbial communities during composting. Since investigations of compost microbial ecology have been done for a variety of purposes, the remainder of the chapter is organized into two sections: a review of the literature describing observations of general trends in microbial communities in composting and a review of the effect of process management on microbial communities.

10.2 MONITORING MICROBIAL COMMUNITIES IN COMPOST

Method selection for monitoring the microbial communities during composting should consider several criteria. First, the method should give reproducible results, which is particularly important given the general complexity of compost. The method should also distinguish between different groups of microorganisms, in particular bacteria, actinomycetes and fungi. Finally, the method should be able to detect a wide variety of microorganisms within these groups, which is important for evaluating the diversity of the microbial community (Fermor *et al.* 1979). Another criterion in method selection might be the identification of viable or active organisms within the composting process.

Several methods have been used for monitoring microbial communities in composting including enrichment plating, phospholipid fatty acid (PLFA) analysis and nucleic acid-based techniques (Figure 10.1). Enrichment plating is considered a cultivation-dependent method while PLFA and nucleic acid-based methods are considered cultivation-independent methods. These methods will be described in the following sections.

10.2.1 Enrichment methods

Enrichment methods, considered traditional microbiological methods, are, to date, the most common methods for monitoring microbial communities in compost. In enrichment methods, the microbial community is removed from the compost, usually by means of liquid extraction and homogenization, and then serial dilutions of extracts are plated and cultured under specific environmental conditions to encourage selection of the greatest number of colonies per plate. As the composting environment can support many different types of microorganisms

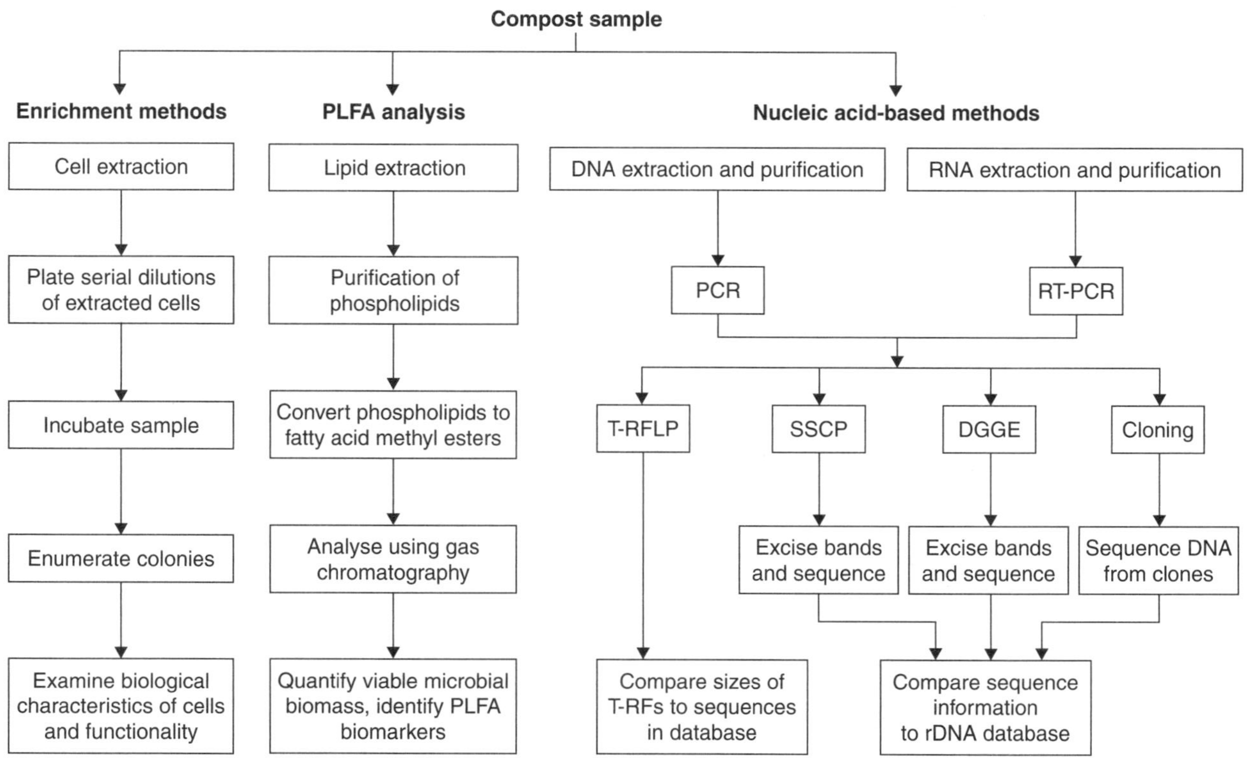

Figure 10.1 Summary of methods used to monitor microbial communities in compost.

with unique growth requirements, compost samples are often examined under several incubation and cultivation conditions to selectively encourage the growth of different groups of microorganisms and capture the diversity of the microbial community.

After several hours or days of incubation, colonies are enumerated. Individual colonies can then be isolated and examined for a variety of characteristics including morphology, nucleic acid and fatty acid composition, growth requirements and tolerances, and functionality. These examinations are done to identify isolated organisms, and, in many cases, to determine their role in the decomposition process.

Enrichment methods have been used to observe the microbial population dynamics during composting of food waste (Pedro *et al.* 1999; Ryckeboer *et al.* 2003), biosolids (Nakasaki *et al.* 1985), animal manures (Fermor *et al.* 1979; Godden *et al.* 1983; Tiquia 2002; Tiquia *et al.* 2002), municipal solid waste (Hassen *et al.* 2001), and woody residues (Davis *et al.* 1991, 1992) and to elucidate the effect of composting factors such as temperature (Waksman *et al.* 1939; Strom 1985), soluble carbohydrate levels (Fermor *et al.* 1979), inoculation (Nakasaki *et al.* 1985; Ichida *et al.* 2001), and nitrogen addition (Davis *et al.* 1991) on microbial communities. Others have used enrichment methods to identify the types of microorganisms present during extreme composting conditions (Beffa *et al.* 1996a, b; Lyon *et al.* 2000).

The advantage of enrichment techniques is that they can provide qualitative and quantitative information about the isolated microbial community. Furthermore, they only detect viable microorganisms in the system. The most significant limitation of enrichment-based methods is that they remove the microbial community from its natural habitat and only isolate organisms that can be cultured under the conditions investigated. Hence, these types of methods are inherently biased since only a small fraction (0.1–10%) of bacteria can be cultivated (Amann *et al.* 1995). In response to this limitation, a number of cultivation-independent methods have been applied to examine the microbial community in compost. These methods are described in the following sections.

10.2.2 PLFA analysis

PLFA analysis is a cultivation-independent method that monitors compost microbial communities by analysis of the phospholipids found in their cell membranes (Vestal and White 1989). Total lipids from a compost sample are first extracted using a mixture of organic solvents and buffer and then purified to obtain only phospholipids (Vestal and White 1989; Lei and VanderGheynst 2000). Phospholipids are converted to fatty acid methyl esters and analysed using a gas chromatograph and identification software. PLFA analysis has been applied to a number of composting studies to monitor microbial community succession throughout the process (Herrmann and Shann 1997; Carpenter-Boggs *et al.* 1998) and to determine the effect of process management on compost microorganisms (Eiland *et al.* 2001; VanderGheynst and Lei 2003).

Table 10.1 Summary of observations regarding PLFA during composting and other biological processes.

PLFA or PLFA group	Observations and interpretations
Cyclopropyl PLFAs	Increased with low pH, low O_2, and nutrient depleted environments. Associated with anaerobic communities and stress.[a] Increased during aerobic composting.[e]
Ratio of iso to anteiso (i/a) branched, saturated PLFAs	Anaerobic i/a ratio greater than aerobic i/a ratio.[b] Increased during aerobic composting.[e]
Polyunsaturated PLFAs	Increased with increasing eukaryotic biomass. Marker for fungal, plant and algal biomass.[c] Increased during the later stages of mesophilic composting. Decreased during thermophilic composting.[e]
Methyl branched, saturated PLFAs	Enriched during the later thermophilic stage of composting and at high pH. Marker for actinomycetes.[d,e]

[a] Findlay and Dobbs (1993); Frostegard *et al.* (1997); Bossio and Scow (1998).
[b] White *et al.* (1996).
[c] Vestal and White (1989); Frostegard and Baath (1996); Eiland *et al.* (2001).
[d] Vestal and White (1989); Hellmann *et al.* (1997); Herrmann and Shann (1997); Carpenter-Boggs *et al.* (1998).
[e] VanderGheynst and Lei (2003).

The advantage of PLFA analysis is that it provides valuable quantitative or semi-quantitative information about compost microbial communities in several ways. Total PLFA content in a compost sample is a quantitative measure of total viable biomass, since phospholipids are rapidly degraded after cell death. Changes in total PLFA during composting have been correlated to other measures of viable biomass, including adenosine triphosphate (ATP) measurement (Eiland *et al.* 2001) and biomass reduction from chloroform fumigation (Hellmann *et al.* 1997). Also, there are PLFA biomarkers associated with specific microbial populations, such as eukaryotes and actinomycetes, and certain metabolic stress conditions, such as oxygen and nutrient limitations, that are relevant to composting systems (Table 10.1). PLFA results can provide overall fingerprints of the compost microbial community that can be compared using principal components analysis (PCA), an ordination technique used to reduce a large data set into fewer uncorrelated variables (Pielou 1984). Despite all the information PLFA analyses can provide, interpretation of PLFA results is sometimes limited by redundancy in biomarkers, by the general nature of the biomarkers, and by extraction of phospholipids from sources other than microorganisms, such as insects or plant material.

10.2.3 Nucleic acid-based methods

Ribosomal RNA (rRNA) is the major component of the ribosome, the site of protein synthesis in microorganisms. The sequence variation of rRNA, especially the small subunit of the ribosome, has been used to infer phylogenetic relationships among microorganisms (Muyzer *et al.* 1993). The small subunit for prokaryotic ribosomes is the 16S subunit and that for eukaryotic ribosomes is the 18S subunit. The internal transcribed spacer (ITS) region, which separates the other regions of rRNA, can also be analysed. The ribosomal DNA (rDNA) is the portion of the genome that encodes for rRNA. Most of the nucleic acid techniques described in this chapter target rDNA, although some techniques target rRNA or specific genes of interest in an organism.

10.2.3.1 Polymerase chain reaction

Although multiple nucleic acid techniques have been developed, the majority of the techniques applied to composting systems involve nucleic acid extraction directly from compost samples, and amplification by polymerase chain reaction (PCR). PCR is a method of amplifying a specific DNA sequence. It requires template DNA, a primer pair that spans the DNA region of interest, a heat stable DNA polymerase, and all four deoxyribonucleoside triphosphates (dNTPs) (Stryer 1995). Heating the DNA to near boiling results in strand separation. Then the primers, designed to bind to regions flanking the sequences of interest, hybridize to each single-stranded DNA molecule, which provides a template for primer elongation. Repeating the process doubles the available template on the first step and exponential amplification of the sequence of interest occurs thereafter. In microbial ecology studies, PCR primers typically target conserved regions of rDNA and flank a variable rDNA region, such that different microorganisms can be identified through the variable sequence PCR products generated. Primers that have been applied to the investigation of composting systems are presented in Table 10.2.

As genomic DNA is a stable molecule in cells, it is not a measure of activity or abundance, rather it is an indication of the presence of a particular organism in the system. If information about the activity of the microorganisms is sought, rRNA should be analysed, because ribosome abundance correlates with metabolic activity. Since PCR requires a DNA template, the extracted rRNA population must first be transcribed into complementary DNA (cDNA) by the enzyme reverse transcriptase. This process is called reverse transcriptase-polymerase chain reaction (RT-PCR).

An advantage of PCR is that microorganisms present in very low numbers can be detected after amplification. Most of the investigations in the literature have used PCR to gain qualitative information about the microbial community. One common limitation for all methods that utilize PCR is that compounds co-purified from compost during DNA extraction, such as humic compounds, can interfere with PCR at levels as low as 0.08–0.64 μg/ml (Tebbe and Vahjen 1993). They can interfere by binding to or denaturing the DNA polymerase, binding to or denaturing the DNA molecules, and preventing the binding of the DNA polymerase to the DNA template (Wilson 1997). Given that primers vary in their susceptibility to inhibitors and determine the DNA region to be amplified, careful PCR primer selection is critical for the successful application of PCR to the analysis of microbial

Table 10.2 PCR primers that have been used in microbial community analyses of compost.

Primer pair name	Target microorganisms	Target DNA region	Nucleic acid method	Reference
Com1, Com2	Eubacteria	16S rDNA	SSCP	(Peters *et al.* 2000; Alfreider *et al.* 2002; Trescher and Tebbe 2002)
8F, 1392R	Eubacteria	16S rDNA	T-RFLP	(LaMontagne *et al.* 2002; Michel Jr. *et al.* 2002; Tiquia and Michel Jr. 2002)
341F, 907R	Eubacteria	16S rDNA	DGGE	(Ishii *et al.* 2000)
357F, 517R	Eubacteria	16S rDNA	DGGE	(Aoshima *et al.* 2001; Pedro *et al.* 2001; Ueno *et al.* 2001; Haruta *et al.* 2002)
PA, PH	Eubacteria	16S rDNA	DGGE	(Kowalchuk *et al.* 1999)
27f, 1492r	Eubacteria	16S rDNA	Cloning	(Dees and Ghiorse 2001)
UNI16S-L, UNI16S-R	Eubacteria	16S rDNA	Cloning	(Blanc *et al.* 1999)
CTO198f-GC, CTO654r	Beta-subgroup proteobacterial ammonia oxidizers	16S rDNA	DGGE	(Kowalchuk *et al.* 1999)
F243, R531	Actinomycetes	16S rDNA	SSCP, cloning	(Peters *et al.* 2000; Dees and Ghiorse 2001)
NS7, NS8	Fungi	18S rDNA	SSCP	(Peters *et al.* 2000)
EF4, fung5, NS2-GC	Fungi	18S rDNA	DGGE	(VanderGheynst *et al.* 2002)
S3EUK, UF1	Fungi	18S rDNA	Cloning	(Dees and Ghiorse 2001)
ITS1-F, ITS4	Fungi	ITS rDNA	T-RFLP	(Roberts *et al.* 2002)
23FPL, 927R	Archaea	16S rDNA	Cloning	Dees and Ghiorse 2001)

communities in compost. Other limitations of PCR include the fidelity of the primers and preferential amplification of certain pieces of template DNA which might bias microbial community analyses in steps downstream of PCR.

PCR has been used in conjunction with cloning, denaturing gradient gel electrophoresis (DGGE), single-strand-conformation polymorphism (SSCP), and terminal-restriction fragment length polymorphism (T-RFLP) to study the microbial ecology of compost, as described later in this chapter. All four of these methods are used to distinguish between PCR products of the same size but with variable sequences.

10.2.3.2 Cloning

The objective of molecular cloning is to make the gene or region of DNA of interest accessible by moving it from a large, complicated genome or mixture of DNA to a small, more manageable one, referred to as a cloning vector, e.g. a plasmid or virus (Madigan *et al.* 2003). The first step in the cloning process is the isolation and possibly fragmentation of the source DNA. Source DNA can include genomic DNA, DNA synthesized by reverse transcriptase, or PCR products. If genomic DNA is used, it is typically cut with a combination of restriction enzymes to yield a mixture of fragments (Madigan *et al.* 2003). The DNA fragments are ligated into a cloning vector, cut at complementary restriction sites, by DNA ligase. The cloning vector is then introduced into a host organism, typically *Escherichia coli* and the clones screened for the desired characteristic(s). To more rapidly detect unique clones, clones are screened by sequencing, restriction profile analysis, and DGGE. The final step typically involves the production of large numbers of the desired clone(s) so that the clone(s) can be further studied. Cloning has been used to study microbial community dynamics of the entire composting process (Alfreider *et al.* 2002) and microbial community diversity in thermophilic composts (Blanc *et al.* 1999; Dees and Ghiorse 2001).

Cloning is probably the most reliable method for separating many pieces of DNA. It is also a good method to employ if precise classification of the microorganisms is required because sequence sizes are not restricted (Alfreider *et al.* 2002). However, it is a resource and time intensive method (Alfreider *et al.* 2002). Furthermore, for a complex system such as compost, the number of clones screened and analysed determines the resolution of the data; higher resolution is necessary to gain useful information about the presence and abundance of particular organisms in the composting system.

10.2.3.3 DGGE

DGGE was first applied to microbial ecology by Muyzer *et al.* in 1993. In DGGE, PCR amplified DNA fragments of the same size are separated on a denaturing gel based on the sequence of the PCR products. The electrophoretic mobility of the PCR products decreases as the products melt, or denature; the location in the gel where the PCR product melts is determined by the sequence of the DNA fragment. Upon completion of electrophoresis, different microbial populations in the same sample are represented by different bands on the gel. Bands in a DGGE gel can be excised, sequenced, and compared to a database of reported sequences for microorganisms (Benson *et al.* 1999). DGGE has been used to monitor the bacterial populations in a variety of composts (Kowalchuk *et al.* 1999; Ishii *et al.* 2000; Aoshima *et al.* 2001; Ueno *et al.* 2001).

DGGE is advantageous because it allows the separation of very similar molecules, even those that differ by only a few base pairs, such as would be seen between rRNA genes from different species and subspecies of the same genera. Community members are resolved as separate bands on the DGGE gel that can then be sequenced and compared to rRNA sequence databases for identification. Another benefit of this method is the fact that multiple samples can be visualized on the same gel, thereby providing an excellent means to profile population dynamics, such as the microbial successions that occur during composting.

A study by Kowalchuk *et al.* (1999) provides an instructive example of how these methods have been utilized to provide insight into composting. Kowalchuk *et al.* (1999) used PCR and RT-PCR to understand the presence and activity, respectively, of ammonia-oxidizing bacteria (AOB) in two types of compost. The authors first amplified 16S rDNA with eubacterial specific primers and then used those PCR products as the template DNA for another amplification with AOB specific primers; this technique is called nested PCR. For the RNA samples, reverse transcriptase was first used to create a template DNA strand and then PCR was performed with the AOB specific primers. Finally, the authors used primers to amplify the structural gene for the active portion of ammonia monooxygenase. All PCR products were then analysed using DGGE. In compost production for mushroom cultivation, *Nitrosospira* DNA and RNA were detected early in the process, however only DNA persisted once the compost temperature increased to 75°C; thus, *Nitrosospira* was only active early in the process. This paper took steps toward understanding the presence and activity of AOB that may affect the net nitrogen balance in compost, and thereby the application of compost in agriculture.

Although DGGE has been a useful technique for examining microbial communities in compost samples, there are several limitations of DGGE in addition to the aforementioned limitations of PCR. Consistent gradient pouring is imperative for making comparisons between gel images. Also, sequencing from DGGE can be complicated by the difficulty of band isolation, the fact that one band may represent more than one PCR product, and that it is a laborious process (Ishii *et al.* 2000). PCR primer specificity is an another important issue, as researchers studying the bacterial community in a garbage composting system with DGGE amplified plant DNA using eubacteria specific primers (Aoshima *et al.* 2001). Finally, the data generated by DGGE is qualitative in nature.

10.2.3.4 SSCP

The SSCP method, originally developed for mutation detection in genetic studies, was recently adapted for analysis of complex microbial communities in the environment (Schwieger and Tebbe 1998). After PCR, the double-stranded PCR products are digested so that only single-stranded regions of DNA remain. Under the appropriate conditions, each single-stranded DNA fragment folds and develops a unique secondary structure (conformation) that depends on the specific DNA sequence and that determines the mobility of the conformation in a gel. These conformations are then separated on a gel in order to develop SSCP profiles representing the targeted compost microbial community. The SSCP method has been used to track changes in populations of bacteria, actinomycetes, and fungi throughout the composting process, with the goals of assessing microbial diversity and identifying important microorganisms (Peters *et al.* 2000; Alfreider *et al.* 2002).

Like DGGE, an advantage of the SSCP method is the ability to excise bands from the gel, obtain the DNA sequence for the excised band, and then attempt to assign the DNA to a particular microorganism through comparison with known sequences archived in a database. An advantage of SSCP over DGGE is that SSCP requires a simple gel, rather than the denaturing gel in DGGE, and that it requires shorter primers than DGGE, which may help to prevent unintended products of amplification. As with all PCR-based methods, PCR primer specificity can be an issue in a mixed microbial community such as compost, where primers targeting bacteria and other primers targeting fungi have been shown to amplify DNA from yeast and maize, respectively (Peters *et al.* 2000).

10.2.3.5 T-RFLP

In T-RFLP, PCR products are digested with a restriction enzyme, which recognizes a specific DNA sequence of four base pairs and cuts the DNA at every recognition site to produce restriction fragments (RFs). The RFs are separated by their size on a gel, generating a pattern that is a function of the DNA sequence of the PCR products. For a complex microbial community such as compost, size separation of all the RFs to yield meaningful results would be very complicated. To simplify analysis of RFs for mixed microbial cultures, one of the primers used in PCR is labelled with a fluorescent dye so that only the terminal-RFs (T-RFs) are detected and an automated sequencer with high resolution can be used to separate the T-RFs (Liu *et al.* 1997). The sequencer results are presented as a chromatogram, identifying the size of each T-RF and the intensity of its fluorescent signal. The T-RFLP method has mainly been applied to investigate bacterial diversity during composting (Michel Jr. *et al.* 2002; Tiquia and Michel Jr. 2002).

As with other nucleic acid-based methods, limitations of T-RFLP stem from nucleic acid extraction and PCR amplification. For instance, inhibitory compounds co-extracted with DNA from compost samples have been shown to decrease the microbial diversity seen in T-RFLP profiles (LaMontagne *et al.* 2002). Also, biases during PCR amplification can complicate the interpretation of T-RFLP results, especially since T-RFs cannot be excised from the gel and sequenced, as with DGGE and SSCP, to confirm their association with particular microorganisms. Typically, three separate digestions of PCR products are performed using three different restriction enzymes. Then, using the sequence database and available analysis programs (Marsh *et al.* 2000), the sizes of all three observed T-RFs are matched with the expected sizes for sequences in the database, allowing some observed T-RFs to be tentatively linked to one or more species. The main advantage of T-RFLP is the use of an automated sequencer, which increases the speed and resolution of the method. However, the higher resolution of the sequencer reveals more variability among replicates (Dunbar *et al.* 2001) and the sequencer detects larger T-RFs less accurately than shorter T-RFs (Liu *et al.* 1997).

10.3 MICROBIAL COMMUNITY DYNAMICS DURING COMPOSTING

Using a variety of the microbiological methods discussed above, a number of investigations have studied the succession of microbial communities in composting. Since many have presented their findings with respect to changes in process temperature, this part of the chapter is organized into three sections corresponding to three commonly termed phases of composting: the self-heating phase, the thermophilic phase, and the cooling and maturation phase.

10.3.1 Self-heating phase

The composting process typically begins in the temperature range of 25–35°C when relatively high levels of readily assimilable carbon, nitrogen, and oxygen are available and the moisture content is favourable for microbial growth. Under these conditions, bacteria tend to proliferate more quickly than other members of the microbial community (Finstein and Morris 1975). The metabolic heat production from this rapid change in microbial activity eventually exceeds latent and sensible cooling and the temperature increases (Figure 10.2). Once the compost temperature reaches 40–45°C, conditions become very unfavourable for the growth of mesophilic organisms and a transition in microbial activity between mesophilic and thermophilic bacteria occurs.

This transition has been observed in respiration rate profiles from the composting process (Figure 10.2). In bench-scale studies by Nakasaki *et al.* (1985), CO_2 evolution rate increased as temperature increased. Once temperature reached about 45°C there was a drop in CO_2 evolution rate and then an increase suggesting a transition in microbial activity from mesophilic to thermophilic organisms. This trend was also observed in their microbial enrichment plating analyses. They found relatively high levels (10^6–10^9 cells/g dry weight) of mesophilic bacteria in the early stages of biosolids composting before temperatures exceeded 45°C. Thermophilic bacteria and actinomycetes also increased during the initial phases of composting but, in general, did not dominate until thermophilic temperatures were reached. Godden *et al.* (1983), in a manure composting study, also noted that thermophilic actinomycetes increased gradually during the transition phase. Furthermore, this study showed that although thermophilic fungi (isolated at 55°C) were present in their system, they were detected at levels 3–4 orders of magnitude lower than fungi isolated at 30°C and 45°C (Godden *et al.* 1983).

These general observations from enrichment studies have also been documented with PLFA studies. As composting transitions from the mesophilic to thermophilic phase, microbial activity, as measured by CO_2 production, viable microbial biomass, as measured by total PLFA, and temperature reach their maxima at approximately the same time (Hermann and Shann 1997; Bolta *et al.* 2003). Branched, monounsaturated PLFAs have been shown to decrease during this transition phase, suggesting their potential as biomarkers for

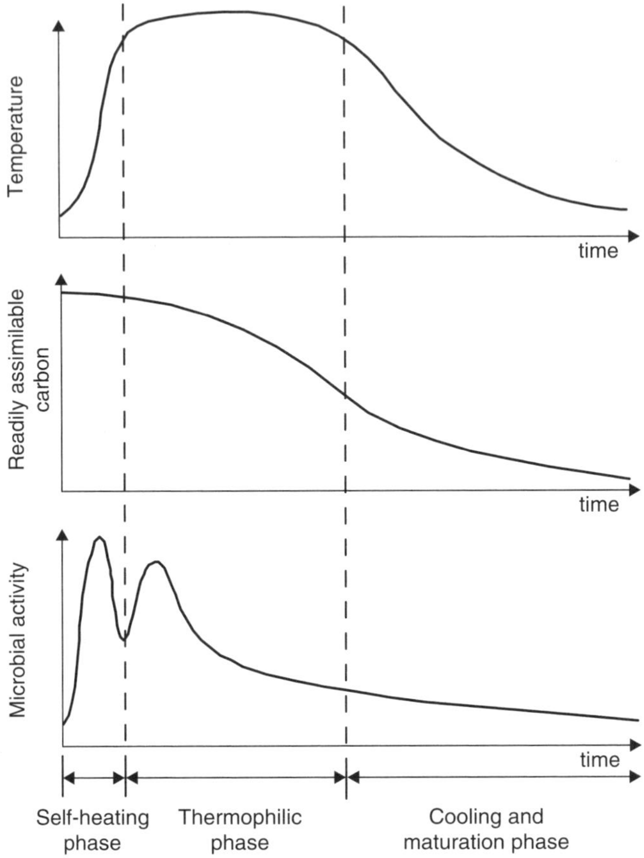

Figure 10.2 General overview of typical temperature, readily assimilable carbon and microbial activity trends during the self-heating, thermophilic, and cooling and maturation phases of composting.

mesophilic bacteria (VanderGheynst *et al.* 2002; VanderGheynst and Lei 2003). During the initial phase of composting, polyunsaturated PLFAs, indicative of eukaryotic microorganisms such as fungi or algae (Table 10.1), have been found at levels up to 14 mol% (Herrmann and Shann 1997; Bolta *et al.* 2003). When temperature increases beyond the typical range for eukaryotic microorganisms, polyunsaturated PLFAs often decrease to levels less than 5 mol% (Herrmann and Shann 1997; Bolta *et al.* 2003).

Microbial activity leads to self-heating and eventually temperatures exceed the tolerance of mesophilic organisms. When temperatures exceeded 45°C, a number of investigators observed that populations of most mesophilic organisms decreased. Nakasaki *et al.* (1985) found that mesophilic bacteria increased rapidly during the initial stages of composting and then declined as the temperature reached 60°C. In an investigation of vegetable, fruit and garden waste compost with enrichment plating techniques, Ryckeboer *et al.* (2003) found that actinomycetes, fungi and yeast decreased. Hassen *et al.* (2001) observed a slow drop in mesophilic bacteria and a rapid decline in yeast and fungi during self-heating. Similarly, Tiquia (2002) observed a rapid decline in numbers of isolated bacteria, actinomycetes and fungi during composting of animal manures. Numbers dropped by a factor of $10–10^3$ as temperature increased from 40°C to 70°C in poultry manure and pig manure composting.

There are exceptions to the general trend of a transition between mesophilic and thermophilic organisms during the self-heating phase that appear to be related to feedstock differences. Nakasaki *et al.* (1985) observed that levels of mesophilic actinomycetes and fungi and thermophilic fungi were low throughout the process including the initial phase when temperatures were low. They attributed this to unfavourable composting conditions for these organisms, such as the lack of a bulking agent. In contrast to the findings from Nakasaki *et al.* (1985), a study investigating the microbial community during manure composting observed relatively high densities of mesophilic and thermophilic bacteria, and mesophilic actinomycetes and fungi early during the process (Godden *et al.* 1983). During the transition from mesophilic to thermophilic temperatures (23–45°C), levels of mesophilic bacteria and mesophilic and thermophilic fungi remained fairly constant while thermophilic bacteria and mesophilic actinomycetes decreased. A likely explanation for this is that the feedstock was initially hot and thermophilic organisms were active before the composting process began. Finally, under some composting conditions fungal PLFA biomarkers have been observed to increase from 5 mol% to nearly 20 mol% as temperatures increased from 25°C to 47°C (VanderGheynst *et al.* 2002). This was likely due to the specific conditions of this composting system that favoured fungi, including low peak temperature and low pH.

During the self-heating phase, the rapid growth of bacteria results in the depletion of soluble carbohydrates and the consumption of oxygen (Fermor *et al.* 1979). A transition in substrate utilization has been observed by the detection of fermentative bacteria, which utilize the remains of easily degradable substrates, and cellulolytic or proteolytic bacteria, which utilize more complex substrates (Figure 10.3 and Table 10.3). In most cases, the presence of either group of bacteria indicates low oxygen conditions.

During the composting of garbage, Ishii *et al.* (2000) detected a number of low GC, non-spore forming gram-positive fermentative bacteria that thrive in the presence of sugars and organic acids using PCR-DGGE. These bacteria included *Staphylococcus* sp. and other relatives of *Lactobacillus*. Fermentative bacteria were also detected in a domestic organic waste composting system between the temperatures of 25–50°C (Alfreider *et al.* 2002). The results for cloning and SSCP of the initial sample showed the presence of enteric bacteria and lactic acid bacteria. *Lactobacillus* sp. was also detected during the self-heating phase of a shredded corn plants, wood chips, and horse manure compost using SSCP (Peters *et al.* 2000). Lactic acid bacteria have been commonly associated with ensilage

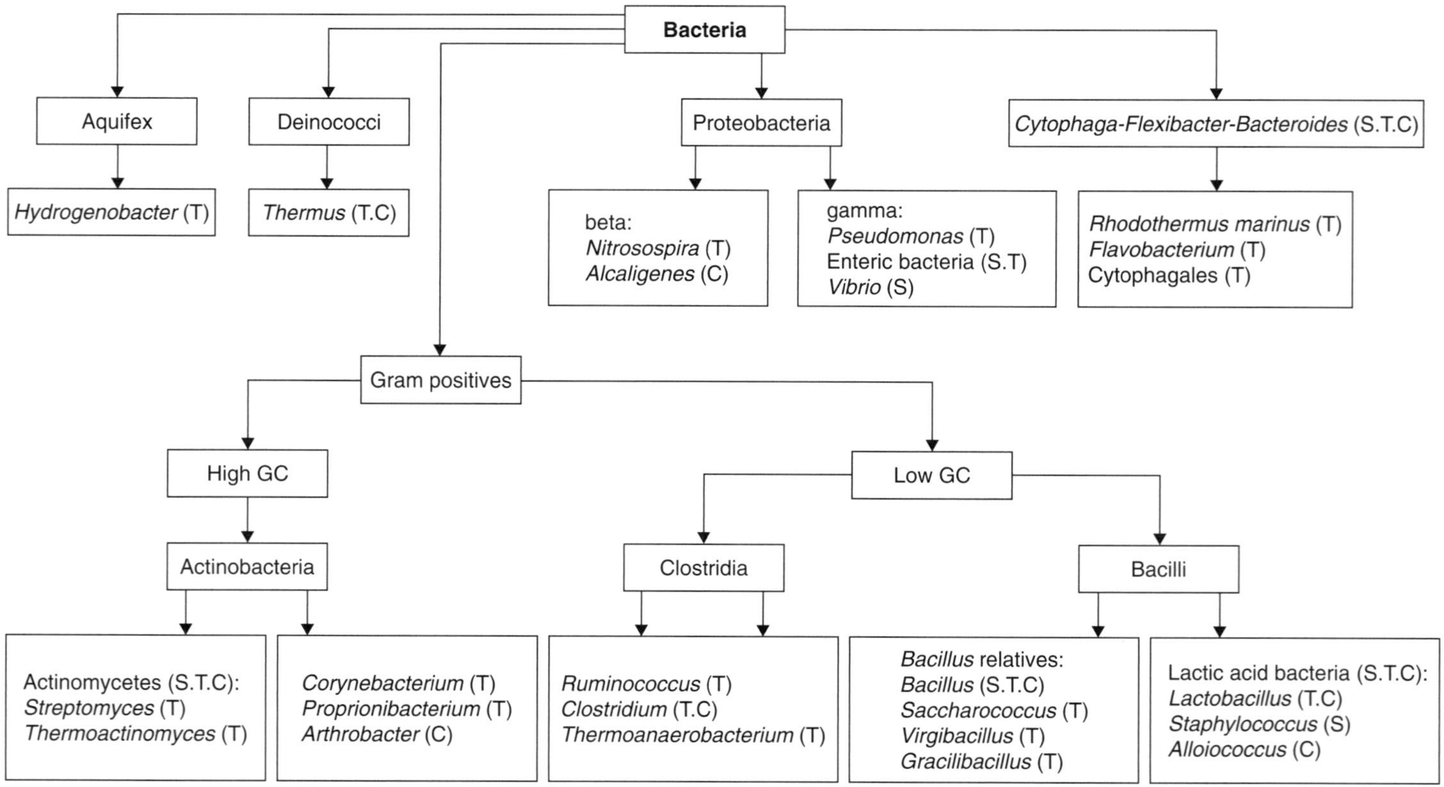

Figure 10.3 Bacterial groups found during composting organized according to Madigan *et al*. (2003). Letters following each group indicate detection during certain phases of composting where "S" indicates detection during the self-heating phase, "T" indicates detection during the thermophilic phase, and "C" indicates detection during the cooling and maturation phase.

Table 10.3 General characteristics of bacterial groups or genera found during composting, according to Madigan *et al.* (2003).

Group or genus name	Oxygen requirement	Spores	Gram status	Significant compounds utilized by organisms
Hydrogenobacter	Aerobe	No	−	Hydrogen, sulphur
Thermus	Aerobe	No	−	Sugars, organic acids
Actinomycetes	Aerobe, facultative aerobe, or anaerobe	Yes/No	+	Sugars, starch, cellulose, hemicellulose, lignin, organic acids, alcohols
Corynebacterium, Arthrobacter	Aerobe	No	+	Sugars, starch
Propionibacterium	Anaerobe	No	+	Sugars, organic acids, alcohols
Clostridia	Anaerobe	Yes	+	Sugars, starch, cellulose, organic acids, alcohols
Bacillus relatives	Aerobe, facultative aerobe	Yes	+	Starch, cellulose, hemicellulose
Lactic acid bacteria	Aerotolerant anaerobe	No	+	Sugars
Nitrosospira	Aerobe	No	−	Ammonia
Alcaligenes	Aerobe	No	−	Hydrogen
Gamma-Proteobacteria	Aerobe, facultative aerobe	No	−	Sugars
Cytophaga-Flexibacter-Bacteriodes (CFB)	Aerobe, anaerobe	No	−	Sugars, starch, cellulose, hemicellulose, chitin

processes, which are also characterized by moderate temperatures, wet conditions, and fresh feedstocks. In addition to these findings, Michel Jr. *et al.* (2002), using the T-RFLP method, detected a number of fermentative Proteobacteria, including *Vibrio* spp. and the enteric bacteria *E. coli.*

Although cellulolytic bacteria have not been observed as often as fermentative bacteria during the self-heating phase, their presence still has been noted. Ishii *et al.* (2000) observed two *Bacillus* species, which are known to be cellulolytic and proteolytic. Similarly, Alfreider *et al.* (2002) observed the presence of a microorganism from the Cytophaga-Flexibacter-Bacteriodes (CFB) group, which is known to contain microorganisms that can degrade cellulose and hemicellulose (Table 10.3). The presence of cellulolytic and proteolytic organisms indicates a shift from the consumption of easily degradable substrates, such as sugars, to the consumption of complex substrates, such as cellulose and proteins.

In summary, during the self-heating phase general trends of decreasing fungal and mesophilic bacterial populations and increasing actinomycete and thermophilic bacterial populations have been found with enrichment plating and PLFA methods. Nucleic acid-based methods have shown the presence of a variety of fermentative bacteria that would consume easily degradable substrates and contribute to the self-heating of the compost. The thermophilic phase of composting continues the trends of heating and decreasing readily assimilable substrate and oxygen. The next section of the chapter describes the microbial community resulting from these changes.

10.3.2 Thermophilic phase

From a process control perspective the thermophilic phase of composting is arguably the most important phase of composting. This is the phase were odours are most commonly emitted from the process (Derikx *et al.* 1991; VanderGheynst *et al.* 1998), potential for ammonia loss is highest (Miller 1993) and human and plant pathogens are inactivated (Bollen 1993; Farrell 1993). Hence, much of the microbial ecology research has emphasized this phase of the process.

The thermophilic phase of composting is characterized by high temperatures and moderate levels of oxygen, moisture and readily available organic substrates depending on process management (Figure 10.2). During this phase there is continued microbial activity, resulting in metabolic heating and substrate and oxygen depletion. In many cases, the process is cooled during the thermophilic phase to maintain temperature between 50°C and 65°C. Methyl branched, saturated PLFAs, generally associated with actinomycetes (Table 10.1), have been observed to increase from negligible levels in initial composting materials to 2–4 mol% during the thermophilic phase (Herrmann and Shann 1997; VanderGheynst and Lei 2003). Also, terminally branched, saturated PLFAs, associated with gram-positive bacteria and enriched in *Thermus* sp. (Wilkinson 1988), have been shown to increase from initial levels of 10 mol% to 30–40 mol% during thermophilic composting (Hellmann *et al.* 1997; Herrmann and Shann 1997; Bolta *et al.* 2003).

A number of reports in the literature have found *Bacillus* spp. and several other gram-positive bacteria to dominate composts maintained in this temperature range. Strom (1985) isolated *Bacillus* spp., two types of non-spore forming bacteria, and two types of actinomycetes, *Thermoactinomyces* sp. and *Streptomyces* sp., during composting of municipal solid waste maintained between 49°C and 65°C. In this study 87% of the organisms isolated were from the genus *Bacillus*. As bench-scale incubation temperature exceeded 61°C, 100% of the isolates were *Bacillus* spp. The predominance of *Bacillus* in enrichment isolates was also observed for mushroom cultivation compost (Fermor *et al.* 1979), vegetable and egg-shell waste compost maintained between 50°C and 60°C (Pedro *et al.* 1999), pig manure and poultry litter compost at 50–70°C (Tiquia 2002) and vegetable, fruit and garden waste compost between 55°C and 75°C (Ryckeboer *et al.* 2003).

Thermophilic compost samples analysed using nucleic acid-based methods have also identified organisms from the Bacilli and Actinobacteria groups (Ishii *et al.* 2000; Peters *et al.* 2000; Dees and Ghiorse 2001). At the peak temperature during the composting of garbage, the strongest band in a DGGE profile presented by Ishii *et al.* (2000) was related to *Bacillus* sp. Other bands were related to *Virgibacillus*, *Gracilibacillus*, and *Corynebacterium* (Figure 10.3). In samples collected from thermophilic compost consisting of shredded corn plants, wood chips, and horse manure, 5 of the 12 bands sequenced from the SSCP gel were most closely related to five different *Bacillus* species. (Peters *et al.* 2000).

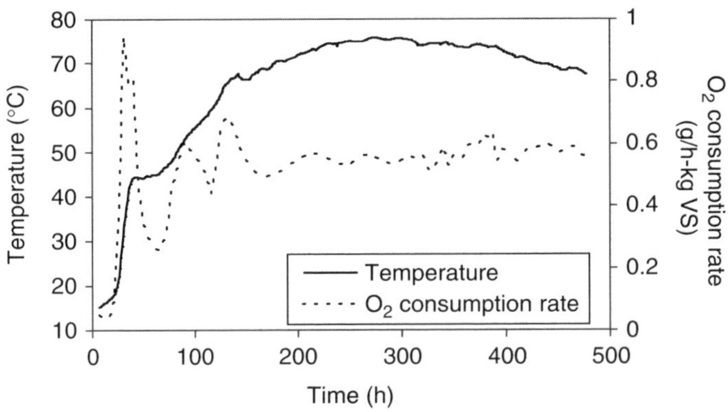

Figure 10.4 Temperature and respiration profiles from aerated static pile composting of synthetic food waste (VanderGheynst *et al.* 1997). Temperature and respiration were measured 75 cm from the air entrance to the pile.

Actinomycetes identified in the same study included two *Streptomyces* species, which were also found by Strom (1985).

Dees and Ghiorse (2001) used a cloning technique to analyse the microbial community in synthetic food waste compost at 60°C and 74°C and dairy manure compost at 50°C. In general, their results were consistent with previous studies since gram-positive bacteria, including *Bacillus* spp. and Actinobacteria, were shown to dominate the hot compost. They detected a number of *Bacillus* spp. and Actinobacteria that were different from species previously isolated from thermophilic compost with cultivation-dependent methods and therefore, revealed greater diversity. This diversity led the authors to believe that the importance of Actinobacteria in previous studies may have been obscured by the abundance of *Bacillus* spp.

In many cases when compost heating is not controlled, temperatures can reach as high as 75°C, yet biological activity can still be detected (Figure 10.4). There have been a number of studies investigating the microbial community under these extreme conditions which are described in the following section.

10.3.2.1 Microbial communities identified at extreme composting temperatures

Using enrichment plating, Beffa *et al.* (1996b) isolated significant numbers of heterotrophic, gram-negative, non-spore forming aerobic thermophilic bacteria closely related to *Thermus thermophilus* during windrow composting of green waste and food waste. These non-spore forming bacteria gradually increased from 10^2–10^4 cells/g dry weight to 10^8–10^9 cells/g dry weight as temperature increased from approximately 25–40°C to 67–78°C. Numbers of these organisms decreased to 10^3–10^4 cells/g dry weight as temperature decreased below 35°C. These investigators also observed levels of spore forming bacteria (Table 10.3) to increase as temperature increased to 65°C. However, as temperature increased from 65°C to 78°C, levels of spore forming bacteria decreased. Furthermore, they found that although spore forming bacteria were present at high temperatures their growth limit was 71°C. These observations suggested that microbial activity at very high temperatures was due to non-spore forming bacteria and these bacteria were responsible for heating the compost above 70°C.

In a similar study, Blanc *et al.* (1999) used a cloning technique to study thermophilic microbial communities in compost. The authors collected compost samples from the hottest spots of two young and two old compost windrows consisting of grass, garden and kitchen waste, and shredded wood. They noted organisms including *Bacillus* sp., *Saccharococcus* sp. and *Rhodothermus marinus* (Figure 10.3) in both young and old compost piles and *T. thermophilus* in young composts only. The authors concluded that *T. thermophilus* and, to a lesser extent, thermophilic *Bacillus* spp. were the dominant organisms in thermophilic composting.

Several other investigations have been carried out to identify thermophilic bacteria present in compost samples ranging from 60°C to 80°C. Beffa *et al.* (1996a) isolated organisms commonly found in geothermal environments that had not been previously identified in composting. One was an obligately autotrophic, non-spore forming, gram-negative, sulphur and hydrogen-oxidizing thermophilic bacterium and was isolated at levels of 10^4–10^6 cells/g dry weight from thermophilic composts. This organism only grew under microaerophilic conditions and was closely related to *Hydrogenobacter*. Also isolated was a facultatively autotrophic gram variable, spore forming thermophilic bacterium at levels of 10^5–10^8 cells/g dry weight. This organism could grow heterotrophically or autotrophically with hydrogen and was most closely related to *Bacillus schlegelii*. Lyon *et al.* (2000) also isolated two xylanolytic *Thermus* strains from a food and green waste windrow composting process at 75°C. These observations are significant because they demonstrate that organisms capable of utilizing complex carbon sources and fermentation products are present at extreme conditions. Certainly, high levels of microbial activity during the thermophilic phase lead to the depletion of readily assimilable carbon so the presence of organisms capable of utilizing lignocellulosic compounds is expected. The presence of such organisms along with organisms that can tolerate low levels of oxygen is discussed in the following section.

10.3.2.2 Impacts of oxygen and nutrient limitations on the microbial community during the thermophilic phase

The high levels of microbial activity during the thermophilic phase often lead to oxygen limitations. Although measured oxygen levels in the pore space might be high enough to support aerobic microbial activity, it is likely microenvironments exist in the compost that can only support anaerobic communities. For instance, as illustrated in Figure 10.5, methane, a product of anaerobic metabolism, was detected during thermophilic composting even when oxygen levels were greater than 10 mole%.

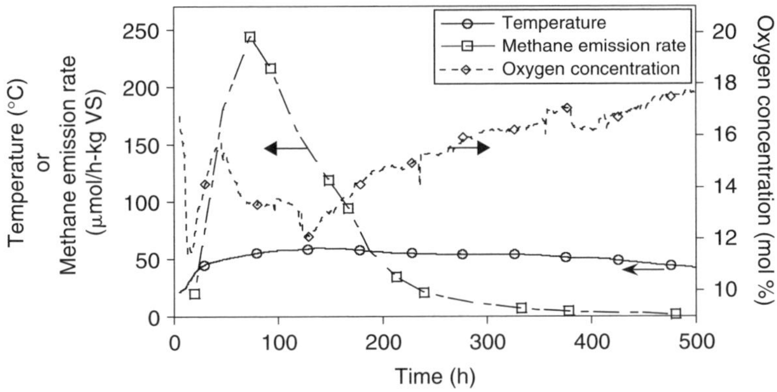

Figure 10.5 Temperature, oxygen concentration and methane emission profiles from an aerated static pile for composting of digested biosolids (VanderGheynst *et al.* 1998). Temperature, oxygen concentration and methane emission were measured 180 cm from the air entrance to the pile.

Evidence for oxygen limitations during composting is also provided by the detection of anaerobic and facultative microorganisms (Figure 10.3 and Table 10.3). During investigations of the microbial community during pig manure composting and poultry litter composting, Tiquia *et al.* (2002) found that levels of isolated denitrifiers were high during the thermophilic phase of composting. Davis *et al.* (1992) found that over 90% of the bacteria isolates from thermophilic pine bark composting could grow under anaerobic conditions. Furthermore, upon analysis of their bark samples using scanning electron microscopy and transmission electron microscopy, Davis and co-workers found many microorganisms were lodged under the surface of the bark and growing within the bark cells, which were likely oxygen and nutrient deprived environments.

As anaerobic organisms are particularly difficult to isolate and culture, nucleic acid-based methods have provided useful tools for detecting such organisms in composting. For instance, *Propionibacterium* sp. and *Clostridium* spp. were detected by several investigators using DGGE and SSCP (Peters *et al.* 2000; Pedro *et al.* 2001). In another study utilizing DGGE, Ueno *et al.* (2001) examined an enriched anaerobic hydrogen-producing microbiota from sludge composts at 60°C. The major organisms detected included *Thermoanaerobacterium* species and *Clostridium* species, which are obligately anaerobic bacteria. Another organism detected was affiliated with *Ruminococcus albus*, which can excrete cellulases that actively degrade cellulose outside of the bacteria (Madigan *et al.* 2003). These findings suggest that anaerobic organisms such as *R. albus* may have an important role in breaking down cellulosic materials in the composting process. *Bacillus* sp. was also detected in this study, although it was not responsible for a major band in the DGGE profiles. This finding is particularly interesting considering *Bacillus* spp. have been found to dominate the thermophilic microbial community in composts managed aerobically.

During windrow composting of yard trimmings, T-RFLP analysis of compost samples indicated there was a distinct shift in the bacterial community from

Proteobacteria, which dominated the initial compost sample, to Proteobacteria and gram-positive bacteria as temperatures exceeded 55°C (Michel Jr. *et al.* 2002). T-RFs associated with Proteobacteria, including enteric bacteria and *Pseudomonas* spp., were also present in thermophilic samples. These are not microorganisms typically associated with high temperatures, but presence of DNA from these microorganisms is not an indication of activity or viability. Nevertheless, *Pseudomonas* spp. have also been detected in hot composts using SSCP and enrichment plating techniques (Fermor *et al.* 1979; Peters *et al.* 2000). During the early thermophilic phase, many matched T-RFs were identified as gram-positive bacteria, possibly related to lactic acid bacteria. As composting proceeded T-RFs related to the genera *Flavobacterium* and *Chryseobacterium* were identified. This finding is significant in that microorganisms from the CFB group have the ability to utilize cellulose, hemicellulose, and chitin (Table 10.3). Similar observations of microorganisms from the CFB group were made during thermophilic composting of shredded corn plants, wood chips, and horse manure using SSCP (Alfreider *et al.* 2002). These authors also identified cellulolytic bacteria such as *Ruminococcus* sp.

In addition to the detection of facultative and anaerobic microorganisms during thermophilic composting, further evidence of oxygen limitations is provided by PLFA biomarkers associated with metabolic stress. Cyclopropyl PLFAs, which have been associated with oxygen limitations (Table 10.1), have been seen at high levels, such as 8 mol%, during thermophilic composting compared to less than 1 mol% during mesophilic composting (Herrmann and Shann 1997). A higher iso/anteiso ratio of branched, saturated PLFAs is another potential biomarker for anaerobic activity (Table 10.1). This ratio has been shown to double as composting transitions from the mesophilic to the thermophilic phase in forced aeration systems (VanderGheynst and Lei 2003), although this ratio may also be affected by increases in temperature (Petersen and Klug 1994). The iso/anteiso ratio also generally increased over the course of 100 days of composting straw and pig slurry; however, the change in the ratio varied depending on the number of carbons in the fatty acid chain (Klamer and Baath 1998). Furthermore, anaerobic conditions during thermophilic composting have been indicated by simultaneous increases in methane emissions and in ether-linked fatty acids, characteristic of anaerobic methanogens (Hellmann *et al.* 1997).

In summary, the thermophilic phase is dominated by a bacterial population that possesses a wide variety of functionalities. Although there are reports of fairly high densities of bacteria at very high temperatures (>70°C), the diversity of the community at these temperatures is likely to be low. With enrichment studies, Strom (1985) found that the species diversity index, related to community stability, was the same for systems incubated between 49°C and 61°C, but dropped dramatically when incubation temperature exceeded 61°C. Furthermore, Ishii *et al.* (2000) found that the number of bands in DGGE decreased as the temperature increased during composting. In contrast to these findings, a study investigating the diversity of actinomycetes using the SSCP method indicated that actinomycete diversity increased as the compost pile reached peak temperatures and maintained thermophilic temperatures near 65–70°C (Peters *et al.* 2000). Unfortunately, interpreting diversity from SSCP and DGGE profiles is problematic due to PCR biases.

While eukaryotes, such as *Aspergillus fumigatus* (Strom 1985) and *Candida* sp. (Peters *et al.* 2000), have been detected during the thermophilic phase of composting, eukaryotes and associated PLFAs and nucleic acids generally have not been found during thermophilic composting. Eukaryotes, however, do recolonize the compost as it cools. This is discussed in the next section.

10.3.3 Cooling and maturation phase

The cooling and maturation phase is characterized by the depletion of nutrients, including readily available carbon and nitrogen sources and moisture. This results in decreased microbial activity and decreased heat production (Figure 10.2). An outcome of these changes in compost physical properties is a very diverse and distinct microbial community.

During the cooling and maturation phase, the reduced microbial activity, resulting in higher oxygen concentrations and lower temperatures, tends to favour aerobic, mesophilic organisms, such as fungi. For instance, enrichment studies have shown an increase in isolated yeast and fungi as compost temperature decreases below 50°C (Tiquia 2000; Hassen *et al.* 2001). In another study using enrichment techniques, the fungal community was compared for matured compost made thermophilically and vermicompost made mesophilically (Anastasi *et al.* 2002). The fungal population was larger in compost compared to vermicompost, however more species were identified in vermicompost than in compost. The genera isolated from both composts included *Acremonium*, *Aspergillus*, *Cladosporium*, *Malbranchae*, *Penicillium* and *Thermomyces*. Recolonization of cooling compost by eukaryotes has also been indicated by an increase in polyunsaturated PLFAs up to 15 mol% (Herrmann and Shann 1997).

However, an increase in fungal populations during the maturation phase is not always observed. Some researchers have reported no increase in polyunsaturated PLFAs, even after the temperature decreased to less than 35°C (Hellmann *et al.* 1997; Bolta *et al.* 2003; VanderGheynst and Lei 2003). This may be due to other factors such as oxygen levels, available substrate levels, pH, or microbial competition. There is some evidence that composts which remain in the thermophilic phase for a longer duration see less recolonization by fungi during the maturation phase (Klamer and Baath 1998; Eiland *et al.* 2001). In one study using DGGE to target eukaryotic rDNA during composting, eukaryotes were only detected early in the process and not during the cooling phase (Ishii *et al.* 2000). The researchers hypothesized that fungal recolonization of the compost was not observed because the compost was too wet, the C/N ratio was too low for fungi, and growth rates of fungi were too slow to compete in a laboratory-scale composting system.

In addition to lower temperatures and likely fungal recolonization, the cooling and maturation phase is also characterized by nutrient limitations, which favour microorganisms capable of hydrolysing cellulose and hemicellulose and tolerating low moisture levels. If the moisture lost from latent cooling is not replenished, the compost moisture content decreases, favouring desiccant-tolerant organisms and the formation of microbial resting structures (Finstein and Morris 1975). For instance, DGGE analysis of compost during the cooling phase detected *Arthrobacter* sp. (Ishii *et al.* 2000), a member of the Actinobacteria group which is commonly found in the soil and is resistant to desiccation and starvation (Madigan *et al.* 2003). Microorganisms capable of hydrolysing cellulose and hemicellulose also thrive during this phase (Figure 10.3 and Table 10.3). For example, enrichment studies have shown an increase in isolated actinomycetes during the cooling and maturation phase (Tiquia 2000; Hassen *et al.* 2001). In addition, levels of actinomycete biomarkers have been shown to remain at thermophilic phase levels or increase further (Hellmann *et al.* 1997; Herrmann and Shann 1997; VanderGheynst and Lei 2003). Microorganisms from the CFB group were also identified with cloning analyses during the maturation phase of domestic waste compost (Alfreider *et al.* 2002).

Even though microbial activity decreases during the cooling and maturation phase, oxygen-limited conditions can exist in regions of the compost. For instance, although the levels of similarity were relatively low (83.4–91.4%), sequences obtained through DGGE analyses were related to fermentative bacteria, including *Clostridium* sp. and *Alloiococcus* sp., a lactic acid bacteria (Ishii *et al.* 2000). After five months of storage, DGGE analyses of matured compost were unable to detect any sequences from aerobic AOB, such as *Nitrosospira*, in samples from the bottom of a pig manure compost pile, likely because of anoxic conditions (Kowalchuk *et al.* 1999). Indicators of metabolic stress, cyclopropyl PLFAs, have also been shown to further increase during the curing phase, even to greater than 20 mol% for one municipal solid waste composting system (Herrmann and Shann 1997). As mentioned previously, these PLFAs can be indicators of anaerobic conditions; however, they can also be associated with the stationary growth phase of microorganisms.

As the thermophilic phase ends, total viable microbial biomass, as measured by total PLFA, decreases as the compost pile cools (Hellmann *et al.* 1997; VanderGheynst *et al.* 2002; Bolta *et al.* 2003). Nevertheless, many studies have found that microbial diversity is extremely high during the cooling phase. The fungal community has been shown to become more diverse during the cooling stage of composting (Fermor *et al.* 1979). T-RFLP analysis of yard trimmings after 136 days of windrow composting also suggested increased bacterial diversity (Michel Jr. *et al.* 2002). Diversity is difficult to assess through PCR methods, but when comparing the number of T-RFs for day 136 compost to the initial compost sample, the number had approximately doubled for each of three restriction enzymes used in the analysis.

Recolonization of the maturing compost by mesophilic microorganisms, the presence of unique microorganisms, and the persistence of microorganisms from the thermophilic phase can contribute to this microbial diversity. When temperatures drop below 50°C, rapid recolonization of food waste compost by yeast, fungi, mesophilic and thermophilic actinomycetes, and cellulolytic bacteria has been observed, with the number of species identified through enrichment methods increasing as the compost cooled (Ryckeboer *et al.* 2003). Also, a hydrogen-oxidizing bacteria from the genus *Alcaligenes* was detected in a garbage composting system using DGGE (Ishii *et al.* 2000) and a dominant microorganism in the thermophilic phase, *Thermus* sp., was identified in the maturation phase of domestic waste compost using cloning methods (Alfreider *et al.* 2002). In addition, the cooling and maturation phase has been marked by the persistence of iso and anteiso branched, saturated PLFAs, indicating the continued importance of gram-positive microorganisms (Hellmann *et al.* 1997; Herrmann and Shann 1997; VanderGheynst and Lei 2003). While many of these microorganisms have also been detected earlier in the composting process, DGGE profiles of the bacterial community during the cooling and maturation phase have been shown to be distinct from the patterns during the initial mesophilic phase (Ishii *et al.* 2000).

In summary, the microbial community has been found to transition along with temperature, substrate complexity and oxygen level during composting. Differences in these general trends between composting systems can be attributed to differences in initial processing conditions such as feedstocks and perhaps the initial inoculum. Other management steps, such as mixing or aeration, can account for the differences in general trends of composting, and certainly affect the microbial community of compost.

10.4 INFLUENCE OF PROCESS MANAGEMENT ON MICROBIAL COMMUNITIES

Many management actions can be made prior to or during the composting process. The actions include, but are not limited to, feedstock composition and structure (e.g. C/N ratio, particle size reduction and bulking), inoculation, aeration, and mixing. In the following sections some of the observations of microbial communities from investigations on compost process management are summarized.

10.4.1 Feedstocks

C/N ratio of compost feedstocks is an important management consideration in composting. A general guideline suggests an initial C/N ratio of between 20 and 40, however this varies with the application (Rynk 1992). An investigation of initial C/N ratios ranging from 11 to 54 was performed using straw and pig slurry as feedstocks (Eiland *et al.* 2001). Composts with lower initial C/N ratios decomposed faster, as indicated by hemicellulose and cellulose degradation, and released excess N, as indicated by nitrate formation. The compost with the lowest initial C/N ratio was enriched in terminally branched, saturated PLFAs, or biomarkers for gram-positive microorganisms, in comparison to the other composts which were enriched in polyunsaturated PLFAs. This likely resulted because of the more active thermophilic phase of the low C/N ratio compost. Interestingly, hemicellulose degradation, which is often associated with fungi, was more rapid in the low C/N ratio compost, which had lower levels of eukaryotic biomarkers than the higher C/N ratio composts. Initial C/N ratio strongly influenced the resultant microbial community, which remained distinct in the various composts even after 12 months of composting, as shown by PCA of PLFA results.

C/N ratio has also been found to be important in increasing viable microbial biomass during composting (Cahyani *et al.* 2002). Windrow composting of rice straw was started with an initial C/N of approximately 60. Within a few days of composting, a peak compost temperature above 60°C was reached. However, total PLFA content of the compost did not increase significantly during this time. After 2 weeks of composting, ammonium sulphate was added to the windrow at the rate of 10 kg/Mg dried rice straw. After nitrogen addition, total PLFA content nearly doubled and continued to increase over the next 50 days of composting.

Davis *et al.* (1991) investigated the effect of nitrogen addition on the microbial community during pine bark composting. They found that urea addition significantly increased the rate of self-heating and increased the numbers of mesophilic and thermophilic bacteria during composting, as measured through enrichment studies. No difference was observed when bark was amended with 4 kg urea/m^3 compared to 8 kg urea/m^3. They also investigated the effect of adding different nitrogen sources. They found that temperatures were higher in bark amended with urea compared to nitrate and also found that numbers of mesophilic and thermophilic bacteria were highest in bark amended with urea followed by KNO_3 and $Ca(NO_3)_2$. They indicated that the piles treated with nitrates were more acidic than urea-treated piles and suggested that the acidic environment favoured fungal growth over the growth of bacteria and actinomycetes.

Since microorganisms are sensitive to pH, initial pH adjustment of compost feedstocks is another process management strategy. PLFA analysis has been used to examine the effect of initial pH adjustment on microbial community structure during in-vessel composting of grape pomace, which has a low pH, and rice straw (Lei and VanderGheynst 2000). Feedstocks had an initial pH of 4.1 or 6, for treatments without or with pH adjustment, respectively. While initial PLFA samples were the same regardless of pH adjustment, samples after 7 days of composting were distinguished by higher levels of terminally branched, saturated PLFAs in the compost with initial pH adjustment and of branched, monounsaturated PLFAs, or mesophilic biomarkers, in the compost without pH adjustment. This was likely due to the longer and hotter thermophilic phase of the compost with higher initial pH, suggesting that higher initial pH was beneficial for growth of thermophiles. Despite differences in temperature profiles and microbial community composition, there was ultimately no difference in cumulative oxygen consumption, or decomposition level, between the two composts after 10 days of composting. One reason for this may have been due to the eventual drop in pH from 6 to 5 for the compost with pH adjustment while the compost with no pH adjustment only changed from a pH of 4.1 to 3.8 during 10 days of composting.

As these studies indicate, feedstock composition, including nitrogen content and pH, can significantly affect the microbial community and temperature profile which develop during composting. However, management of these factors may not necessarily speed up the composting process.

10.4.2 Temperature

Based on the observations from community dynamics studies, temperature is probably the most dominant factor affecting the microbial community during composting. A few studies have investigated the role of temperature on the microbial community by performing constant temperature incubations without allowing the composting to self-heat.

Waksman *et al.* (1939) used enrichment methods to monitor the microbial community during horse manure composting in bench-scale systems held at 28°C, 50°C, 65°C and 75°C for 61 days. They found that although decomposition initially proceeded more rapidly at 28°C, after 61 days of incubation the most decomposition occurred at 50°C followed by 28°C and 65°C, which had very similar results, and then 75°C. Many different types of microorganisms dominated the compost incubated at 28°C, the majority of which were bacteria followed by actinomycetes in the early stages of composting and fungi during the later stages. This was the only incubation temperature that supported organisms such as protozoa and nematodes. At 50°C, bacteria and actinomycetes dominated the microbial community toward the beginning of the process while thermophilic fungi dominated toward the end. Similar observations of bacteria and actinomycetes were made at 65°C, however no fungi were detected. Finally at 75°C, only spore forming bacteria were isolated.

In another constant temperature study, the bacterial community was analysed during the composting of garbage at 40°C with and without added soya bean oil over a period of 22 days (Aoshima *et al.* 2001). Using DGGE, the compost with oil addition was shown to include a high abundance of lactic acid bacteria, including *Staphylococcus* sp. and *Lactobacillus* sp. There were some shifts in microbial community noted with respect to time, but most of the variation was in the intensity of the bands, not in their presence or absence. This was possibly because the system was operated in a semi-continuous manner, as garbage with oil was supplied to the reactor every other day. As indicated by cloning, the bacterial community in the compost without oil addition was dominated by lactic acid bacteria, but also included *Bacillus* spp.

In general, microorganisms observed in these constant temperature studies correlated with those observed at particular temperatures during the composting process, as previously discussed. At temperatures less than 50°C, fungi, mesophilic bacteria, and fermentative bacteria were prevalent, while at higher temperatures, actinomycetes, thermophilic fungi, and ultimately spore forming bacteria were present.

10.4.3 Inoculation

In many systems, compost is seeded with material containing a mixed microbial community in order to accelerate the process. Nakasaki *et al.* (1985) investigated biosolids compost amended with different amounts of compost seed. Respiration rate profiles indicated a smoother transition from mesophilic to thermophilic conditions and temperature increased slightly faster as seeding increased from 0% to 133% of the initial dry weight of biosolids. While initial microbial populations were different between treatments, no trend in final populations could be observed with respect to seeding percentage, suggesting other factors such as initial readily assimilable carbon and nitrogen levels or duration of the thermophilic phase, which were different for each treatment, may have played a larger role in the microbial community than inoculation.

In another study, inoculation was again seen to affect initial compost microbial populations, but not final populations. Grape pomace and rice straw feedstocks were composted for 1 week in forced aerated reactors, where one system was inoculated with 5% (g dry inoculum/g total dry solids) mature compost and compared with a non-inoculated system (Lei and VanderGheynst 2000). After inoculation, microbial communities in the initial samples of inoculated and non-inoculated composts were different, as indicated by PLFA analysis. Both composts achieved peak temperatures of 40°C or slightly higher, but the inoculated compost remained at peak temperature for more than 2 days relative to 1 day for the non-inoculated compost. Regardless of inoculation, both composts showed decreases in branched, monounsaturated PLFAs, likely indicating decreases in mesophilic bacteria, during the transition from the self-heating to thermophilic phase. Other than the differences in initial microbial community and slight variation in temperature profiles, inoculation did not significantly impact composting, as the cumulative oxygen consumption, or decomposition level, of both composts was the same after 10 days.

In other studies, inoculation has proven to be an effective management strategy for increasing decomposition rate. For instance, Ichida *et al.* (2001) investigated inoculating poultry waste compost with a mixed culture of bacteria to facilitate decomposition of feathers. Feathers were soaked in a mixed culture of *Bacillus licheniformis* and *Streptomyces* sp. and then composted with a mixture of poultry litter and straw. Their composts containing treated feathers had higher initial levels of bacteria than composts containing untreated feathers, and treated feathers decomposed faster than untreated feathers.

In another example of effective inoculation, a mixture of wood chips and organic household waste was inoculated with 0% or 14% (by volume) 1-week-old compost (Bolta *et al.* 2003). The inoculated compost had higher initial levels of eukaryotic microorganisms, as indicated by polyunsaturated PLFAs. These organisms may have been significant in degrading cellulose in the feedstocks and releasing more readily degradable substrates that resulted in rapid self-heating. The inoculated compost reached 70°C, the peak temperature, 5 days before the non-inoculated compost. After mixing and watering on day 8 of composting, only the inoculated compost returned to thermophilic temperatures and maintained a thermophilic microbial community. PLFA biomarkers for actinomycetes and gram-positive bacteria decreased to half of previous levels in the non-inoculated compost. While inoculation did not have a long-term effect on the compost microbial communities as indicated by similar PLFA biomarker composition after 198 days of composting, it had a timely effect that resulted in more rapid decomposition. The inoculated compost mineralized nearly four times more organic matter.

Rather than inoculating with a mixed microbial community, specific organisms may be inoculated to confer the compost with specific biological characteristics, e.g. disease suppression (Hoitink and Boehm 1999). Important issues associated with this type of inoculation include when to inoculate during the composting process and how to manage the process after inoculation.

In one study, the fungal biological control agent, *Trichoderma harzianum*, was inoculated into three different grape pomace and rice straw compost samples taken throughout the process and then incubated for 1 week at 26, 30, or 35°C (VanderGheynst *et al.* 2002). *T. harzianum* only colonized the compost sample from the end of the thermophilic phase, which had been exposed to temperatures greater than 45°C for 4 days. This compost sample was most suitable for fungal colonization, as indicated by relatively lower pH and lower respiration rate, or less microbial competition, than other compost samples. *T. harzianum* population density was 4–5 orders of magnitude higher after incubation at temperatures of 26°C and 30°C than at 35°C, stressing the importance of process management after inoculation. *T. harzianum* in the inoculated composts was monitored with a selective enrichment method since DGGE targeting the 18S rDNA failed to detect *T. harzianum* due to the inhibitory effect of non-target DNA on its PCR amplification. Not all monitoring methods for microorganisms are applicable in all situations, and frequently, a combination of methods is needed. For the purposes of this study, the combination of selective enrichment plating and PLFA techniques allowed for the detection of the *T. harzianum* and insight into the microbial community present at the time of inoculation.

10.4.4 Composting method

Another process management issue is whether there is a measurable benefit from mixing during composting or from increased aeration. One study addressed this question by examining composting of dairy manure and rice straw in either static or mixed reactors with low (210 ml wet air/min-kg) or high (620 ml wet air/min-kg) aeration rates (VanderGheynst and Lei 2003). Reactors with the high aeration rate had higher oxygen concentrations in the effluent air by 2–4% than reactors with the low aeration rate. Nevertheless, PLFA biomarkers still suggested anaerobic activity since cyclopropyl PLFAs increased from negligible levels to 4 mol% in the low aeration treatment and 7 mol% in the high aeration treatment. If this increase in cyclopropyl PLFAs was due to oxygen limitations rather than nutrient limitations, this study indicates that a high aeration rate might not have any impact on decreasing anaerobic activity during composting.

Mixing and aeration were also found to have implications for the management of eukaryotic populations in compost (VanderGheynst and Lei 2003). Reactors with the low aeration rate had higher temperatures by as much as 5°C, relative to the reactors with the high aeration rate. After the thermophilic composting phase, polyunsaturated PLFAs returned to initial levels of 7 mol% in reactors with the high aeration rate. However, in reactors with the low aeration rate, the eukaryotic biomarkers remained near 1 mol%. This lack of eukaryotic recolonization may have been due to higher temperatures achieved during composting, lower oxygen levels, or differences in microbial community composition. Furthermore, polyunsaturated PLFAs reached slightly higher levels in static bed reactors, possibly indicating damage to fungal hyphae during mixing. This potentially negative effect of mixing might be a consideration if a beneficial fungus was inoculated into compost. In addition, this study may indicate a possible method for managing compost microbial communities, such as eukaryotes, through aeration.

10.5 CONCLUSIONS

Compost can support a diverse microbial community which, based on the literature, is highly related to temperature, oxygen concentration, and substrate composition. PLFA studies confirmed the observations from enrichment plating studies and were able to provide quantitative information on total microbial biomass and semi-quantitative information on microbial community structure during several types of composting processes. More recently, nucleic acid-based methods and very specific selective enrichment studies were able to provide qualitative information on the microbial community, but most importantly, were able to identify specific microorganisms that had not been previously identified using general enrichment plating methods.

Although nucleic acid-based methods were, in general, better at capturing a more diverse population of organisms compared to enrichment plating, they did have limitations, especially when the DNA from more dominant microorganisms masked the presence of other members of the microbial community. An additional limitation has been that the nucleic acid-based techniques rely on sequence databases for matching isolated DNA to known microorganisms. These databases are continually under development and for several composting studies were unable to match sequence information to isolated DNA. Given the limitations of the methods presented in this chapter, a study utilizing a combination of methods might be the most thorough approach for understanding the microbial community during composting.

There were far more studies that examined general trends in the microbial community during composting compared to studies examining the link between process management and the microbial community. Studies of the latter type would help improve our understanding of the role of microbial communities in composting, which could have an impact on controlling compost emissions that impact environmental health and on better reproducibility with respect to composting process performance and product quality.

Finally, a variety of microorganisms with unique abilities were isolated from composting operations. For instance, fungi and anaerobic bacteria capable of degrading complex carbohydrates and hyperthermophiles that thrived at temperatures as high as 80°C have been observed. The role of anaerobic organisms in a process that is intended to be aerobic is an issue that should be studied in detail in the future. Furthermore, the tolerances of the hyperthermophiles and their enzymes to high temperatures and the diversity of the microbial community could make compost a rich source of organisms and enzymes for industrial and biotechnical applications.

REFERENCES

Alfreider, A., Peters, S., Tebbe, C.C., Rangger, A. and Insam, H. (2002) Microbial community dynamics during composting of organic matter as determined by 16S ribosomal DNA analysis. *Compost Sci. Util.* **10**, 303–312.

Amann, R.I., Ludwig, W. and Schleifer, K. (1995) Phylogenetic identification and *in situ* detection of individual microbial cells without cultivation. *Microbiol. Rev.* **59**, 143–169.

Anastasi, A., Varese, G.C., Voyron, S., Scannerini, S. and Marchisio, V.F. (2002) Systemic and functional characterization of fungal biodiversity in compost and vermicompost. *Proceedings of the 2002 International Symposium on Composting and Compost Utilization*, Columbus, OH.

Aoshima, M., Pedro, M.S., Haruta, S., Ding, L.X., Fukada, T., Kigawa, A., Kodama, T., Ishii, M. and Igarashi, Y. (2001) Analyses of microbial community within a composter operated using household garbage with special reference to the addition of soybean oil. *J. Biosci. Bioeng.* **91**, 456–461.

Beffa, T., Blanc, M. and Aragno, M. (1996a) Obligately and facultatively autotrophic, sulphur- and hydrogen-oxidizing thermophilic bacteria isolated from hot composts. *Arch. Microbiol.* **165**, 34–40.

Beffa, T., Blanc, M., Lyon, P.-F., Vogt, G., Marchiani, M., Fischer, J.L. and Aragno, M. (1996b) Isolation of *Thermus* strains from hot composts (60–80°C). *Appl. Environ. Microbiol.* **62**, 1723–1727.

Benson, D.A., Boguski, M.S., Lipman, D.J., Ostell, J., Ouellette, B.F., Rapp, B.A. and Wheeler, D.L. (1999) GenBank. *Nucl. Acids Res.* **27**, 12–17.

Blanc, M., Marilley, L., Beffa, T. and Aragno, M. (1999) Thermophilic bacterial communities in hot composts as revealed by most probable number counts and molecular (16S rDNA) methods. *FEMS Microbiol. Ecol.* **28**, 141–149.

Bollen, G.J. (1993) Factors involved in inactivation of plant pathogens during composting. In: *Science and Engineering of Composting: Design, Environmental, Microbiological and Utilization Aspects* (eds. H.A.J. Hoitink and H.M. Keener), pp. 301–318. Renaissance Publications, Worthington, OH.

Bolta, S.V., Mihelic, R., Lobnik, F. and Lestan, D. (2003) Microbial community structure during composting with and without mass inocula. *Compost Sci. Util.* **11**, 6–15.

Bossio, D.A. and Scow, K.M. (1998) Impacts of carbon and flooding on soil microbial communities: phospholipid fatty acid profiles and substrate utilization patterns. *Microb. Ecol.* **35**, 265–278.

Cahyani, V.R., Watanabe, A., Matsuya, K., Asakawa, S. and Kimura, M. (2002) Succession of microbiota estimated by phospholipid fatty acid analysis and changes in organic constituents during the composting process of rice straw. *Soil Sci. Plant Nutr.* **2002**, 733–743.

Carpenter-Boggs, L., Kennedy, A.C. and Reganold, J.P. (1998) Use of phospholipid fatty acids and carbon source utilization patterns to track microbial community succession in developing compost. *Appl. Environ. Microbiol.* **64**, 4062–4064.

Davis, C.L., Donkin, C.J., Hinch, S.A. and Germishuizen, P.J. (1991) The microbiology of pine bark composting: an electron-microscope and physiological study. *Biores. Technol.* **40**, 195–204.

Davis, C.L., Hinch, S.A., Donkin, C.J. and Germishuizen, P.J. (1992) Changes in microbial population numbers during the composting of pine bark. *Biores. Technol.* **39**, 85–92.

Dees, P.M. and Ghiorse, W.C. (2001) Microbial diversity in hot synthetic compost as revealed by PCR-amplified rRNA sequences from cultivated isolates and extracted DNA. *FEMS Microbiol. Ecol.* **35**, 207–216.

Derikx, P.J.L., Simons, F.H.M., Op Den Camp, H.J.M., Van Der Drift, C., Van Griensven, L.J.L.D. and Vogels, G.D. (1991) Evolution of volatile sulphur compounds during laboratory-scale incubations and indoor preparation of compost used as a substrate in mushroom cultivation. *Appl. Environ. Microbiol.* **57**, 563–567.

Dunbar, J., Ticknor, L.O. and Kuske, C.R. (2001) Phylogenetic specificity and reproducibility and new method for analysis of terminal restriction fragment profiles of 16S rRNA genes from bacterial communities. *Appl. Environ. Microbiol.* **67**, 190–197.

Eiland, F., Klamer, M., Lind, A.M., Leth, M. and Baath, E. (2001) Influence of initial C/N ratio on chemical and microbial composition during long term composting of straw. *Microb. Ecol.* **41**, 272–280.

Farrell, J.B. (1993) Fecal pathogen control during composting. In: *Science and Engineering of Compost: Design, Environmental, Microbiological and Utilization Aspects* (eds. H.A.J. Hoitink and H.M. Keener), pp. 282–300. Renaissance Publications, Worthington, OH.

Fermor, T.R., Smith, J.F. and Spencer, D.M. (1979) The microflora of experimental mushroom composts. *J. Horticul. Sci.* **54**, 137–147.

Findlay, R.H. and Dobbs, F.C. (1993) Quantitative description of microbial communities using lipid analysis. In: *Handbook of Methods in Aquatic Microbial Ecology* (ed. P.F. Kemp), pp. 271–284. Lewis Publishers, Boca Raton, FL.

Finstein, M.S. and Morris, M.L. (1975) Microbiology of municipal solid waste composting. *Adv. Appl. Microbiol.* **19**, 113–151.

Frostegard, A. and Baath, E. (1996) The use of phospholipid fatty acid analysis to estimate bacterial and fungal biomass in soil. *Biol. Fert. Soils* **22**, 59–65.

Frostegard, A., Petersen, S.O., Baath, E. and Nielsen, T.H. (1997) Dynamics of a microbial community associated with manure hot spots as revealed by phospholipid fatty acid analysis. *Appl. Environ. Microbiol.* **63**, 2224–2231.

Godden, B., Penninckx, M., Pierard, A. and Lannoye, R. (1983) Evolution of enzyme activities and microbial populations during composting of cattle manure. *Eur. J. Appl. Microbiol. Biotechnol.* **17**, 306–310.

Haruta, S., Kondo, M., Nakamura, K., Aiba, H., Ueno, S., Ishii, M. and Igarashi, Y. (2002) Microbial community changes during organic solid waste treatment analysed by double gradient-denaturing gradient gel electrophoresis and fluorescence *in situ* hybridization. *Appl. Microbiol. Biotechnol.* **60**, 224–231.

Hassen, A., Belguith, K., Jedidi, N., Cherif, A., Cherif, M. and Boudabous, A. (2001) Microbial characterization during composting of municipal solid waste. *Biores. Technol.* **80**, 217–225.

Hellmann, B., Zelles, L., Palojarvi, A. and Bai, Q. (1997) Emission of climate-relevant trace gases and succession of microbial communities during open-windrow composting. *Appl. Environ. Microbiol.* **63**, 1011–1018.

Herrmann, R.F. and Shann, J.F. (1997) Microbial community changes during the composting of municipal solid waste. *Microb. Ecol.* **33**, 78–85.

Hoitink, H.A.J. and Boehm, M.J. (1999) Biocontrol within the context of soil microbial communities: a substrate-dependent phenomenon. *Ann. Rev. Phytopathol.* **37**, 427–446.

Ichida, J.M., Krizova, L., LeFevre, C.A., Keener, H.M., Elwell, D.L. and Burt Jr., E.H. (2001) Bacterial inoculum enhances keratin degradation and biofilm formation in poultry compost. *J. Microbiol. Meth.* **47**, 199–208.

Ishii, K., Fukui, M. and Takii, S. (2000) Microbial succession during a composting process as evaluated by denaturing gradient gel electrophoresis analysis. *J. Appl. Microbiol.* **89**, 768–777.

Klamer, M. and Baath, E. (1998) Microbial community dynamics during composting of straw material studied using phospholipid fatty acid analysis. *FEMS Microbiol. Ecol.* **27**, 9–20.

Kowalchuk, G.A., Naoumenko, Z.S., Derikx, P.J.L., Felske, A., Stephen, J.R. and Arkhipchenko, I.A. (1999) Molecular analysis of ammonia-oxidizing bacteria of the beta subdivision of the class Proteobacteria in compost and composted materials. *Appl. Environ. Microbiol.* **65**, 396–403.

LaMontagne, M.G., Michel Jr., F.C., Holden, P.A. and Reddy, C.A. (2002) Evaluation of extraction and purification methods for obtaining PCR-amplifiable DNA from compost for microbial community analysis. *J. Microbiol. Meth.* **49**, 255–264.

Lei, F. and VanderGheynst, J.S. (2000) The effect of microbial inoculation and pH on microbial community structure changes during composting. *Proc. Biochem.* **35**, 923–929.

Liu, W., Marsh, T.L., Cheng, H. and Forney, L.J. (1997) Characterization of microbial diversity by determining terminal restriction fragment length polymorphisms of genes encoding 16S rRNA. *Appl. Environ. Microbiol.* **63**, 4516–4522.

Lyon, P.-F., Beffa, T., Blanc, M., Auling, G. and Aragno, M. (2000) Isolation and characterization of highly thermophilic xylanolytic *Thermus thermophilus* strains from hot composts. *Can. J. Microbiol.* **46**, 1029–1035.

Madigan, M.T., Martinko, J.M. and Parker, J. (2003) *Brock Biology of Microorganisms.* Prentice Hall, Upper Saddle River, NJ.

Marsh, T.L., Saxman, P., Cole, J. and Tiedje, J. (2000) Terminal restriction fragment length polymorphism analysis program, a web-based research tool for microbial community analysis. *Appl. Environ. Microbiol.* **66**, 3616–3620.

Michel Jr., F.C., Marsh, T.J. and Reddy, C.A. (2002) Bacterial community structure during yard trimmings composting. In: *Microbiology of Composting* (eds. H. Insam, N. Riddech and S. Klammer), pp. 25–42. Springer-Verlag, Berlin.

Miller, F.C. (1993) Minimizing odour generation. In: *Science and Engineering of Compost: Design, Environmental, Microbiological and Utilization Aspects* (eds. H.A.J. Hoitink and H.M. Keener), pp. 219–241. Renaissance Publications, Worthington, OH.

Muyzer, G., Dewaal, E.C. and Uitterlinden, A.G. (1993) Profiling of complex microbial populations by denaturing gradient gel electrophoresis analysis of polymerase chain reaction-amplified genes coding for 16S ribosomal RNA. *Appl. Environ. Microbiol.* **59**, 695–700.

Nakasaki, K., Sasaki, M., Shoda, M. and Kubota, H. (1985) Change in microbial numbers during thermophilic composting of sewage sludge with reference to CO_2 evolution rate. *Appl. Environ. Microbiol.* **49**, 37–41.

Pedro, M.S., Haruta, S., Hazaka, M., Shimada, R., Yoshida, C., Hiura, K., Ishii, M. and Igarashi, Y. (2001) Denaturing gradient gel electrophoresis analyses of microbial community from field-scale composter. *J. Biosci. Bioeng.* **91**, 159–165.

Pedro, M.S., Hayashi, N.R., Mukai, T., Ishii, M., Yokota, A. and Igarashi, Y. (1999) Physiological and chemotaxonomical studies on microflora within a composter operated at high temperature. *J. Biosci. Bioeng.* **88**, 92–97.

Peters, S., Koschinsky, S., Schwieger, F. and Tebbe, C.C. (2000) Succession of microbial communities during hot composting as detected by PCR-single-strand-conformation-polymorphism-based genetic profiles of small-subunit rRNA genes. *Appl. Environ. Microbiol.* **66**, 930–936.

Petersen, S.O. and Klug, M.J. (1994) Effects of sieving, storage, and incubation temperature on the phospholipid fatty acid profile of a soil microbial community. *Appl. Environ. Microbiol.* **60**, 2421–2430.

Pielou, E.C. (1984) *The Interpretation of Ecological Data.* John Wiley & Sons, Inc., New York.

Roberts, M.S., Klamer, M., Frazier, C. and Garland, J.L. (2002) Community profiling of fungi and bacteria in an in-vessel composter for the NASA advanced life support program. *Proceedings of the 2002 International Symposium on Composting and Compost Utilization*, Columbus, OH.

Ryckeboer, J., Mergaert, J., Coosemans, J., Deprins, K. and Swings, J. (2003) Microbiological aspects of biowaste during composting in a monitored compost bin. *J. Appl. Microbiol.* **94**, 127–137.

Rynk, R. (ed.) (1992) *On-Farm Composting Handbook.* Northeast Regional Agricultural Engineering Service, Ithaca, NY.

Schwieger, F. and Tebbe, C.C. (1998) A new approach to utilize PCR-single-strand-conformation polymorphism for 16S rRNA gene-based microbial community analysis. *Appl. Environ. Microbiol.* **64**, 4870–4876.

Strom, P.F. (1985) Identification of thermophilic bacteria in solid-waste composting. *Appl. Environ. Microbiol.* **50**, 900–913.

Stryer, L. (1995) *Biochemistry.* W. H. Freeman and Company, New York.

Tebbe, C.C. and Vahjen, W. (1993) Interference of humic acids and DNA extracted directly from soil in detection and transformation of recombinant DNA from bacteria and yeast. *Appl. Environ. Microbiol.* **59**, 2657–2665.

Tiquia, S.M. (2002) Evolution of extracellular enzyme activities during manure composting. *J. Appl. Microbiol.* **92**, 764–775.

Tiquia, S.M. and Michel Jr., F.C. (2002) Bacterial diversity in livestock manure composts as characterized by terminal restriction fragment length polymorphisms (T-RFLP) of PCR-amplified 16S rRNA gene sequences. In: *Microbiology of Composting* (eds. H. Insam, N. Riddech and S. Klammer), pp. 65–81. Springer-Verlag, Berlin.

Tiquia, S.M., Wan, J.H.C. and Tam, N.F.Y. (2002) Microbial population dynamics and enzyme activities during composting. *Compost Sci. Util.* **10**, 150–161.

Trescher, K. and Tebbe, C.C. (2002) Molecular microbial community analysis: genetic profiles as a tool to characterize composting processes. *Proceedings of the 2002 International Symposium on Composting and Compost Utilization*, Columbus, OH.

Ueno, Y., Haruta, S., Ishii, M. and Igarashi, Y. (2001) Microbial community in anaerobic hydrogen-producing microflora enriched from sludge compost. *Appl. Microbiol. Biotechnol.* **57**, 555–562.

VanderGheynst, J.S., Gossett, J.M. and Walker, L.P. (1997) High-solids aerobic decomposition: pilot-scale reactor development and experimentation. *Process Biochem.* **32**, 361–375.

VanderGheynst, J.S., Cogan, D.J., DeFelice, P.J., Gossett, J.M. and Walker, L.P. (1998) Effect of process management on the emission of organosulphur compounds and gaseous antecedents from composting processes. *Environ. Sci. Technol.* **32**, 3713–3718.

VanderGheynst, J.S., Dooley, T.M. and Marshall, M.N. (2002) The influence of process management and microbial community structure on the cultivation of a biological control agent in compost. *Proceedings of the 2002 International Symposium on Composting and Compost Utilization*, Columbus, OH.

VanderGheynst, J.S. and Lei, F. (2003) Microbial community structure dynamics during aerated and mixed composting. *Trans. ASAE* **46**, 577–587.

Vestal, J.R. and White, D.C. (1989) Lipid analysis in microbial ecology. *BioScience* **39**, 535–541.

Waksman, S.A., Cordon, T.C. and Hulpoi, N. (1939) Influence of temperature upon the microbiological population and decomposition processes in composts of stable manure. *Soil Sci.* **47**, 83–98.

White, D.C., Stair, J.O. and Ringelberg, D.B. (1996) Quantitative comparisons of *in situ* microbial biodiversity by signature biomarker analysis. *J. Indust. Microbiol.* **17**, 185–196.

Wilkinson, S.G. (1988) Gram-negative bacteria. In: *Microbial Lipids* (eds. C. Ratledge and S.G. Wilkinson), Vol. 1, pp. 299–488. Academic Press, London.

Wilson, I.G. (1997) Inhibition and facilitation of nucleic acid amplification. *Appl. Environ. Microbiol.* **63**, 3741–3751.

11

Overview of resource recovery technologies for biowastes

W.H. Rulkens

11.1 INTRODUCTION

Solid and semi-solid wastes cause a severe worldwide environmental problem (Tchobanoglous *et al.* 1993). This problem is characterised by the fact that the origin, amount and composition of these wastes vary strongly, not only regarding the region where the problem exists, but also with respect to time. The origin of these wastes is from household activities, from activities of the municipalities, industrial production systems, as well as from agricultural production processes. In fact all wastes are originating from activities of human beings. Abatement of these wastes occurs via several scenarios, varying between prevention, recovery of valuable products from wastes, waste destruction and final disposal of waste residues. The optimal approach depends on many factors. Among others these factors include the origin, amount, type and composition of the wastes, the available technologies to treat the waste, costs of waste treatment, market value of the recovered valuable products, costs of disposal, available technical expertise, environmental policy at local, regional and national level, social

 Resource Recovery and Reuse in Organic Solid Waste Management. Edited by Piet Lens, Bert Hamelers, Harry Hoitink, Werner Bidlingmaier. ISBN: 1 84339 054 X. Published by IWA Publishing, London, UK

acceptance of the proposed solution and the local, regional and national circumstances in general. Often a combination of different treatment processes is most attractive.

Solid and semi-solid waste can be subdivided in several groups. One group comprises the so-called biowastes. These wastes mainly consist of organic material, more or less originating from biological origin. As examples of these wastes three types can be identified: the organic fraction of municipal solid wastes (MSW), sewage sludge and pig manure. The latter two wastes also contain substantial amounts of water and can also be characterised as liquid wastes. In this chapter, these three types of wastes will be discussed in more detail regarding those treatment options that are focused on the recovery of valuable products from these types of wastes.

In general a treatment process for a biowaste consists of a combination of several single treatment steps, often indicated as unit operations. All these unit operations are based on four types of processes: phase separation (and phase contact) processes, molecular separation processes, (micro)biological conversion processes and chemical conversion processes. When dealing with biowastes and the focus is on recovery of valuable products, the general opinion is that (micro)biological treatment steps, especially microbiological conversion steps are more favourable than chemical/thermal conversion steps. However, though this is often the case, recovery of valuable products from waste streams is always part of a complete waste treatment scenario focused on the abatement of the environmental problem caused by that waste. Thus even in cases where bioconversion is the dominant process, physical/chemical treatment steps are also required. And for some types of wastes, it is even more efficient to use only physical/chemical treatment steps in a treatment scenario.

The aim of this chapter is to present a brief overview of the most important resource recovery technologies for solid wastes and especially for biowastes. These recovery technologies consist in general of one or more basic treatment steps (core treatment steps) in combination with pre-treatment and post-treatment steps. The overview will be given based on a brief discussion of possible treatment scenarios for the three types of biowastes mentioned. Preceding this discussion, a short outline will be given regarding the role and application of environmental technologies in the abatement of environmental problems caused by wastes in general. The role of environmental biotechnology will be highlighted in that context.

11.2 ENVIRONMENTAL TECHNOLOGY

Environmental technology comprises those technologies that are primarily focused on the abatement of environmental pollution in general, and, in the context of this chapter, the environmental problems caused by biowastes.

In general an integral treatment process (or treatment scenario) for a waste stream exists of a large number of fundamental treatment steps connected to each other in a logical way. In agreement with the definition of unit operations, there are three types of such fundamental treatment steps:

- Physical treatment steps, aimed at separation and concentration of the contaminants without changing their composition. Physical treatment steps can be subdivided in phase separation processes, in which phases are separated (e.g. membrane separation and sieving), and molecular separation processes, in which molecules are transferred from one phase to another phase (e.g. adsorption at activated carbon and stripping). Physical treatment steps can also be used to bring compounds, which are present in different phases, into contact with each other.
- Biological conversion steps, focused on a biological conversion of contaminants into less or non-polluting compounds or into compounds that can be removed and concentrated more easily than the original contaminants. Applicable biological conversion steps in the area of environmental technology are:
  - Degradation of organic contaminants into compounds which are non-polluting or which are substantially less polluting than the original compounds. Examples are the conversion of organic contaminants present in municipal wastewater into carbon dioxide, water and a small amount of biomass, or the conversion of nitrogen containing organic compounds into molecular nitrogen.
  - Conversion of organic contaminants into valuable products. Examples are the conversion of organics into biogas, the production of amino acids from nitrogen and organic carbon containing wastes, and the production of single-cell protein (e.g. algae).
  - Conversion of contaminants into compounds that can be separated and concentrated more easily than the original compounds. Examples are the conversion of nitrogen containing compounds into ammonia or into nitrate, the partial oxidation of hydrogen sulphide into elementary sulphur and the mobilisation of insoluble heavy metal sulphides into soluble heavy metal sulphates. Also the reverse, the reduction of sulphates in volatile sulphides, could be of interest.
- Chemical conversion steps, aimed at a chemical conversion of contaminants into less or non-polluting compounds. Important chemical conversion processes are oxidation at high or low temperatures, wet oxidation, gasification, pyrolysis, hydrolysis, dechlorination, electrochemical detoxification and reduction.

Biological and chemical conversion steps are always applied in connection with physical treatment steps. Often all three types of basic treatment steps are applied in an integrated treatment process for a waste stream. The number of possible physical, biological and chemical treatment steps is considerable. Consequently, for treating a certain type of waste there are several alternatives. In case that pollutants can be converted biologically or chemically, there is always the question what type of conversion step is most attractive. The answer to this question cannot be given by single comparison of the separate conversion steps but requires comparison of the integrated treatment process, including all physical, chemical and biological treatment steps.

Depending on the type of treatment steps, environmental technology can be subdivided into environmental biotechnology and chemical/physical environmental technology. Both types of environmental technology are partly complementary and partly overlapping.

In the abatement of environmental pollution, biological methods have already been in use for a long time. For over a century, micro-organisms have been used for the purification of municipal wastewater. Also composting of various sorts of organic waste has already been applied in practice for a long time. In the last decades the practical application of biological treatment processes to abate environmental pollution has shown a strong expansion, both with respect to the number and the types of applications. Additionally, biological treatment of industrial wastewater, purification of polluted air and off gases, treatment of organic wastes and decontamination of soil has been increasingly implemented.

The microbiological conversion process of a pollutant depends on several microbiological and technological factors:

- The molecular structure of the contaminant.
- The availability of the contaminant to the micro-organisms. Factors that are important in that context are the physical state of the contaminant (as particle, adsorbed or absorbed to non-polluting compounds of the waste stream, dissolved in water, etc.).
- Environmental factors (presence of electron acceptors or electron donors, temperature, moisture content, pH and presence of nutrients).
- The availability of the appropriate micro-organisms.
- The concentration of micro-organisms in the reactor system.
- Mass transfer limitations.
- The removal of intermediates produced during microbiological degradation.

Microbiological conversion processes (or treatment steps) have certain characteristics compared with chemical conversion steps (Rulkens 1992):

- Microbiological processes are especially suitable to convert a large number of organic pollutants to non- or only slightly polluting compounds such as nitrogen, carbon dioxide and water.
- In contrast with chemical conversion processes such as incineration, wet oxidation and pyrolysis, microbiological processes occur at relatively low temperatures, usually between 0°C and 60°C.
- Microbiological processes require an aqueous environment. The presence of a water phase often means a physical or physical/chemical post-treatment of this water phase.
- The conversion rates of microbiological processes are in general very low. This means long residence times in the conversion reactor and large reactor volumes.
- Microbiological processes are sensitive to toxic compounds.
- Micro-organisms are often able to produce very specific and valuable compounds from mixtures of organic pollutants. Chemical conversion processes are often much less specific. However, chemical processes that are carried out

at a high temperature such as incineration, pyrolysis or gasification have the possibility to transfer recalcitrant organic compounds into valuable compounds such as oil, gas or charcoal.
- Long adaptation periods are often necessary to start microbiological processes and this is especially common for anaerobic processes.

11.3 TREATMENT OF MSW

11.3.1 Introduction

MSW can be defined as waste that primarily originates from households and other human activities at a more centralised level (commercial or industrial activities and activities at municipality level) that produce comparable types of wastes. This type of waste can be characterised as follows:

- The production of MSW is strongly related to the (partly non-sustainable) lifestyle of people and the level of welfare and convenience that is pursued.
- It is a continuously ongoing production of waste, without an outlook that changes in lifestyle and attitude will prevent the production of this type of waste.
- The amount and composition of this waste is changing continuously.
- The amount and composition varies strongly, not only between the group of highly industrialised countries, developing countries, and countries that are in an economic transition period, but also between separate countries in each group.
- There is an increasing environmental awareness of the problems caused by this waste and of the potentials to recover valuable products from this waste.
- The increasing environmental awareness has strongly changed the focus to solve this problem. Currently landfill and simple incineration are becoming less and less acceptable, also due to the change in environmental policy and legislation. Attention is forced into the development and application of recovery technologies. An important aspect in a successful recovery of valuable products from MSW is the application of source separation approaches.
- A large part of MSW consists of organic waste fractions. Therefore, MSW is often also categorised under the name biowaste, although a large part of the waste also consists of inorganic material and inerts.

This section gives a short overview of the most interesting technologies for recovery of valuable products from this type of waste. These technologies are aerobic composting, anaerobic composting, incineration, pyrolysis and gasification.

11.3.2 Composition of MSW

MSW consists of organic and inorganic substances (Tchobanoglous *et al.* 1993; Henry and Heinke 1996). The organic fraction is comprised of waste paper,

Table 11.1 Typical composition of MSW in the United States (Tchobanoglous *et al.* 1993).

Waste component	Weight (%)
Paper	38.1
Yard waste	12.1
Food waste	10.9
Plastics	10.5
Metals	7.8
Rubber, leather and textiles	6.6
Glass	5.5
Wood	5.3
Miscellaneous inorganic wastes	3.2

cardboard, plastics, textiles, rubber, yard wastes, food wastes and wood. Part of these organic wastes decomposes rapidly causing offensive odours. The inorganic fraction contains glass, clay, sand, dirt, ferrous compounds, aluminium and tin cans. The MSW may also contain a small fraction of hazardous waste compounds originating from, for example, used batteries, residual amounts of pesticides and paint. Table 11.1 gives a typical composition of residential MSW in the US (1990), excluding recycled materials and food wastes discharged with wastewater. MSW also contains relatively large quantities of water, and the size of the waste particles varies over a broad range.

In practice the composition of MSW can vary much more than is indicated in Table 11.1. One reason is that the type and composition of waste are also strongly dependent on factors such as level of industrial development, standards of living, consumer environmental awareness, consumption patterns and changes in these patterns, introduction of new types of products, package material, and national, regional and local environmental policy. A second reason, in fact also strongly linked with the first reason, is the impact of the various types of source reduction programmes. One approach in these programmes is the separate collection of wastes at the source, in this case primarily the households. Separate collection of glass, paper, cardboard, vegetables, food and yard wastes exists in many regions. Where their use is common, kitchen food waste grinders result in the discharge of most food wastes with the wastewater, causing a strong change in composition of the residual solid waste. Thus, the collection system has a strong impact on the quantity and composition of MSW.

It will be clear that the typical composition of MSW and the continuous change in its quantity and composition sets special requirements to waste management systems and always requires a tailor made solution at a regional level. It has to be taken into account that separately collected waste streams often can be treated more efficiently both in technical and economic respects, and with a higher degree of environmental sustainability as they are more concentrated and have a more homogeneous composition.

11.3.3 Separation and pre-processing of MSW

The recovery of reusable and recyclable materials from MSW requires the separation of MSW. In general this separation process can take place either at the source of the generation of the waste, such as households or individual industries, or, at a central location when dealing with a collection system that generates unseparated MSW (Tchobanoglous *et al.* 1993; Henry and Heinke 1996). Even separately collected wastes often require a further separation, for example to remove toxic or specific components that hinder reuse or recycling of the material. In addition to an adequate separation of the waste stream, generally a pre-processing step is also necessary for the subsequent production of valuable products from the various waste fractions. These pre-treatment processes are sometimes necessary to make an efficient separation process technically feasible. Dependent on the types of useful products or types of energy that are aimed to be produced, mandated goals have to be set for the various waste fractions. As many types of useful products and energy sources can be produced, a large variety in required separation goals and pre-treatment goals is possible, resulting in many scenarios for separation and pre-treatment.

The most important separation and pre-treatment processes are size reduction, screening, density separation and electromagnetic separation. Size reduction is a process in which the waste material particles are mechanically reduced in size. There are several modifications of this process such as shredding, grinding or milling. From each modification various types exist. Which modification and which type is most feasible depends on the type of waste that has to be reduced in size and on the qualifications set for the product of the size reduction system. Screening is a unit process aimed at the separation of a granular mixture of different materials into two or more size fractions. The process can be applied either on wet or dry material. Sometimes it is combined with a size reduction system. Several modifications of the screening process exist, including vibrating screens, rotary screens and disc screens.

Density separation is aimed at separation of light materials, such as paper and plastics, from heavier materials, such as sand particles and ferrous materials. The process is based on differences in the settling velocity of these materials in an airflow, in this case in a so-called zig-zag zifter. A gas cyclone is most appropriate for the separation of fine granular material into a fine granular light fraction, mainly consisting of organic material, and a fine granular heavy fraction, mainly consisting of inorganic material. Density separation for materials with a high specific weight can also be obtained by using liquid phases with specific densities. Magnetic separation is a unit operation in which ferrous materials or other magnetisable materials can be separated from non-magnetic substances such as organic materials. Eddy current separation can similarly be used to separate non-ferrous metals like copper and aluminium.

Besides these four specific groups of separation and pre-processing systems many other unit operations are available. Their usefulness depends on the typical aim that has to be achieved in the separation and pre-processing steps. In general

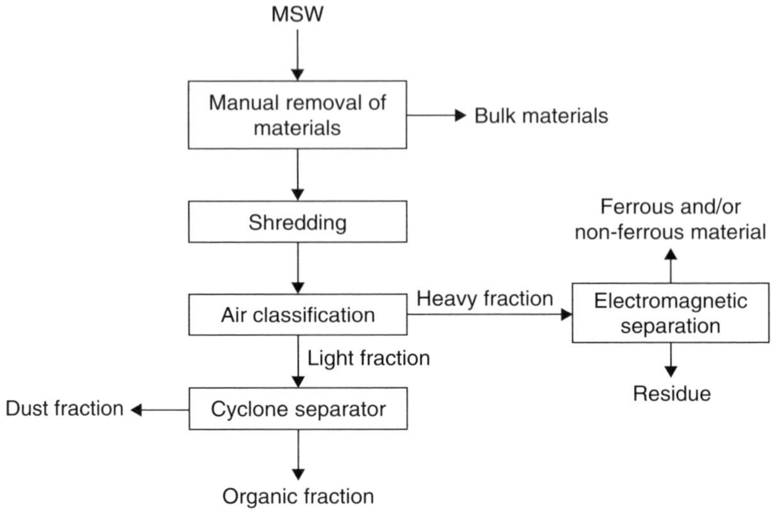

Figure 11.1 Separation and pre-processing of MSW.

the various separation and pre-processing steps are linked together as can be illustrated in a simplified process flow diagram as shown in Figure 11.1 (Tchobanoglous *et al.* 1993).

Within the framework of this chapter this will not be discussed in further detail.

11.3.4 Composting of MSW

11.3.4.1 Introduction

Composting is considered as a biological transformation process in which MSW is converted into a product, compost, which can be beneficially used as fertiliser or as soil conditioner (Tchobanoglous *et al.* 1993; Henry and Heinke 1996; Evans 2001; Cheremisinoff 2003). The general aim of a composting process is the transformation of a biodegradable unstable waste such as MSW into a stable useful product, to reduce the amount of waste and to destroy unwanted compounds present in the waste, such as pathogens, insect eggs and weed seeds. It is a pre-requisite that this process is executed in a technically, financially and sustainably efficient way. The transformations involved in the composting process mainly occur in a natural way by organisms such as bacteria, fungi, yeasts and antimonycetes. These microbiota are primarily dealing with the conversion of compounds such as proteins, amino acids and carbohydrates. Depending on the use of oxygen, two types of composting processes can be distinguished: aerobic composting (in general considered as the standard composting process) and anaerobic composting, the latter is a high-solids anaerobic digestion.

11.3.4.2 Aerobic composting

The aerobic composting process is characterised by three temperature phases. If the appropriate pre-treatment is applied with the appropriate process conditions, the first phase is characterised by a rapid rise in temperature due to the activities of the organisms present. If a certain temperature level is achieved then the second phase starts. In this phase, at a temperature of 50–60°C, the microbiological activity is highest. This phase continues as long as optimal process conditions exist and the waste still contains sufficient easily biodegradable material. After this phase the temperature decreases and the third phase starts. In this phase, slowly biodegradable materials are converted and the temperature gradually decreases to ambient. The total processing time for these three phases is about 3 weeks, but can vary depending on the feedstock types and process management. To obtain complete stabilisation an additional extensive post-treatment of 6–8 weeks or more is necessary. After that period a final mechanical/physical treatment is necessary to prepare a high-quality compost. A complete, but very schematical illustration of the aerobic composting process is given in Figure 11.2.

The composting process is in principle a simple process using natural biological conversion processes. However, the actual design of the process and the operational performance is quite complex because in general the waste to be composted varies strongly in time with respect to both composition and physical appearance. The most relevant process parameters in the composting process are temperature, moisture content, C/N ratio, pH, particle size and particle size distribution. Regarding the practical performance of the aerobic composting system many methods are practised. In general the various methods can be classified as agitated methods (in which the compost is agitated periodically) and static methods (where the compost remains static). Without further discussion of the specific

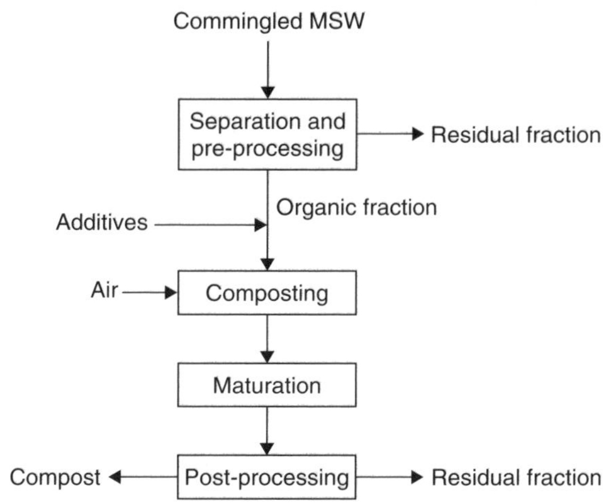

Figure 11.2 Schematical representation of the aerobic composting process.

differences between all these systems the main systems are windrow composting systems, aerated static pile composting systems and in-vessel composting systems. Each of these systems can be subdivided into several modifications, differing from each other with respect to scale, complexity and technological level of operation.

The specific aspects that have to be taken into account in a composting process and in the production of a compost of high quality are:

- Proper process design and operation. In this respect the appropriate pre-treatment of the compost, the aeration, effective control of temperature and oxygen level and moisture content has to be considered.
- Proper control of the process regarding the elimination and prevention of the production of odours.
- Elimination of pathogenic organisms and other non-desirable organisms, insect eggs and weed seeds.
- Quality regarding the presence of heavy metals and eventually pesticides. Product quality is a key factor in the public acceptance and the market value of the product. In case of yard wastes, collected separately, a high-quality compost with low levels of heavy metals in the compost can be obtained. However, in composting the organic fraction that has been separated mechanically from commingled MSW, the resulting compost often is still unacceptable due to the high level of heavy metals and toxic organic pollutants.

11.3.4.3 Anaerobic composting/digestion

The organic fraction of MSW can also be treated anaerobically in order to convert organic components into biogas. Two variations of this process exist. The first is a low-solids anaerobic process. In this process, the organic fraction of MSW is (after an appropriate pre-treatment process) diluted with water. In some cases nutrients have to be added as well. This slurry can be anaerobically digested in a continuous flow ideally mixed tank reactor or in a cascade of mixed tank reactors. Digestion temperature can be either at a thermophilic or a mesophilic level. After the digestion process the slurry is separated into a dewatered sludge and a water phase. For both phases a final destination has to be found, often after further treatment. The use of the water phase and resulting large reactor volumes makes the process less cost attractive.

Instead of a low-solids anaerobic digestion process, a high-solids anaerobic digestion process can also be applied. The first step in this process is an anaerobic digestion of the solid granular organic phase under thermophilic conditions. The retention time amounts to about 25 days. After this period the residual solid phase can be dried by using heat from a part of the produced biogas. Another possibility to reduce the moisture content is a subsequent aerobic composting process. The result is a relatively dry product that, after a mechanical/physical post-treatment, can be converted into high-quality compost if contaminant levels are low. It is evident that the problem of the water phase, as was mentioned in the low-solids anaerobic treatment process, can now fully be avoided.

11.3.4.4 Brief evaluation of the composting process

Composting of the organic fraction of MSW is worldwide applied on a large scale. The majority of these processes are aerobic. A typical aspect of the composting process is the relatively long treatment time (retention time), more than 20 days, as the main treatment step, followed by an extensive post-treatment process during 6–8 weeks. High-quality compost, usable as fertiliser, can in fact only be obtained from separately collected yard and food waste. Compost obtained from the mechanically separated organic fraction from commingled MSW often has an unacceptable quality for use as fertiliser or soil amendent. In that case however, composting can be useful to make a stabile product (in case of landfill, if allowed), to reduce the volume, or to remove water to produce a dry product that eventually can be incinerated in combination with the production of heat or electrical energy. It is also possible to co-compost sewage sludge obtained from municipal wastewater treatment plants, with the organic fraction of MSW.

Composting is in fact an attractive natural process to produce valuable products such as organic fertiliser, soil conditioner and biogas (in case of anaerobic composting). However, it will also be clear that for an efficient composting process, in addition to the biological conversion process, many physical and mechanical processes have to be applied in an appropriate way. Also if the role of a composting process in the management of MSW is considered, it will be clear that non-biological treatment steps are of crucial importance before and after the composting step.

11.3.5 Thermal conversion processes of MSW

11.3.5.1 Introduction

MSW contains a substantial percentage of organic (carbonaceous) wastes, not only organic wastes such as food wastes, but also wastes such as paper waste, plastics and cardboard. This fraction of MSW can be transformed by a variety of thermal processes (Henry and Heinke 1996; Stessel 1996; Tchobanoglous *et al.* 2001; Cheremisinoff 2003). Dependent on the type of thermal treatment either energy or valuable (mainly) gaseous components can be produced. In addition to the production of energy or valuable gaseous components the thermal processes also have the advantage of a strong reduction in the total amount of waste, so that only a relatively small residual amount has to be landfilled. The most important thermal processes for MSW are direct combustion (with energy recovery), production of refuse derived fuel for incineration in combination with energy production, pyrolysis and gasification.

11.3.5.2 Combustion with energy recovery

There are two systems that can be used for the incineration of MSW. A simplified schematic representation of the predominant type of combustion process with

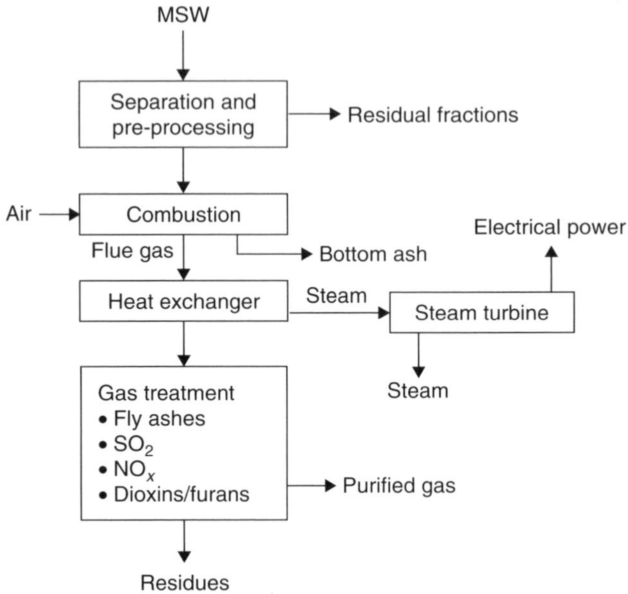

Figure 11.3 Schematical representation of a combustion process with energy recovery.

energy recovery is shown in the process scheme given in Figure 11.3. The first step of the process is the removal of large pieces and also part of the non-combustable materials as far as this is possible in a simple way. Then the waste is fed to a grate furnace where the combustion takes place. Combustion air is introduced at the bottom of the incinerator. Energy from the combustion gases is recovered by means of a heat exchanger with water as heat exchange medium. The hot water is fed to a boiler where steam is produced that is used for the production of electricity by means of a steam turbine. It is also possible to directly use the hot water as an energy source. The combustion gases contain various types of pollutants such as NO_x, SO_2 and fly ash (particulate pollutants). With appropriate pollutant control equipment NO_x can be destroyed by injection of NH_3, SO_2 and other acid inorganic pollutants such as HCl can be removed by absorption on lime and particulate material can be removed either by bag house filtration or by electrostatic precipitation. Specific equipment to abate possible emissions of dioxins and furans can be installed in the exhaust gases processing system. At the bottom of the combustion chamber furnace ash is removed. The advantage of this system is that unseparated commingled MSW can be used directly after large pieces have been removed.

A more efficient combustion and energy recovery system can be obtained if only the organic fraction of commingled MSW is used in the incineration process. Therefore, this organic fraction has to be separated from the commingled MSW. In fact this fraction, called refuse derived fuel (RDF), can directly be used in the incineration process. However, often further processing, for example drying and densification, takes place before this fraction is fed to an incinerator. Due to more or less homogeneous composition of RDF but also due to the higher energy content of RDF compared with unseparated commingled MSW, the incinerator is simple and the incineration process is more efficient and can be better controlled. However, it is clear that despite these advantages in the use of RDF, there are additional costs related with the production of RDF.

There are many modifications of the waste combustion processes which are too numerous to discuss in this general overview. In general all these systems are characterised by the fact that they are very capital intensive, and that for treatment of the incineration gases advanced equipment is necessary to satisfy all emission standards. Technically this is not a problem, however, costs are in general substantial. In all incineration processes a lot of ashes are produced. Depending on the quality of these ashes, especially with respect to heavy metals, these ashes can be used elsewhere, for example as building material, or may have to be disposed in a controlled landfill.

11.3.5.3 Pyrolysis

Pyrolysis is a process in which an organic waste (such as the organic waste fraction of MSW) is heated under pressure in an oxygen-free atmosphere to a temperature between 350°C and 500°C (Tchobanoglous *et al.* 1993). As most organic substances are instable at these temperatures, thermal cracking and condensation reactions take place. This is also the reason that pyrolysis, which is basically an endothermic process, is often called destructive distillation. The product of the pyrolysis process consists of three fractions:

- A gaseous fraction containing a large variety of components such as methane, hydrogen, carbon monoxide and carbon dioxide.
- A liquid brown fraction consisting of a tar and pyrolytic oil, also containing components such as methanol and acetone with a relatively high-caloric value.
- A char consisting of pure carbon and inorganic material such as ash.

A schematic representation of the pyrolysis process is given in Figure 11.4. The exact composition of the various products of the pyrolysis process strongly depends on the composition and physical state of the waste, the type of pyrolysis system and the process conditions. All products of the pyrolysis can basically be used in a beneficial way as energy source or as feedstock for the production of specific chemicals. However, subsequent treatment will be necessary, especially in case of feedstock production.

Pyrolysis is applied worldwide as an industrial process for the production of charcoal from wood, coke and coke gas from coal. In the petrochemical industry pyrolysis is used for cracking heavy petroleum fractions. Up to now the application of pyrolysis for treating organic wastes originating from MSW has not been successful. The main reasons for this lack of success are the complex composition of the starting material and the poor quality of the end products. However, a lot of

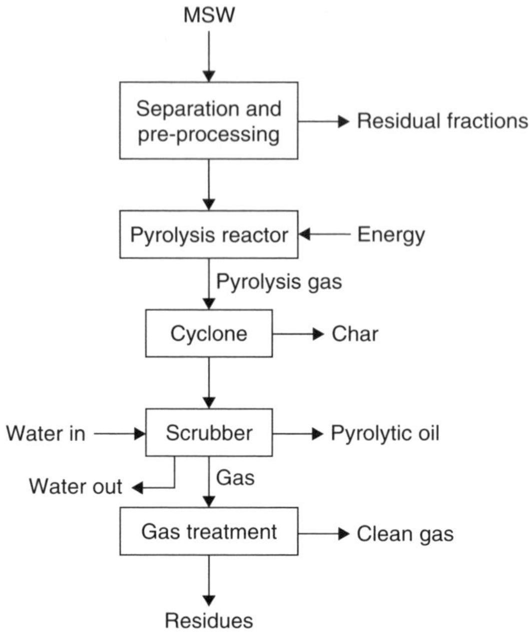

Figure 11.4 Schematical representation of the pyrolysis process.

attention is currently paid to the pyrolysis of biomass, focused on the production of energy rich fuels.

11.3.5.4 Gasification

Gasification is a process in which an organic waste (such as the organic waste fraction of MSW) is partly combusted and heated (800–900°C) in an atmosphere with an oxygen amount substantially lower than the stochiometric amount required for complete combustion (Tchobanoglous *et al.* 1993). It is an exothermic process in which the heat to sustain the process is derived from exothermic reactions. The products of the gasification process are:

- A low-caloric gas containing carbon dioxide, hydrogen, carbon monoxide, methane and other hydrocarbon components.
- A charcoal, also containing inerts originally present in the waste.

The process scheme of gasification is for a large part identical to that of the pyrolysis process with the exception that in the gasification process a certain amount of oxygen is supplied. The applied process temperature is higher than in the pyrolysis process. The gaseous and solid products obtained are strongly dependent on the composition and physical nature of the waste, the gasification system and the process conditions applied. Basically the obtained products, especially the gases, can be beneficially used as fuels or as feedstocks for the production of certain types of chemicals. However, it will be clear that the use as feedstock requires additional treatment of the gas, especially focused on quality improvement of the product. The obtained charcoal and inerts can also be used as a fuel. Similar to pyrolysis, gasification cannot presently be considered as a waste conversion technology that is technically and economically feasible to treat organic fractions of MSW. The main reasons are the complex composition of the material that has to be treated, the quality of the end products and the complexity of the process. Also similar to pyrolysis, gasification is getting more and more attention as a prospective technology for production of fuel from biomass. Compared to complete incineration of the organic fraction of MSW, pyrolysis and gasification have the advantage that the exhaust gas treatment required to prevent emission of pollutants to the atmosphere is less complex and therefore less expensive.

11.3.6 Discussion

The previous analysis has made it clear that from an environmental point of view it is not evident which technology for treating MSW is most sustainable. Besides the specific sustainability aspects also other aspects have to be taken into account in a complete evaluation. The most important aspects are:

- Technical feasibility and reliability.
- Financial feasibility (strongly dependent on the market value of the products).
- Public acceptability (based on rational and irrational arguments).

An important aspect in this evaluation is that the environmental problem caused by MSW has to be solved completely. In that respect the complete chain of the waste treatment, also including the final destination of the non-reusable residual fractions has to be assessed.

When comparing the aerobic and anaerobic composting system, the anaerobic system looks more sustainable although the majority of the current systems is based on an aerobic conversion process. Among the thermal processes, combustion with energy recovery is in fact the only one that is at present technically feasible. Comparison of biological and thermal treatment technology illustrates that biological treatment is much simpler than thermal treatment. However, with only biological treatment a complete solution of the waste problem is not possible. Mechanical and physical processes are always required and the quality of these processes strongly influences the quality and therefore the revenues of the final products. Combustion also requires the application of some mechanical/physical treatment. Another typical difference between biotechnological and thermal treatment is the difference in residence time of the waste in the process installation, which is 8–10 weeks for a complete composting process and only a few hours or even substantially less for the thermal process. The possible combination of composting and thermal treatment is interesting and may have considerable potential, especially when dealing with a low-quality compost and an incineration process that is suitable to incinerate compost with a high level of energy recovery.

11.4 TREATMENT OF SEWAGE SLUDGE

11.4.1 Introduction

Developments in municipal wastewater treatment during the past 30 years are characterised by continuous improvement of the quality of the effluent and of the efficiency of the treatment process. This development proceeded parallel with an increasing awareness of the problems associated with sewage sludge. These problems are the high costs of sludge treatment, which are often more than 50% of the total treatment costs of municipal wastewater, and the risks for the environment and human health caused by toxic pollutants and pathogens in the sludge.

Due to this increasing awareness regarding the risks for the environment and human health, the original application of the sludge as a fertiliser in agriculture has come under strong pressure. This is also caused by the increasing insight into the possible effect of these toxic pollutants and pathogens. Parallel to this development, the government policy and regulations regarding the application of sludge in agriculture have changed remarkably. However, it has also become clear that policy and legislation regarding sludge application and sludge management in general are strongly dependent on local, national and regional circumstances and conditions.

Environmental sustainability is becoming increasingly important in the development and application of innovative sludge treatment systems. A large part of the current research on sludge treatment is focused on the recovery of energy from sludge, eventually in combination with a beneficial use of the inorganic compounds in the sludge.

The primary aim of this section is to give a short overview of the various sewage sludge treatment methods and principles, mainly focused on the recovery of energy from sludge, and to evaluate these methods very briefly (Rulkens 2003). The various sludge treatment processes (or unit operations) that are relevant and that will be discussed further are subdivided according to the following groups:

- Anaerobic digestion.
- Incineration, gasification, pyrolysis, oil from sludge.
- Subcritical oxidisation, supercritical oxidisation and hydrothermal treatment.
- Integrated use of sewage sludge as energy and valuable material resource.

As sustainable management of sludge includes the abatement of the direct environmental problem caused by the sludge, some attention is also paid to the composition of the sludge as well.

11.4.2 Characterisation of sewage sludge

To identify potential scenarios for a more sustainable treatment it is useful to briefly assess the composition of the sewage sludge. As a rough guide this composition is characterised by five groups of components:

- Non-toxic organic carbon compounds (approximately 60% on dry basis) and nutrients containing nitrogen and phosphorous.

- Toxic pollutants:
  - Heavy metals, such as Zn, Pb, Cu, Cr, Ni, Cd, Hg, As (concentrations vary from more than 1000 ppm to less than 1 ppm).
  - PCBs, PAHs, dioxins, pesticides, endocrine disrupters, linear-alkyl-sulphonates, nonyl-phenols, etc.
- Pathogens and other microbiological pollutants.
- Inorganic compounds such as silicates, aluminates, calcium and magnesium containing compounds.
- Water, varying from a few percent to more than 95%.

The fundamental problem of the sludge is that all these compounds are present in one mixture. Organic carbon, phosphorous and nitrogen containing compounds can be considered as valuable compounds. This often holds for the inorganic compounds as well. Sustainable treatment involves the recovery and useful reuse of the valuable products and the minimisation of the possible adverse impact of sewage sludge on the environment and on human beings. For reasons of transport, disposal or efficient treatment it is also often required to remove the water. With respect to the amount of Kjeldahl-N in the sludge it has to be noted that this amount is small compared with the amount of Kjeldahl-N present in the wastewater. Kjeldahl-N is mostly present as ammonia, a potential fertiliser component. It has to be noted that both the industrial production of ammonia from an energy source, such as natural gas, and the biological destruction in the wastewater treatment requires energy. In modern wastewater treatment systems most of the Kjeldahl-N is volatilised as molecular nitrogen, using also a certain amount of the valuable organic carbon compounds present in the wastewater. The water in the sludge has in fact no real value, but is very often an obstacle and an important cost factor in sludge management.

11.4.3 Anaerobic digestion of sewage sludge

Anaerobic digestion is used to stabilise sludge and convert part of the volatile compounds into biogas. The biogas can be applied as an energy resource either at the wastewater treatment plant itself or elsewhere. Currently, anaerobic digestion is mainly applied at large- and medium-sized treatment plants. However, there is a growing interest in the application of anaerobic treatment in small-sized plants.

The vast majority of the anaerobic processes applied in practise are mesophilic (30–35°C). The sludge retention time in the anaerobic digestor amounts to approximately 20 days. Biogas production strongly depends on the type of sludge and also on the operating conditions of the digestor. The gas production from a mixture of primary and secondary (biological) sludge roughly amounts to 1 m^3 biogas per kilogram volatile solids biodegraded. Typical reductions in the volatile solids content are in the range of 20–40%, depending on the type of sludge. After the digestion process the sludge is dewatered. The supernatant of this dewatering process may contain high concentrations of phosphate and ammonia and needs further treatment. The result of the anaerobic digestion is a reduction in the final amount (dry matter) of sludge of 20–30%.

It is also possible to digest sludge thermophilically at a temperature of about 55°C. Compared to mesophilic digestion, thermophilic treatment has some advantages, such as a somewhat higher biogas production and a larger reduction in the amount of volatile solids. Also, the retention time of the sludge in the reactor can be reduced. The dewaterability of the sludge is also improved. In general more coagulant/flocculant is required for an appropriate dewatering process. The smell can be a disadvantage. Thermophilic digestion is also successfully applied in practise.

The biogas production from sludge can be increased by disintegration of the sludge, especially focused on disintegration of the bacterial mass in order to release organic substances that are easily accessible to a subsequent anaerobic biodegradation process. In the literature many pre-treatment methods to disintegrate sludge have been presented and discussed (Chiu *et al.* 1995; Tiehm *et al.* 1997; Scheminski *et al.* 2000; Goel *et al.* 2002; Müller 2002). The most important methods are:

- *Mechanical disintegration*. This process uses primarily mechanical/physical methods to disintegrate the sludge, especially secondary sludge. Mechanical disintegration techniques that have been investigated are, among others, ballmills, high pressure homogenisers (consisting of a high pressure pump and a homogenising valve), ultrasonic disintegraters, mechanical jet smash techniques (in which the sludge is forced at high pressure to flow through a nozzle, followed by a collision of the sludge flow at an inert surface) and high performance pulse techniques (in which shockwaves in the sludge are induced by application of a high voltage pulse of up to 10 kV).
- *Chemical treatment*. Three types of techniques can be mentioned: alkali hydrolyses, acid hydrolyses and ozone treatment. As prior to the anaerobic digestion step the chemically pre-treated sludge has to be neutralised, alkali and acid treatment require a relatively large amount of chemicals and cause, in general, a strong increase of the mineral content of the sludge liquid. That is the main reason that these two methods are not attractive. Ozone treatment is the most interesting chemical pre-treatment method. It causes a hydrolysis and solubilisation of cell walls resulting in a release of biodegradable cell material. Also, recalcitrantly biodegradable compounds are transferred into easily biodegradable substances. To obtain a substantial improvement of the biogas production, relatively high amounts of ozone are necessary: up to 5% of the total solids content of the sludge. Compared to the biogas production from sludge that has not been pre-treated, the production of biogas can be doubled. Also the reduction in volatile solids content of the sludge can be more or less doubled. A disadvantage of the use of ozone is that the dewatering properties of the sludge, in general, deteriorate.
- *Thermal hydrolyses*. This process aims at disintegrating and partially solubilising the biological cells in the sludge by heating the sludge to a temperature of 130–180°C during a period of 30 min. After cooling, the treated sludge can be digested. This principle is applied in the Cambi process. Compared to conventional anaerobic digestion an increase in biogas production of about 50% is

observed in combination with a strong reduction of the amount of volatile solids. An additional advantage is that the hydraulic retention time in the digestor can be reduced. In combination with a higher concentration of dry matter in the digestor this results in a much smaller digestor volume. Also a substantial improvement of the mechanical dewaterability is obtained, resulting in a lower volume of sludge. By means of an efficient heat exchange system and co-generation of heat and electrical power the net energy need of the treatment system can be minimised.

- *Enzymatic treatment.* This treatment process involves the use of enzymes to degrade cell walls. Costs of the use of enzymes are in general relatively high, while the effect on the biogas production is very limited.

The potential of the various pre-treatment processes to increase the anaerobic biodegradation rate and to produce a much larger amount of biogas is substantial. Also the lower residual amount of the dewatered sludge, which may have to be disposed of in a controlled landfill or further treated, is an advantage. Most systems, however, are still in the developing stage. To make a quantitative assessment with respect to the technical, economical and environmental feasibility, more data is necessary. Regarding the assessment of the feasibility of a pre-treatment method, the extra biogas production, total energy balance, final amount of sludge, amount of chemicals used, costs of sludge disposal and investment costs have to be taken into account and need to be analysed.

11.4.4 Incineration, gasification and pyrolysis of sewage sludge

Incineration of sludge is applied worldwide to reduce the sludge volume. Incineration is especially applicable when controlled disposal, agricultural or another beneficial use of sludge is not possible due to the heavy metal content of the sludge or due to the large quantities of sludge. Potential environmental problems related to sludge incineration include the emissions of pollutants with exhaust gases to the atmosphere and the quality of the ashes. However, there is a lot of standard technology available to abate the gaseous emissions very efficiently so that the stringent air quality standards can be met. Also the ash quality, especially with respect to heavy metals in the ash, is not a serious environmental problem. Due to the high temperatures applied in the incineration process and the composition of the inorganic compounds in the sludge, the heavy metals are very well immobilised and resistant to leaching. As in general the incineration process deals with large quantities of polluted exhaust gases, the costs of an efficient and adequate gas treatment system are very high. This is the main reason that sludge incineration is rather expensive. For mechanically dewatered sludge it is possible to recover energy from the combustion process. This energy can, for example, be used, if necessary, for drying of the mechanically dewatered sludge cake prior to the incineration process, or can be used for production of electricity. In practise, however, the recovery of energy from sludge incineration is seldom the primary aim. It will be clear, however, that the potential of sludge incineration to produce a surplus of energy, is substantial, provided that there is a suitable application for the beneficial use of this energy surplus and the entire process has been designed for both the incineration of the sludge and the chosen application for the energy surplus. To improve the recovery of energy from sludge incineration plants, several options can be mentioned:

- Application of more efficient heat exchange systems.
- Application of more efficient mechanical dewatering systems.
- Application of more advanced drying systems.
- Use of heat pumps.
- Co-generation of heat and power.

It will be clear that the use of additional equipment for energy recovery also results in higher operating costs.

To avoid the high costs of a stand-alone incineration plant for sludge and also to improve the energy recovery efficiency, the possibilities of incinerating dried sludge in a coal-fired power plant have been investigated. In that situation beneficial use can be made of the existing coal combustion installation and the existing exhaust gas treatment system. As the amount of sludge that is incinerated is small compared to the amount of coal, the effect of the incineration of the sludge on the quality of the exhaust gases and on the quality of the ash can be neglected. Also the net energy recovery from the incineration of the sludge can probably be strongly increased by improvement of the dewatering and drying process of the sludge and by the use of the low-caloric waste heat from the exhaust gases of the power plant. Use of the polluted drying gas in the incineration process can also contribute to better total process performance.

It is also possible to treat sludge by pyrolysis or gasification, comparable to pyrolysis and gasification of MSW (Stolarek and Ledakowicz 2001). In the pyrolysis process the sludge is heated under pressure to a temperature of 350–500°C in the absence of oxygen. The sludge is then converted into char, ash, pyrolysis oils, water vapour and combustible gases. Part of the solid and/or gaseous products of the pyrolysis process are incinerated and used as heating energy in the pyrolysis process. Depending on the equipment used and the applied operating conditions several modifications of this process exist. Gasification involves the breakdown of dried sludge in an ash and in combustible gases at temperatures usually about 1000°C in an atmosphere with a reduced amount of oxygen (Jaeger and Mayer 2000). In addition, thermal treatment techniques exist that combine the specific aspects of both pyrolysis and gasification.

Pyrolysis and gasification of sewage sludge have potential advantages compared to incineration. One advantage is that the conversion of the combustible gases of both systems into electrical power can be achieved more efficiently. In addition, valuable gases can be produced as basic chemicals or as fuel. Also the treatment of the waste gases is easier, mainly due to the much lower quantities of gas that have to be treated. However, the process performance of pyrolysis and gasification is much more complicated than incineration. From a technological point of view, these processes are still in their development phase. In order to offer a complete solution of the sludge problem, an adequate and sustainable solution (market) for both the valuable and the waste products has to be developed. It can therefore be concluded that, on basis of the existing technological development, gasification and pyrolysis are not short-term options to solve the sludge problem. However, this can change in the future.

An interesting process for efficient use of the organic carbon containing compounds, somewhat comparable to pyrolysis, is thermal conversion of these compounds to liquid fuel such as oil (Skrypsi-Mäntele *et al.* 2001). In this process, which consists of a large number of unit operations, and which is applied in practise, dried sludge pellets are heated in the absence of oxygen in a gas generator to a temperature of approximately 450°C. The organic compounds are converted into char, oil and non-condensable gases. These gases are brought into contact with the char and converted into straight chain hydrocarbons, which are then condensed into oil. The char is used as energy source to provide the energy for the gas reactor. The quality of the oil, important for the application of the product, can be controlled by using catalysts and by application of the right process condition. It will be clear that the economic feasibility of this process strongly depends on the market value of the oil.

11.4.5 Subcritical oxidation, supercritical oxidation and hydrothermal treatment of sewage sludge

Subcritical and supercritical oxidation involve the oxidation of sludge in the water phase at high temperatures and pressures (Shanableh and Shimizu 2000; Lendormi *et al.* 2001). Subcritical oxidation, also including wet air oxidation (WAO), occurs at temperatures and pressures below the critical point of water (374.2°C and 22.1 Mpa) at temperatures between 120°C and 300°C. In this process two major reactions take place: (a) hydrolysis of complex organic compounds into simpler molecules, such as carbohydrates, amino acids, and low molecular volatile fatty acids which are soluble in water and (b) oxidation of organic compounds into oxidation products such as carbon dioxide and water. Nitrogen containing organic compounds are in general converted into ammonia. With subcritical oxidation only a partial oxidation of the organic compounds can be obtained. The process is mainly applied in practise as a sludge conditioning process aimed at improving of the dewaterability of the sludge and reduction of the final amount of sludge. The degree of oxidation of the sludge (the degree of conversion of the organics), the dewaterability of the sludge, and the composition of the supernatant strongly depend on the process parameters applied such as temperature and retention time of the sludge in the reactor. Several technical modifications of this process exist, most of which are applied in practise. With respect to energy recovery from the sludge in the subcritical oxidation process, two possibilities are relevant. The first possibility is the direct recovery of heat from the wet oxidation process by heat exchange. The second possibility is the use of the volatile fatty acids (such as acetic acid) and other organic soluble or colloidal compounds in the supernatant as a resource for the production of biogas.

The use of these volatile fatty acids as a carbon source in the denitrification process is also an option.

Supercritical oxidation occurs at temperatures and pressures above the supercritical point of water. Supercritical water has special properties such as a superior ability to dissolve oxygen and organic compounds. But it has a low solubility for mineral compounds such as sodium chloride. With supercritical oxidation the organic compounds are completely oxidised. Nitrogen, present in nitrogen containing compounds such as ammonia and amino acids, is converted into pure nitrogen. The oxidation rate at supercritical conditions is much higher than at subcritical conditions. The required retention time of the sludge in a supercritical reactor is in the order of a few seconds to 1 minute. In subcritical oxidation this retention time amounts to approximately 30 minutes. Also with supercritical oxidation superior dewatering properties of the sludge can be obtained. Energy recovery from this oxidation process can occur directly by heat exchange in the reactor or from the exit flow from the reactor. Compared with subcritical oxidation processes, supercritical oxidation processes are much more complex with respect to reactor design, reactor materials and reactor operation. It is much more expensive than subcritical oxidation. The process takes place at high temperatures, which together with the presence of oxygen causes an aggressive oxidising environment. The presence of chlorides in the sludge can cause corrosion. Minerals can cause scaling and precipitation at the reactor walls. The presence of inert inorganic particles requires special measures with respect to heat exchange as well.

Compared with sludge incineration, supercritical oxidation has the advantage that off gas treatment is very simple so that the costs of off gas treatment can be neglected. However, as already mentioned, supercritical treatment is much more complicated. In fact the process is still in its developing phase. A lot of research is still necessary to develop this process for large-scale treatment of bulk waste materials, such as sludge.

Thermal decomposition of sludge in the water phase at high temperatures (between 150°C and 450°C) can also occur without using oxygen or another oxidant. This process is called hydrothermal treatment and is also still in the developing phase (Shanableh and Jomaa 2001). In the absence of oxygen or another oxidant, the main reaction mechanism of hydrothermal treatment is hydrolysis of sludge compounds, resulting in the production and accumulation of high concentrations of dissolved organic compounds in the sludge water phase. For example, thermal decomposition of proteins results in amino acids, the decomposition of lipids produces fatty acids and the decomposition of fibre material and hydrocarbons leads to low molecular hydrocarbon substances such as sugars. The composition of the soluble compounds in the water phase strongly depends on temperature and treatment time. Use of oxidants in the hydrothermal treatment, which makes the process more comparable to wet oxidation, has a positive effect on the formation of volatile fatty acids. These compounds are very interesting as a carbon resource, not only in the production of biogas but also in the denitrification process and the biological phosphorous removal from wastewater.

11.4.6 Integrated use of sewage sludge as energy source and construction material

In sewage sludge both the organic carbon containing compounds and the inorganic compounds represent valuable materials. There are several possibilities to use these compounds simultaneously in a useful and beneficial way. A lot of effort has been put into the manufacturing of valuable products by thermal solidification of the inorganic sludge compounds, especially in Japan. The starting point in this production process is either incinerator ash, obtained after combustion of sludge, or dried sludge. The solidification process occurs at high temperatures, up to 1000°C. Depending on the specific process modifications and the applied operating conditions, various types of products can be made, such as artificial light weight aggregates, slags and bricks. Even with recovery of the energy represented by the organic compounds in the sludge, each process route (also process routes including the prior incineration step and/or drying step), requires always a net input of external energy. A lot of experience exists with all types of processes, mainly in Japan (Endo *et al.* 1997; Okuno *et al.* 2002).

Another beneficial way to use the inorganic and organic compounds of the sludge, is through the use of sludge in the production of Portland cement (Taruya *et al.* 2002). In this process, which is also applied in practise, either the ash or the dried sludge can be used. In general the amounts of ash or dried sludge that are utilised as raw materials are only a few percent of the total amount of the raw materials used. Specific requirements set to the sludge quality are a sufficiently low concentration of phosphate and mercury in the sludge.

With respect to the above-mentioned application of sludge, it is concluded from the current practise that the organic carbon containing compounds in the sludge can provide a substantial part of the total energy need. However, in most cases an additional amount of external energy is still necessary. There are several options to reduce these external energy requirements. The most important options are:

- Improvement of the mechanical sludge dewatering before sludge incineration or sludge drying.
- Improvement of the energy efficiency of the sludge drying process, for example by using waste heat, use of heat pumps, or use of advanced drying processes.

11.4.7 Mechanical dewatering and drying of sewage sludge

In current sludge treatment processes the dewatering of sewage sludge is a key process. Two types of dewatering processes are relevant: mechanical sludge dewatering and sludge drying. Mechanical dewatering occurs by means of centrifuges, sieve belt presses or filter presses. With these processes sludge can be dewatered to a final dry solids concentration of 20–50%. Mechanical dewatering is a crucial step if the final destination of sludge is disposal in a controlled disposal site. It is very often also an important treatment step if the sludge has to be incinerated. Sludge drying is aimed at a complete removal of the water from the sludge. The drying process is often preceded by a mechanical dewatering process. Sludge drying is an important pre-treatment step in case of incineration (if the mechanically dewatered sludge is not directly fed to the incinerator), pyrolysis or gasification of the sludge. Sludge drying is often also applied if the inorganic compounds of the sludge are used for the production of valuable products such as cement, bricks, light weight aggregates and the organic material in the sludge is used as energy source. Costs of mechanical dewatering and sludge drying contribute significantly to the total operational costs of sludge treatment and therefore to the total costs of the entire wastewater treatment process.

11.4.8 Discussion

It will be clear from the given overview that many options/systems exist to recover energy from sewage sludge. Some of these options are already applied on a practical scale, others are still in a stage of development or in a very preliminary research phase. Which options are most beneficial to be applied or to be developed in the future not only depends on specific local and regional circumstances, but also on the national and international policy and legislation with respect to sewage and sludge treatment. A solution for the sludge problem can also be developed in combination with other industrial activities, such as the production of electrical energy from coal-fired power plants, the production of Portland cement, or the treatment of other types of wastes, such as MSW. Therefore, it can be expected that future developments in sludge treatment are also strongly influenced by the developments in these areas.

The previous analysis also made clear that the sludge problem cannot be solved by biological processes only. Physical and chemical processes will always be necessary. One reason is the fact that the final destination of sludge requires a final product with low water content. This means that mechanical dewatering, eventually in combination with a drying step or incineration step has to be applied in all cases. The other reason is that sludge contains toxic organic and inorganic compounds and that these compounds require appropriate treatment such as immobilisation or destruction. The presence of toxic compounds in sludge is another reason that the beneficial use of biological treatment steps is mainly limited to the application of anaerobic digestion processes. Other biological treatment processes are of course possible, but these processes are much more complicated and require complex post-treatment steps, both for recovery of the valuable products originating from the biological treatment step, and for further treatment of the sludge.

It has to be noted that the conclusions of this discussion are mainly based on the assumption that energy recovery from sludge is the primary aim of the sludge treatment process. If the recovery of phosphate is desired in addition to energy, the conclusions are more or less the same. However, in that case also biological processes intended to concentrate or release phosphate may also become important tools, depending on the chosen scenario for recovery of phosphates.

11.5 TREATMENT OF PIG MANURE

11.5.1 Introduction

The tremendous growth in livestock farming during the last decades has in several countries resulted in significant amounts of animal wastes. These wastes can cause serious environmental problems such as phosphate supersaturation of soils, eutrophication of surface waters, nitrate leaching to groundwater, and emission of ammonia and other odour compounds to the atmosphere (ten Have 1991; Taiganides 1992; El-Ahraf and Willis 1996). In the Netherlands, the most problematic agricultural waste is liquid pig manure or pig slurry. Legislation restricting the use of fertiliser minerals on agricultural land has resulted in annual pig slurry surpluses that amounted to several million tonnes in the year 2000. Problems associated with pig slurry are caused by the high concentrations of nutrients, such as nitrogen and phosphorous compounds, and minerals, for example potassium and the presence of odour compounds.

Solving environmental problems associated with pig slurry requires a broad integrated strategy, including:

- Feed measures such as reduction of phosphorus and nitrogen containing compounds in the feed, the use of essential amino acids (e.g. lysine) and addition of the enzyme phytase to improve the digestibility of phosphorus containing compounds resulting in a more efficient phosphorus uptake.
- Emission control to reduce gaseous odour and ammonia emissions from livestock buildings, storage facilities and land application.
- Optimisation of pig slurry distribution and utilisation on agricultural land.
- Processing of pig manure slurries, focused on the recovery of valuable products for application inside or outside the agricultural sector or for improvement of nutrient management efficiency in land application. With respect to the scale of processing, a distinction may be made between single-farm scale and regional scale (conglomerations with high concentrations of livestock farms).

For a relatively long period, it has been believed in the Netherlands that large-scale central processing would contribute significantly to the solution of the pig-slurry surplus problem. However, central processing of slurries requires high investments and a sound organisational and financial basis, as well as a well-organised network for the distribution and marketing of the products. Due to lack of such a sound basis, combined with the predicted effect of farm-scale measures on the future volume and composition of pig manure, the introduction of large-scale central animal-slurry processing failed. Nevertheless, the experience gained with central-processing concepts together with the development of techniques for the treatment of pig slurry on a farm scale have resulted in more insight, a growing awareness, and a new approach to the animal-waste problem. This new approach involves a full integration of the relevant processes in livestock farming: feed and way of feeding to the animals, processing of the solid and liquid wastes that can be reused at farm scale, recovery of energy from ventilation air, and aspects dealing with animal health and welfare.

On a single farm, the possibilities for animal-slurry processing seemed at first limited. However, on such a farm environmental problems can often be solved in a more sustainable way by integration of the technological solution of the manure problem into the livestock farming process. Based on this approach research is going on into the development of an innovative new animal friendly housing system for pigs that also produces more environmentally friendly organic fertilisers replacing the traditional pig slurry.

The compounds in pig slurry that are environmentally most problematic are undoubtedly the nitrogen containing compounds, mainly ammonia. These compounds may contribute to the pollution of all three environmental compartments: soil, surface water and groundwater, and the atmosphere. However, ammonia also represents a very valuable component because it can be used as basis for an artificial fertiliser or as a chemical to be applied in denoxification processes of incineration gases.

This section will focus on two aspects. The first aspect is the role of physical/chemical environmental technologies and environmental biotechnologies in a central-processing system for pig slurry. The second aspect focuses on the significance of these type of technologies for the recovery of nitrogen compounds from pig manure.

11.5.2 Composition of pig manure

The characteristic composition of pig manure in the Netherlands is given in Table 11.2. From this table it is clear that pig manure mainly consists of water. In fact, pig manure should be considered as agricultural wastewater.

Table 11.2 shows that pig slurry contains many different compounds that may be regarded as valuable substances when recovered in a pure form. This provides a sound basis for its reuse as valuable products. The four main categories of compounds in pig slurry are:

- Organic compounds, which are mainly present as suspended solids.
- Nitrogen containing compounds, including ammonia, which are mostly dissolved in the liquid phase.
- Phosphorus containing compounds, which for the most part are present in the suspended fraction.
- Minerals (e.g. potassium, sodium, chloride, calcium, magnesium, sulphate) most of which are dissolved in the liquid phase.

Nitrogen is excreted by pigs as urea in the urine and as part of organic compounds in the faeces. When in contact with manure or manure-stained surfaces, urea is rapidly converted to ammonium as a result of an enzymatic reaction, and this ammonium volatilises to form ammonia, which spreads in animal houses. During long-term storage, organic nitrogen compounds are degraded anaerobically, leading to a slow increase in ammonium concentration in the slurry.

Table 11.2 Characteristic composition of pig manure (indicative values).

Compound	Concentration (kg/m^3)
Water	900
Dry solids	100
Volatile solids	70
Non-volatile ash	30
Suspended solids	80
Chemical oxygen demand	120
N-Kjeldahl	8
N-ammonia	6
P (as P_2O_5)	5
K (as K_2O)	8
Ca (as CaO)	4
Mg (as MgO)	2
Cl^-	2
S (as SO_4^{2-})	2
pH	7

The values given in Table 11.2 are only indicative. In the future, pig slurry will probably change in composition as a consequence of technobiological developments, leading to more concentrated slurries containing less total nitrogen and total phosphorus.

11.5.3 Centralised processing of pig manure

11.5.3.1 Introduction

Large-scale central processing of manure is aimed at the separation of manure into clean water and valuable products that can be reused. A large-scale central plant for processing pig manure has to meet environmental, technical and financial constraints. This means that the treatment process should meet present and future standards with respect to the emission of pollutants into the atmosphere, surface water and soil. Several centralised-processing systems for pig manure have been developed in the Netherlands (ten Have 1993; Rulkens and ten Have 1994). Two processes will be discussed which, however, are currently out of operation because of reasons already mentioned in the previous paragraph. One process is the so-called Promest system, which is partly based on biotechnological process steps. The other process is the so-called Memon process, which is based on physical/chemical process steps.

11.5.3.2 The Promest system

A simplified process diagram of the Promest system is presented in Figure 11.5. The process starts with a screening step to remove large objects from the raw manure. Next, the manure is fed to an anaerobic reactor in which biogas is produced. This biogas is used as an energy source. Another purpose of the anaerobic treatment is to convert volatile acids, which otherwise would be emitted in the drying step. The anaerobic treatment facilitates the mechanical separation applied to the slurry leaving the anaerobic digester. In this mechanical separation step, a centrifuge is used, which separates the manure in a cake with a dry matter content of about 25% and in a liquid effluent with a dry matter content of about 3%.

The liquid from the centrifuge is passed to an aerobic biological treatment step, in which the volatile ammonia is converted into nitrate and part of the organic compound is oxidised. To prevent pH reduction lime is added in this step. The effluent resulting from the biological nitrification step is concentrated in a multi-effect evaporator. The concentrate of this evaporator, which has a dry solid content of about 25%, is fed into an indirect dryer system, together with the cake of the centrifuge. As a result of the anaerobic and aerobic biological treatment processes, the condensate from the evaporator is practically pollutant-free and can be discharged into surface water.

The vapour from the dryer, which contains water, dust particles, ammonia and other odour compounds, is used as a heat source for the evaporation process. For that purpose the vapour is fed to the heat exchanger of the evaporator. The condensate of the dryer, containing some volatile pollutants such as ammonia, is treated in the aerobic biological treatment step. The non-condensables are treated in an incinerator. The product of the dryer is pelletised. To meet the required percentages of N, P and K in the products, minerals can be added to the product.

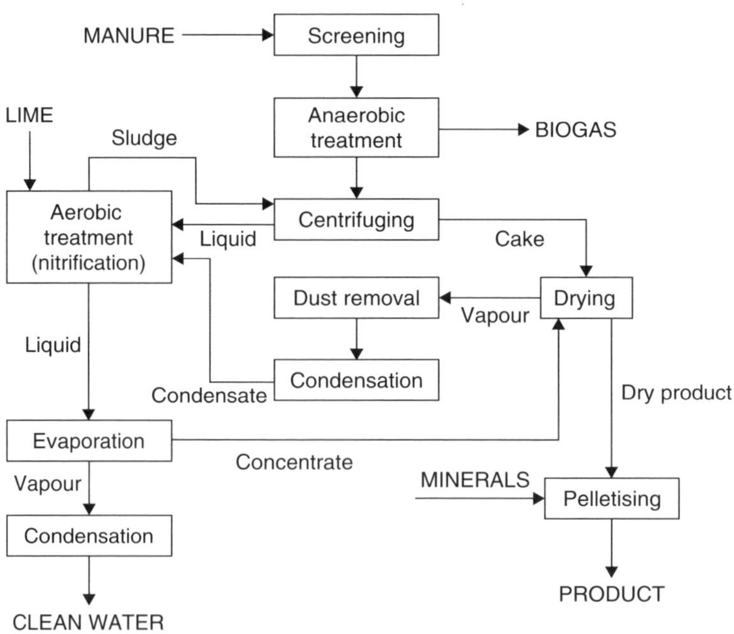

Figure 11.5 Simplified process diagram of the Promest system.

11.5.3.3 The Memon system

A simplified process diagram of the Memon system is shown in Figure 11.6. The Memon system is based on the so-called Carver-Greenfield drying process (Rashidi *et al.* 1986). The process starts with a screening step to remove large objects from the raw manure. Next, the manure is acidified with sulphuric and/or nitric acid in order to lower the volatility of the ammonia. The acidified manure is mixed up with a non-toxic biodegradable carrier oil with a boiling point far above that of water and a negligible solubility in water.

Using a carrier oil with a high boiling point, in this case a paraffin oil, makes it possible to dehydrate the slurry completely in a multi-effect evaporator. After dewatering, the dried manure particles are separated from the carrier oil in a centrifuge. Part of the separated carrier oil is fed into the mixer and the remaining part is treated in a stripper, in which fatty manure compounds, dissolved in the carrier oil, are separated from the carrier oil. The stripped carrier oil is fed back into the evaporator, while the fatty manure compounds are removed as a so-called "pig oil".

The remaining carrier oil in the dewatered manure cake is removed in a stripper consisting of a dryer in which steam is used as stripping gas. The mixture of oil vapour and water vapour from the stripper is returned to the evaporator. Due to the acidification, the major part of the ammonia remains in the manure during the evaporation and drying process.

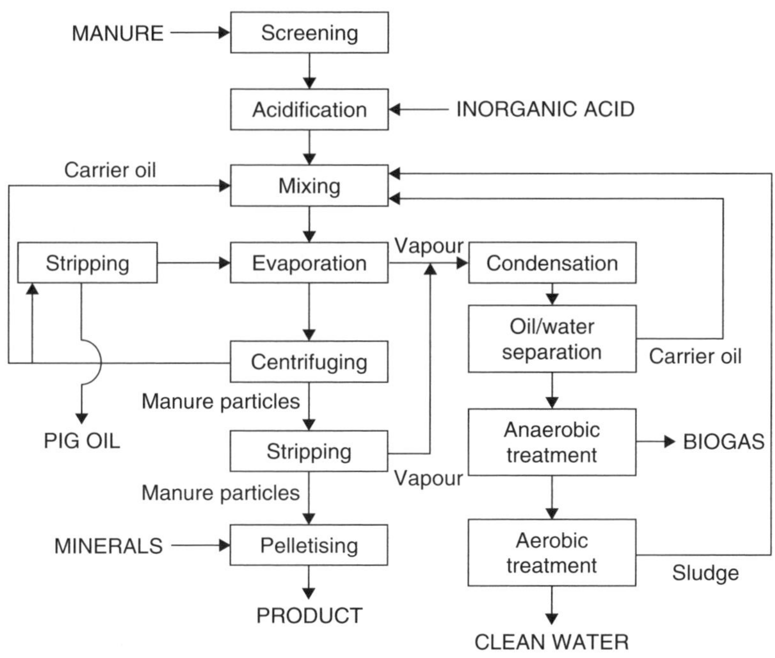

Figure 11.6 Simplified process diagram of the Memon system.

The vapour from the evaporation process consists of water, some carrier oil, volatile organic compounds such as fatty acids, and also a small amount of ammonia. The amount of ammonia strongly depends on the amount of acids used for acidification. After condensation, the carrier oil and water phase can be separated easily. The separated carrier oil is reused. The water phase has to be treated in order to remove the fatty acids, ammonia and traces of carrier oil. For that purpose, the water phase is fed into an anaerobic system, followed by a combined nitrification/denitrification process. Also physical/chemical treatment is possible.

11.5.3.4 Evaluation

Comparison of the two treatment processes shows several major differences. The most important are:

- The Promest system comprises two essential biological treatment steps: anaerobic treatment and aerobic treatment (nitrification). Due to the anaerobic step, the residence time of the manure in the Promest system is substantially higher than in the Memon system. This also means that a much larger treatment plant is required.
- The Promest system needs more energy due to the separate manure dryer, however it also produces a part of the energy needed.
- In both processes the nitrogen is recovered as a valuable component. However to that aim in the Memon system a lot of inorganic acid is necessary in order to achieve that goal. These inorganic acids end up in the dried product.
- The Memon system starts with purely physical/chemical process steps and is therefore less sensitive to components that may disturb biological processes.

Though not fully illustrated by the brief process schemes in Figures 11.5 and 11.6, it should be clear that both treatment processes are rather complex. What type of treatment process is most attractive can be derived from an integrated evaluation of these processes. In this evaluation both technical and financial aspects have to be considered. In the financial evaluation not only are the total treatment costs important but also the revenues from the products. It is very difficult to make an exact estimate of the revenues, because the market for these products has to be developed. Therefore, the general conclusion from this comparison is that at present it is impossible to postulate that a more biologically oriented central treatment process is preferable to a more physically/chemically oriented central treatment process.

11.5.4 Recovery of nitrogen compounds from pig manure

11.5.4.1 Introduction

Nitrogen compounds in pig manure are valuable compounds which can be recovered for beneficial use. Recovery of nitrogen compounds can be part of an integrated complete treatment system for pig manure. However, it is also possible to recover nitrogen compounds, directly or after a specific conversion step, without the direct aim to solve the complete environmental problem of the manure. Such a treatment may of course contribute indirectly to a sustainable solution. To that aim various types of physical/chemical and biological treatment steps are available (Rulkens *et al.* 1998).

11.5.4.2 Physical/chemical steps to recover ammonia

Several physical/chemical treatment steps are available to directly recover nitrogen compounds, in this case especially ammonia, from pig manure. As these processes have been extensively elaborated in literature and handbooks, they will be mentioned only very briefly.

Stripping of ammonia in combination with absorption In stripping, the fact that ammonia has a high relative volatility at high pH values is utilised. The process is characterised by the following process steps:

- Lime or sodium hydroxide is added to the slurry to increase the pH. As a result, the equilibrium in the reaction $NH_4^+ + OH^- \rightleftharpoons NH_3 + H_2O$ shifts to the right. This shift promotes the transfer of ammonia to the gaseous phase.
- The slurry is stripped with air or steam at a high temperature. A higher temperature raises the volatility of ammonia.
- Ammonia resulting from the gas-stripping phase is absorbed into a liquid phase, for example water or an acid solution (HCl, H_2SO_4, H_3PO_4, or HNO_3).

Transmembrane chemo-sorption In this process, hydrophobic, microporous, hollow-fibre membranes with gas-filled pores are used. The gas phase enables transport of gaseous ammonia through the membrane, while it forms a barrier for other components (Klaassen and van Voorneburg 1995). The process is characterised by the following process steps:

- The slurry, free of particles, is led through the hollow-fibre membranes.
- Lime or sodium hydroxide is added to raise the pH of this liquid and thus increase the concentration of volatile gaseous NH_3.
- An aqueous solution of an inorganic acid is led along the external area of the membrane.
- Ammonia diffuses from the slurry through the gas-filled pores of the membrane to this acid solution.

Precipitation of ammonia from an aqueous solution Precipitation of ammonia can occur because certain types of ammonia salt are only moderately or barely soluble in water. The most important mineral is magnesium ammonium phosphate hexahydrate, $MgNH_4PO_4.6H_2O$ (struvite) (Buchanan *et al.* 1994). The solubility of this mineral in water is about 98.7 mg/l. The precipitation process is characterised by the following process steps:

- A sufficient amount of soluble magnesium and/or phosphate ions is added to liquid manure (magnesium hydrophosphate solution or magnesium oxide is a suitable reagent).
- The precipitate is separated from the manure.
- The precipitate is heated to evaporate ammonia and to recover magnesium hydrophosphate for reuse.
- The evaporated ammonia is absorbed in water or an aqueous solution of a mineral acid.

11.5.4.3 Physico-chemical and biological pre-treatment processes increasing the ammonia content of liquid agricultural wastes

Organic nitrogen can be converted by physico-chemical or biological pre-treatment. During this process, the organic matter is degraded, resulting in an increase in ammonia in the liquid phase. Two examples of this kind of treatment are wet oxidation at a high temperature and anaerobic digestion.

Wet oxidation During wet oxidation organic matter is oxidised in a water phase by adding oxygen at a high temperature and a high pressure. In this process, organic components are generally degraded to form carbon dioxide and water. During wet oxidation of agricultural waste slurry, much energy is produced which can be used in other treatment steps (e.g. drying and evaporation).

At subcritical conditions (temperature $<374°C$), organic nitrogen is converted to ammonia. The high temperature of the medium offers an ideal opportunity for stripping or evaporating ammonia from the liquid and to recover it from the resultant vapour in the form of a mineral salt, such as ammonium sulphate, ammonium phosphate or ammonium nitrate.

Anaerobic digestion During the anaerobic digestion of pig manure, part of the organic compounds in the manure are converted into biogas. The latter may provide energy for other treatment steps, such as drying and evaporation. The organic nitrogen present as amino acids, urea and protein and in fibres in the slurry is for the major part converted to ammonia. Experience with this process has been gained on a large scale.

11.5.4.4 Biological conversion of nitrogen compounds aimed at production of valuable products

Microbiological processes can be used to convert nitrogen compounds in agricultural wastes to valuable and reusable products (Taiganides 1992). Basically these products can be removed from the treated slurry to be used as a separate product.

It is also possible that after biological conversion the nitrogen containing product is further processed simultaneously with other components of the manure. The most important biological conversion processes are briefly presented below.

Microbiological conversion of liquid animal wastes into single-cell proteins Nitrogen in animal slurries may be present in fibre solids, in bacteria, or in some soluble form like proteins, ammonia and free amino acids. The primary aim of microbiological conversion of these nitrogen compounds is the conversion to useful products that can be used as animal feed or as a source for the production of other valuable compounds. Much research on biological conversion processes of manure nitrogen has been carried out in the past in which algae, yeast, fungi, bacteria, earthworms and insects are grown on animal excreta for protein syntheses. The most interesting systems are algae growth, fungal fermentation and bacterial growth.

Algae are grown in diluted slurry ponds. They use solar energy to synthesise cellular material from carbon dioxide, ammonia, organic nitrogen compounds and other essential nutrients. In algae ponds aerobic bacteria breakdown organic matter and release carbon dioxide. This carbon dioxide is consumed during the day by algae that release oxygen during the night. The oxygen promotes the growth of bacteria. This symbiotic existence accelerates the total conversion rate.

Single-cell protein can also be produced from liquid animal wastes by fungal fermentation. These processes require closed sterilised systems to prevent interfering growth of unwanted micro-organisms. Fungi are able to produce a wide range of enzymes so they can adapt to many substrates. This makes many fungi suitable for liquid fermentation of animal slurries.

Conversion of nitrogen compounds present in liquid animal wastes into single-cell protein is also possible by the use of aerobic bacteria. A standard biological wastewater treatment system like an activated sludge reactor may be used for this purpose. The product is a mixture of bacterial protein mass and organic solids which have to be harvested simultaneously.

Comparison of the three types of biomass acquired from algae, fungi and bacteria leads to the conclusion that conversion of nitrogen compounds and resources of total Kjeldahl nitrogen into valuable products is the highest for fungi. However, the overall quality of the treated liquid waste is the highest for algae and bacteria. It is further observed that all these conversion processes are strongly influenced by many process factors.

Nitrification of liquid pig manure During research and development of large-scale manure processing plants in the Netherlands much attention has been paid to biological nitrification. In this process bacteria oxidise ammonia to nitrate. As opposed to ammonia, nitrate is not volatile and can therefore be concentrated in evaporation and drying steps. The most common way of nitrification is achieved in the activated sludge process. As acid is formed during nitrification, a neutralising agent like lime ($Ca(OH)_2$) has to be added in order to maintain the optimum pH for the nitrification to proceed. An additional aspect of nitrification is the occurrence of spontaneous denitrification that reduces the concentration of the valuable compound nitrate to worthless nitrogen gas.

Manufacturing of lysine from pig manure For more than 20 years the amino acid lysine has been used as a feed additive for pigs. Sanders (1993) investigated a process for the fermentative production of lysine using ammonia from pig manure as a raw material. This has been investigated on a small pilot plant scale. In this process anaerobically fermented manure is separated in a solid and a liquid phase. The liquid phase is used for lysine production. It is first sterilised by a heat-shock system and then sterilised molasses is added. Lysine is produced by a bacterial strain of the species *Corynebacterium*. After removal of suspended particles by centrifuging, lysine is recovered from the liquid phase by an ion exchange system. It appears that the lysine produced using pig slurry is very similar to the lysine that has been on the market for years as a feed ingredient. With this fermentation process approximately 75% of the nitrogen containing compounds present in the separated liquid phase is converted to lysine. The rest of the nitrogen is still present in an insoluble form or has been converted to bacterial cell matter that can be harvested easily by centrifugation.

The commercial value of lysine may contribute to a reduction of the total treatment costs of animal slurries. An additional advantage of lysine production as a step in a pig slurry treatment process is also that wastewater and other wastes from the lysine production can easily be treated in the pig manure treatment plant. Though laboratory scale research was successful, this process has not been applied at full scale yet.

Production of duckweed Another option is the cultivation of duckweed in ponds containing manure and wastewater. This process is applied on a large scale in South-East Asia and North and South America, in areas with a moderate climate. This duckweed can be used as feed for fish. During an investigation carried out in the Netherlands, it was found that this process can also be applied in cold climates, provided that the manure is heated to the appropriate temperature (Beaujean 1997). In this investigation, it was also found that not only N and P but also large amounts of minerals (e.g. potassium) can be removed from manure in this way.

11.5.5 Discussion

Liquid livestock wastes contain valuable products, including nitrogen and phosphorus containing compounds and organic carbon compounds. The challenge of how to manage this type of waste is primarily focused on the way to recover these valuable compounds in combination with solving the environmental problem caused by these wastes. From the previous analysis it will be clear that a complete solution of this environmental problem, that will also be sustainable with respect to energy use and recovery of valuable products, requires a combination of several physical, chemical and microbiological treatment steps. Microbiological steps have a high potential to produce specific products with a high intrinsic value from the agricultural waste. However, the direct contribution of such high value-adding biotechnologies to the solution of the environmental problems is limited. Physical treatment steps focused on the separation of various compounds from the manure are always necessary. There are in general many scenarios to solve the environmental problem caused by manure in a sustainable way. Each of these scenarios results in various valuable products. The products from a treatment process can be used both inside and outside the agricultural sector. The market value depends not only on the type of product but also on the local situation, the governmental policy, the quality of the product and the acceptance of the product.

It will also be clear from the previous discussion that the analysis and implementation of strategies to manage the manure problem is different from that applied in MSW or sludge treatment. The main reason is of course the absence (currently) of very toxic compounds such as heavy metals and pesticides in the manure. This provides more flexibility in the selection of a treatment scenario. However, from a market point of view the quality of the products from a manure treatment process should be high. In the case of manure this means awareness of the problem of pathogens and weed seeds in products that will return to the agricultural sector.

11.6 GENERAL DISCUSSION

In the previous sections it has been made clear how environmental technologies can be applied to treat three different types of biowastes in such a way that not only the problem caused by these wastes is eliminated, but also that valuable components present in these wastes are recovered for beneficial reuse or energy production. Further it has been shown that a complete elimination of the environmental problem caused by these wastes always requires a combination of various types of environmental technologies. It is also obvious that for each type of (bio)waste several scenarios exist that can be applied to satisfy the conditions of the elimination of the biowaste simultaneous with the beneficial recovery of valuable products from that waste. Which of these scenarios is most optimal depends on many factors including technological, financial and marketing factors, as well as factors dealing with social acceptability of the chosen solution and with the local, regional and national environmental policies. In fact the solution of a biowaste problem always requires a tailor-made approach. And it can also be expected that with increasing insight in the environmental problem and progress in waste prevention methods, the development of new, innovative, environmental technologies and optimal scenarios is in fact a very dynamic process, changing continuously. This is one of the uncertainties decision makers working on waste treatment have to deal with.

The previous analysis has also illustrated the role of biotechnological processes in the treatment of these biowastes. Biotechnological processes are attractive at places where we have to deal with conversion processes, as part of the integrated treatment scenario. As already discussed in paragraph 11.2, biological conversion processes are characterised by the fact that conversion of components present in the waste can be carried out very specifically. They are comparable with natural conversion processes and the biological closed loop systems taking place in nature. Hence, the potential of these biological conversion processes is higher than those of chemical/thermal conversion processes. However, it is also evident from the previous analysis that chemical/thermal processes can sometimes be more attractive and more environmental sustainable. This is especially the case when dealing with (bio)wastes that also contain toxic organic pollutants that are recalcitrant or not biodegradable.

In all cases biological or chemical conversion processes have to be combined with physical processes focused on separation of components. Therefore, the discussion which type of environmental technology is more attractive to treat a certain type of biowaste, has to be focused on the core process as well, especially the conversion process. This process strongly affects the choices of pre-treatment and post-treatment steps in a complete integrated treatment scenario.

Finally, it is obvious that in the previous analysis many types of environmental treatment steps and treatment scenarios have been discussed and elaborated. In that respect this chapter provided a general overview of the most important biological, chemical and physical treatment steps. However, of course, this overview must not be considered as complete. There are potentially many more environmental technologies which may become attractive for practical application in the future.

REFERENCES

Beaujean, J.M.E. (1997) *Duckweed in the "Peel Region" (in Dutch)*. Bogey Venlo BV, The Netherlands.

Buchanan, J.R., Mote, C.R. and Robinson, R.B. (1994) Thermodynamics of struvite formation. *Trans. ASAE* **37**(2), 617–621.

Cheremisinoff, N.P. (2003) *Handbook of Solid Waste Management and Waste Minimization Technologies*. Elsevier Science, USA. ISBN 0-7506-7507-1.

Chiu, Y.-C., Chang, C.-N., Lin, J.-G. and Huang, S.J. (1997) Alkaline and ultrasonic pretreatment of sludge before anaerobic digestion. *Water Sci. Technol.* **36**(11), 155–162.

El-Ahraf, A. and Willis, W.V. (1996) *Management of Animal Waste, Environmental Health Problems and Technological Solutions*, 185 pp. Praeger, London.

Endo, H., Nagayoshi, Y. and Suzuki, K. (1997) Production of glass ceramics from sewage sludge. *Water Sci. Technol.* **36**(11), 228–235.

Evans, G. (2001) *Biowaste and Biological Waste Treatment*. James & James (Science Publishers) Ltd, London, UK. ISBN 1-902916-08-5.

Goel, R., Tokutomi, T. and Yasui, H. (2002) Anaerobic digestion of excess activated sludge with ozone pretreatment. *IWA World Water Congress 2002 Melbourne*. Paper no. e21393a.

Henry, G.J. and Heinke, G.W. (1996) *Environmental Science and Engineering*, 2nd edn. Prentice Hall, Upper Saddle River, NJ. ISBN 0-13-120650-8.

Jaeger, M. and Mayer, M. (2000) The Noell conversion process – a gasification process for the pollutant-free disposal of sewage sludge and the recovery of energy and materials. *Water Sci. Technol.* **41**(8), 37–44.

Klaassen, R. and van Voorneburg, F. (1995) *Removal of Ammonia from Liquid Manure and Waste Waters with Membrane Chemo-Sorption*. TNO report Ref. No. R95-118, TNO, Apeldoorn, the Netherlands (in Dutch).

Lendormi, T., Prévot, C., Doppenberg, F., Spérandio, M. and Debellefontaine, H. (2001) Wet oxidation of domestic sludge and process integration: the Mineralis® process. *Water Sci. Technol.* **44**(10), 163–169.

Müller, J.A. (2002) Sludge pre-treatment processes – results and applications. *IWA World Water Congress 2002 Melbourne*. Paper no. e21094a.

Okuno, N., Isikawa, Y. and Andow, K. (2002) Evaluation of thermal sewage-sludge-solidification process by LCA–CO_2. *IWA World Water Congress 2002 Melbourne*. Paper no. e20481a.

Rashidi, A.S., Horil, R.S., Greenfield, C. and Smith, D.L. (1986) The hyperion Carver-Greenfield facility: innovative technology for municipal sludge dehydration. *59th Annual WPCF Conference*, October 1986.

Rulkens, W.H. (1992) Environmental biotechnology versus chemical/physical environmental technology. *Med. Fac. Landbouww. Univ. Gent.* **57/4a**, 1437–1449.

Rulkens, W.H. (2003) Energy from sludge: an overview and brief evaluation of most important options. In: *Proceedings of the 2nd International and 13th National Conference on Renewable Energy Sources – The New Insights into Sewage Sludge* (ed. G. Malina), pp. 33–47, 3–5 February 2003, Technical University of Czestochowa, Czestochowa, Poland. ISBN 83-7193-197-6.

Rulkens, W.H. and ten Have, P.J.W. (1994) Central processing of pig manure in the Netherlands. *Water Sci. Technol.* **30**(7), 157–165.

Rulkens, W.H., Klapwijk, A. and Willers, H.C. (1998) Recovery of valuable nitrogen compounds from agricultural liquid wastes: potential possibilities, bottlenecks and future technological challenges. *Environ. Pollut.* **102**(S1), 727–735.

Sanders, J.P.M. (1993) Development of an industrial process technology for the manufacturing of lysine from pig manure. In: *Proceedings of the First International Symposium on Nitrogen Flow in Pig Production and Environmental Consequences* (eds. M.W.A. Verstegen, L.A. Den Hartog, G.J.M. Van Kempen and J.H.M. Metz), pp. 398–403. Pudoc Scientific Publishers, Wageningen, The Netherlands.

Scheminski, A., Krull, R. and Hempel, D.C. (2000) Oxidative treatment of digested sludge with ozone. *Water Sci. Technol.* **42**(9), 151–158.

Shanableh, A. and Jomaa, S. (2001) Production and transformation of volatile fatty acids from sludge subjected to hydrothermal treatment. *Water Sci. Technol.* **44**(10), 129–136.

Shanableh, A. and Shimizu, Y. (2000) Treatment of sewage sludge using hydrothermal oxidation – technology application challenges. *Water Sci. Technol.* **41**(8), 85–92.

Skrypsi-Mäntele, S., Bridle, T.R., Freeman, P., Luceks, A. and Ye, P.D. (2000) Production of high quality fuels using the enhanced Enersludge (TM) process. *Water Sci. Technol.* **41**(8), 45–52.

Stessel, R.I. (1996) *Recycling and Resource Recovery Engineering: Principles of Waste Processing*. Springer-Verlag, Berlin/Heidelberg/New York. ISBN 3-540-61100-2.

Stolarek, P. and Ledakowicz, S. (2001) Thermal processing of sewage sludge by drying, pyrolysis, gasification and combustion. *Water Sci. Technol.* **44**(10), 333–340.

Taiganides, E.P. (1992) *Pig Waste Management and Recycling, The Singapore Experience*. IDRC, Ottawa, Ont. 368 pp.

Taruya, T., Okuno, N. and Kanaya, K. (2002) Reuse of sewage sludge as raw material of Portland cement in Japan. *Water Sci. Technol.* **46**(10), 255–258.

Tchobanoglous, G., Theisen, H. and Vigil, S. (eds.) (1993) *Integrated Solid Waste Management: Engineering Principles and Management Issues*. McGraw-Hill Inc. ISBN 0-07-112865-4.

ten Have, P.J.W. (1991) Technical and economic aspects of manure processing. In: *Treatment and Use of Sewage Sludge and Liquid Agricultural Wastes* (ed. L' Hermite, P.), pp. 70–78. Elsevier Applied Science, London/New York.

ten Have, P.J.W. (1993) Nitrogen and the industrial processing of pig manure. In: *Proceedings of the First International Symposium on Nitrogen Flow in Pig Production and Environmental Consequences* (eds. M.W.A. Verstegen, L.A. Den Hartog, G.J.M. Van Kempen and J.H.M. Metz), pp. 386–397. Pudoc Scientific Publishers, Wageningen, The Netherlands.

Tiehm, A., Nickel K. and Neis, U. (1997) The use of ultrasound to accelerate the anaerobic digestion of sewage sludge. *Water Sci. Technol.* **36**(11), 121–128.

12

Fundamental parameters of aerobic solid-state bioconversion processes

Tom L. Richard

12.1 INTRODUCTION

Aerobic bioprocessing of solid waste can be accomplished using a range of approaches and technologies (Rynk and Richard 2001). This chapter will discuss the essential parameters used to manage these processes, including substrate composition, oxygen, temperature, moisture, pH and matrix characteristics such as mechanical strength, bulk density, porosity and permeability. The importance of these parameters is indicated by their use in defining and classifying systems, such as aerobic vs. anaerobic, thermophilic vs. mesophilic, liquid vs. solid state, etc., as illustrated in Figure 12.1 for oxygen and moisture. While aerobic, thermophilic solid-state composting is the most widely used strategy used for solid-waste bioconversion at present, many of these same basic principles also apply to vermicomposting, as well as other biological bioconversion and stabilization processes. With the success or failure of any technology largely dependant on effective

Edited by Piet Lens, Bert Hamelers, Harry Hoitink, Werner Bidlingmaier. ISBN: 1 84339 054 X. Published by IWA Publishing, London, UK

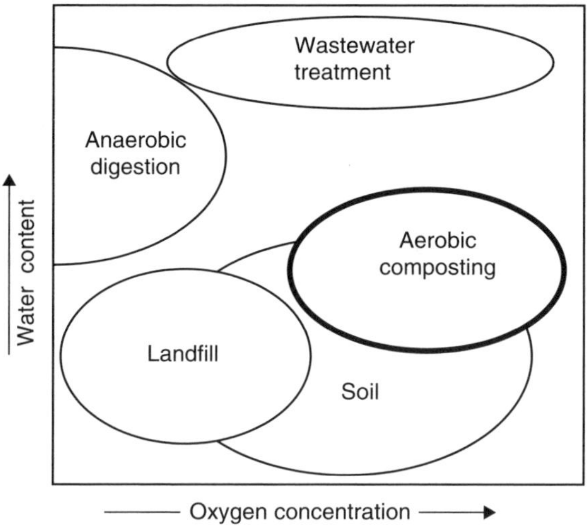

Figure 12.1 Biodegradation processes classified by H_2O and O_2 conditions.

process control, a fundamental understanding of these parameters is essential for both researchers and practitioners of organic waste management systems.

The primary goal for most aerobic bioprocessing of solid wastes is stabilization of the organic substrates. Biodegradation occurs through a complex web of biochemical reactions by which microorganisms catabolize substrates for their growth and metabolic needs (Alexander 1977; Paul and Clark 1996). Multiple cycles of biodegradation result in increasingly recalcitrant residues, which are less likely to produce odors, attract vectors, or support pathogens that can cause disease. Not surprisingly, a diverse organic wastestream cycling through various stages of biodegradation normally supports a diverse microbial ecosystem. Both wide species diversity (Chang and Hudson 1967; de Bertoldi *et al.* 1981, 1983; Strom 1985a, b; Amner *et al.* 1988) and fluctuating population dynamics (Hardy and Sivasithamparam 1989; Fogarty and Tuovinen 1991; Miller 1991; Joshua 1997) are observed in composting processes. Under certain circumstances the initial microbial species composition can effect substrate degradation rates (Matthur *et al.* 1986; Nakasaki *et al.* 1994a), although not always at statistically significant levels (de Bertoldi *et al.* 1983; Magan *et al.* 1989; Faure and Deschamps 1991). Even with similar initial populations, differences in environmental conditions will influence the evolving microbial ecosystem (Alexander 1977). Although individual microbial species each have their own specific requirements, management of composting and other solid-state waste treatment processes typically focus on the ecosystem as a whole.

In composting, this microbial biodegradation occurs in a matrix of insoluble solids, and the term "solid state" is used to differentiate this bioprocess from more common submerged fermentations such as wastewater treatment (Losane *et al.* 1985; Saucedo-Castañeda *et al.* 1990; Metcalf and Eddy 1991; Shuler and Kargi

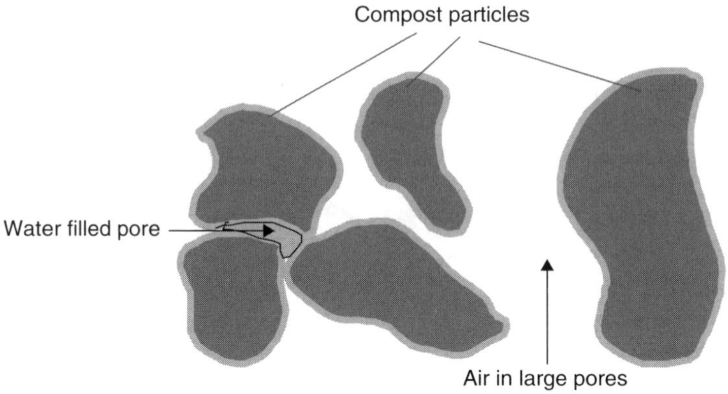

Figure 12.2 Macroscopic solid-state matrix, with solid, aqueous, and gas phases.

1992). Both types of bioprocesses generally also contain both liquid and gaseous components, with the major difference being the high concentrations (>30%) of insoluble solids in the solid-state matrix (see Figure 12.2). Without an aqueous phase there would be little microbial activity (Griffin 1981), while in the absence of a gas phase there would be limited oxygen transfer (Miller 1989) and anaerobic reactions would dominate. This three-phase matrix requires the exchange of gaseous molecules, such as CO_2, H_2O, and O_2 across phase boundaries, and the transport of energy from the particle to the air and from particle to particle. Such effects create a linkage between particle-scale transport processes and microbial degradation.

12.2 COMPOSTING PARAMETERS

12.2.1 Substrate composition

Of the many elements required for microbial decomposition, carbon and nitrogen are both the most important and the most commonly limiting, although phosphorus can also be limiting in certain organic solid wastes. Carbon is both an energy source, liberated by the oxidation of carbon–carbon bonds, and the basic building block for most biomolecules, making up about 50% of the mass of microbial cells.

Nitrogen is a crucial component of proteins and nucleic acids. Microorganisms, whose biomass is often over 50% protein, need adequate supplies of nitrogen for rapid growth. When there is too little nitrogen, the microbial population will not grow to its optimum size, and biodegradation rates slow down. In contrast, too much nitrogen allows rapid microbial growth and accelerates decomposition, but this can create serious odor problems as oxygen is used up and anaerobic conditions occur. In addition, some of this excess nitrogen will be given off as NH_3, N_2O, and N_2 gases, generating odors and greenhouse gas emissions while allowing valuable nitrogen to escape. Therefore, materials with a high nitrogen content, such as food wastes, biosolids, and manures, require careful management to insure adequate oxygen transport, as well as thorough blending with a high carbon waste like paper, woodchips, or leaves. For most materials, a carbon/nitrogen (C/N) ratio of about 30 : 1 (by weight) will keep these elements in approximate balance, although other factors such as substrate bioavailability and specific surface area can also come into play.

A generalized formula for the C/N ratio can be written as:

$$R = \frac{Q_1(C_1 \times (100 - M_1)) + Q_2(C_2 \times (100 - M_2)) + Q_3(C_3 \times (100 - M_3)) + \cdots}{Q_1(N_1 \times (100 - M_1)) + Q_2(N_2 \times (100 - M_2)) + Q_3(N_3 \times (100 - M_3)) + \cdots} \tag{12.1}$$

where
R = C/N ratio of compost mixture,
Q_n = mass of material n ("as is", or "wet weight"),
C_n = carbon (%) of material n,
N_n = nitrogen (%) of material n,
M_n = moisture content (%) of material n.

If direct carbon data is unavailable but C/N ratio and N data are, the carbon content can easily be calculated:

$$C_n = N_n \times \frac{\mathrm{C}}{\mathrm{N}} \tag{12.2}$$

Or alternately, %C can be estimated from the volatile solids (VS) content (Rynk 1992) by:

$$C_n = \frac{VS_n}{1.8} \tag{12.3}$$

By simplifying for a two material mixture, and rearranging the general equation (Richard and Trautmann 1996), the mass of the second material required to balance any particular feedstock would be:

$$Q_2 = \frac{Q_1 \times N_1 \times \left(R - (C_1/N_1)\right) \times (100 - M_1)}{N_2 \times \left((C_2/N_2) - R\right) \times (100 - M_2)} \tag{12.4}$$

C/N ratios may need to be adjusted depending on the bioavailability of these elements. This is commonly an issue with high carbon materials, which are often derived from wood and other lignified plant materials, as increased lignin content reduces biodegradability. Particle size is also an important factor, with smaller particles degrading more quickly than large particles of the same material. Bioavailability can also be a factor with nitrogen sources, especially fertilizer nitrogen, where nearly instant availability can exceed the assimilative capacity of the microbial community and be lost as odorous ammonia gas and nitrate in leachate and runoff.

12.2.2 Oxygen

Oxygen – its presence or absence – plays a dominant role in organic waste management, although in reality the separation between aerobic and anaerobic systems is far less distinct than Figure 12.1 implies. Aerobic respiration dominates whenever significant amounts of molecular oxygen are present, as microorganisms maximize their energy release when they use oxygen as the electron acceptor during biodegradation.

Oxygen consumption rates and CO_2 and moisture evolution can be calculated directly from stoichiometric relationships, although this estimate should not be used to design aeration rates, which must also provide for the much higher heat removal requirement (see below). The generalized formula for an aerobic biodegradation reaction (Richard 1997b) can be written as:

$$C_aH_bO_cN_d + \left(a + \frac{b}{4} - \frac{c}{2} - \frac{3d}{4}\right)O_2 \rightarrow aCO_2 + \left(\frac{b}{2} - \frac{3d}{2}\right)H_2O + dNH_3 \quad (12.5)$$

Equation (12.5) greatly simplifies the aerobic respiration process, and only applies to the net organic substrate that is actually mineralized to CO_2 and NH_3. Internal cycling of organic matter into microbial biomass or recalcitrant organic compounds would undergo different reactions.

In solid-state systems completely aerobic conditions are difficult to achieve. This difficulty is compounded by transport limitations at both microscopic and macroscopic scales. At a microscopic scale, the impact of oxygen limitations occurs primarily in the liquid films surrounding particles and the solid particles themselves, where microbial biomass and biodegradable substrates are concentrated. Within these liquid and solid phases oxygen transport is driven by diffusion, with the oxygen concentration dropping toward the interior of aqueous films and particles, and corresponding increases in the production of odorous gases as illustrated in Figure 12.3. These relationships have been modeled using biofilm theory by Hamelers (1993, 2001) and Tseng *et al.* (1995). These simulations predict that anaerobic conditions will occur within the interior of substrate particles greater than 0.25 mm in diameter, even with high oxygen concentrations in the gas phase surrounding the particles.

At a macroscopic scale, oxygen transport through the substrate matrix is dependent on convection and diffusion processes, which in some cases are inadequate to maintain high oxygen concentrations throughout the pore space. McCauley and Shell (1956) measured oxygen diffusion rates in a mixed food scrap substrate and concluded that diffusion alone is generally not adequate to maintain high oxygen concentrations throughout a mass of material, particularly with materials experiencing rapid degradation or with excessive moisture. With slower degrading substrates, oxygen diffusion may be adequate for intraparticle oxygen transport, as was reported by Nakasaki *et al.* (1987) for sewage-sludge cake. Natural convection can complement diffusion in passive composting systems (Veeken *et al.* 2002), although this process has proven difficult to measure experimentally (Miller *et al.* 1989).

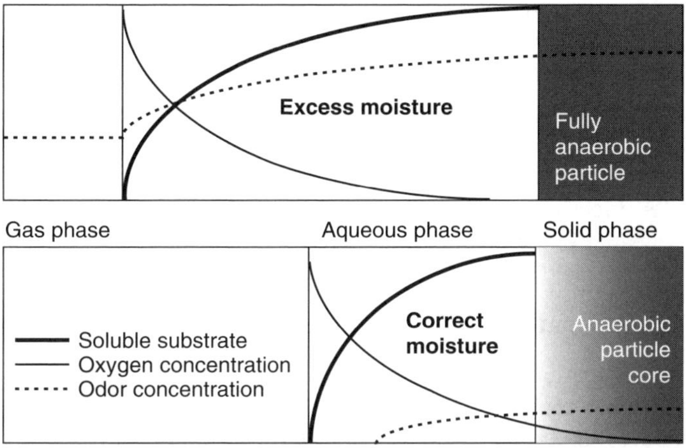

Figure 12.3 Relationships between soluble substrate, oxygen, and odorous gas concentrations in aqueous films and particles. Shading to the right illustrates the intensity of anaerobic activity.

Forced aeration systems can also be subject to oxygen transfer limitations on a macroscopic scale. Variations in packing, particle size or moisture within the matrix can act initiate preferential flow paths, which cause short circuiting in some regions and leave other zones with inadequate flow. Even a relatively homogeneous mixture will have a variably sized and tortuously connected pore structure. Frictional resistance by the pore walls will affect the gas velocity distribution, so that one can expect convection to dominate in the very large pores, while diffusion dominates in the very small pores. Results from bench-scale experimental reactors indicated that even at high oxygen concentrations (generally above 18%), a doubling of flow rates increased degradation rates by 20–30% (Miller 1984).

These theoretical considerations are born out in full-scale composting systems, where low oxygen conditions are indeed quite commonplace. Randle and Flegg (1978) and Miller *et al.* (1989) report the oxygen concentration distributions in mushroom composting stacks, which illustrate the predominance of low oxygen concentrations through much of the windrow interior. Although turning windrows or applying forced aeration can temporarily replenish oxygen levels, oxygen depletion can occur rapidly. In sewage-sludge composting, interstitial oxygen concentration can be reduced to less than 5% (v/v) within 15 min (USEPA 1985), while in poultry waste composting complete depletion can occur in as little as 2 min (Galler *et al.* 1978).

In composting and similar aerobic solid-state microbial ecosystems, oxygen limitations can significantly reduce substrate degradation (Suler and Finstein 1977; Nakasaki *et al.* 1990), with rates at 2–3% O_2 about half those at 20.9% (Richard *et al.* 1999). These studies reported oxygen concentration as measured in the gas phase of the matrix or exhaust stream. Such measurements indicate the mean concentration flowing past the sensor at a single point, and give no indication of the distribution of concentration within the variously sized pores. In particular, the impact of the aeration rate on the gas phase oxygen transport into small pores, and the impact of particle size and substrate degradation rate on oxygen transport within the liquid/solid matrix, may influence the relationship between measured oxygen concentration in the gas phase and the redox potential or dissolved oxygen concentration actually experienced by the microorganisms. Such mass transfer effects are the presumed mechanism behind the reported influence of airflow rate on biodegradation rate (Wiley 1956; Miller 1984, 1991).

12.2.3 Temperature

Aerobic decomposition is an exothermic process, and the limited surface to volume ratio of most composting systems limits heat loss and causes temperatures to rise (Haug 1980, 1993; Finstein *et al.* 1986; Miller 1991). These higher temperatures result in an acceleration in the degradation process, reflecting increases in microbial growth rates, enzymatic production, and the reaction rates of the enzymes produced. Higher degradation rates liberate additional energy, creating a positive feedback loop and increasing both temperatures and degradation rates up to an optimum temperature, above which degradation rates decline. This decline results from a number of factors, both on a molecular scale as enzymes required to catalyze degradation reactions are denatured, and at an ecological scale as individual microorganisms and entire species experience thermal death (Moats 1971; Tansey and Brock 1978). This decrease in reaction rate at higher temperatures has been observed by many investigators, with the temperature peak usually occurring within the temperature range of 50–62°C in small-scale systems (Haug 1980, 1993; Lasaridi *et al.* 1996; Richard 1997a) (Figure 12.4). Slightly higher optimum temperatures, up to near 70°C, have been recorded from investigations

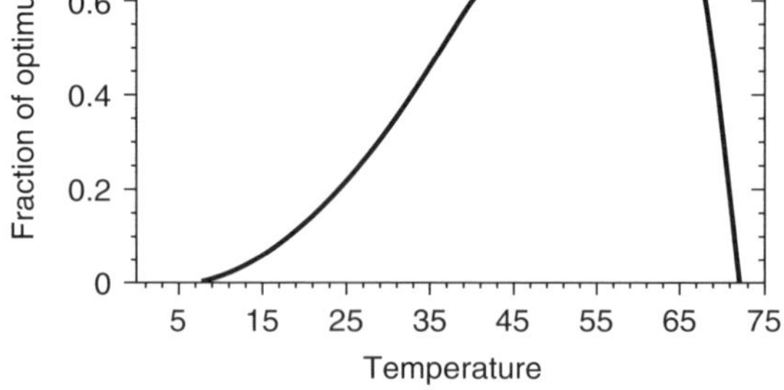

Figure 12.4 Effect of temperature on biodegradation rate.

in pilot-scale reactors (Schultze 1962; Adani *et al.* 2002). These presumably result from temperature gradients in these systems relative to the temperature measurement point. Since similar challenges to temperature homogeneity and measurement occur in full-scale composting systems, set points of 60–70°C may be appropriate for process control.

A critical factor in compost system design is insuring adequate airflow to remove excess heat. Aerobic biodegradation releases about 14 kJ/g of O_2 consumed (Finstein *et al.* 1986; Haug 1993). Thus, for a known rate of decomposition of a well defined organic substrate, the energy removal requirements can be calculated from the oxygen demand derived in equation (12.5), or by measuring changes in chemical oxygen demand (COD) through a mass balance analysis. Up to 90% of this energy is removed by convection (Finstein *et al.* 1986), so that in mechanically ventilated systems blower size requirements can be estimated from changes in enthalpy between the inlet and exhaust airstreams (Haug 1993).

$$\Delta\text{Enthalpy} = \text{Airflow}(h_{\text{exhaust}} - h_{\text{inlet}}) \tag{12.6}$$

where
ΔEnthalpy = the change in enthalpy (kJ/day),
Airflow = mass of dry air (kg/day),
h = enthalpy (kJ/kg dry air) under inlet or exhaust conditions.

Note that equation (12.6) includes units of time, which would need to be matched to the time scale of the biodegradation used to determine the oxygen demand. As the biodegradation rate varies dramatically during the first few days of a composting process, accurate estimates are essential for proper aeration system design (Hamelers 2001).

12.2.4 Moisture

Moisture content is widely recognized as a important control parameter for aerobic solid-state biodegradation processes (Gray *et al.* 1971a; Haug 1980, 1993; Finstein *et al.* 1983; de Bertoldi *et al.* 1983; Miller 1991; Golueke 1991; Richard 1992). Moisture management requires a balance between two functions: encouraging microbial activity and permitting adequate oxygen supply. Water is essential to the decomposition process, and water stress is among the most common limitations for microbial activity on solid substrates (Griffin 1977, 1981; Smith 1978; Griffin and Luard 1979). Low moisture conditions can also restrict the movement of bacteria, so that physical dispersal allows mixed composting systems to function at lower moisture contents than static systems (Miller 1989). However, mixtures can be too wet, thereby reducing the oxygen supply. Excess moisture will fill many of the pores between particles with water, enhancing opportunities for microbial decomposition while simultaneously limiting oxygen transport (Miller 1989, 1991).

Moisture content is usually measured in terms of gravimetric water content (% H_2O on a wet basis (w.b.)) or its complement: dry matter (% dry basis (d.b.)). These will be the conventions in this chapter except as otherwise noted. However, these measurements are only a proxy for the actual biophysical parameters. Microbial activity in composting systems varies more precisely with matric water potential (Miller 1989; Richard 1996; Bloom and Richard 2002), while oxygen transfer depends on aqueous surface area and film thickness (McCauley and Shell 1956; Hamelers 1993; Tseng *et al.* 1995).

Matrix structure and water content change dynamically during the composting process. The decomposition process reduces particle size and increases matrix dry bulk density, leading to a reduction in total porosity (van Ginkel *et al.* 1999). Although 0.5–0.6 g of metabolic water is produced per g VS decomposed (Miller 1991), the heat and airflow generated during composting evaporate significantly more water than is produced and tend to dry the material out (Nakasaki *et al.* 1987). This biodrying process allows the sequential addition of wet feedstocks to the system to maximize decomposition with minimal bulking amendments (Choi *et al.* 2001). During the active composting phase, additional water is sometimes added to prevent premature drying and incomplete stabilization (Finstein *et al.* 1983; Keener *et al.* 1994; Atkinson *et al.* 1996). These and similar management strategies can help avoid unfavorable moisture and air-filled porosity conditions, but achieving optimal conditions is more difficult as these are generally not well defined.

Regan *et al.* (1973) reviewed literature values reported for optimum moisture content, which ranged from 25% to 80% H_2O with most values in the 50–70% range. Similar optima have been observed more recently by Murwira *et al.* (1990) and Nakasaki *et al.* (1994b). Hamelers and Richard (2001) developed a deductive model of the affects of moisture on composting kinetics, defining these relationships based on fundamental physical properties and biological mechanisms. Richard *et al.* (2002) applied this model to experimental data from a manure and papermill-sludge composting system (see Figure 12.5). The results indicated that the optimum moisture content for biodegradation can vary widely for different compost mixtures and times in the composting process, ranging from near 50% to over 70% (w.b.). Variations in the relationship between moisture content and

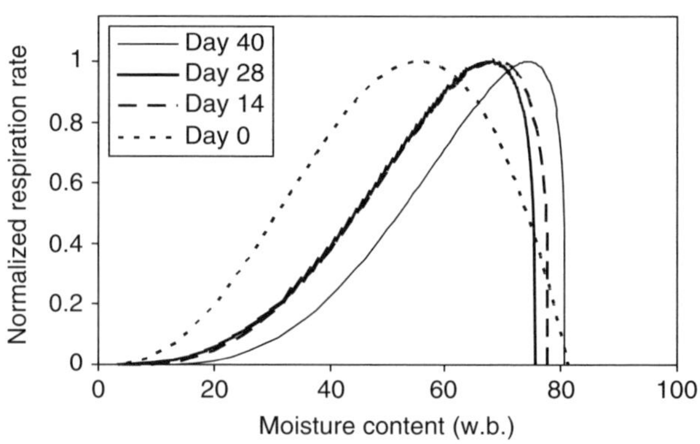

Figure 12.5 Effect of moisture on biodegradation rates for a papermill-sludge–manure mixture.

matric water potential may have contributed to the breadth of the optimum moisture range, which appears to vary with substrate as well as with time.

Understanding the caveat that optimum moisture is not well defined, in practice most composting operations will start with a moisture content near 60% (w.b.). The general formula for percent moisture is:

$$G = \frac{(Q_1 \times M_1) + (Q_2 \times M_2) + (Q_3 \times M_3) + \cdots}{Q_1 + Q_2 + Q_3 + \cdots} \quad (12.7)$$

where
Q_n = mass of material n ("as is", or "wet weight"),
G = moisture goal (% w.b.),
M_n = moisture content (% w.b.) of material n.

By simplifying for a two material mixture, and rearranging the general equation, the mass of the second material Q_2 required would be:

$$Q_2 = \frac{(Q_1 \times G) - (Q_1 \times M_1)}{M_2 - G} \quad (12.8)$$

Solutions for three or more materials can be solved simultaneously for both moisture (equation (12.7)) and the C : N ratio (equation (12.1)), with the caveat that an explicit solution of these two equations requires only two unknowns. Examples for three and four material mixtures are provided by Richard and Trautmann (1996).

The drying that occurs during composting is almost all lost via convection, and the change in the mass of water in a system can be calculated if airflow rate and inlet and outlet conditions are known:

$$\Delta\text{Moisture} = \text{Airflow}(H_{\text{exhaust}} - H_{\text{inlet}}) \quad (12.9)$$

where
Moisture = the change in water mass (kg/day),
Airflow = mass of dry air (kg/day),
H = absolute humidity, kg water/kg dry air, under inlet or exhaust conditions.

When calculating changes in the water mass balance, it is important to remember that the water loss predicted by equation (12.9) is somewhat offset by water gains, which includes the metabolic water produced during the decomposition of the organic solid waste, as indicated by equation (12.5), and may also include precipitation in outdoor systems.

12.2.5 The effect of pH

The biochemical reactions of aerobic decomposition can modify the pH of the substrate matrix, with potentially important impacts on the composting process. Several researchers have observed an initial drop in pH in batch composting systems, particularly with poorly buffered or highly degradable substrates, which can reduce degradation rates and delay thermophilic conditions until the pH recovers to the neutral range (usually within a few days to 1 week) (Wiley and Pearce 1957; Gray *et al.* 1971b; de Bertoldi *et al.* 1983; Miller 1991; Joshua 1997). Lime has sometimes been used to raise pH into the neutral or alkaline range when this acidic phase is severe or prolonged (Wiley 1956; Wiley and Pearce 1957; Joshua 1997). However, this approach can have negative impacts, as deamination and ammonification often raise pH within 1 or 2 weeks to approximately 8.5 (Miller 1991), so that lime additions early on could ultimately raise pH to levels that inhibit microbial activity, as well as exacerbate ammonia losses (Wiley and Pearce 1957; Miller 1991).

A few studies have attempted to control pH and examined the effect on the composting process. Wiley (1956) and Wiley and Pearce (1957) observed a pH drop to 4.5–5.0 within 12–24 h when composting a municipal refuse mixture. This acidic phase was followed within 2–4 days by a rapid rise in pH to 8.0–9.0, accompanied by a transition to thermophilic decomposition and increased CO_2 production. Addition of $Ca(OH)_2$ in two doses totaling 16–27% (d.b.) eliminated this acidic lag period, with peak temperatures achieved within 36 h. Addition of acetic acid to control pH between 5.0 and 6.0 prevented or retarded heat production and prolonged the degradation process. Also working with mixed refuse, (Regan *et al.* 1973, Jeris and Regan 1973) observed a dramatic increase in degradation rates between pH 6.0 and 7.0, with peak degradation rates observed at pH 8.2. Phosphate salts or hydrochloric acid were used to control pH in these experiments. Nakasaki *et al.* (1993) performed laboratory experiments composting food scraps with and without lime addition, and also observed more rapid degradation when pH was maintained above 7.0. All three of these experimental studies observed significantly greater biodegradation rates at pH $\geqslant 7.0$ than at pH $\leqslant 6.0$, and severe limitations on degradation at pH $\leqslant 5.0$. Degradation rates appear relatively stable in the range of 7.0–8.5, with decreases above 9.0. Although models of the effect of pH on microbial activity have been developed for pure cultures (Rosso *et al.* 1995), they have not yet been applied to composting systems.

12.2.6 Matrix considerations

Compost is characterized by non-uniform distributions of substrates, microorganisms, intraparticle porosity, aqueous film thickness, etc. At a particle scale, access to substrate by microorganisms or even extracellular enzymes may be limited by the size of internal pores (Ladisch *et al.* 1983; Lin *et al.* 1985; Neuman and Walker 1992). At a cellular scale, high lignin levels in cell walls may restrict access to cellulose and other more readily degradable compounds (Haug 1993; Van Soest 1994). Compost is frequently subject to mechanical mixing which can shear particles or aggregates and expose new surfaces to microbial attack. Thus, substrate degradation rates can be expected to change over time not only as a function of substrate concentration, but also as bioavailability of carbon and nutrients changes due to particle structure and composition.

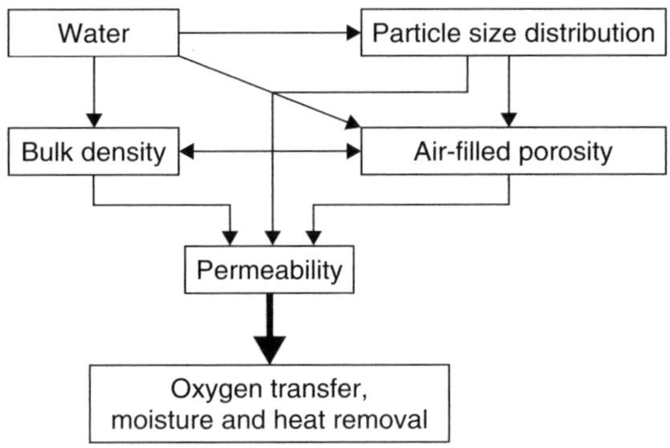

Figure 12.6 Interacting physical relationships in a solid-state matrix.

The physical changes in the composting matrix also interact to influence oxygen transfer, moisture, and heat removal (see Figure 12.6). Changes in air-filled porosity can be predicted from easily measured characteristics such as bulk density, water content, and VS content (Richard *et al.* 2002). In full-scale composting systems pile height and density cause compaction of the lower regions, reducing porosity and increasing bulk density. The mechanical strength of the substrate can reduce this compaction, maintaining structural integrity of the pile. These matrix properties also have effects on permeability and thus mass and energy transfer. Physical properties thus have an important impact on system design and performance, influencing the maximum acceptable pile size, pressure drop, and blower requirements in mechanically ventilated systems, and convective air movement in naturally ventilated systems.

12.3 CONCLUSIONS

Although a wide variety of composting technologies and other aerobic bioprocessing strategies have been developed for managing solid waste, all are bound by the same physical, chemical, and biological constraints. Physical properties of the substrate determine porosity, permeability, and the impact of compaction at different locations in a pile. Its chemical composition determines bioavailability and maximum biodegradation rates. Achieving rapid biological stabilization requires optimization of temperature, moisture, oxygen, and pH, all of which can work in concert in a well operating system.

Airflow is a particularly important means of controlling several of these parameters, including temperature, moisture, oxygen, and pH. In most composting configurations, temperature regulation dominates airflow requirements, providing adequate oxygen and moisture removal, and allowing the aerobic degradation of organic acids to prevent dramatic drops in the system pH. When airflow is not adequate in the early stages of composting, the system can overheat and essentially pasteurize itself. Inadequate airflow can also lead to anaerobic odors, excess water, and accumulations of organic acids that can cause a drop in pH. Thus control of airflow, whether through active or passive means, often means the difference between success and failure of the system as a whole.

Various approaches to implementing these principles allow differing degrees of process control, space requirements, capital cost, and management expertise. A better understanding of the fundamental process parameters in aerobic solid-waste management should lead to improved implementation, and to even more effective systems in the years to come.

REFERENCES

Adani, F., Baido, D., Calcaterr, E. and Genevini, P. (2002) The influence of biomass temperature on biostabilization-biodrying of municipal solid waste. *Biores. Technol.* **83**, 173–179.

Alexander, M. (1977) *Introduction to Soil Microbiology*, 2nd edn. John Wiley and Sons, New York.

Amner, W., McCarthy, A.J. and Edwards, C. (1988) Quantitative assessment of factors affecting the recovery of indigenous and released thermophilic bacteria from compost. *Appl. Environ. Microbiol.* **54**(12), 3107–3112.

Atkinson, C.F., Jones, D.D. and Gauthier, J.J. (1996) Biodegradabilities and microbial activities during composting of municipal solid waste in bench-scale reactors. *Compost Sci. Util.* **4**(4), 14–23.

Bloom, E. and Richard, T.L. (2002) Relative humidity and matric potential constraints on composting microbial activity. ASAE Paper No. 027010. ASAE, St. Joseph, MI.

Chang, Y. and Hudson, H.J. (1967) The fungi of wheat straw compost, I. *Trans. Br. Mycol. Soc.* **50**(4), 649–666.

Choi, H.L., Richard, T.L. and Ahn, H.K. (2001) Composting high moisture materials: biodrying poultry manure in a sequentially fed reactor. *Compost Sci. Util.* **9**(4), 303–311.

de Bertoldi, M., Citernese, U. and Griselli, M. (1981) Microbial populations in compost process. In: *Composting: Theory and Practice for City, Industry and Farm*, pp. 26–33. J.G. Press, Emmaus, PA.

de Bertoldi, M., Vallini, G. and Pera A. (1983) The biology of composting: a review. *Waste Manage. Res.* **1**, 157–176.

Faure, D. and Deschamps, A.M. (1991) The effect of bacterial inoculation on the initiation of composting of grape pulps. *Biores. Technol.* **37**(3), 235–238.

Finstein, M.S., Miller, F.C., Strom, P.F., MacGregor, S.T. and Psarianos, K.M. (1983) Composting ecosystem management for waste treatment. *Bio/Technology* **1**, 347–353.

Finstein, M.S., Miller, F.C. and Strom, P.F. (1986) Waste treatment composting as a controlled system. In: *Biotechnology: A Comprehensive Treatis* (eds. H.-J. Rehm and G. Reed), *8 vol.*; *Microbial Degradations* (W. Schönborn (Vol. Ed.)), pp. 363–398. VCH, Weinheim.

Fogarty, A.M. and Touvinen, O.H. (1991) Microbiological degradation of pesticides in yard waste composting. *Microbiol. Rev.* **55**(2), 225–233.

Galler, W.S., Davey, C.B., Meyer, W.L. and Airan, D.S. (1978) *Animal Waste Composting with Carbonaceous Material*, 95 pp. EPA-600/2-78-154. USEPA, Cincinnati, OH.

Golueke, C.G. (1991) Principles of composting. In: *The BioCycle Guide to the Art and Science of Composting* (eds. BioCycle staff), pp. 14–39. J.G. Press, Emmaus, PA.

Gray, K.R., Sherman, K. and Biddlestone, A.J. (1971a) Review of composting – part 2: the practical process. *Process Biochem.* **6**(10), 22–28.

Gray, K.R., Sherman, K. and Biddlestone, A.J. (1971b) Review of composting – Part 1. *Process Biochem.* **6**(6), 32–36.

Griffin, D.M. (1977) Water potential and wood-decay fungi. *Ann. Rev. Phytopathol.* **15**, 319–329.

Griffin, D.M. (1981) Water and microbial stress. In: *Advances in Microbial Ecology* (ed. M. Alexander), pp. 91–136. Plenum Press, New York/London.

Griffin, D.M. and Luard, E.J. (1979) Water stress and microbial ecology. In: *Strategies of Microbial Life in Extreme Environments* (ed. M. Shilo), pp. 49–63. Dahlem Konferenzen, Berlin.

Hamelers, H.V.M. (1993) A theoretical model of composting kinetics. In: *Proceedings of the International Composting Research Symposium* (eds. H.A.J. Hoitink and H. Keener), pp. 36–58. Renaissance Publications, Columbus, OH.

Hamelers, H.V.M. and Richard, T.L. (2001) The effect of dry matter on the composting rate: theoretical analysis and practical implications. ASAE Paper No. 017004. ASAE, St. Joseph, MI.

Hamelers, H.V.M. (2001) *A Mathematical Model for Composting Kinetics*. Ph.D. Dissertation, Wageningen University.

Hardy, G.E.St.J. and Sivasithamparam, K. (1989) Microbial, chemical and physical changes during composting of a eucalyptus (*Eucalyptus calophylla and Eucalyptus diversicolor*) bark mix. *Biol. Fertil. Soils* **8**, 260–270.

Haug, R.T. (1980) *Compost Engineering: Principles and Practice*, 665 pp. Ann Arbor Science, Ann Arbor, MI.

Haug, R.T. (1993) *The Practical Handbook of Compost Engineering*, 718 pp. Lewis Publishers, Boca Ratan, FL.

Jeris, J.S. and Regan, R.W. (1973) Controlling environmental parameters for optimum composting III. *Compost Sci.* **14**(3), 16–22.

Joshua, R.S. (1997) *Biorecovery of Food Effluent Residues.* Ph.D. Thesis. La Trobe University, Melbourne, Australia.

Keener, H.M., Elwell, D.L., Das, K.C. and Hansen, R.C. (1994) Remix frequency of compost based on moisture control. ASAE Paper No. 944066. ASAE, St. Joseph, MI.

Ladisch, M.R., Lin, K.W., Voloch, M. and Tsao, G.T. (1983) Process considerations in the enzymatic hydrolysis of biomass. *Enzyme Microbial. Technol.* **5**(2), 82–101.

Lasaridi, K.E., Papadimitriou, E.K. and Balis, C. (1996) Development and demonstration of a thermogradient respirometer. *Compost Sci. Util.* **4**(3), 53–61.

Lin, K.W., Ladisch, M.R., Boloch, M., Patterson, J.A. and Noller, C.H. (1985) Effect of pretreatments and fermentation on pore size in cellulosic materials. *Biotechnol. Bioeng.* **27**(10), 1427–1433.

Losane, B.K., Ghildyal, N.P., Budiatman, S. and Ramakrishna, S.V. (1985) Engineering aspects of solid state fermentation. *Enzyme Microb. Technol.* **7**(6), 258–265.

McCauley, R.F. and Shell, B.J. (1956) Laboratory and operational experiences in composting. Proceedings of the 11th Industrial Waste Conference. *Purdue Univ. Eng. Bull. Series No.* **91**, 436–453.

Magan, N., Hand, P., Kirkwood, I.A. and Lynch, J.M. (1989) Establishment of microbial inocula on decomposing wheat straw in soil of different water contents. *Soil Biol. Biochem.* **21**(1), 15–22.

Matthur, R.S., Magu, S.P., Sadasivam, K.V. and Gaur, A.C. (1986) Accelerated compost and improved yields. *BioCycle* **27**(2), 42–44.

Metcalf and Eddy (1991) *Wastewater Engineering: Treatment, Disposal and Reuse*, (Edited/Revised by G. Tchobanoglous and F.L. Burton), 1334 pp. McGraw-Hill, Inc., New York.

Miller, F.C. (1984) *Thermodynamic and Matric Water Potential Analysis in Field and Laboratory Scale Composting Ecosystems*, 564 pp. Ph.D. Thesis. Rutgers University, New Brunswick, NJ.

Miller, F.C. (1989) Matric water potential as an ecological determinant in compost, a substrate dense system. *Microb. Ecol.* **18**, 59–71.

Miller, F.C., Harper, E.R. and Macauley, B.J. (1989) Field examination of temperature and oxygen relationships in mushroom composting stacks: consideration of stack oxygen based on utilisation and supply. *Aust. J. Exp. Agricul.* **29**, 741–750.

Miller, F.C. (1991) Biodegradation of solid wastes by composting. In: *Biological Degradation of Wastes* (ed. A.M. Martin), pp. 1–31. Elsevier Applied Science, London.

Moats, W.A. (1971) Kinetics of thermal death of bacteria. *J. Bacteriol.* **105**(1), 165–171.

Murwira, H.K., Kirchmann, H. and Swift, M.J. (1990) The effect of moisture on the decomposition rate of cattle manure. *Plant Soil* **122**, 197–199.

Nakasaki, K., Hakano, Y., Akiyama, T., Shoda, J. and Kubota H. (1987) Oxygen diffusion and microbial activity in the composting of dehydrated sewage sludge cakes. *J. Ferment. Technol.* **56**(1), 43–48.

Nakasaki, K., Yaguchi, H., Sasaki, Y. and Kubota, H. (1990) Effect of oxygen concentration on composting of garbage. *J. Ferment. Bioeng.* **70**(6), 431–433.

Nakasaki, K., Yaguchi, H., Sasaki, Y. and Kubota, H. (1993) Effects of pH control on composting of garbage. *Waste Manage. Res.* **11**, 117–125.

Nakasaki, K., Fujiwara, S. and Kubota, H. (1994a) A newly isolated thermophilic bacterium, *Bacillus lichniformis* HA1 to accelerate the organic matter decomposition in high rate composting. *Compost Sci. Util.* **2**(2), 88–96.

Nakasaki, K., Aoki, N. and Kubota, H. (1994b) Accelerated composting of grass clippings by controlling moisture level. *Waste Manage. Res.* **12**, 13–20.

Neuman, R.P. and Walker, L.P. (1992) Solute exclusion from cellulose in packed columns: experimental investigation and pore volume measurements. *Biotechnol. Bioeng.* **40**(2), 218–225.

Paul, E.A. and Clark, F.E. (1996) *Soil Microbiology and Biochemistry*, 2nd edn. Academic Press, San Diego, CA.

Randle, P. and Flegg, P.B. (1978) Oxygen measurements in a mushroom composting stack. *Sci. Horticul.* **8**, 315–323.

Regan, R., Jeris, J.S., Basser, R., McCann, K. and Hudek, J. (1973) *Cellulose Degradation in Composting*, 142 pp. EPA-R3-73-029, PB 215 722. USEPA, Washington, DC.

Richard, T.L. (1992) Municipal solid waste composting: physical and biological processing. *Biomass Bioener.* **3**(3–4), 163–180.

Richard, T.L. (1996) Capillary theory and matric potential. URL: http://compost.css.cornell.edu/oxygen/capillary.html. Accessed December 17, 2003.

Richard, T.L. (1997a) *The Kinetics of Solid-State Aerobic Biodegradation*, 357 pp. Ph.D. Thesis. Cornell University, Ithaca, NY.

Richard, T.L. (1997b) *Stoichiometry and Kinetics*. A E 573 class website. http://www.abe.iastate.edu/Ae573_ast475/stoichiometry_notes.htm. Accessed December 17, 2003.

Richard, T.L. and Trautmann, N. (1996) Solving the moisture and carbon–nitrogen equations simultaneously (including embedded links). URL: http://compost.css.cornell.edu/calc/cn_ratio.html. Accessed December 17, 2003.

Richard, T.L, Walker, L.P. and Gossett, J.M. (1999) The effects of oxygen on solid-state biodegradation kinetics. *Proceedings of the Institution of Biological Engineering* **2**, A10–A30. IBE Publications, Athens, GA.

Richard, T.L., Hamelers, H.V.M., Veeken, A.H.M. and Silva, T. (2002) Moisture relationships in composting processes. *Compost Sci. Util.* **10**(4), 286–302.

Rosso, L., Lobry, J.R., Bajard, S. and Flandrois, J.P. (1995) Convenient model to describe the combined effects of temperature and pH on microbial growth. *Appl. Environ. Microbiol.* **61**(2), 610–616.

Rynk, R. (ed.) (1992) On-farm Composting Handbook. Northeast Regional Agricultural Engineering Service, Ithaca, NY.

Rynk, R. and Richard, T.L. (2001) Commercial composting production systems, pp. 51–93. In: *Compost Utilization in Horticultural Cropping Systems* (eds. P.J. Stoffella and B.A. Kahn), 414 pp. Lewis Publishers, Boca Raton, FL, USA.

Saucedo-Castañeda, G, Gutiérrez-Rojas, M., Bacquet, G. and Raimbault, M. (1990) Heat transfer simulation in solid state fermentation. *Biotechnol. Bioeng.* **35**, 802–808.

Schultze, K.L. (1962) Continuous thermophilic composting. *Appl. Microbiol.* **10**(2), 108–122.

Shuler, M.L. and Kargi, F. (1992) *Bioprocess Engineering: Basic Concepts.* Prentice Hall, Englewood Cliffs, NJ.

Smith, D.W. (1978) Water relations of microorganisms in nature. In: *Microbial Life in Extreme Environments* (ed. D.J. Kushner), pp. 369–380. Academic Press, New York.

Strom, P.F. (1985a) Effect of temperature on bacterial species diversity in thermophilic solid-waste composting. *Appl. Environ. Microbiol.* **50**, 899–905.

Strom, P.F. (1985b) Identification of thermophilic bacteria in solid-waste composting. *Appl. Environ. Microbiol.* 50, 906–913.

Suler, D.J. and Finstein, M.S. (1977) Effect of temperature, aeration, and moisture on CO_2 formation in bench-scale, continuously thermophilic composting of solid waste. *Appl. Environ. Microbiol.* **33**(2), 345–350.

Tansey, M.R. and Brock, T.D. (1978) Microbial life at high temperatures: ecological aspects. In: *Microbial Life in Extreme Environments* (ed. D.J. Kushner), pp. 159–216. Academic Press, London.

Tseng, D.Y., Chalmers, J.J., Tuovinen, O.H. and Hoitink, H.A.J. (1995) Characterization of a bench-scale system for studying the biodegradation of organic solid wastes. *Biotechnol. Progr.* **11**, 443–451.

van Ginkel, G.T., Raats, P.A.C. and van Haneghem, I.A. (1999) Bulk density and porosity distributions in a compost pile. *Neth. J. Agricul. Sci.* **47**, 105–121.

Van Soest, P.J. (1994) *Nutritional Ecology of the Ruminant*, 2nd edn., 476 pp. Cornell University Press, Ithaca, NY.

Veeken, A., de Wilde, V. and Hamelers, H.V.M. (2002) Passively aerated composting of straw-rich pig manure: effect of compost bed porosity. *Comp Sci. Util.* **10**(2), 114–128.

Wiley, J.S. (1956) Progress report on high-rate composting studies. Proceedings of the 11th Industrial Waste Conference. *Purdue Univ. Eng. Bull. Series No.* **91**, 334–341.

Wiley, J.S. and Pearce, G.W. (1957) A preliminary study of high-rate composting. *Trans. ASCE* **122**, 1009–1034.

USEPA (1985) *Composting of Municipal Wastewater Sludges.* Design Manual Number 44, WWBLDM44. EPA 625/4-85/014. Available from: National Small Flows Clearinghouse. Morgantown, WV. 68 pp.

13

Trends in solid-waste management through composting in the US

Harry A.J. Hoitink and Frederick C. Michel Jr

13.1 INTRODUCTION

Enforcement of regulations governing the concentrations of priority pollutants in environmental samples has had major impacts on the development of solid-waste management practices in the US (Chaney *et al.* 2001). Diversion of priority pollutants from the solid-waste stream has allowed composting to become the preferred alternative process for treatment of an ever-increasing diversity of solid wastes (Michel Jr. *et al.* 2002). Composting of commingled municipal solid wastes (MSW) has all but been abandoned in the US because the quality of the compost produced cannot consistently meet the expectations of compost users. Landscapers, as described in more detail below, have been unwilling to accept risks associated with the distribution of plastics and glass in such contaminated composts to homeowners even though some batches of MSW composts can meet US-EPA guidelines governing the concentrations of priority pollutants. As a result, waste separation became adopted widely for recycling of MSW in the US Composting of

Edited by Piet Lens, Bert Hamelers, Harry Hoitink, Werner Bidlingmaier. ISBN: 1 84339 054 X. Published by IWA Publishing, London, UK

yard wastes now is practiced widely (Goldstein and Madtes 2001). Furthermore, composting of food and restaurant wastes is increasing rapidly. Municipal, paper mill and other industrial wastewater sludges have been composted for decades in the US (reviewed in Brown *et al.* 1998). This trend is increasing as priority pollutants are removed from these waste streams and where land application of the raw material poses unacceptable problems. A brief review presented in this chapter reveals that the composting process also is used widely for bioremediation of soils contaminated with biodegradable pollutants.

Concerns over the release of odors, pathogens and of nutrients from animal manures into the environment that surrounds farms has emphasized problems associated with utilization of these wastes. These pressures, coupled with better knowledge of compost mineralization rates (Gilmour 1998), have allowed composting to become a mainstream agricultural practice. An *On-Farm Composting Handbook* published in the early 1990s (Rynk *et al.* 1992) greatly facilitated the introduction of appropriate composting technologies to farmers. A direct spin-off from this trend has been the emergence of novel animal housing and manure management systems which avoid or minimize the generation of liquid manures and include a composting system as an integral component of their design (Elwell *et al.* 1998; Keener *et al.* 2002). The latest technologies generate composts cost effectively from manures if factors described below are considered. This development has been most effective on poultry farms. It has significantly increased the opportunity for the adoption of organic farming practices on grain farms that surround mega poultry farms.

A second factor that has had major impacts on acceptance of composting as a waste treatment option in the US is the value-added nature of high quality composts produced from the best operating systems. The most value added aspect of high quality composts generally is its peat substitute value in potting mixes and during out planting of ornamental shrubs in landscapes (Gouin 1998; Fitzpatrick 2001; Alexander 2001). The second most valuable aspect is the plant disease suppressive nature of composts (Cohen *et al.* 1998; Hoitink and Boehm 1999). Utilization of composts has facilitated the phase-out of methyl bromide and it also significantly curtailed soil fungicide use (De Ceuster and Hoitink 1999). Biological control agent-fortified composts, which are now becoming available and that provide more predictable disease control than natural composts (Zhang *et al.* 1998; Pharand *et al.* 2002), promise to further increase the value of composts.

Guidelines governing composting and compost quality standards as well as standardized methods for analysis of these compost properties that have been sanctioned by the US Composting Council, in addition to educational programs for compost producers and utilizers (Kashmanian and Rynk 1998), all have supported the new directions described above. Due to continuous progress in the field of composting technologies as a result of feedback from utilization industries on issues related to compost quality, market values for composts have increased. As a result, subsidies provided by governmental agencies to the composting infrastructure in the US are declining as compost prices increase due to market demand. In conclusion, solid-waste management through composting has had major impacts on soil and water quality as well as on plant health management practices. This chapter highlights the most recent trends in composting and in compost utilization in the US.

13.2 TRENDS IN MUNICIPAL AND INDUSTRIAL SOLID WASTES

13.2.1 Acceptance of composting for treatment of MSW

Estimates suggest that the US generates at least 400 million tons of MSW each year (Goldstein and Madtes 2001). This database reveals that the estimated amount of MSW recycled has grown from only 8% in 1990 to 32% in 2001. Yard wastes (leaves, brush, grass clippings, sod, etc.) represent a substantial portion of this recycled material. Today, more than 3000 such composting facilities exist in the US, with the greatest number located in the Great Lakes Region (Goldstein and Madtes 2001). An example of a low technology facility is presented in Figure 13.1. The total tonnage of yard wastes composted in the US now exceeds 38 million tons per year. It has become a major industry that cannot meet market demands. Some of these plants also compost food wastes and a few are composting food industry wastes in addition to agricultural wastes. This trend is becoming more common in the US.

Composting of mixed MSW in the US lost the momentum that it had in the early 1990s (Satkofsky 2001b). The primary reason for the decline is product quality. Compost produced from mixed MSW cannot consistently meet US-EPA 503 rules governing concentrations of heavy metals and other priority pollutants (Cook and Beyea 1998). Glass and plastics in this product pose hidden problems, which typically are not recognized until the organic matter has decomposed in the field

Figure 13.1 Municipal yard trimmings composting facility in Michigan (USA).

(Krause *et al.* 1996; Skimina 1998). An effort is underway in the US to reduce landfilling costs by utilizing mixed MSW compost as daily or as final cover.

As mentioned earlier, composting of source-separated MSW has increased rapidly in the US. The concentrations of potential toxic and carcinogenic chemical contaminants in separated composted municipal wastes typically are below those levels causing a concern (Cook and Beyea 1998). Products that cannot meet US-EPA regulations governing compost quality must be landfilled. The concentrations of cadmium, lead, mercury, PCDDFs, PCBs and polyaromatic hydrocarbons (PAHs) in these source-separated composts that meet the regulations are low enough so that their utilization at a rate of $10\,Mg\,ha^{-1}y^{-1}$ would not exceed soil levels permitted by US-EPA's 503 rule for these pollutants until after at least 200 years of application at that rate. The plow depth assumed in these calculations was 25 cm. They determined, therefore, that adequate time would be available for monitoring of treated soils to avoid the development of potential problems associated with utilization of these composts in the US.

13.2.2 Bioremediation through composting

Utilization of the composting process for bioremediation of soils and industrial wastes contaminated with toxic organics expanded rapidly during the 1990s (Cole 1998; Reddy and Michel Jr. 1998). In general, composting is less costly than incineration or other remediation alternatives. The treated product can be used to generate topsoil for reclamation of such disturbed sites. Contaminants such as diesel fuel, motor oil, and to a lesser extent PAHs and pentachlorophenol (Laine and Jorgensen 1997), are now treated via composting/bioremediation. Composting also is used for bioremediation of soils contaminated with explosives, including trinitrotoluene, nitrocellulose, RDX, and HMX (Williams *et al.* 1992; Pennington *et al.* 1995). Some success has been reported for bioremediation of recalcitrant chlorinated compounds such as polychlorinated biphenyls via composting (Hogan *et al.* 1989; Michel Jr. *et al.* 2001a). However, highly chlorinated PCB congeners often are not degraded during composting (Michel Jr. *et al.* 2001a).

Considerable research has been performed on bioremediation of pesticides during composting (Michel Jr. *et al.* 1995, 1997; Buyuksonmez *et al.* 1999). Pesticides that readily biodegrade in soil in general are also destroyed rapidly during composting (Buyuksonmez *et al.* 2000). This efficacy of composting for bioremediation of pesticides has led to its use for adsorption and treatment of pesticide rinsate wastewater (Rao *et al.* 1995; Kuo and Regan 1999). Unfortunately, not all pesticides are degraded adequately during composting. Recent literature from the US and New Zealand reports composts contaminated with the herbicides clopyralid and picloram at concentrations high enough to affect the most sensitive crops (Bezdicek *et al.* 2001; Fietje 2001; Cotton *et al.* 2001). Although these herbicides do substantially decompose during the composting process, they remain present in composts under some conditions at concentrations high enough (10–100 ppb) to pose problems to tomato and other sensitive crops (Vandervoort *et al.* 1997; Bezdicek *et al.* 2001). Several states have banned the use of clopyralid in landscape settings and lawns to avoid this problem.

13.3 TRENDS IN COMPOSTING OF MANURES

13.3.1 Background information

During the past decade, the US-EPA and state environmental agencies began emphasizing reduction of non-point pollutants in surface waters. This emphasis focused principally on pathogens and nutrients (N and P) in water draining from farmland (Millner 1998; Boersma and Murarka 1995; Sutton and Power 1996). Odors, flies and manure spilled on public roads in rural areas have become major issues also.

Several states require that quantities of available essential plant nutrients in manures applied to cropland not exceed those required for the crop on the treated field. This has motivated farmers to distribute manures among farms resulting in utilization of nutrients in more environmentally sound and beneficial ways. Composting is beginning to play a role in this process because it allows for lower costs and storage of products with fewer problems than manures. More importantly, composts can be directed into value-added markets (McCoy and Cogburn 2001; De Ceuster and Hoitink 1999).

The magnitude of potential manure management problems in the US ranges from minor for small farms to immense for facilities housing millions of chickens or thousands of hogs or cows. Some farms produce as much solid waste as a mid-sized municipality. Each animal industry poses unique problems and the nature of problems varies by region in the US.

13.3.2 Poultry industry

The poultry industry has made significant progress recently by adopting novel housing and composting facilities and manure marketing strategies (Elwell *et al.* 1998). Egg producing farms have installed composting technology equipped with air scrubber facilities which control ammonia emissions (Figure 13.2). Such facilities produce stabilized composted manures with a high nitrogen content (4.25–5.0%). The manure in these facilities is mixed frequently so that flies cannot complete life cycles. Thus, it has become possible for these facilities to become certified as organic farms.

These new technologies increasingly are being constructed in minimum tillage grain producing regions. Application of this type of compost at a low rate of 2.5 tons per hectare increases the rate of decomposition of littler during the winter which in turn decreases survival of plant pathogens and the severity of diseases caused by soilborne plant pathogens. This in turn results in improved grain yields and quality. This trend has increased organic farming in these regions.

Broiler facilities have made significant improvements in "sheet" composting of manures (Miller 1999). The sawdust-bedded chicken litter used in many of these systems is composted in between flocks for 1 week at a time through four to seven flock cycles. The manure from these farms is stabilized adequately by this process so that the final product removed from barns can be composted more easily or applied directly to farmland. An interesting side benefit from repeated reuse of composted bedding in broiler facilities is the decrease in animal mortalities observed in

Figure 13.2 Poultry manure composting at a 800,000 bird layer facility in Ohio (USA).

each broiler production cycle over time (Miller 1999). The food industry is beginning to assist this trend in the poultry industry by signing preferential contracts with meat and or egg producers that compost the manure. This in part pays for the cost of composting on such farms.

13.3.3 Hog industry

Perhaps the most interesting development in manure management through composting is the High Rise Hog system (Figure 13.3). This novel technology allows production of hogs on slatted floors under controlled conditions (Stowell *et al.* 1999; Keener *et al.* 1999; Mescher *et al.* 1999). The manure is composted underneath this floor with bulking agents such as sawdust and wood chips that are positioned on an aerated floor. Liquids are evaporated during this process. When managed properly, the concentrations of NH_3 and H_2S in the system are very low resulting in better animal performance than observed in the traditional liquid hog manure systems where health problems often are caused by these emissions. The manure produced after several cycles of hogs can be composted further and then marketed beneficially in the horticultural industries (Keener *et al.* 2001; Michel Jr. *et al.* 2001b). This approach to swine production may allow further separation of this industry from traditional farming systems.

As odors are managed in this new system, large facilities could be constructed in regions deficient in organic by-products as has already occurred for egg farms. These new technologies, therefore, may pave the way for the development of more cost effective and environmentally friendly animal production technologies based on composting, perhaps preferentially to be located in regions where methyl bromide still is used due to the absence of organic amendments, as described in more detail below.

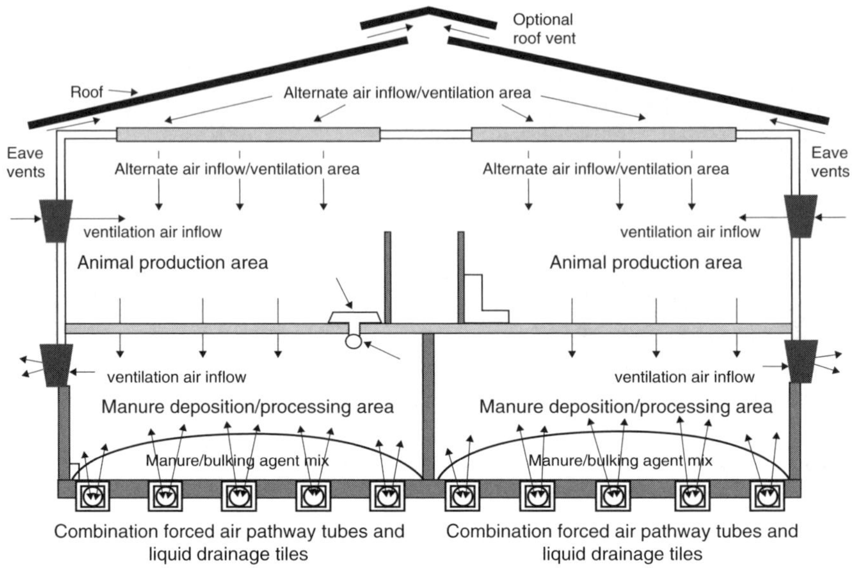

Figure 13.3 Diagram of the High Rise Hog system.

13.3.4 Composting of animal mortalities

The impact of composting on agriculture has been most significant perhaps for treatment of animal mortalities. The number of farms that compost poultry mortalities in the US exceeds the number of farms that compost manure and the number of yard waste composting systems (Kashmanian and Rynk 1996). Methods for composting of mortalities are now well established for swine, cattle, horses, sheep, fish, as well as some other animal carcasses (Stanford *et al.* 2000). These methods effectively eliminate common animal pathogens (Kashmanian and Rynk 1996; Stanford *et al.* 2000). A possible exception is the causative agent for BSE. The fate of prions during composting needs to be examined (Sreevatsan and Michel Jr. 2002).

13.4 TRENDS IN UTILIZATION OF COMPOSTS

13.4.1 Compost quality impacts

The most critical factor affecting compost utilization is product quality. As already mentioned above, MSW composts that are contaminated with glass or plastics cannot be marketed successfully, except for in regions not familiar with the consequences of application of such materials to soil. After the problems become recognized, the contaminated product is refused. The concentration of these materials and of heavy metals has not been an issue for composts prepared from source-separated wastes as long as adequate quality control procedures are practiced. Residual concentrations of pesticides in composts prepared from separated wastes

Figure 13.4 Effect of compost age (days) on plant growth response in potting media. Composts were made from a 3 : 1 mixture (wet weight) of dairy manure and wheat straw.

typically also do not pose problems to most markets (Rao *et al.* 1995; Michel Jr. *et al.* 1997). Limits on concentrations for heavy metals set by US-EPA's 503 rule continue to be reviewed. On the other hand, high concentrations of heavy metals detected in some commercial sources of inorganic fertilizers have become a concern (Kranert and Fruth 2002). It seems likely, therefore, that this entire issue will continue to be under review for some time.

Standard methodologies for the determination of physical, chemical and biological properties of composts are in the process of being adopted in the US (http://tmecc.org). Most states adhere to limits set by the US-EPA's 503 rule for priority pollutants but some enforce lower limits. One factor that often is overlooked is compost stability and the response of plants to compost applications. Inadequate and excessive compost stability can cause utilization problems. Although a significant amount of information is available on the degree of stability required for a particular type of compost to avoid the generation of major utilization problems by composts (Brewer and Sullivan 2003; Changa *et al.* 2003; Chen 2003; Cooperband *et al.* 2003), quantitative guidelines for their utilization in soils or as mulches or as components in potting mixes are just now being characterized. Figure 13.4 demonstrates the response of cucumber seedlings to compost maturity/stability. After 74–88 days of composting growth in the compost-amended mix (30% compost on a volume basis) was equivalent to that in a standard peat mix. Rational standards for utilization of different types of composted manure products cannot yet be proposed because it differs for each type of bedding, manure and utilization pathway. For this reason, composts typically are utilized on the basis of past experiences with a specific product from a given supplier as opposed to on the basis of standard physical, chemical, and/or biological properties that define quality.

13.4.2 Product blending and performance

In spite of the difficulties mentioned above, most composts are utilized beneficially in the US. Blends of various types of composts tailored to suit the needs of specific niche markets are becoming available (Gouin 1998). Across the US, landscapers have all but completely replaced peat with composts. Blends of composted leaves and yard wastes are being sold as "garden composts" which has increased revenues for landscapers and improved the quality of gardens for homeowners. Composted yard wastes also are blended with tree bark to produce products that do not immobilize nitrogen and yet provide the longer lasting mulch effects typically associated with high lignin content tree bark (Hoitink and Boehm 1999). Composts release nutrients in slow release form and also improve soil structure. However, management of essential plant nutrients, in particular of N, remains a difficult problem associated with utilization of composts because it is difficult to accurately predict (1) fertility values for composts and (2) the quantity of nutrients released from the residual organic fraction in soil. The entire fertility aspect associated with these organic products requires basic research as well as more extension-related activities.

13.4.3 Environmental applications

Novel environmental applications for compost continue to emerge. The use of compost is being promoted to help soils retain water and nutrients, reduce runoff and erosion, filter storm water, replace pesticides and biodegrade pollutants in the natural environment. A prominent public-supported effort is the Soils for Salmon program in the Pacific Northwest region (Block 2001; Satkofsky 2001a). This program encourages residents and land developers to use compost so that degrading effects of urban development on water quality are reversed while the habitat for endangered salmon is improved. State Departments of Transportation across the US increasingly use composts for erosion control on steep banks along highways. The composting industry, therefore, is beginning to market composts for environmental performance-related properties. Other environmental uses that emerged during the past three decades include the use of compost for remediation of disturbed lands such as wetlands, abandoned mine lands, brownfields and landfills (Chaney *et al.* 2001; Henry 2002).

Composts have had significant impacts on pesticide use in the US. In the nursery industry, methyl bromide has been replaced by composts that naturally suppress diseases caused by soilborne plant pathogens such as *Pythium* and *Phytophthora* spp. (Quarles and Grossman 1995; De Ceuster and Hoitink 1999). Biocontrol agents can be inoculated into composts to provide even greater suppression of diseases (Cohen *et al.* 1998; Hoitink and Boehm 1999). Much remains to be accomplished, however, to fully commercialize these concepts. Finally, composts also may induce systemic resistance in plants to diseases (Zhang *et al.* 1998; Pharand *et al.* 2002). This type of environmentally induced resistance may affect diseases as well as pests (Heil 1999). During the next decade, better knowledge will undoubtedly be developed on how composts regulate these beneficial effects in plants. Better knowledge undoubtedly will allow production of value-added products from composts and thus facilitate composting as a waste treatment option in applications where it is not yet cost effective today.

REFERENCES

Alexander, R. (2001) Compost utilization in landscapes. In: *Compost Utilization in Horticultural Cropping Systems* (eds. P.J. Stofella and B.A. Kahn), pp. 151–176. Lewis Publishers.

Bezdicek, D., Fauci, M., Caldwell, D., Finch, R. and Lang, J.C. (2001) Persistent herbicides in compost. *BioCycle* **42**(7), 25–30.

Block, D. (2001) Composting advances in Oregon and Washington. *BioCycle* **42**(2), 28–36.

Boersma, L.L. and Murarka, I.P. (eds.) (1995) *Waste Management and Utilization in Food Production and Processing*. CAST Task Force Report, no. 12, p. 125.

Brewer, L.J. and Sullivan, D.M. (2003) Maturity and stability evaluation of composted yard trimmings. *Compost Sci. Util.* **11**(2), 96–112.

Brown, S., Angle, J.S. and Jacobs, L. (1998) In: *Beneficial Co-Utilization of Agricultural, Municipal and Industrial By-Products*, p. 430. Kluwer Academic Publishers.

Buyuksonmez, F., Rynk, R., Hess, T. and Bechinski, E. (1999) Occurrence, degradation and fate of pesticides during composting. *Compost Sci. Util.* Part I, **7**(4), 66–82.

Buyuksonmez, F., Rynk, R., Hess, T. and Bechinski, E. (2000) Occurrence, degradation and fate of pesticides during composting. *Compost Sci. Util.* Part II, **8**(1), 61–81.

Chaney, R.L., Ryan, J.A., Kukier, U., Brown, S.L., Siebielec, G., Malik, M. and Angle, J.S. (2001) Heavy metal aspects of compost use. In: *Compost Utilization in Horticultural Cropping Systems* (eds. P.J. Stofella and B.A. Kahn), pp. 323–359. CRC Press, Boca Raton, FL.

Changa, C.M., Wang, P., Watson, M.E., Hoitink, H.A.J. and Michel Jr., F.C. (2003) Assessment of the reliability of a commercial maturity test kit for composted manures. *Compost Sci. Util.* **11**(2), 125–143.

Chen, Y. (2003) Nuclear magnetic resonance, infra-red and pyrolysis: application of spectroscopic methodologies to maturity determination of composts. *Compost Sci. Util.* **11**(2), 152–168.

Cole, M.A. (1998) Remediation of soils contaminated with toxic organic compounds. In: *Beneficial Co-Utilization of Agricultural, Municipal and Industrial By-Products* (eds. S. Brown, J.S. Angle and L. Jacobs), pp. 175–194. Kluwer Academic Publishers, Dordrecht, The Netherlands.

Cohen, R., Chefetz, B. and Hadar, Y. (1998) Suppression of soil-borne pathogens by composted municipal solid waste. In: *Beneficial Co-Utilization of Agricultural, Municipal and Industrial By-Products* (eds. S. Brown, J.S. Angle and L. Jacobs), pp. 113–130. Kluwer Academic Publishers, Dordrecht, The Netherlands.

Cook, J. and Beyea, J. (1998) Potential toxic and carcinogenic chemical contaminants in source-separated municipal solid waste composts: review of available data and recommendations. *Tox. Environ. Chem.* **67**, 27–69.

Cooperband, L.R., Stone, A.G., Fryda, M.R. and Ravet, J.L. (2003) Relating compost measures of stability and maturity to plant growth. *Compost Sci. Util.* **11**(2), 113–124.

Cotton, M., Storelli, S. and Levenson, H. (2001) Assessing California's organic materials management structure. *BioCycle* **42**(7), 42–29.

De Ceuster, T.J.J. and Hoitink, H.A.J. (1999) Prospects for composts and biocontrol agents as substitutes for methyl bromide in biological control of plant diseases. *Compost Sci. Util.* **7**, 6–15.

Elwell, D.L., Keener, H.M., Carey, D.S. and Schlak, P.P. (1998) Composting unamended chicken manure. *Compost Sci. Util.* **6**, 22–35.

Fietje, G. (2001) New Zealand city confronts contamination. *BioCycle* **42**(7), 31.

Fitzpatrick, G.E. (2001) Compost utilization in ornamental and nursery crop production systems. In: *Compost Utilization in Horticultural Cropping Systems* (eds. P.J. Stofella and B.A. Kahn), pp. 135–150. Lewis Publishers.

Gilmour, J.T. (1998) Carbon and nitrogen mineralization during co-utilization of biosolids and composts. In: *Beneficial Co-Utilization of Agricultural, Municipal and Industrial By-Products* (eds. S. Brown, J.S. Angle and L. Jacobs), pp. 89–112. Kluwer Academic Publishers, Dordrecht, The Netherlands.

Goldstein, J. and Madtes, C. (2001) The state of garbage in America. *BioCycle* **42**(12), 42–54.

Gouin, F.R. (1998) Using compost in the ornamental horticulture industry. In: *Beneficial Co-Utilization of Agricultural, Municipal and Industrial By-Products* (eds. S. Brown,

J.S. Angle and L. Jacobs), pp. 131–138. Kluwer Academic Publishers, Dordrecht, The Netherlands.

Heil, M. (1999) Systemic acquired resistance: available information and open ecological questions. *J. Ecol.* 87, 341–346.

Henry, C. (2002) Swamp restoration project Website: http://faculty.washington.edu/clh/wet.html. University of Washington.

Hogan, J.A., Toffoli, G.R., Miller, F.C., Hunter, J.V. and Finstein, M.S. (1989) Composting physical model demonstration: mass balance of hydrocarbons and PCBs. In: *International Conference on Physical, Chemical and Biological Detoxification of Hazardous Waste*, pp. 742–757. Technomic Publishing Co, Lancaster, PA.

Hoitink, H.A.J. and Boehm, M.J. (1999) Biocontrol within the context of soil microbial communities: a substrate-dependent phenomenon. *Annu. Rev. Phytopathol.* **37**, 427–446.

Kashmanian, R. and Rynk, R. (1996) Agricultural composting in the United States-trends and driving forces. *J. Soil Water Conser.* **51**(3), 194–201.

Kashmanian, R. and Rynk, R. (1998) Creating positive incentives for agricultural composting. *Am. J. Alternative Agricul.* **13**(1), 40–45.

Keener, H.M., Elwell, D.L., Menke, T. and Stowell, R. (1999) Design and management of a High-Rise ™ Hog facility manure drying bed. Paper no. 994108. ASAE Meeting Presentation. Toronto, Ont., Canada, July 18–21, 1999.

Keener, H.M., Stowell, R.R., Elwell, D.L., Michel Jr., F.C. and Menke, T. (2001) Operating performance of a High-Rise Hog facility. *Proceedings of the 2001 Animal and Poultry Waste Management Center Annual Meeting*, October 3–5, 2001. pp. 508–516.

Kranert, M. and Fruth, F. (2002) Comparison of the introduction of heavy metals into soils by fertilizers and compost. In: *Proceedings of the 2002 Symposium on Composting and Compost Utilization* (eds. F.C. Michel Jr., R.F. Rynk and H.A.J. Hoitink), pp. 968–983. JG Press, Emmaus, PA.

Krause, C.R., Horst, L.E. and Hoitink, H.A.J. (1996) Morphological and chemical properties of plastic residues in composts. *Scanning Microsc.* **10**, 997–1004.

Kuo, W.S. and Regan, R.W. (1999) Removal of pesticides from rinsate by adsorption using agricultural residuals as medium. *J. Environ. Sci. Health*, Part B – Pesticides, Food Contaminants and Agricultural Wastes **34**, 431–447.

Laine, M.M. and Jorgensen, K.S. (1997) Effective and safe composting of chlorophenol-contaminated soil in pilot scale. *Environ. Sci. Technol.* **31**, 371–378.

McCoy, S. and Cogburn, B. (2001) Texas makes inroads with highway use of compost. *BioCycle* **42**(2), 67–70.

Mescher, T., Menke, T., Stowell, R., Veenhuizen, M. and Keener, H. (1999) Design, performance, and economics of a High-Rise swine finishing building. Paper no. 994107. ASAE Meeting Presentation. Toronto, Ont., Canada, July 18–21, 1999.

Michel Jr., F.C., Reddy, C.A. and Forney, L.J. (1995) Microbial degradation and humification of 2,4-dichlorophenoxyacetic acid during the composting of yard trimmings. *Appl. Environ. Microbiol.* **61**, 2566–2571.

Michel Jr., F.C., Reddy, C.A. and Forney, L.J. (1997) Fate of [-2-^{14}C]-labeled diazinon during the composting of yard trimmings. *J. Environ. Quality* **26**, 200–205.

Michel Jr., F.C., Quensen, J. and Reddy C.A. (2001a) Bioremediation of a PCB contaminated soil via composting. Bioremediation. *Compost Sci. Util.* **9**, 274–284.

Michel Jr., F.C., Pecchia, J., Sun, H. and Keener, H.M. (2001b) Use of high rise building and composting to manage swine manure. *Proceedings of the 2001 Animal and Poultry Waste Management Center Annual Meeting*, October 3–5, 2001. pp. 728–732.

Michel Jr., F.C., Rynk, R.F. and Hoitink, H.A.J. (2002) *Proceedings of the 2002 Symposium on Composting and Compost Utilization*, 1598 pp. JG Press, Emmaus, PA.

Miller, F.C. (1999) Poultry farm pioneers low-rate composting. *BioCycle* **40**, 59–61.

Millner, P.D. (1998) Microbial effects of environmental health and product quality aspects of recovery and co-utilization of bio-mineral products. In: *Beneficial Co-Utilization of*

Agricultural, Municipal and Industrial By-Products (eds. S. Brown, J.S. Angle and L. Jacobs), pp. 247–257. Kluwer Academic Publishers, Dordrecht, The Netherlands.

Pennington, J.C., Hayes, C.A., Myers, K.F., Ochman, M., Gunnison, D., Felt, D.R. and McCormick, E.F. (1995) Fate of 2,4,6 trinitrotoluene in a simulated compost system. *Chemosphere* **30**, 429–438.

Pharand, B., Carisse, O. and Benhamou, N. (2002) Cytological aspects of compost-mediated induced resistance against Fusarium crown and root rot in tomato. *Phytopathology* **92**, 424–438.

Quarles, W. and Grossman, J. (1995) Alternatives to methyl bromide in nurseries – Disease suppressive media. *IPM Pract.* **17**, 1–13.

Rao, N., Grethlein, H.E. and Reddy, C.A. (1995) Mineralization of atrazine during composting with untreated and pretreated lignocellulosic substrates. *Compost Sci. Util.* **3**, 38–46.

Reddy, C.A. and Michel Jr., F.C. (1998) Fate of xenobiotics during composting. *Proceedings of the International Symposium of Microbial Ecology*, Halifax NS, August 20–25, 1998.

Rynk, R., van de Kamp, M., Willson, G.B., Singley, M.E., Richard, T.L., Kolega, J.J., Gouin, F.R., Laliberty Jr., L., Kay, D., Murphy, D.W., Hoitink, H.A.J. and Brinton, W.F. (1992) *On-Farm Composting Handbook*, 186 pp. Northeast Regional Agricultural Engineering Service, Cornell Extension, Ithaca, NY 14853.

Satkofsky, A. (2001a) Organic recycling initiatives spawned by salmon recovery. *BioCycle* **42**(9), 68–72.

Satkofsky, A. (2001b) MSW composting developments in the U.S. *BioCycle* **42**(12), 58–64.

Skimina, C.A. (1998) Costs and environmental impacts of co-utilization of by-products in nursery operations. In: *Beneficial Co-Utilization of Agricultural, Municipal and Industrial By-Products* (eds. S. Brown, J.S. Angle and L. Jacobs), pp. 149–155. Kluwer Academic Publishers, Dordrecht, The Netherlands, 430 pp.

Stanford, K., Larney, F.J., Olson, A.F., Yanke, L.J. and McKenzie, R.H. (2000) Composting as a means of disposal of sheep mortalities. *Compost Sci. Util.* **8**, 135–146.

Stofella, P.J. and Kahn, B.A. (2001) *Compost Utilization in Horticultural Cropping Systems*. CRC Press, Boca Raton, FL.

Stowell, R., Elwell, D., Strobel, B., Keener, H. and Menke, T. (1999) Air flow and air quality within a High-Rise ™ Hog facility. Paper no. 99–4109. *ASAE Meeting Presentation*, Toronto, Ont., Canada, July 18–21, 1999.

Sreevatsan, S. and Michel Jr., F.C. (2002) Prion diseases (Spongiform Encephalopathies): an overview. In: *Proceedings of the 2002 Symposium on Composting and Compost Utilization* (eds. F.C. Michel Jr., R.F. Rynk and H.A.J. Hoitink), pp. 1304–1324. JG Press, Emmaus, PA.

Sutton, A.L. and Power, J.F. (eds.) (1996) *Integrated Animal Waste Management*. CAST Task Force Report, ISSN 0194–4088, no. 128. p. 87.

Vandervoort, C., Zabik, M.J., Braham, B. and Lichfeldt, D.W. (1997) Fate of selected herbicides applied to turf grass: effect of composting on residuals. *Bull. Environ. Contam. Toxicol.* **58**, 38–45.

Williams, R.T., Ziegenfuss, P.S. and Sisk, W.E. (1992) Composting of explosives and propellant contaminated soils under thermophilic and mesophilic conditions. *J. Indust. Microbiol.* **9**, 137–144.

Zhang, W., Han, D.Y., Dick, W.A., Davis, K.R. and Hoitink, H.A J. (1998) Compost and compost water extract-induced systemic acquired resistance in cucumber and Arabidopsis. *Phytopathology* **88**, 450–455.

14

Anaerobic bioprocess concepts

J. Mata-Alvarez and S. Macé

14.1 INTRODUCTION

Anaerobic digestion (AD) is a process which decomposes organic matter in the absence of oxygen. The main products are biogas (a mixture of approximately 65% methane and 35% carbon dioxide) and a rather low amount of bacterial biomass (around 0.05 g volatile suspended solids (VSS)/g chemical oxygen demand (COD) removed).

$$\text{Organic matter} \rightarrow CH_4 + CO_2 + \text{Bacterial biomass} \quad (14.1)$$

The biogas composition depends on the substrate composition. Some stoichiometric relations have been proposed (McCarty 1964). If the organic compound has the formula $C_xH_yO_zN_t$, the ratio CH_4/CO_2 can be given approximately by:

$$\text{Ratio}\frac{CH_4}{CO_2} = \frac{4-T}{4+T} \quad (14.2)$$

where T is a parameter which is given by the compound formula:

$$T = \frac{(2z + 3t - y)}{x} \quad (14.3)$$

 Resource Recovery and Reuse in Organic Solid Waste Management. Edited by Piet Lens, Bert Hamelers, Harry Hoitink, Werner Bidlingmaier. ISBN: 1 84339 054 X. Published by IWA Publishing, London, UK

	Liquids	Solids Semi-solids
Industrial	Food industry Paper industry	Food industry Food markets
Urban	Domestic wastewaters	Sewage sludge Organic fraction MSW
Agriculture	Pig slurries Catty slurries Poultry slurries	Farm yard manure Bedding material Energy crops

Figure 14.1 Broad field of applications of AD processes.

Grease yields high percentages of CH_4 whereas oxalic or formic acids give low methane yields.

Figure 14.1 presents the broad field of application of AD processes. Traditionally, AD has been used to treat liquid wastes with or without suspended solids, such as manures, domestic or industrial wastewaters, sludges from biological or physico-chemical treatments, etc. References ranging from scientific studies to full-scale experiences are very extensive in this field (Mata-Alvarez *et al.* 2000; Van Lier *et al.* 2001). However, there are large quantities of solid wastes such as agricultural and municipal (MSW), which have attracted comparatively little attention on the part of AD specialists until recently. Due to their large organic matter content, these wastes offer great potential for biogas production.

One of the first studies on the topic of AD of solid wastes was conducted by Golueke and McGauhen (Golueke and McGauhen 1970; Golueke 1971) at the University of California. These researchers examined the technical feasibility of the AD of MSW mixed with other wastes such as manure and sewage sludge. Following this study, many others studies were published (see, for instance, Kispert *et al.* 1975) devoted to the AD of MSW as a whole, i.e. prior to any sorting. The studies mainly addressed the problem of landfill biomethanization. In Europe, the beginning of the study of AD of solid waste took place later, in the 1980s, and was dedicated directly to the organic fraction of municipal solid waste (OFMSW).

Figure 14.2 shows a basic COD balance in a digester treating an organic waste. As can be seen, 90% of the COD is removed in the process. Of course this figure shows an example and real parameters depend on the type of waste and on the digester operating conditions. Ten per cent of the initial COD remains in the effluent, whereas 5% is used to synthesize biomass.

AD of complex wastes may be employed to optimize organic matter removal (measured, for instance, by the COD) and/or methane production. The degree of

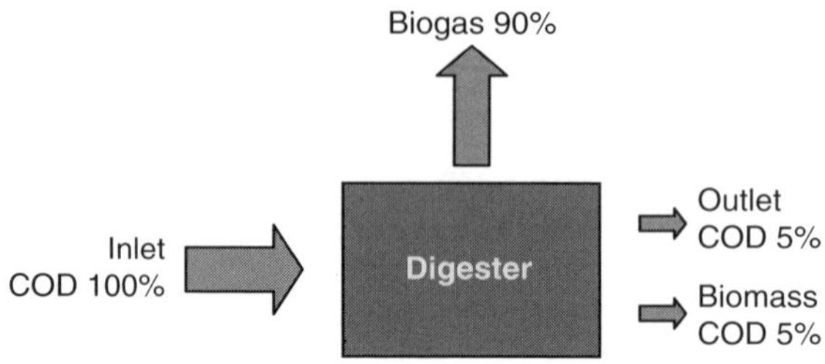

Figure 14.2 Typical COD digester balance.

organic matter reduction required is a function of the final use of the treated waste, and must be in accordance with the link that follows in the waste treatment chain. That is, the requirements of the process depend on the destination of the effluent. Methane production may be regarded as an alternative source of energy or a means by which energy can be saved in the whole plant. Both objectives condition the design criteria. In addition, either an AD model or accumulated experience from pilot (or full-scale) tests on different wastes are required for the proper design of the anaerobic digester.

Comparing aerobic processes with anaerobic ones points out a number of advantages as well as some drawbacks. For instance, anaerobic processes exhibit the advantage of a positive energetic balance and, moreover, the energy produced is renewable. A second advantage is that the biomass yield coefficient is much lower. Among the drawbacks are the rate of the process, which together with the low bacterial production, leads to a slow start-up. Moreover, reduced species such as ammonia are not removed by anaerobic reactors. The combination of both aerobic and anaerobic treatments usually offers the best solution.

14.2 MAIN STEPS IN AD PROCESSES

AD of organic wastes is a complex system involving several bacterial groups living in syntrophic association (Mata-Alvarez 2002). The process takes place by virtue of many simultaneous reactions catalysed by the microorganisms. The compounds present exhibit different oxidation states until finally being converted into carbon dioxide and methane.

Figure 14.3 shows a simplified diagram where substrate degradation together with microorganism growth takes place. This process also includes cell decay and is a process, which commonly accompanies the aerobic one. It can be represented by the following rate equation:

$$\frac{dX}{dt} = -Y\frac{dS}{dt} - b. \tag{14.4}$$

where S is the substrate concentration, X the microorganisms concentration, Y the yield coefficient and b the decay coefficient. Typical values for the yield coefficient

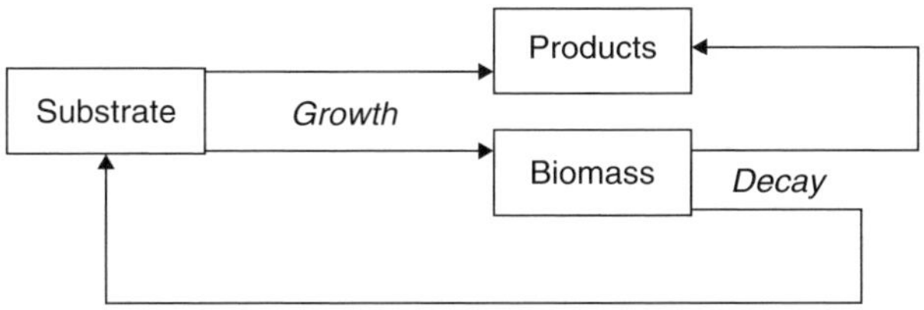

Figure 14.3 A simple scheme for microorganism growth and decay.

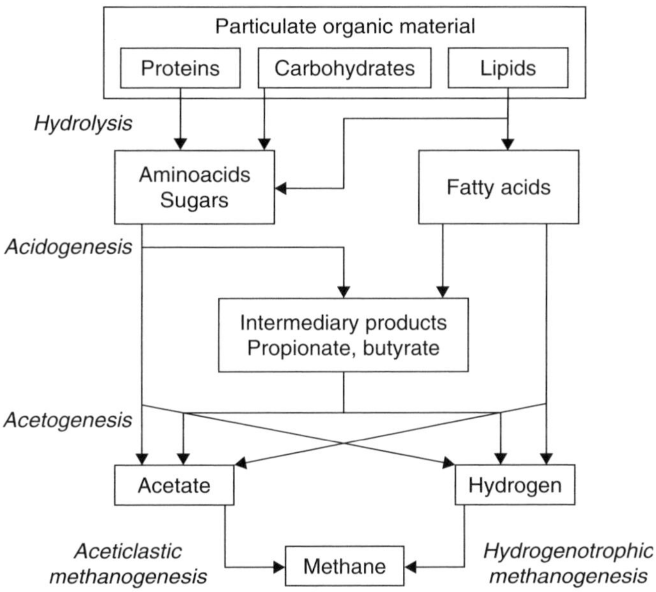

Figure 14.4 A complex diagram for AD considering five simultaneous processes (hydrolysis, acidogenesis, acetogenesis, aceticlastic methanogenesis and hydrogenotrophic methanogenesis) (Gujer and Zehnder 1983).

Y in aerobic systems are around 10 times higher than in anaerobic ones. This points out the reduced biomass production mentioned above.

AD processes are far more complex than aerobic ones, as can be seen in Figures 14.4 and 14.5, which present the main biochemical processes classified in four groups: hydrolysis, fermentation of monomers, acetogenesis/hydrogenotrophic methanogenesis and aceticlastic methanogenesis (Gujer and Zehnder 1983; Pohland 1992; Siegrist *et al.* 1993; García-Heras *et al.* 1999). There are, in addition, two other processes: The first is the action of homoacetogenic bacteria, which produce acetate from bicarbonate and hydrogen (acetogenic respiration of bicarbonate). The second is nitrate and sulphate reduction to respectively ammonium and sulphide by nitrate and sulphate-reducing bacteria. This process oxidizes reduced organic products, e.g. acetate and hydrogen at the expense of the electron acceptors nitrate and sulphate (Pohland 1992).

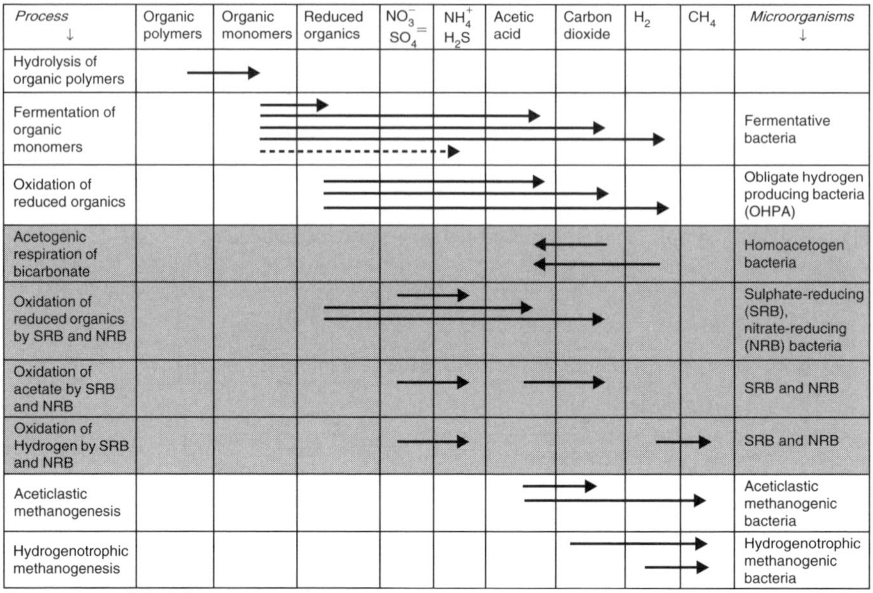

Organic polymers: Carbohydrates, lipids, proteins
Organic monomers: Sugars, organic acids, aminoacids
Reduced organics: Volatile fatty acids (propionic, butiric, valeric).

Figure 14.5 A complex diagram for AD in a matrix form (Pohland 1992; García-Heras *et al.* 1999). Each arrow indicates the reactants and products (first row) of a process (first column) catalysed by a group of bacteria (last column).

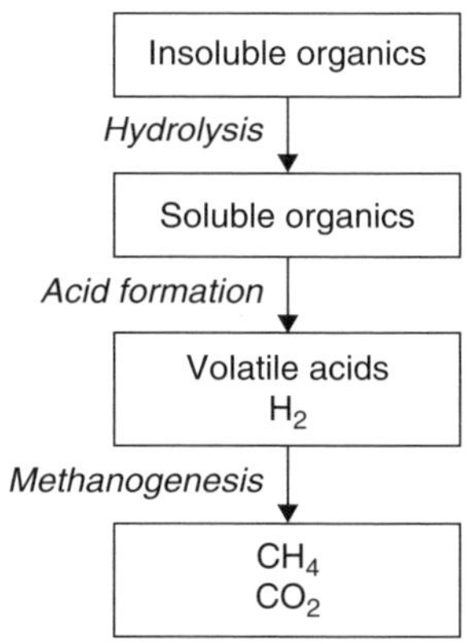

Figure 14.6 AD described by three processes: hydrolysis, acidogenesis and methanogenesis.

For practical applications, a simplified approach can be used (Figure 14.6). Only three single processes are considered here: (a) the hydrolysis of non-soluble compounds, which solubilize organic matter; (b) the acidogenesis, which converts soluble compounds to volatile fatty acids (VFA); and finally (c) the methanogenesis which converts acids to methane. This last step is the critical one, as methanogens are very sensitive to environmental conditions. Moreover, methane-formers have much slower growth rate than acidogenic bacteria. Thus, acid-formers outgrow methane-formers thus inducing acidic conditions in digesters. A pH below 6 affects the activity of methanogenic bacteria by inhibition. On the contrary, acidogens are not affected by this low pH, and can continue to produce acids, which makes the pH decrease. This, at the same time affects the methanogens more and inhibits the removal of these acids. As a consequence, the pH decreases still further. This negative circle slows down the activity of methanogenic bacteria and eventually totally inhibits their activity, causing the digester to fail. Digester failure must be avoided, as start-up is quite a slow process. Adequate control of the alkalinity and VFA concentration can avoid such a situation (Cecchi *et al.* 2002).

14.3 ENVIRONMENTAL FACTORS CONTROLLING THE AD PROCESS

14.3.1 Nutrients

In addition to the obvious presence of organic carbon to be degraded, there are many other substances, both organic and inorganic, which are necessary for an AD process to run optimally. Namely, there is the requirement of phosphorous and nitrogen (generically called nutrients), sulphur, vitamins and some traces of minerals (Fe, Ni, Mg, Ca, Na, Ba, Tu, Mo, Se and Co, generically called micronutrients). The presence of micronutrients in small quantities can stimulate growth (Kelly and Switzenbaum 1984; Speece *et al.* 1986). However, it must be taken into account that if a certain threshold level is surpassed, the presence of micronutrients can be inhibitory.

Recalling equations (14.1) and (14.4) and considering the low yield of biomass production of an anaerobic process, the contents of nutrients and micronutrients of an amount of waste is usually enough for digestion to proceed. Thus, nutrient requirements for anaerobic processes are much less exigent than for aerobic ones. Nevertheless, it is advisable to check the availability of nutrients, as a deficit of these compounds could exist.

As nutrients and micronutrients are necessary to synthesize cellular matter, the quantitative requirement is a function of the operating conditions of the digester. Taking the following empirical formula $C_5H_9O_3N$ (Speece and McCarty 1964) as representative of the composition of proteinic matter, the ratio COD/N is 11.4. Similarly, it can be deduced that for cellular matter the ratio P/N is between 1/5 and 1/7. Moreover, taking into account yields for cellular matter production (Mosey 1983) and the digester load, values between 400/7 (high load) and 1000/7 (low load) have been reported for the COD/N ratio (Henze and Harremoes 1983). Consequently, an average ratio COD/N/P of around 600/7/1 can be recommended for a substrate to be anaerobically digested.

14.3.2 Toxic substances

Methanogenic bacteria are especially sensitive to the presence of some compounds that at a given concentration can inhibit their activity (Speece 1987). These substances are called inhibitors or, more generally, toxicants. The problem of toxicity is complex as sometimes it is difficult to predict. To begin with, there are no general agreement on threshold limits and, as mentioned before, at low concentration some toxicants (e.g. metals) can become stimulants for AD. The inhibitory capability is not only dependent on the concentration of the substance but also on the environmental conditions, i.e. pH, temperature and the concentration of other substances, which can behave synergistic or antagonistic with a given inhibitor (Kugelman and Chin 1971).

14.3.2.1 Common inhibitors of anaerobic digesters

There are a number of common substances that can affect the AD process and that are considered toxic or inhibitory at a given threshold level. VFA, pH, free ammonia and hydrogen sulphide are the most frequent. Others are salinity or some xenobiotic compounds.

VFA are intermediary compounds of organic matter anaerobic degradation. Hence, they are present in variable amounts, depending on the existing environmental conditions in the digestion system. The undissociated species have been reported as more toxic because they can more easily diffuse to the inner parts of the cell (Andrews 1969; Pohland and Martin 1969). Consequently, pH together with the alkalinity level will in a way determine the VFA toxicity and the threshold level will depend on these parameters. Among VFA, propionic and butyric acids have been described as the most inhibitory. In accordance with Boone and Xum (1987), propionic acid concentrations over 3000 mg/l are definitively toxic and cause digestion failure.

Another common compound, ammonium, considered to be a nutrient, can become inhibitory at some concentrations, depending on the pH (Koster 1986). As in the case of VFA, the toxic species is the undissociated one, i.e. free ammonia. Solubility and temperature also have a large effect on ammonia toxicity. For instance pig manure, which has high ammonia-N concentrations (around 4 g/l and even more) cannot be directly digested at thermophilic conditions because of ammonium toxicity (Angelidaki and Ahring 1994; Hansen *et al.* 1998).

Hydrogen sulphide presents a similar behaviour: its toxicity is also very dependent on the pH and alkalinity conditions of the system. The literature describes toxicity threshold values from 200 mg/l to 1500 mg/l (O'Flaherty *et al.* 1998). In addition to the environmental factors, these differences and – this is general for other inhibitors – can also be attributed to bacteria acclimation. In this case, the iron ion can also influence these values, as it can remove sulphur anion by precipitation of FeS (Suschka *et al.* 2001). This precipitation can be used as a method to overcome toxicity problems derived from the presence of H_2S. Generally, the presence of metal ions in anaerobic environments can act either as a stimulant or an inhibitor. As reference values, concentrations over 1 mg/l for heavy metals or 5–8 g/l for metals of Group II, can be toxic, in accordance with an EPA report (EPA, USA 1974). These values are to be considered with care as they are dependent on environmental factors.

Finally, it should be noted that values for alkalinity over 1.5 g/l and a pH in the range of 6–7.5 are advisable to avoid problems with acids (VFA and H_2S).

14.3.2.2 Xenobiotics

Compounds known as xenobiotics are substances not present in the natural environment but are produced by human activities. Some of them are classified as priority pollutants as they are known for their acute toxicity. As a consequence, they can be a problem for the AD of some residual streams. Typical examples are solvents, including alcohols, organic acids, esters, ketones, etc., frequently present in industrial wastewaters, which can be very toxic for methanogenic bacteria (Leisinger 1983; Madsen and Rasmussen 1996). Together with these solvents, other inorganic substances and also small quantities of industrial products such as pesticides, dyes, drugs and other synthetic compounds can become inhibitors or toxic to anaerobic environments (Johnson and Young 1983). In addition, other compounds used in the synthesis of industrial substances, especially those containing halogens, sulphur, hydroxyl, amines, nitrocompounds, ether, ketones and other functional groups should also be included in this classification. However, it must be taken into account that some xenobiotics can be biodegraded in certain conditions, including by the genetic manipulation of microrganisms.

14.3.3 Temperature effects

The influence of temperature on AD processes is very pronounced, both from the kinetic and the thermodynamic point of view (Kim *et al.* 2002). In particular, methanogenesis is strongly influenced by this parameter, degradation rates and yields increasing, as usual, with temperature. Figure 14.7 presents a scheme in which the rate of the AD process is compared with the temperature. As can be seen, over this range, two optimal ranges, with maximum activity, have been identified: mesophilic (around 35°C) and thermophilic (around 55°C). Most of the digestion systems are operated within these temperature ranges.

Although the kinetics are faster at thermophilic temperatures, some imbalances may occur due to the complexity of the several steps involved in the process. Thus, acidogenesis can produce more acids than methanogenesis can convert at a higher temperature. This problem can affect the digestion of wastes with a high biodegradability as has been observed by Bernal *et al.* (1992) with market wastes if the process is not carried out in separate reactors, i.e. one for hydrolysis and fermentation and another for methanogenesis.

To conclude, although biogas production yields and bioreaction kinetics can be more favourable at thermophilic temperature, optimal conditions depend on the type of substrate – concentration and biodegradability – and type of system used.

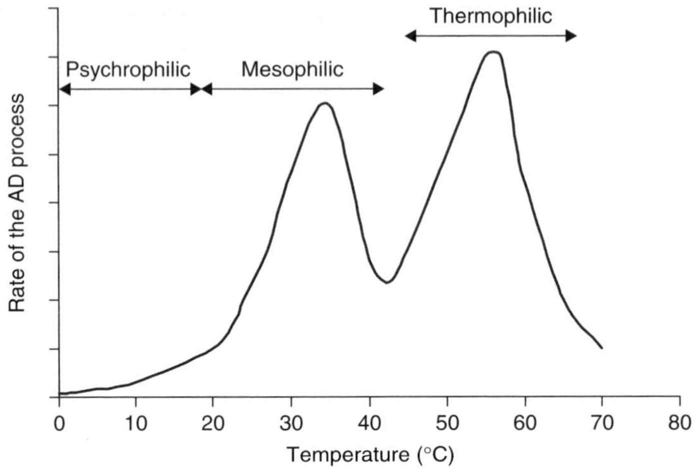

Figure 14.7 Temperature ranges for AD. Optima are for mesophilic and for thermophilic bacteria, around 30–35°C and 55–60°C, respectively.

14.4 IMPORTANT PARAMETERS IN ANAEROBIC DIGESTERS

A number of parameters are used to design or to monitor the behaviour of anaerobic digesters (Cecchi *et al.* 2002). Before introducing them, it is convenient to recall the basic variables concerning concentrations:

- *Substrate concentration:* S ($kg_{substrate}/m^3_{reactor}$) generally expressed in terms of:
  - *Total solids (TS)*: This is the residue upon water evaporation, and the content of dry matter in the substrate sample, after 48 h drying at 105°C (constant weight). This is a raw estimation of all the organic and inorganic matter content in the original sample.
  - *Total volatile solids (TVS)*: This is the fraction of solid matter which can be oxidized and driven off as gas at 550°C in 24 h (constant weight). As an approximation, it is the organic fraction of the dry matter determined at 105°C, the TS.
  - *COD*: This is a measure of all the organic matter in the substrate. More specifically, it is the oxygen equivalent of the organic matter that can be oxidized. It is measured by using a strong chemical oxidizing agent in an acidic medium.
- *Microorganisms or biomass concentration:* X. This parameter is generally expressed in terms of TVS or in VSS although it can be also referred to as COD.

14.4.1 Main design and operating parameters

Main design and operating parameters are:

- *Hydraulic retention time:* HRT (d). The HRT is defined as the ratio of the reactor volume to the influent substrate flow rate. Thus, it is a measure of the time that the substrate spends inside the digester:

$$\mathrm{HRT} = \frac{V}{Q} \tag{14.5}$$

where V is the reactor volume (m^3) and Q the flow rate (m^3/day). Hence, HRT is expressed in days.

Microorganisms have to digest the organic matter that enters into the digester. Thus, they need some time and an appropriate concentration. This means that the designer must allow enough HRT and a concentration not too high, to prevent the digester from being overloaded. The parameters that are used to measure the load of a digester are the following.

- *Solids retention time:* SRT (d). The SRT is a measure of the sludge age and usually expressed in days.

$$\mathrm{SRT} = \frac{V \cdot X}{Q \cdot X_e} = \mathrm{HRT}\frac{X}{X_e} \tag{14.6}$$

where X is the microorganisms concentration inside the digester and X_e in the effluent stream. In attached biomass systems (see next section), X is very high and so, consequently, is the SRT. In completely mixed systems, it coincides with the HRT, as X is equal to X_e.

- *Organic loading rate:* OLR ($kg_{substrate}/m^3_{reactor}$ day), is the amount of substrate introduced into the digester volume in a given time:

$$\mathrm{OLR} = \frac{Q \cdot S}{V} = \frac{S}{\mathrm{HRT}} \tag{14.7}$$

Usually, the substrate concentration S is expressed in terms of TVS or COD. As can be seen, OLR can be increased by decreasing the HRT or by increasing the substrate concentration S. However, the latter approach affects the digester performance in an additional way: viscosity of the digester content increases and this can negatively affect mixing efficiency for a stirred digester. In a single-stage anaerobic stirred digester, the upper limit in concentration is around 12–15% TS.

- *Specific organic loading rate:* SOLR ($kg_{substrate}/kg_{SVS}$ day). This parameter is not so common in anaerobic digester design, although it is very common in activated sludge systems (known there as F/M ratio). It expresses the OLR per unit of active biomass, and thus is more comprehensive than the simple OLR. It can also explain why attached biomass systems support much higher OLR than do suspended growth systems:

$$\mathrm{SOLR} = \frac{S}{\mathrm{HRT} \cdot X} \tag{14.8}$$

Again, $kg_{substrate}$ can be put in terms of TVS, COD, etc. A special difficulty comes from the fact that X is quite difficult to estimate because of the impossibility of distinguishing the active biomass and the substrate in the reactor by means of the volatile solids (VS) measurement.

All these described parameters are very important for the design of stirred digesters, i.e. to find the volume which is the main design parameter. However, it must be taken into account that the effective reactor volume is affected by scum and grit accumulation due to the presence of inorganic matter. Poor mixing performance also negatively affects the required volume. The effective digester volume required depends on the sizing criterion chosen. Two criteria are usually considered. One fixes the HRT, and the other fixes the OLR, although, as can be seen from equation (14.7), that both criteria are related.

14.4.2 Yield parameters

On the other hand there are parameters that measure the yield of the anaerobic process. Basically there are three:

- *Specific gas production:* SGP ($m^3_{biogas}/kg_{substrate\ fed}$) is the amount of biogas produced per unit of substrate fed or in some cases per unit of substrate converted into biogas:

$$SGP = \frac{Q_{biogas}}{Q \cdot S} \tag{14.9}$$

  where Q_{biogas} is the biogas flow rate (m^3/day). For instance, it can be set in terms of the TVS in the feed, as $m^3_{biogas}/kg_{substrate\ fed}$. SGP is often used to compare the performance of different anaerobic processes.
- *Gas production rate:* GPR ($m^3_{biogas}/m^3_{reactor}$day). This is the biogas produced per unit of reactor volume and time:

$$GPR = \frac{Q_{biogas}}{V} \tag{14.10}$$

  As can be seen from equations (14.7), (14.9) and (14.10), both the GPR and SGR are related through the OLR:

$$GPR = SGP \cdot OLR \tag{14.11}$$

- *Substrate removal yield:* η(%). This is a dimensionless parameter defined as the percentage of substrate removed in the digester:

$$\eta = \frac{S - S_e}{S} \cdot 100 \tag{14.12}$$

  where S_e is the concentration of TVS in the effluent (kg/m^3). Units for substrate concentration are usually TVS or COD.

14.5 BIODEGRADABILITY AND DEGREE OF BIODEGRADATION

At a given temperature, a given amount of VS of a particular waste can be converted to a maximum amount of biogas, provided optimal conditions are prevalent. This conversion can be measured through what has been called "ultimate biogas yield" (B_0). Obviously, the higher the ultimate methane yield, the higher the biodegradability. However, the actual amount of biogas (SGP) that is obtained in a digester at this given temperature is not this yield, but a function of it and the operating conditions of the digester. The ratio:

$$f_B = \frac{B_0}{\mathrm{SGP}} \tag{14.13}$$

is a measure for the degree of biodegradation, which depends on the one hand on the biodegradability at a specified temperature that is, its ultimate biogas yield. This factor is intrinsic to the waste itself. On the other hand it depends on the digester operating conditions, kinetics, the flow pattern within the digester, digestion phases, etc. See Mata-Alvarez (2002) and Sanders *et al.* (2001) for a more thoroughgoing discussion on the biodegradability concept.

Assuming a first-order model for the biodegradation rate, with a kinetic constant k (d^{-1}), the following expression can be obtained for the specific biogas production in a stirred digester (Mata-Alvarez 2002):

$$\mathrm{SGP} = \frac{B_0}{1 + 1/(k \cdot \mathrm{HRT})} \tag{14.14}$$

which links SGP with both the biodegradability and kinetic factors. Although, this expression is of limited use, because it is based on a first-order model, the qualitative trends that can be deduced are of interest (see Figure 14.8). Thus, it is clear that, the higher the HRT, the more biogas is produced and the closer the SGP

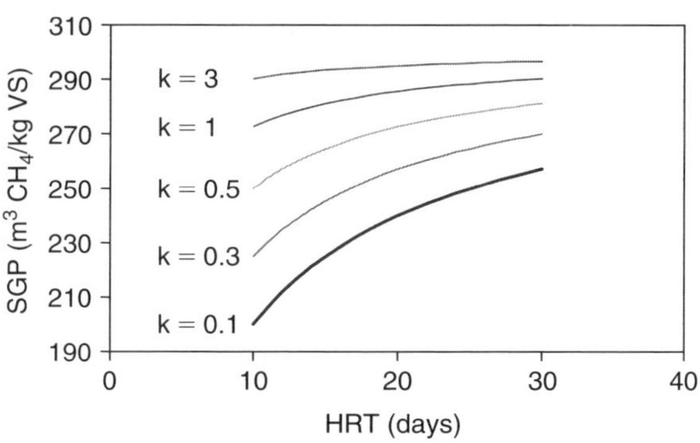

Figure 14.8 A simulated example (using equation (14.14)) for a substrate with $B_0 = 300\,\mathrm{m^3\ CH_4/kg\ VS}$. Kinetic constant k values are expressed in d^{-1}.

will approach B_0. This approach is more noticeable at high values of the kinetic constant. Thus, for low kinetic rates, it is necessary to operate the digester at longer HRT if a higher SGP is desired. When the kinetic constant is high, even at low HRT (obviously, over the washout value), the yields are high and the profile SGP vs. HRT is much flatter.

14.6 TYPES OF DIGESTERS

In this section, different types of digesters are described. First, Section 14.6.1 describes the continuous stirred tank reactor (CSTR), which is the most used and is quite appropriate to digest slurries. Section 14.6.2 deals with reactors with high biomass concentration, which can be used only with liquids without suspended solids, although they are also used in solid-waste treatment in two-phase reactors as described at the end of this section. Finally, Section 14.6.3 describes some commercial types of digesters for the treatment of the OFMSW.

14.6.1 Flow-through digesters

Flow-through stirred tank digesters consist of a void recipient generally heated (different devices are used to this end) and stirred by different systems (mechanical, biogas recirculation and liquid recirculation). In these reactors (known as continuous stirred digester, CSTR), the liquid and the microorganisms have the same residence time (see equation (14.6)) for the liquid and for the microorganisms (HRT = SRT). Active biomass is inside the digester for the same amount of time as the liquid, as they are completely mixed with it, forming a suspended growth system (Figure 14.9). In other words, the growth rate is the same as the dilution rate and consequently the HRT must be high enough to prevent the microorganisms to washout. In practice, this means that stirred tank digesters should be designed with residence times normally above 10 days, which in turn means large volumes (see equation (14.2)).

Flow-through stirred tank digesters have been in use for decades for the anaerobic stabilization of sewage sludge produced in wastewater treatment plants (Chynoweth and Pullammanappallil 1996). In addition, piggery waste is another typical example of the type of waste suited for this digester (Zeeman *et al.* 1988).

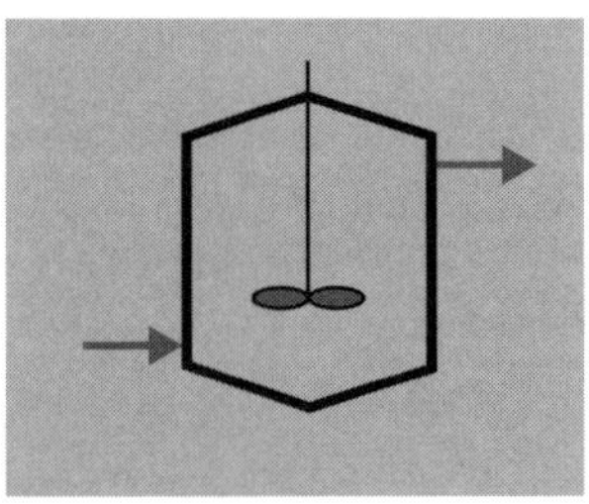

Figure 14.9 Sketch of the continuous stirred digester. HRT is equal to SRT and is in the range 10–25 days. The digester can work with slurries.

Biogas produced is used as a fuel source to provide hot water for the piggery and to keep the digester temperature around 35°C. Electricity is also produced if the piggery is large enough.

A distinctive problem of these reactors is that slurried wastes such as the OFMSW do not keep a homogenous consistency because heavier fractions and contaminants sink and a floating scum layer forms during the digestion process, resulting in the formation of three layers of distinct densities, or phases, in the reactor. The floating layer accumulates at the top of the reactor and if it is significant may hamper effective mixing. The heavier layer accumulates at the bottom of the reactor, where it may damage the propellers. It is therefore necessary to prevent the substrate to contain such particles before feeding it into the digester.

Another technical drawback of the complete mixed reactor is the possible occurrence of short-circuiting, i.e. the passage of a fraction of the feed through the reactor with a shorter retention time than the average retention time of the bulk stream. Not only does short-circuiting diminish the biogas yield, it can impair the proper hygienization of the wastes, i.e. the kill-off of microbial pathogens which requires a minimum retention time. It is therefore very important to provide adequate conditions at the inlet and outlet ports, together with some internal devices that minimize the potential danger associated with pathogen washout due to short-circuiting.

14.6.2 Digesters with large active biomass concentration

SRT is a significant factor in biological reactors, particularly for anaerobic systems and in view of the relatively long mean generation times of the methanogens and acetogens. Consequently, attempts were made to design digesters in which active biomass could be retained and washout of the slower growing microorganisms is avoided. Two types of digesters were designed that meet this goal:

(a) Digesters based on keeping suspended biomass inside the reactor.
(b) Attached growth systems.

14.6.2.1 Digesters based on keeping suspended biomass inside the reactor

Among the first type are the contact digester and the upflow anaerobic sludge blanket (UASB) reactor. The first one is quite similar to the activated sludge system used in aerobic wastewater treatment: it comprises two units. The first is a CSTR digester and the second a settler (Figure 14.10). Between both units a degasser is usually included to improve the settleability of the anaerobic sludge, as this is the main drawback of the system. Thickened sludge is then recirculated to the first unit to maintain a high biomass concentration, while a clarified effluent is discharged. A small amount of waste sludge is withdrawn from the system (to balance the slow anaerobic growth rate). Depending on the strength of the waste, high biomass concentrations can be obtained (up to more than 25 g/l).

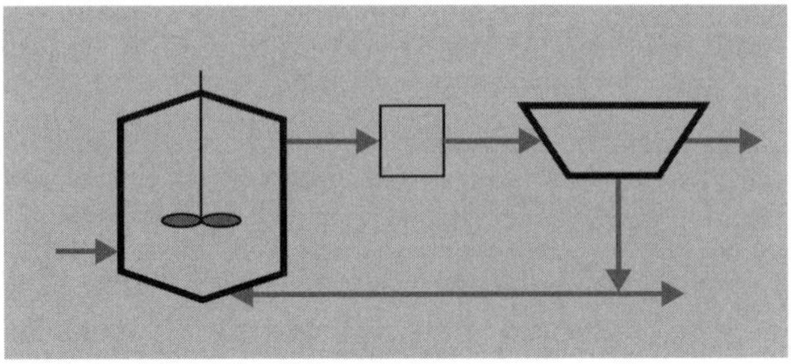

Figure 14.10 Sketch of the anaerobic contact digester.

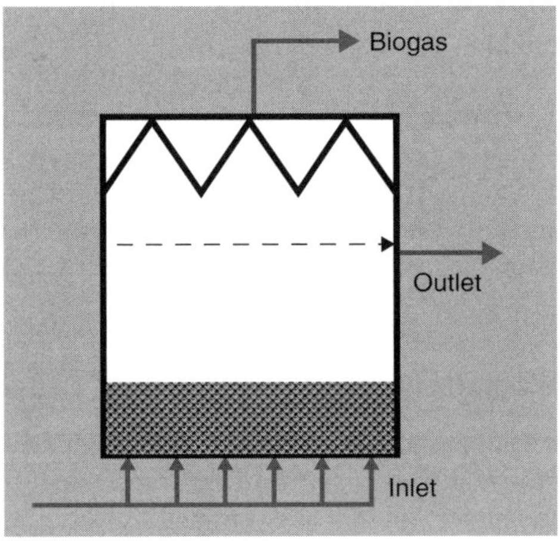

Figure 14.11 Sketch of the UASB digester.

Anaerobic contact systems have been installed in pulp and paper wastewater plants, and are the oldest system of this type (since 1955). Today, they are not generally used because there are many problems with the settler, as biomass still produces gas and tends to float.

An improvement to keep biomass inside the digester was achieved in 1976 in a sugar factory located in Breda (the Netherlands) in a pilot plant of 6 m^3 upflow anaerobic granular sludge reactor (Lettinga and Vinken 1980). It is based on the development of a highly settable flocs, called granules due to their form. These granules have a considerable diameter (up to 5 mm), and due to their density are kept in the bottom of the reactor. Settling velocities of granular sludge of approximately 60 m^3/h are common, whereas the superficial upflow velocities in UASB reactors (Figure 14.11) usually remain below 2 m^3/h. This allows the reactor operation with SRT of over 200 days at HRT of only 6 h. In other words, a consequence of the high biomass concentration that can be achieved inside a UASB reactor (up to 30 g/l) is the high loading rates that can be applied. Other advantages of this system are that there is a low loss of solids, and that no mechanical mixing is necessary (biogas generation is enough to ensure adequate mixing). In addition, if granular seed comes from another reactor, start-up can be completed in a few days.

Formation of granules is due to microbiological and biochemical factors, substrate characteristics and reactor operating conditions (so that filamentous bacteria are washed out). Feedstocks with a large proportion of VFA do not prevent the granulation process. Extensive literature is available on the appropriate conditions allowing granule formation (Daffonchio *et al.* 1995; Thaveesri *et al.* 1995; Yu *et al.* 2001; Gonzalez-Gil *et al.* 2001).

An important design consideration is that the influent should be distributed uniformly at the base of the reactor (in industrial reactors a typical value is one feed inlet per 10 m^2 cross section).

Given these advantages, this system is probably one of the most successful in industry, and hundreds of industrial plants are installed worldwide. Notwithstanding the high efficiency of UASB reactors, there are some aspects of its operation and stability that do not always meet the expectations. Examples are factors contributing to the granulation of the sludge, the effect of operating conditions and effluent characteristics on the granules or optimum biomass hold-up. Therefore, intense research has also been done to improve UASB reactors, with new developments such as the expanded granular sludge bed (EGSB) reactors (Wiegant and Maaskant 2000) and internal recirculation (IC) reactors (Hack *et al.* 1987; Pereboom and Vereijken 1994).

14.6.2.2 Anaerobic attached growth systems

Attached growth-based anaerobic digesters constitute another way of keeping active biomass inside the reactor. Basically there are three types: the downflow reactor, the anaerobic filter and the expanded or fluidized bed reactor. In all of these, a biofilm is developed with biomass fixed on a support. This is of plastic material, although other supports made of clay or even stones have been described (Hobson *et al.* 1981). In stationary downflow reactors, the support material is formed by tubes or sheeting arranged to achieve a maximum specific surface, so as to favour the contact between microorganisms and substrate. Normally they leave enough volume to avoid any clogging problem. These fixed reactors were developed in the NRC in Canada (Kennedy and van den Berg 1982), although problems related to cost and clogging have prevented wide use. More successful has been the "anaerobic filter", which is similar to the trickling filters used in aerobic wastewater treatment. However, unlike the trickling filter the system is contained in an airtight vessel and the liquid movement is from the bottom upwards (Figure 14.12).

In anaerobic filters most of the biomass is fixed to the walls of the support, but a significant fraction grows suspended in its interstitial spaces. Active biomass concentrations can be also very high and loads up to 20 kg/m^3d have been reported (Hobson *et al.* 1981). Especially significant for its size was the installation of a 13,000 m^3 anaerobic filter in the distillery that Bacardi has in Puerto Rico, to treat its wastes (Mata-Alvarez and Llabrés 1984). Costs of packing material,

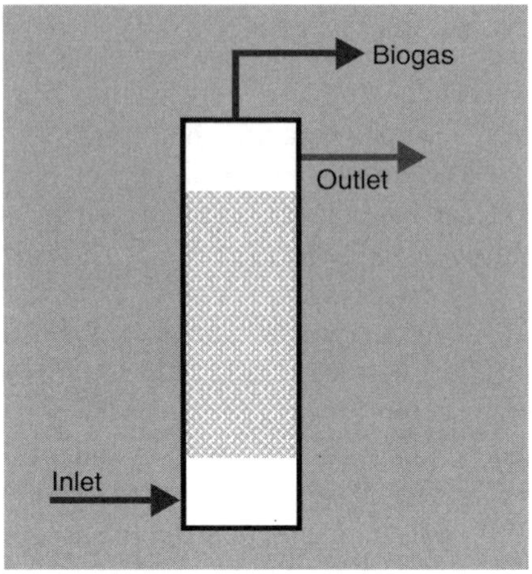

Figure 14.12 Sketch of the anaerobic filter.

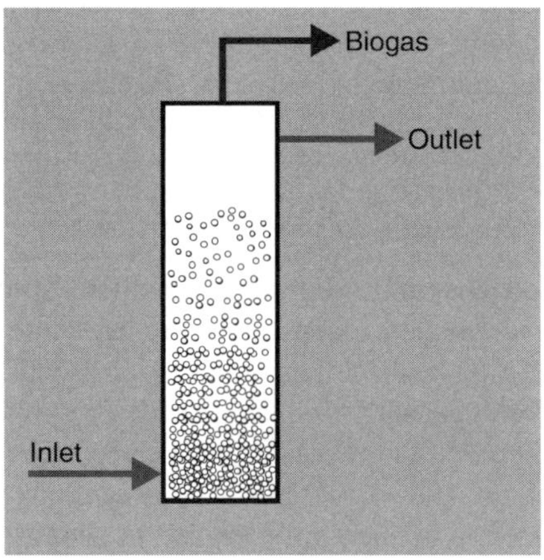

Figure 14.13 Sketch of the expanded bed reactor.

restrictions imposed by suspended solids (as they clog the filter) and the need for periodic backwashes have considerably limited the use of this technology.

Probably, the most successful attached growth system has been the anaerobic expanded bed (AEB) or the anaerobic fluidized bed (AFB). These simply apply the principles of expanded and fluidized beds to anaerobic systems (Figure 14.13). No clear border between the two systems exists: it is commonly accepted that an expansion of the bed beyond 30–50% can be considered fluidized. Shear forces create a thin, dense biofilm, preventing diffusion limitations to take place.

In these digesters, biomass is attached to the surface of the particles of the bed. Size and type of medium used as support are important factors, as they are directly related to the fluidization energy, the most significant process cost. Sand and, especially, granular activated carbon (GAC) are the most popular. Usually, particle sizes are between 0.2 and 1 mm.

Advantages of this system are linked to the free movement of the particles, which prevents the digester from clogging. The high biomass concentrations that can be achieved and which ensures high SRT also aids the stability of the reactor in the event of pulse presence of toxic compounds or sudden overloadings and allows short HRT. This results in cost-effective installations. Drawbacks are related to the complexity of the system design and the cost of the energy required for the expansion/fluidization process.

14.6.2.3 Two-phase digesters

Two-phase digesters are a combination of the digesters described above in such a configuration, which ideally optimizes the biological reactions taking place inside. Optimizing these reactions separately in different phases carried out in different reactors may lead to a larger overall reaction rate and biogas yield (Ghosh *et al.* 2000). Typically, two stages are used: in the first phase, the liquefaction–acidification reactions take place; in the second, acetogenesis and methanogenesis are the prevalent reactions. With these two steps occurring in distinct reactors, it becomes possible to increase the rate of methanogenesis which can be made the rate-limiting step by using a fixed biomass digester (Weiland 1992; Kübler and Wild 1992). Additionally, it is possible to increase the rate of hydrolysis in the first stage by using microaerophilic conditions, lower pH or other means (Capela *et al.* 1999; Wellinger *et al.* 1999).

The increased technical complexity of two-stage relative to single-stage systems has not, however, always been translated in the expected higher rates and yields (Weiland 1992). In fact, the main advantage of two-stage systems is not their putative higher reaction rate, but rather a greater biological reliability for wastes which cause unstable performance in one-stage systems (Vandevivere *et al.* 2002).

Two-phase systems provide some protection against the fluctuations of OLR or against the presence of inhibitory substances. Unstable performance can be caused either by fluctuations of OLR (due to waste heterogeneity or discontinuous feeding) or by wastes excessively charged with inhibiting substances such as nitrogen (especially with those two-stage systems with biomass attached growth digesters in the second stage). In order to increase rates and resistance to shock loads or inhibiting substances, it is desirable to achieve high-cell densities of the slowly growing methanogenic consortium in the second stage.

At industrial scale, there are not many two-phase systems. The main reason for this is the increased cost derived from the increased complexity of these systems. One industrial system using this approach is the BTA (Biotechnische Abfallverwertung Gmbh & Co.) process designed to treat the OFMSW. In the BTA process, illustrated in Figure 14.14, the 10% TS pulp exiting the pasteurization step is dewatered and the liquor directly sent to the methanogenic reactor (Kübler and Wild 1992). The solid cake is resuspended in process water and

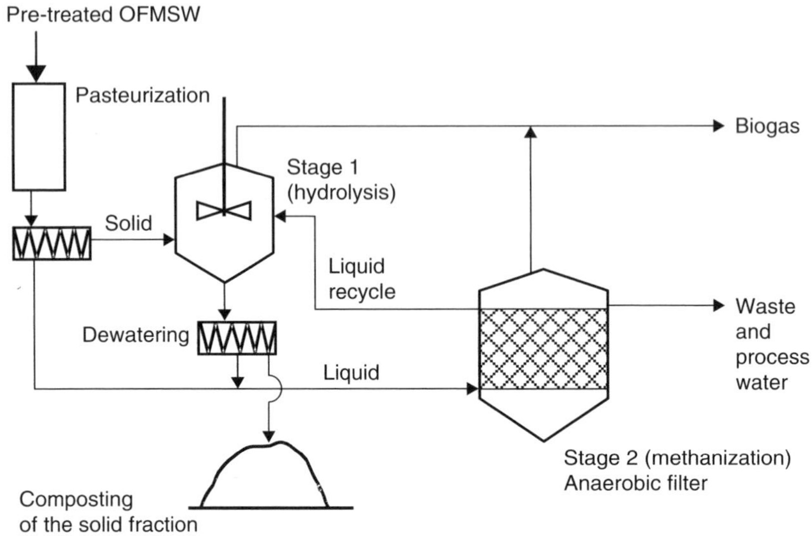

Figure 14.14 Two-phase system used in the BTA process to treat OFMSW.

hydrolysed in a complete mix reactor under mesophilic conditions (HRT 2–3 d). The pH within the hydrolysis reactor is maintained in the range 6–7 by recirculating processed water from the methanogenic reactor. The output stream of the hydrolysis reactor is once more dewatered and the liquor fed to the methanogenic reactor. The latter, receiving only liquid effluents, is designed as a fixed film loop reactor in order to increase biomass concentration and age.

14.6.3 Commercial digesters for solid-waste treatment

Different types of anaerobic reactors to treat solid wastes were specifically developed to treat the OFMSW, which with circa 3,000,000 tons/year of installed capacity is the most significant application of this technology for solid substrates. However, it must be taken into account that in this type of treatment not only the reactor is important but also the unit operations before and after the anaerobic bioreaction. These include magnetic separation, comminution in a rotating drum or shredder, screening, pulping, gravity separation or pasteurization, substrate conditioning for reaction, etc. As post-treatment steps, the typical sequence involves mechanical dewatering, aerobic maturation, and wastewater treatment (Figure 14.15). However, this section deals only with the digester, which is the heart of the plant.

A typical classification divides digesters in two main digester types: dry (when the substrate has a solid content comprised between 25% and 35% or even 40% TS) and wet (when TS is below 15%). Dry type digesters can approximately be considered as plug-flow, whereas wet digesters are mixed flow (they are variations of the CSTR described in Section 14.6.1).

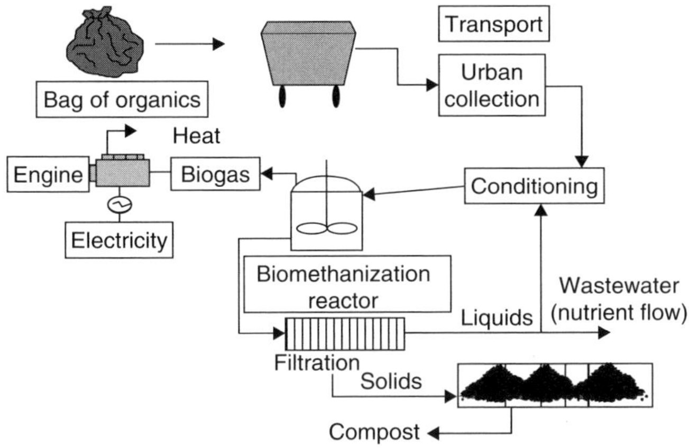

Figure 14.15 Basic operations of a plant to treat the OFMSW in a "wet" type digester.

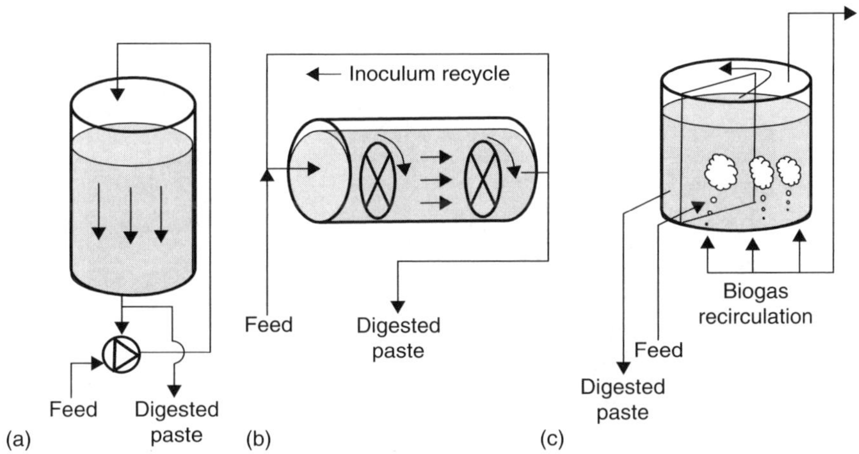

Figure 14.16 Three types of "dry" type digesters: (a) Dranco technology, (b) Kompogas and BRV technologies and (c) Valorga technology (Vandevivere *et al.* 2002).

Wet type digesters require a previous step of substrate conditioning where it is mainly diluted to the required solid content (as stated before, below 15% TS). One of the first "wet-type" full-scale plants for the treatment of OFMSW was built in Vaasa (Finland) in 1989, using the so-called WAASA (commercial name) technology.

Dry type digesters are plug-flow, and the substrate advances through the reactor while it converts to biogas. Different approaches have been developed, as shown in Figure 14.16 taken from Vandevivere *et al.* (2002) since 1984 (Mata-Alvarez 2002).

Table 14.1 shows some of the present market technologies, classified as dry and wet systems. Many others exist, but only the ones with a larger installed capacity (over 15,000 tons/year) have been included in Table 14.1. Figure 14.17 shows one

Table 14.1 Some of the present market AD technologies for organic solid waste.

	Technology	Owner
Dry Technologies	BRV	LINDE
	DRANCO	OWS
	KOMPOGAS	KOMPOGAS
	VALORGA	TECMED
Wet Technologies	BIOSTAB	ROS ROCA
	BTA	BTA
	KCA	LINDE
	WAASA	ALCYON/CYTEC

Figure 14.17 In the background, the stirred tank digester of 3500 m^3 (ROS ROCA technology) in Boden, Germany.

of the reactors of the Boden plant in Germany built with the ROS ROCA technology. Stirring is achieved by biogas injection in this case. Figure 14.18 shows the Valorga reactor in La Coruña (Spain), one of the last most recent plants built with this technology.

14.7 FINAL REMARKS

Anaerobic digestion is a mainstream technology for treating sewage sludge, and in Europe its application to the treatment of OFMSW has lately received great attention, due to the fossil-fuel derived energy savings. Industrial high strength wastewater applications of AD as a pre-treatment prior to sewer discharge have also increased in the last years to meet the need to reduce energy costs.

Figure 14.18 The four 4300 m^3 vertical reactors of Valorga technology built in La Coruña (Spain) treating MSW.

In this chapter a review of some of the concepts of AD has been presented. Anaerobic digestion takes place in several steps and has definite optimal temperatures to maximize rate and yields. Some of the parameters used to design and monitor the digester performance have been described. Main digester configurations have also been reviewed.

Expectations for the near future of AD technology are optimistic as it can be considered a sustainable technology, especially because of the very important factor of the greenhouse gas emission reduction agreed at the Kyoto Summit (Tilche and Malaspina 1998).

Another factor that has to contribute to the consolidation of AD as a mainstream technology for OFMSW is the fact that the digested residue can be considered quite stable organic matter with a very slow turnover of several decades (over 50 years), given adequate soil conditions. In this way, the natural imbalance in CO_2 can be adjusted by restoring or creating organically rich soil. The removal of CO_2 constitutes an extra benefit that could help place AD among the most relevant technologies in this field (Verstraete *et al.* 2000).

14.8 CONCLUSIONS

After the development observed during the last 15 years of the AD process applied to solid wastes, it can be concluded that today it can be considered as an optimal technology from the environmental point of view. The future of anaerobic technology should be sought in the integration of this process into an overall sustainable waste treatment perspective. Life cycle assessment applied to this technology offers a number of interesting features as can be seen in the paper of Edelmann (2002).

REFERENCES

Andrews, J.F. (1969) Dynamic model of the anaerobic digestion process. *J. Sanit. Eng. Div.*, ASCE, **95**, SA1, 95–116.

Angelidaki, I. and Ahring, B.K. (1994) Anaerobic thermophilic digestion of manure at different ammonia loads: effects of temperature. *Water Res.* **28**, 727–731.

Bernal, O., Llabrés, P., Cecchi F. and Mata-Alvarez, J. (1992) A comparative study of the thermophilic biomethanization of putrescible organic wastes. *Odpadní vody/ Wastewaters* **1**(1), 197–206.

Boone, D.R. and Xum, L. (1987) Effects of pH, temperature and nutrients on propionate degradation by a methanogenic enrichment culture. *Appl. Environ. Microbiol.* July 1987, 1589–1592.

Capela, I.F., Azeiteiro, C., Arroja, L. and Duarte, A.C. (1999) Effects of pre-treatment (composting) on the anaerobic digestion of primary sludges from a bleached kraft pulp mill. In: *Proceedings of the II International Symposium on Anaerobic Digestion of Solid Wastes*, Barcelona, June 15–17, 1999 (eds. J. Mata-Alvarez, A. Tilche and F. Cecchi), Vol. 1, pp. 113–120. International Association on Water Quality.

Cecchi, F., Traverso, P., Pavan, P., Bolzonella, D. and Innocenti, L. (2002) Characteristics of the OFMSW and behaviour of the anaerobic digester process. In: *Biomethanization of the Organic Fraction of Municipal Solid Wastes* (ed. J. Mata-Alvarez), pp. 1–20. IWA Publishing Company, London.

Chynoweth, D.P. and Pullammanappallil (1996) Anaerobic digestion of solid wastes. In: *Microbiology of Solid Waste* (eds. A.C. Palmisano and M.A. Barlaz). CRC Press, Inc., Boca Raton.

Daffonchio, D., Thaveesri, J. and Verstraete, W. (1995) Contact angle measurement and cell hydrophobicity of granular sludge from upflow anaerobic sludge bed reactors. *Appl. Environ. Microbiol.* **61**, 3676–3680.

Edelmann, W. (2002) Products, impacts and economy of anaerobic digestion of OFMSW. In: *Biomethanization of the Organic Fraction of Municipal Solid Wastes* (ed. J. Mata-Alvarez), pp. 265–302. IWA Publishing Company, London.

EPA (1974) Process design manual for sludge treatment and disposal. EPA 635/1-74 006. Technology Transfer, Washington DC.

García-Heras, J.L., Salaberría, A., Prevot, C. and Sancho, L. (1999) Increase of organic loading rate and specific methane production by splitting phases in anaerobic digestion of sewage sludge. In: *Proceedings of the II Internat. Symposium on Anaerobic Digestion of Solid Wastes*, Barcelona 15–18 June 1999 (eds. J. Mata-Alvarez, A. Tilche, and F. Cecchi), Vol. 2, pp. 77–80. Printed by Gràfiques 92.

Ghosh, S., Henry, M.P., Sajjad, A., Mensinger, M.C. and Arora, J.L. (2000) Pilot-scale gasification of MSW by high-rate and two-phase anaerobic digestion. *Water Sci. Technol.* **41**(3), 101–110.

Golueke, C.G. and McGauhen, P.H. (1970) *Comprehensive Studies of Solid Waste Management*. First and Second Annual Reports. Bureau of Solid Waste Management, Washington DC.

Golueke, C.G. (1971). *Comprehensive Studies of Solid Waste Management*. Third Annual Report. US EPA, Washington DC.

Gonzalez-Gil, G., Seghezzo, L., Lettinga, G. and Kleerebezem, R. (2001) Kinetics and mass-transfer phenomena in anaerobic granular sludge. *Biotechnol. Bioeng.* **73**(2), 125–134.

Gujer, W. and Zehnder, A.J.B. (1983) Conversion processes in anaerobic digestion. *Water Sci. Technol.* **15**(8–9), 127–167.

Hack, P., Vellinga, S. and Hubets, L. (1987) Growth of granular sludge in the Biopaq IC-reactor. *Gasmat Workshop* 26–27 October 1987, Lunteren, The Netherlands.

Hansen, K.H., Angelidaki, I. and Ahring, B.K. (1998) Anaerobic digestion of swine manure: inhibition effect by ammonia. *Water Res.* **32**, 5–12.

Henze, M. and Harremoes, P. (1983) Anaerobic treatment of wastewater in fixed film reactors – a literature review. *Water Sci. Technol.* **15** (8–9), 1–101.

Hobson, P.N., Bousfield, S. and Summers, R. (1981) Types of digesters being constructed and the operation of digesters. In: *Methane Production from Agricultural and Domestic Wastes* (ed. A. Porteous), 1st edn, pp. 91–163. Applied Science Publishers Ltd., Ripple Road, Barking, Essex, England.

Johnson, L.D. and Young, J.C. (1983) Inhibition of anaerobic digestion by organic priority pollutants. I. *Water Pollut. Contr. Fed.* **55**, 1441–1450.

Kelly, C.R. and Switzenbaum, M.S. (1984) Anaerobic treatment: temperature and nutrient effects. *Agric. Wastes* **10**, 135–154.

Kennedy, K.J. and van der Berg, L. (1982) Anaerobic digestion of piggery waste using a stationary fixed film reactor. *Agric. Wastes* **4**, 151–158.

Kim, M., Ahn Y.-H. and Speece, R.E. (2002) Comparative process stability and efficiency of anaerobic digestion; mesophilic vs. thermophilic. *Water Res.* **36**(17), 4369–4385.

Kispert, R.G., Sadek, S.E. and Wise, D.L. (1975) An economic analysis of fuel gas production from solid waste. *Res. Recov. Conserv.* **1**, 95.

Koster, I. (1986) Characteristics of the pH influenced adaptation of methanogenic sludge to ammonium toxicity. *J. Chem. Technol. Biotechnol.* **36**, 445–455.

Kübler, H. and Wild, M. (1992) The BTA-process high rate biomethanisation of biogenous solid wastes. In: *Proceedings on the International Symposium on Anaerobic Digestion of Solid Waste*, Venice, 14–17 April 1992 (eds. F. Cecchi, J. Mata-Alvarez and F.G. Pohland), pp. 535–538. International Association on Water Pollution Research and Control.

Kugelman, I.J. and Chin, K.K. (1971) Toxicity synergism, and antagonism in anaerobic waste treatment processes. Anaerobic biological treatment processes. *Adv. Chem. Ser.* 105. American Chemical Society, Washington DC.

Leisinger, T. (1983) Microorganisms and xenobiotics compounds. *Experientia* **39**, 1183–1191.

Lettinga, G. and Vinken, J.N. (1980) Feasibility of the up-flow anaerobic sludge blanket (UASB) process for the treatment of low strength wastes. *Proceedings of the Purdue Industrial Waste Conference* **35**, 625–634.

Madsen, T. and Rasmussen, H.B. (1996) A method for screening the potential toxicity of organic chemicals to methanogenic gas production. *Water Sci. Technol.* **33**(6), 213–220.

Mata-Alvarez, J. and Llabrés, P. (1984) Digestores anaerobios: Evolución y tendencias. *Afinidad* **391**(5–6), 318–322.

Mata-Alvarez, J., Llabrés, P. and Macé, S. (2000) Anaerobic digestion of solid wastes. An overview of research achievements and perspectives. *Biores. Technol.* **74**(1), 3–16.

Mata-Alvarez, J. (2002) Fundamentals of the anaerobic digestion process. In: *Biomethanization of the Organic Fraction of Municipal Solid Wastes* (ed. J. Mata-Alvarez), pp. 1–20. IWA Publishing Company, London.

McCarty, P.L. (1964) Anaerobic waste treatment fundamentals. *Publ. Works* **107**, September.

Mosey, F.E. (1983) Mathematical modelling of the anaerobic digestion process: regulatory mechanisms for the formation of short-chain volatile acids from glucose. *Water Sci. Technol.* **15**(8–9), 209–232.

O'Flaherty, V., Mahony, T., O'Kennedy, R. and Colleran, E. (1998) Effect of pH on growth kinetics and sulphide toxicity thresholds of a range of methanogenic, syntrophic and sulphate-reducing bacteria. *Process Biochem.* **33**(5), 555–569.

Pereboom, J. and Vereijken, T. (1994) Methanogenic granule development in full scale internal circulation reactors. *Water Sci. Technol.* **30**(8), 9–21.

Pohland F.G. (1992) *Design of Anaerobic Processes for the Treatment of Industrial and Municipal Wastes* (eds. J.F. Malina and F.G. Pohland), Vol. 7. Water Quality Management. Technomic Publishing Co., Lancaster, PA.
Pohland, F.G. and Martin, J.C. (1969) Dynamic model of the anaerobic digestion process – a discussion. *J. Sanit. Eng. Div.*, ASCE **95**, SA6, 1197–1202.
Sanders, W.T.M., Geerink, M., Zeeman, G. and Lettinga, G. (2001) Anaerobic hydrolysis kinetics of particulate substrates. *Water Sci. Technol.* **41**(3), 17–24.
Siegrist, H., Renggli, D. and Gujer, W. (1993) Mathematical modelling of anaerobic mesophilic sewage treatment. *Water Sci. Technol.* **27**(2), 25–36.
Speece R.E. and McCarty, P.L. (1964) Nutrient requirements and biological solids accumulation in anaerobic digestion. *Adv. Water Pollut. Res.* **2**, 305–322.
Speece, R.E., Parkin, G.F., Bhattacharya, S. and Takashima, M. (1986) Trace nutrient requirements of anaerobic digestion. In: *Proceedings of the European Water Pollution Control Association* pp. 177–188. Conference on Anaerobic Waste Water Treatment 1986. Industrial Presentations B.V., Schiedam, The Netherlands.
Speece, R.E. (1987) Toxicity. In: *Anaerobic Digestion of Biomass* (eds. D.Y. Chynowth and R. Isaacson), Elsevier applied science Ltd.
Suschka, J., Machnicka, A. and Poplawski, S. (2001) Phosphates recovery from iron phosphates sludge. *Environ. Technol.* **22**, 1295–1301.
Thaveesri, J., Daffonchio, D., Liessens, B., Vandermeeren, P. and Verstraete, W. (1995) Granulation and sludge bed stability in upflow anaerobic sludge bed reactors in relation to surface thermodynamics. *Appl. Environ. Microbiol.* **61**, 3681–3686.
Tilche, A. and Malaspina, F. (1998) Biogas production in Europe. In: Paper presented at the *10th European Conference Biomass for Energy and Industry*, Würzburg, Germany, 8–11 June, 1998.
Van Lier, J.B., Tilche, A., Ahring, B.K., Macarie, H., Moletta, R., Dohanyos, M., Pol, Hulshoff, L.W., Lens, P. and Verstraete, W. (2001) New perspectives in anaerobic digestion. *Water Sci. Technol.* **43**(1), 1–18.
Vandevivere, P., De Baere, L. and Verstraete, W. (2002) Types of anaerobic digesters for solid wastes. In: *Biomethanization of the Organic Fraction of Municipal Solid Wastes* (ed. J. Mata-Alvarez), pp. 111–140. IWA Publishing Company, London.
Verstraete, W., Van Lier, J., Pohland, F., Tilche, A., Mata-Alvarez, J., Ahring, B., Hawkes, D., Cecchi, F., Moletta, R. and Noike, T. (2000) Developments at the Second International Symposium on Anaerobic Digestion of Solid Waste (Barcelona, 15–19 June 1999). *Biores. Technol.* **73**, 287–289.
Weiland, P. (1992) One- and two-step anaerobic digestion of solid agroindustrial residues. In: *Proceedings of the International Symposium on Anaerobic Digestion of Solid Waste*, Venice, 14–17 April 1992 (eds. F. Cecchi, J. Mata-Alvarez and F.G. Pohland), pp. 193–199. International Association on Water Pollution Research and Control.
Wellinger, A., Widmer, C. and Schalk, P. (1999) Percolation – a new process to treat MSW. In: *Proceedings of the II International Symposium on Anaerobic Digestion of Solid Waste*, Barcelona, 15–17, June 1999 (eds. J. Mata-Alvarez, A. Tilche and F. Cecchi), Vol. 1, pp. 315–322. International Association on Water Quality.
Wiegant, W. and Maaskant, W. (2000) Unified design concept for UASB, EGSB and other modified UASB reactors. Mededelingen – Faculteit Landbouwkundige en Toegepaste Biologische Wetenschappen (Universiteit Gent), **65**(3a), 23–29.
Yu, H., Tay, J. and Fang, H. (2001) The roles of calcium in sludge granulation during UASB reactor start-up. *Water Res.* **35**(4), 1052–1060.
Zeeman, G., Sutter, K., Koster, V.T., Koster, M. and Wellinger, A. (1988) Psychrophilic digestion of dairy cattle and pig manure: start up procedures of batch, fed-batch and CSTR-type digesters. *Biol. Wastes* **26**, 15–31.

PART FOUR: Resources from organic solid waste

15

Use of compost as suppressor of plant diseases

D. De Clercq, L. Vandesteene, J. Coosemans and J. Ryckeboer

15.1 INTRODUCTION

Soilborne plant pathogens can cause serious losses to both agricultural and horticultural crops.

Until the 1950s, the primary measures used to control soilborne diseases were the application of organic soil amendments derived from animal and green manures, crop rotation, and, for some crops, steam sterilization (Baker and Cook 1974). Synthetic pesticides have been used since to control soilborne diseases. Health and environmental public concern on the use of these chemicals has brought about an important change of mentality, which has led to environmental regulations. The use of many of the commonly used pesticides has been prohibited or severely restricted during the last two decades. In addition, for several minor crops such as woody ornamental and floricultural crops and several major

 Resource Recovery and Reuse in Organic Solid Waste Management. Edited by Piet Lens, Bert Hamelers, Harry Hoitink, Werner Bidlingmaier. ISBN: 1 84339 054 X. Published by IWA Publishing, London, UK

crops, resistant varieties are not available. All this has sparked substantial interest in alternative biological control of plant diseases, as part of an integrated crop management.

Nowadays, several alternatives to the use of pesticides are under development. One approach represents a return to the old practice, which consists in replacing conventional pesticide use and soil sterilization procedures with soil organic matter management. Examples include application of manures, green manures, composts, or biocontrol-agent-fortified composts, which can provide effective control of diseases and insect pests.

Since millennia, composting has been an empirically used method in agriculture. For example, amending compost to soil instead of animal manure circumvents odor problems. In addition, provided that the composting process was performed in an approved manner, compost is usually free of plant and human pathogens as well as weed seeds, which most probably cannot be presumed for animal manure (Ryckeboer 2001, 2002). Only recently, composting has regained substantial interest because of the re-valorization of what is now called "maintenance of biomass". In the European Union (EU), 50 million tons of compostable waste is generated each year. Presently 30% of this waste (15 million tons) is composted, resulting in 9 million tons of compost, and used mainly in agriculture, horticulture, or hobby gardens. The remaining 35 million tons of compostable waste is incinerated or dumped.

The composting process is one of the most interesting technologies for recycling waste of animal and plant origin. The end product of the composting process, the compost, has not only beneficial effects on soil erosion, improves water infiltration and retention, and supplies essential nutrients to support sustained plant growth, but also provides suppression to plant diseases. From the late 1970s to the 1980s (Chef *et al.* 1983; Nelson *et al.* 1983; Schüler *et al.* 1989), throughout the 1990s (Hoitink *et al.* 1991; Ryckeboer and Coosemans 1997a, b; Tuitert *et al.* 1998) until now (Krause 2001; Ryckeboer 2001; Blok *et al.* 2002; Bruns and Schüler 2002) evidence has accumulated that composts have suppressive properties to several plant pathogens. In some cases control of root rot with composts can be as effective as that obtained with fungicides (Hardy and Sivasithamparam 1991; Hoitink *et al.* 1991; Ownley and Benson 1991).

From another point of view, the annual use of peat in horticulture in the EU is about 30 million m^3. Peat is relatively cheap and highly suitable as a growing substrate because of its high stability and porosity, low salinity and "relatively inert" biological properties (Hoitink and Fahy 1986). Most peats are, however, low in available energy reserves and most reports state that potting mixes solely containing peat as organic substrate are conducive to development of soilborne diseases, particularly if they are heated or steamed first. Soilborne diseases caused by *Phytophthora* spp., *Pythium* spp., *Rhizoctonia solani*, *Fusarium* spp., as well as some nematode diseases are not suppressed in peat media of different decomposition levels (Hoitink and Fahy 1986). A few authors report, however, that especially less decomposed or "light" peat harvested from near the surface of the peat bog are suppressive to *Pythium* and *Rhizoctonia* damping-off and *Fusarium* wilt of tomato (Tahvonen 1982a, b; Wolffhechel 1988; Le Quéré *et al.* 1990). This peat could contain a larger diversity of biocontrol agents (Tahvonen 1982a, b; Boehm *et al.* 1997). The suppressive effect of some "light" peat sources is however of short duration (Boehm and Hoitink 1992).

In addition, peat is a non-renewable source, originating from Northern America, Scandinavia, the Baltic States, Poland, Ireland, UK, and Germany. In Germany and Ireland, and also in other countries, peat bogs increasingly are getting appreciated as natural resources that need to be protected. The use of certain composted organic wastes as full or partial substitutes for peat in potting mixes is likely to reduce grower dependence on imported peat.

Increasing the opportunities to use compost in horticulture and agriculture as a (potting) soil amendment and nutrient source for plants and the soil microbial community would contribute to recycling of waste and reduce the use of artificial fertilizers. Growers can also benefit from an increased use of compost by the fact that many types of compost have disease-suppressive properties and, therefore, use of disease-suppressive compost will reduce fungicide use and crop losses caused by plant pathogens.

15.2 SOILBORNE PLANT DISEASES SUPPRESSED BY COMPOST

Numerous reports describe suppressive effects of composts against plant diseases. Some selected examples presented below illustrate the diversity of soilborne plant diseases, which can be suppressed by a wide number of compost types.

Some of the pioneer reports on the suppression of soilborne pathogens after compost amendment involved the suppression of *Pythium* damping-off and *Phytophthora cinnamomi* and *Poinsetta* crown rot after amendment of composted hardwood bark (Hoitink *et al.* 1977; Moustafa and Hoitink 1977; Daft *et al.* 1979). In particular, the suppression of diseases caused by *Pythium* spp. has been well documented (Table 15.1). Root rot caused by *Rhizoctonia* diseases has been investigated to a similar extent (Table 15.1). Tuitert *et al.* (1998) reported that growth of *R. solani* was suppressed in potting mixes amended with non-cured composts or composts cured for 5–7 months, whereas the potting mix amended with compost cured for 1 month was conducive to *Rhizoctonia* damping-off of cucumber.

Composted hardwood bark, sufficiently mature, suppressed chrysanthemum, and flax *Fusarium* wilts (Chef *et al.* 1983). Other experiments demonstrated suppression of wilt diseases of tomato caused by *Fusarium oxysporum* f.sp. *lycopersici* and *Verticillium dahliae* after addition of biowaste compost (33% v/v) to the potting mix (Ryckeboer and Coosemans 1997b). Bollen (1993) demonstrated that amending a potting mix with 20% v/v mature source-separated household waste (referred to hereafter as biowaste compost) resulted in suppression of *Phytophthora* root rot of tomato (c.a. *Phytophthora nicotianae*) and *Fusarium* wilt of carnation (c.a. *F. oxysporum* f.sp. *dianthi*). Tränkner (1995) observed a reduction in primary scab

Table 15.1 Suppressiveness of composts to fungal plant diseases: a literature survey.

Disease causing agent	Plant	Compost type	References
Phytophthora cinnamomi	Lupine	Hardwood bark compost	Hoitink *et al.* (1977)
Pythium spp.	Tomato	Hardwood bark compost	Moustafa and Hoitink (1977)
Poinsetta crown rot	*Poinsetta*	Hardwood bark compost	Daft *et al.* (1979)
Rhizoctonia solani	Radish and celosia	Hardwood bark compost	Nelson and Hoitink (1983)
Fusarium wilt	Chrysanthemum and flax	Hardwood bark compost	Chef *et al.* (1983)
Aphanomyces root rot and *Pythium* damping-off	Pea	Municipal sludge compost	Lumsden *et al.* (1983)
Rhizoctonia	Bean, cotton, and radish		
Fusarium wilt	Cucumber		
Phytophthora crown rot	Pepper		
Pythium aphanidermatum	Cucumber	Liquorice root compost	Hadar and Mandelbaum (1986); Mandelbaum *et al.* (1988); Mandelbaum and Hadar (1990)
Pythium ultimum	Cucumber	Hardwood bark compost	Chen *et al.* (1987); Chen *et al.* (1988a, b)
Pythium ultimum and *Rhizoctonia solani*	Ornamental plants	Municipal sludge compost	Kuter *et al.* (1988)
Pythium ultimum and *Rhizoctonia solani*	Pea, bean, and beetroot	Household waste compost	Schüler *et al.* (1989)
Phytophthora spp.	Waratah and *Banksia occidentalis*	Eucalyptus bark compost	Hardy and Sivasithamparam (1991)
Rhizoctonia solani, *Pythium aphanidermatum* and *Sclerotium rolfsii*	Radish, cucumber, bean and chickpeas	Grape marc compost and separated cattle manure compost	Gorodecki and Hadar (1990); Hadar and Mandelbaum (1986)
Damping-off	Peas and cotton	Sewage sludge compost	Lewis *et al.* (1992)
Phytophthora nicotianae and *Fusarium oxysporum* f.sp. *dianthi*	Tomato	Mature biowaste compost (20% v/v)	Bollen (1993)
Venturia inaequalis	Apple	Biowaste compost	Tränkner (1995)
Pythium graminicola	Creeping bentgrass	Brewery sludge compost and biosolids compost	Craft and Nelson (1996)
Pythium spp., *Fusarium oxysporum*, and *Verticillium dahliae*	Tomato	Biowaste compost (25% v/v) biowaste compost (33% v/v)	Ryckeboer and Coosemans (1996b; 1997b)
Pythium ultimum and *Rhizoctonia solani*	Cucumber and radish	Biowaste compost and yard waste compost	Ryckeboer *et al.* (1998)
Rhizoctonia solani	Cucumber	Non-cured composts or composts cured for 5–7 months	Tuitert *et al.* (1998)

infections of apple (c.a. *Venturia inaequalis*) after amending soil with biowaste compost. Several articles report effects of composts on severity of nematode diseases (Hoitink and Fahy 1986; Ryckeboer and Coosemans 1996a, b).

All these examples summarized in Table 15.1 demonstrate the suppressive properties of different composts to several soilborne plant pathogens and a few even affect foliar diseases. As many as there are success stories, there are examples where compost addition did not lead to a reduced disease severity (e.g. Craft and Nelson 1996). One of the most critical limitations of the use of composts as a biocontrol product is the fact that the level of protection and its consistency are generally not sufficient to compete with conventional methods of disease control. Failure or success of a compost treatment depends on the plant species, the pathogen and on a complex interaction of physical, chemical, and biological factors, which in turn are determined by the raw starting material, compost maturity (decomposition level or the microbial-carrying capacity), and the process of composting used.

Several possible factors which can help to decrease this variability are composting with selected and homogenous starting material, fortifying composts with beneficial microorganisms, or combining compost amendments with pesticides in an Integrated Pest Management (IPM) program.

15.3 FACTORS IN DISEASE SUPPRESSION

15.3.1 Chemical and physical factors

Chemical and physical properties have an important impact on the suppressiveness of substrates to soilborne plant diseases. Particle size, air-filled pore space, salinity, nutrient availability, carbon–nitrogen ratio, moisture content, pH, pathogen inhibitors released by compost, are among the chemical and physical parameters that can affect the suppressiveness of composts (Ryckeboer 2001).

Small particles can harbor a lot of microorganisms because their activity is merely situated on the particle surface. However, if particle size is too small this can lead to compaction of the soil with oxygen deficiency as a result. Different plants demand varying optimal oxygen levels in the root environment. Therefore, air-filled pore size can not only directly affect the plants, but it can also influence root rot caused by *Phytophthora* (Hoitink and Fahy 1986). Low oxygen levels in the root environment can also be engendered by compost still undergoing high rates of decomposition (Hoitink and Fahy 1986).

Excessive salt concentrations can cause stress to some plants, weakening them. High salinity composts aggravate *Pythium* and *Phytophthora* root diseases unless they are applied months ahead of planting to permit sufficient leaching (Hoitink *et al.* 1997). The non-availability of a nutrient can make the plant weaker and more susceptible to pathogens. It has been recognized for decades that high ammonium and low nitrate nutrition aggravates *Fusarium* wilts (Schneider 1985). Perhaps for this reason predominantly ammonium-releasing composts prepared from municipal sewage sludges (low C/N ratio) aggravate *Fusarium* wilt (Hoitink *et al.* 1987;

Hoitink *et al.* 1996b). On the other hand, composts made from high C/N materials such as tree barks, immobilize N and suppress *Fusarium* wilt diseases if colonized by an appropriate microorganisms (Trillas-Gay *et al.* 1986).

Water retention capacity is affected upon compost amendment (Dick and McCoy 1993). It permits good moisture content which favors soil microorganisms and plants. Soil pH can also influence microbial growth and plant disease severity (Trillas-Gay *et al.* 1986).

The concentration of a diverse array of chemicals present in various compost substrates can influence suppressiveness to plant diseases (Hoitink *et al.* 1991). Some composts, such as 6-month old hardwood bark compost, release inhibitors of plant pathogens, whereas 2-year-old hardwood bark compost does not. These inhibitors primarily affect *Phytophthora* spp. and possibly also some nematodes (Malek and Gartner 1975), but they do not affect *R. solani* (Hoitink and Fahy 1986). According to Hoitink *et al.* (1993) utilization of composts prepared from heterogeneous wastes varying in stability (degree of decomposition), salinity, and nitrogen availability can lead to extreme opposite effects in disease suppression.

15.3.2 Biological factors

The rhizosphere microorganisms plays a crucial role in suppression of soilborne pathogens. This can be explained by several mechanisms, from which the competition for nutrients between the pathogens and their antagonists is the classic example. Antibiosis, parasitism, and induction of systemic resistance are the three other defined mechanisms that contribute to suppressiveness (Ryckeboer 2001).

Adding compost to (potting) soil can contribute to biological suppressive effects by the introduction of (i) a diversity or specific group of compost microorganisms, and of (ii) organic substrate that can alter activity of the edaphic microbial population. Three levels of specificity in biological control by compost have been identified (Hoitink and Boehm 1999), and will be discussed hereafter. Most potting mixes formulated with sufficiently stabilized composts, that also establish optimum physical properties relative to drainage and aeration as well as nutrient concentrations, naturally suppress *Pythium* and *Phytophthora* root rots. In contrast, only about 20% of the bark composts tested naturally suppress *Rhizoctonia* damping-off (Krause *et al.* 2000), and a very narrow range of composts is able to induce systemic resistance in plants (Krause 2001).

For a variety of composts, heating (60°C) destroys suppressiveness to *Pythium*, *Rhizoctonia*, and *Fusarium* diseases (reviewed by Hoitink and Fahy 1986). Samples removed from the high-temperature center of compost piles are conductive to the plant pathogens. Addition of small amounts of the edge compost, that harbors beneficial microorganisms, restores the suppressiveness. Incubation at 25°C of these suppressive center compost samples for a few days restored the suppressiveness to *Pythium ultimum* (Chen *et al.* 1988b). Chen *et al.* (1988a) also demonstrated a high correlation between high-microbial activity in the suppressive compost-amended medium, measured by fluorescein diacetate (FDA) hydrolysis, and induction of suppression to damping-off of *P. ultimum*. As the pathogen density was not decreased in this suppressive medium, it was proposed that the microorganisms were actively taking up the available nutrients creating a nutrient sink and microbiostasis for the pathogen.

Microbial diversity contributes to efficient biocontrol provided by compost: it seems more advantageous to have a combination of different antagonists because of the protection received against a larger and more diverse spectrum of plant pathogens under different conditions (Punja 1997). However, the composition of the microorganisms active in the control changes, as the organic matter content decomposes. Consequently, the microbial-carrying capacity of the amendment declines, and suppression is lost (Hoitink and Fahy 1986). The effect of the decomposition level of compost will be discussed.

15.3.2.1 Mechanisms of disease control with composts

As mentioned above, disease control with composts has been attributed to four possible mechanisms:

(i) successful competition for nutrients by beneficial microorganisms,
(ii) antibiotic production by beneficial microorganisms (antibiosis),
(iii) successful parasitism (or predation) against pathogens by beneficial microorganisms,
(iv) activation of disease-resistance genes in plants by beneficial microorganisms.

Competition occurs when there is a demand by two or more microorganisms for the same resource. Competition between beneficial microorganisms and the pathogen may lead to disease control. Carbon, nitrogen, iron, infection sites, and oxygen are the most important factors in this competition (Paulitz 1990; Tronsmo 1996). For example, competition seems to be the most potent mechanism in the control of *F. oxysporum* by *Trichoderma harzianum* T-35 (Sivan and Chet 1989). Siderophores, for example low-molecular-weight ferric-specific ligands, are produced under iron-limiting conditions to supply iron to the cell, which may be produced by antagonists, in such a way limiting the iron availability for some pathogens, like *Pythium* spp. (Leong and Zhiqiang 1997).

Antibiosis occurs when the production of specific and non-specific toxic metabolites or antibiotics by one organism has a direct effect on another organism. Such compounds may be volatile or non-volatile (Tronsmo 1996). Thomashow and Weller (1988) were the first able to provide indirect *in vivo* proof, by comparing the biological control effect of two *Pseudomonas fluorescens* isolates, one that produced the antibiotic phenazine-1-carboxylate, and the other a phenazine-negative Tn5-mutant. The mutants that did not produce phenazine showed very little suppressiveness, but replacing the phenazine gene restored the biological control activity. The importance of the toxin gliotoxin has also been proven in the antagonism of *Gliocladium virens* against *P. ultimum* (Roberts and Lumsden 1990). Take-all decline in barley caused by *Gaeumannomyces graminis* var. *tritici* results from the build-up of fluorescent *Pseudomonas* spp. that produce an antifungal metabolite 2,4-diacetylphloroglucinol (Weller *et al.* 2002).

Successful parasitism or predation of plant pathogens by beneficial microorganisms in composts represents another important mechanism of suppression of plant diseases by composts (Hoitink and Boehm 1999). Mycoparasitism covers a multitude of different interactions (Chet 1990). Four stages can be distinguished, that is chemotrophic growth, recognition, attachment, and degradation of the host wall by production of lytic enzymes, chitinases, and β-1,3-glucanases (Elad *et al.* 1983; Cherif and Benhamou 1990). For example, antagonists such as *Trichoderma hamatum*, *T. harzianum* and *T. viride* can parasite pathogens by enzymatic degradation of the cell walls of *R. solani*, *Sclerotium rolfsii*, *Phytophthora* spp., and *Pythium* spp. (Dennis and Webster 1971).

In 1991 a fourth resistance mechanism triggered by microorganisms became apparent. Three research groups demonstrated independently from each other that plant growth promoting rhizobacteria could induce a reduced pathogen infection through a systemic plant-mediated effect (Alström 1991; Van Peer *et al.* 1991; Wei *et al.* 1991). Meera *et al.* (1994) demonstrated a plant growth promoting and systemic-induced resistance effect with fungi. This phenomenon is triggered by beneficial microorganisms and is called "induced systemic resistance" (ISR). Since, ISR has proven to be effective in many plant species against a broad spectrum of plant pathogens and resistance is held up for long time (Van Loon *et al.* 1998). ISR shows some phenotypic similarities with pathogen induced "systemic acquired resistance" (SAR), but differences become apparent when unraveling the underlying mechanisms. Depending on the initial stimulus (necrotic pathogen, beneficial microorganisms, chemicals, etc.), a complex network of interconnecting signal transduction pathways, in which salicylic acid, jasmonic acid, and ethylene play dominant roles, are set in motion and renders the plant more resistant (Hammerschmidt *et al.* 2001; Pieterse *et al.* 2001; Kuc 2002). Although studying the signal transduction pathways is an interesting subject, the underlying pathways remain a mystery today.

It was recently demonstrated that composts can induce systemic responses in plants leading to a reduced disease severity, as described in more detail below.

15.3.2.2 General suppression

The composting process is often divided into three phases. During the second phase, known as peak heating, temperatures can reach 60–75°C and most pathogens and beneficial microorganisms are killed at this time. As the temperature declines below 40–45°C, bacteria and fungi, including biocontrol agents recolonize the compost (Hoitink and Fahy 1986). It is during this "curing" phase that composts become suppressive to soilborne plant diseases caused by *Pythium* and *Phytophthora* spp. This mechanism is referred as "general suppression" (Cook and Baker 1983) and owes its activity to the total microbial biomass. Not one single microorganism is responsible for general suppression, but diverse mesophilic microorganisms recolonizing the compost from the outer low-temperature layer into the compost pile are. General suppression results from competition for nutrients and niches by numerous bacterial taxa that either alter the activity of or induce microbiostasis of these pathogens (Chen *et al.* 1988a, b; Mandelbaum and Hadar 1990; Hardy and Sivasithamparam 1991; Boehm *et al.* 1993). For example, composted pine or hardwood bark- and light peat-amended potting mixes are naturally suppressive to *Pythium* damping-off of cucumber via the general suppression mechanism (Boehm *et al.* 1993; 1997; Hoitink and Boehm 1999).

General suppression is a kinetic phenomenon. Though fully colonized by microorganisms, compost that has been through the pasteurization phase but not the curing phase is not suppressive. Raw compost still contains high amounts of free glucose and other easily metabolized products. Suppressiveness to root diseases caused by *Pythium* spp. and *Phytophthora* spp. does not occur until after the soluble substances have been utilized. At this point sugars are depleted, but enough slow release sources of energy such as cellulose and lignin remain to sustain high-microbial activity. Eventually, as composts decompose further, all available energy has been utilized by the microorganisms and humic substances incapable of sustaining biocontrol agents remain. At this point, pathogens increase in populations and diseases develop (Boehm *et al.* 1997). Another factor important during the curing that affects suppressiveness is the moisture content. It should be 40–50% w/w so that fungal as well as bacterial biocontrol agents can colonize the compost (Hoitink *et al.* 1997).

15.3.2.3 Specific suppression

"Specific suppression" owes its activity to the effects of individual or selected groups of microorganisms on pathogens and this suppression is transferable (Cook and Baker 1983). A minor group of specific microorganisms, for example *Trichoderma* spp., is capable of giving suppression to *R. solani* (Ryckeboer and Coosemans 1997a) and *S. rolfsii* (Wells *et al.* 1972) which both produce nutrient independent sclerotia.

Suppression of *Rhizoctonia* damping-off with compost amendment is, however, variable. This variability is due, in part, to random recolonization of compost by effective biocontrol agents after peak heating (Hoitink *et al.* 1997). Whether or not the appropriate biocontrol agents colonize compost depends partly on the location and duration of compost curing (Kuter *et al.* 1983, 1988). For instance, bark compost cured near a forest area was suppressive to *R. solani*, but the same compost produced in an enclosed reactor was much lower in species diversity and conducive (Kuter *et al.* 1983).

During the curing phase, many composts in an open environment are colonized by biocontrol agents of pathogens including strains of the genera *Bacillus*, *Enterobacter*, *Pseudomonas*, *Pantoea*, *Streptomyces*, and *Trichoderma* (Hoitink and Fahy 1986; Chung and Hoitink 1990; Hardy and Sivasithamparam 1991; Hoitink *et al.* 1997; Ryckeboer 2001; Ryckeboer *et al.* 2002).

The most important biocontrol agents of *R. solani* are members of the genus *Trichoderma* (Mihuta and Rowe 1986), including isolates of *T. hamatum* and *T. harzianum* (Harman *et al.* 1980; Kuter *et al.* 1983; Nelson *et al.* 1983; Lewis and Papavizas 1985). These biocontrol agents interact with a number of bacterial species to suppress *Rhizoctonia* crown rot and *Fusarium* wilt (Trillas-Gay *et al.* 1986; Kwok *et al.* 1987; Alabouvette 1990; Kok *et al.* 1996; Tuitert *et al.* 1998; Ryckeboer *et al.* 2002).

15.3.2.4 Induced systemic resistance

Some indications led to the assumption that induced resistance plays a role in control of soilborne pathogens with compost. Field observations show that mulches applied under parts of trees somehow control the entire root system (Hoitink *et al.* 1997). In addition, the observation that compost also affected foliage diseases corroborated this assumption. Tränkner (1992) reported that compost incorporated into soil reduced the severity of powdery mildew of wheat. In 1995, Clulow *et al.* reported that compost coated onto tubers protected potatoes from late blight. Zhang *et al.* (1996), Vaes (2000), and Lievens *et al.* (2001) demonstrated that composts can induce systemic responses in cucumber plants leading to reduced *Pythium* root rot using a split-root system. These cucumber plants grown on compost-amended substrates proved also more resistant to anthracnose, caused by the foliar pathogen *Colletotrichum orbiculare*, compared to a non-amended, conducive peat mix (Zhang *et al.* 1996). Autoclaving of the compost-amended potting mixes destroyed the effect whereas fortification of a sterilized mix with a small quantity of suppressive mix restored the effect. This suggested that beneficial microorganisms played a role in the systemic effect. The increased resistance level could be correlated, after challenge infection with the pathogen, with an increased activity of pathogenesis related or PR proteins (Zhang *et al.* 1998). Krause (2001) observed that *T. hamatum* strain 382 induces systemic resistance in radish against bacterial spot caused by *Xanthomonas campestris*.

Further research is necessary to understand how and when compost can induce resistance to plants, because this opens up perspectives to control not only soilborne pathogens with compost, but also a broad spectrum of foliage pathogens. As aforementioned only a few composts are able to induce systemic resistance. Therefore, fortifying compost with special beneficial microorganisms could assure more consistent disease suppression (Ryckeboer 2001; Ryckeboer *et al.* 2002).

15.4 BIOCONTROL AGENTS AND COMPOST MATURITY

The mere presence of a known antagonist in the compost does not ensure suppression of *Rhizoctonia* diseases (Chung and Hoitink 1990). A factor often overlooked by specialists working with biocontrol agents is the decomposition level of organic matter in compost-amended potting mixes (Hoitink *et al.* 1997; Hoitink and Boehm 1999). Fresh undecomposed organic matter that is not fully colonized by soil microorganisms capable of inducing microbiostasis not only stimulates soil microorganisms but also supports populations of plant pathogens such as *Pythium* spp., which result in increased disease (Hoitink *et al.* 1997). However, when the incorporated organic matter has been adequately colonized by soil microorganisms and microbiostasis has been established, the disease is suppressed even though the population of the pathogen is elevated (Hoitink and Boehm 1999). Fresh organic matter also does not support biocontrol of *Rhizoctonia* diseases because the high concentrations of free nutrients (glucose, amino acids, etc.) repress production of enzymes by biocontrol agents such as *Trichoderma* spp. required for parasitism (Nelson *et al.* 1983; Chung *et al.* 1988a, b). Composts must be stabilized well enough for soluble carbohydrates to be metabolized. At this stage, *Trichoderma* spp. shift from saprophytic activity to being parasitic to *R. solani*, secreting lytic enzymes and attacking this pathogen (Hoitink *et al.* 1997). Furthermore, *R. solani* is highly competitive as a saprophyte (Garrett 1962). It can utilize cellulose and therefore colonize cellulose-rich fresh organic wastes but not cellulose-poor mature compost (Chung *et al.* 1988b).

In contrast, excessively humified organic matter, such as dark Sphagnum peat, does not support the activity of biocontrol agents either, because this medium no longer has the microbial-carrying capacity required for growth and activity of biocontrol agents, although high salinity and poor soil structure can also explain the problem (Boehm *et al.* 1993, 1997; Krause 2001; Ryckeboer 2001). In the ideal situation, there is a slow but continuous release of nutrients, which is probably the case for mature compost that is not too mature. Researchers do not agree yet on practical guidelines that define the critical stage of decomposition. Composting industry presently controls decomposition level by maintaining constant conditions during the entire process and adhering to a given time schedule (Hoitink *et al.* 1997).

15.5 CONTROLLED FORTIFICATION

Although compost-amended potting mixes can be produced to provide consistent natural suppression of damping-off and root rot diseases caused by *Phytophthora* and *Pythium* spp., variable results are obtained for suppression of other plant diseases. Fortifying the potting mixes with specific biocontrol agents could be a solution to provide a more consistent suppression of diseases with compost (Hoitink and Boehm 1999).

Several factors may contribute to the variability of composts to induce suppression of other soilborne diseases. The compost maturity, its microbial-carrying capacity, and the variability of colonization of composts after peak heating by the appropriate microorganisms are important factors, as explained above. The latter is especially important for diseases that are specifically suppressed (Kuter *et al.* 1983, 1988). The diversity of microorganisms that colonize bark composts after peak heating is sufficient to provide consistent suppression of soilborne diseases caused by *Phytophthora* and *Pythium* spp. However, before complete recolonization by mesophilic microorganisms occurs, these “generally suppressive” composts could be fortified with biocontrol agents that trigger suppression and induce resistance to other plant diseases.

Fortification of mature composts with biocontrol agents has improved the consistency as well as the spectrum of the suppression induced, particularly for *Rhizoctonia* damping-off (Nelson *et al.* 1983; Phae and Shoda 1990; Kok *et al.* 1996; Hoitink *et al.* 1997; Ryckeboer *et al.* 1998; Ryckeboer 2001; Ryckeboer *et al.* 2002). Isolates of several *Trichoderma* spp. can provide effective control of this disease, particularly if applied in combination with any of several bacterial biocontrol agents (Kwok *et al.* 1987). This procedure seems to be helpful for composts produced in closed composting systems where colonization by biocontrol agents is less effective than the colonization of composts produced near a forest (Kuter *et al.* 1983). This controlled fortification strategy promises to stimulate the adoption of this approach to biocontrol as a common agricultural practice. The establishment of inoculated organisms requires that abiotic factors be at least within their tolerance range, and that the appropriate niche be either vacant or occupied with compatible microorganisms (Finstein and Morris 1975). Therefore, these specific inoculants must be fortified into composts at a specific time after peak heating (Hoitink *et al.* 1991; Hoitink and Boehm 1999).

Specific biocontrol agent-fortified compost-amended potting mixes were developed by Hoitink (1990). Potting mixes amended with composted hardwood bark and composted municipal sludge can be made suppressive to both *Rhizoctonia* and *Pythium* damping-off diseases when they are fortified with specific biocontrol agents such as *T. hamatum* 382 and *Chryseobacterium gleum* 299 (formerly *Flavobacterium balustinum* 299). These biocontrol agents are introduced into compost after peak heating, but before significant recolonization by mesophiles has occurred. Compost-amended potting mixes have been produced that are 2.5–3.2 times as suppressive compared to non-fortified natural compost-amended mixes (Hoitink and Fahy 1986; Ryckeboer 2001).

In Japan, suppressive composts were produced using a strain of *Bacillus subtilis* (strain N4-1), a strain that induces biological control of *R. solani* in composts and produces endospores that are viable after peak heating (Phae and Shoda 1990; Nakasaki *et al.* 1996).

Ryckeboer developed biowaste and yard waste composts suppressive to soilborne diseases caused by *Pythium* spp., *R. solani*, and *F. oxysporum* f.sp. *lycopersici* (Ryckeboer 2001; Ryckeboer *et al.* 2002). Fortification of compost or compost-amended potting mixes with single or binary inocula of biological control agents (*Trichoderma* sp. and/or bacterium) improved disease control (Figure 15.1). Recent work resulted in the selection of a high number of promising bacterial and fungal biocontrol agents (including strains of *P. fluorescens*, *Bacillus* spp., ...) that are able to colonize mature biowaste and yard waste composts (Ryckeboer, unpublished work).

15.6 RESTRICTIONS

Quality control of composts is of primary importance for biological control of soilborne diseases. Success or failure of any compost treatment for disease control

Figure 15.1. *Pythium* damping-off of cucumber plants after addition of compost to the potting mix. A: potting mix containing conducive compost, B: potting mix containing suppressive compost, A+ and B+: potting mixes A and B, respectively, fortified with binary inoculum of biocontrol agents (*Trichoderma* sp. and bacterium).

depends on the nature of the raw feedstock from which it was prepared, the maturity of the compost, as well as the method of composting used. Composts that vary in stability, salinity, and nitrogen availability cannot be used predictably, unless application rates are very low and none of the chemical properties of the composts exceed loading rates for a particular crop (Hoitink *et al.* 1993). Failure to appreciate the importance of compost quality contributes to lack of suppressiveness to plant diseases. The maturity level of composts critically affects their successful utilization in agricultural and horticultural crops. Immature composts induce robust microbial activity in the soil for a period of time after their incorporation, resulting in oxygen depletion, a variety of indirect toxicity problems to plant roots and, consequently, greater susceptibility to invasion by pathogens. Although needs exist for rapid assays that predict maturity and suppressive properties of composts (Ryckeboer 2001), chemical and biological stability of the finished product seem to be difficult to define with a single chemical or biological method of analysis (Zucconi *et al.* 1981; Inbar *et al.* 1990). Due to the poor knowledge base regarding compost maturity, current successful use of compost in agricultural and horticultural crops is based on knowledge of the process by which each compost was prepared and of the utilization history of the compost.

15.7 PREDICTION OF SUPPRESSIVENESS

Due to the link between compost maturity and disease suppressiveness, two approaches to prediction of suppressiveness can be distinguished. The first considers estimation of compost maturity and the second focuses on microbial activity.

A rapid assay that has been used to predict compost stability or maturity is the self-heating test (BGK 1994). The rate of respiration is one of several procedures that can be used to monitor stability of composts (Iannotti *et al.* 1994). Maturity of compost can be correlated to non-destructive analysis of soil organic matter, using “diffuse reflectance Fourier transformed infra red spectroscopy” (DRIFT) and "cross polarization magic angle spinning ^{13}Carbon nuclear magnetic resonance spectroscopy", which allows the characterization of biodegradable components of soil organic fractions (Chen 2003).

A simple, rapid, and sensitive enzyme assay that determines microbial activity based on the rate of hydrolysis of FDA is used to predict suppressiveness of potting mixes to *Pythium* diseases via the general suppression phenomenon (Chen *et al.* 1988a, b; Mandelbaum and Hadar 1990; Boehm and Hoitink 1992; Ryckeboer 2001). FDA is hydrolyzed by a number of enzymes in living cells such as proteases, cellulases, lipases and esterases (Schnürer and Rosswall 1982). As FDA activity in suppressive light Sphagnum peat substrates declines to $<192\ \mu g$ FDA hydrolyzed per hour per gram dry weight of mix, the population of *P. ultimum* increases and root rot develops (Inbar *et al.* 1991; Boehm and Hoitink 1992). During this collapse in suppressiveness, bacterial species diversity in the rhizosphere or the potting medium did not change, though the species composition was affected (Hoitink *et al.* 1996a, b; Boehm *et al.* 1997). Biocontrol agents decline in populations and microorganisms abundant in low quality erodable soils increase.

In an attempt to predict disease suppressiveness of biowaste compost against *Rhizoctonia* damping-off, Tuitert *et al.* (1998) proposed that FDA hydrolysis might not sufficiently distinguish the relative microbial activity for conducive low-maturity composts from that for suppressive mature composts.

Vaes (2000) concluded that high-microbial activity in biowaste and yard waste composts is a prerequisite to achieve sufficient protection against *Pythium* damping-off, but high-microbial activity is not necessarily predictive for disease suppression. Craft and Nelson (1996) found that suppression of *Pythium* diseases of creeping bentgrass in batches of brewery sludge and biosolid composts is directly related to the microbial activities in the composts. On the other hand, the mechanisms of *Pythium* disease suppression in turkey litter are not related directly to the compost microbial activity. Therefore, the microbiological properties of *Pythium* root disease-suppressive composts may differ substantially, and measurements of microbial populations and activity may not be predictive of the level of disease suppression in all composts.

According to Hoitink and Boehm (1999), the decomposition level of organic matter critically affects population activities of biocontrol agents as well the composition of bacterial taxa (e.g. *Pseudomonas* and *Pantoea* spp.). Studying microbial diversity and population shifts, in addition to microbial activity, might therefore be a useful and surplus predictive parameter for disease suppression. Recently, several new culture independent and polymerized chain reaction (PCR)-based methods have been developed to study microbial diversity and spatiotemporal shifts. Owing to its high throughput, automated and semi-quantitative feasibilities, terminal restriction fragment length polymorphism (t-RFLP), developed by Liu *et al.* (1997), has, for example, successfully been used in analyzing the temporal changes in bacterial community structure during composting of yard trimmings (Michel *et al.* 2002). Linking this type of data to disease-suppressive activities could possibly reveal new insights.

15.8 CONCLUSIONS

The development of disease-suppressive substrates is one of the major challenges for durable agricultural and horticultural systems. In horticulture for example, most potting soils prepared with "dark" Sphagnum peat are known as substrates with low-microbial-carrying capacity that exhibit practically no disease-suppressive activity. The addition of compost to potting soil is known to have prominent disease-suppressive activity. However, the majority of research has been performed on lignin-rich types of compost, while future research will include biowaste and yard waste composts as re-uses of biomass (Ryckeboer 2001). Suppressive effects of compost-amended substrates, caused by biotic factors, have been due to natural recolonization of composts by biocontrol agents after peak heating. Pathogens are killed when the composting process is managed properly. Beneficial microorganisms are also killed but may survive in the outer low-temperatures zones of compost piles. A reinvasion of beneficial microorganisms from the environment in which composts are produced can also occur. Recontamination with biocontrol agents of composts, either natural or controlled, is necessary to render the composts suppressive. The chemical and physical properties of the compost are inevitably linked to the biological factor and, therefore, also determine structure and activity of the microbial populations during compost curing as well as in the rhizosphere.

Several composts and compost-amended potting mixes were reported to be excellent carriers of specific biocontrol agents (Hoitink and Boehm 1999; Ryckeboer 2001; Ryckeboer *et al.* 2002). Available data suggest that mixtures of bacterial and fungal biocontrol agents are more effective than single biocontrol agents in triggering suppression of, for example, *Rhizoctonia* and *Pythium* damping-off. Success or failure of any compost treatment for disease control depends on the nature of the raw product from which it was prepared, the maturity of the compost, as well as the process of composting used. The lack of appreciation for the importance of compost quality may be responsible for some of the reported failures (Hoitink and Fahy 1986; Hoitink and Boehm 1999). In the future, new techniques might prove to be helpful in the prediction of disease suppressiveness with compost amendment. It can be assumed that as a predictive tool for disease suppressiveness of the end product the compost will demand chemical, physical as well as biological criteria, and requires the analysis of a huge number of compost samples from different origins and processes. This is a challenge for the further research.

ACKNOWLEDGMENTS

The authors gratefully acknowledge the EU (Quality of Life and Management of Living Resources, Key action 5, grant QLK5-CT-2001-01442) for financial support. The contents of this publication are the sole responsibility of its publisher(s) and in no way represent the view of the Commission or its services nor anticipates its future policy in this area.

REFERENCES

Alabouvette, C. (1990) Biological control of *Fusarium* wilt pathogens in suppressive soils. In: *Biological Control of Soil-Borne Plant Pathogens* (ed. D. Hornby), pp. 27–43, CAB International, Wallingford, UK.

Alström, S. (1991) Induction of disease resistance in common bean susceptible to halo blight bacterial pathogen after seed bacterization with rhizospere pseudomonads. *J. Gen. Appl. Microbiol.* **37**(6), 495–501.

Baker, K.F. and Cook, R.J. (1974) *Biological Control of Plant Pathogens*. W.H. Freeman and Company, San Francisco, USA.

BGK (1994) *Methodenbuch zur Analyse von Kompost*, 3rd edn. Bundesgütegemeinschaft Kompost e.V., Køln, Germany.

Blok, W.J., Coenen, T.C.M., Pijl, A.S. and Termorshuizen, A.J. (2002) Disease suppression and microbial communities of potting mixes amended with biowaste compost. In: *Proceedings of the 2002 International Symposium: Composting and Compost Utilization, Columbus, Ohio* (eds. F.C. Michel, R.F. Rynk and H.A.J. Hoitink), pp. 630–644, The JG Press Inc., Emmaus, Pennsylvania, USA.

Boehm, M.J. and Hoitink, H.A.J. (1992) Sustenance of microbial activity in potting mixes and its impact on severity of *Pythium* root rot of *Poinsettia*. *Phytopathology* **82**(3), 259–264.

Boehm, M.J., Madden, L.V. and Hoitink, H.A.J. (1993) Effect of organic matter decomposition level on bacterial species diversity and composition in relationship to *Pythium* damping-off severity. *Appl. Environ. Microbiol.* **59**, 4171–4179.

Boehm, M.J., Wu, T., Stone, A.G., Kraakman, B., Iannotti, D.A., Wilson, G.E., Madden, L.V. and Hoitink, H.A.J. (1997) Cross-polarized magic-angle spinning 13C nuclear magnetic resonance spectroscopic characterization of soil organic matter relative to culturable bacterial species composition and sustained biological control of *Pythium* root rot. *Appl. Environ. Microbiol.* **63**, 162–168.

Bollen, G.J. (1993) De ziektewerende werking van GFT-compost. In: *Brief verspreid door IB-DLO, Haren*, i.v.m. Wetenschappelijke bijeenkomst over Compostering op 20 januari 1994.

Bruns, C. and Schüler, C. (2002) Suppressive effects of composted yard wastes against soil borne plant diseases in organic horticulture. *Proceedings of the 2002 International Symposium: Composting and Compost Utilization, Columbus, Ohio* (eds. F.C. Michel, R.F. Rynk and H.A.J. Hoitink), pp. 620–629, The JG Press Inc., Emmaus, Pennsylvania, USA.

Chef, D.G., Hoitink, H.A.J. and Madden, L.V. (1983) Effects of organic components in container media on suppression of *Fusarium* wilt of chrysanthemum and flax. *Phytopathology* **73**, 279–281.

Chen, Y. (2003) Nuclear magnetic resonance, infra-red and pyrolysis: application of spectroscopic methodologies to maturity determination of composts. *Comp. Sci. Util.* **11**, 152–168.

Chen, W., Hoitink, H.A.J. and Schmitthenner, A.F. (1987) Factors affecting suppression of *Pythium* damping-off in container media amended with composts. *Phytopathology* **77**, 755–760.

Chen, W., Hoitink, H.A.J. and Madden, L.V. (1988a) Microbial activity and biomass in container media for predicting suppressiveness to damping-off caused by *Pythium ultimum*. *Phytopathology* **78**, 1447–1450.

Chen, W., Hoitink, H.A.J., Schmitthenner, A.F. and Tuovinen, O.H. (1988b) The role of microbial activity in suppression of damping-off caused by *Pythium ultimum*. *Phytopathology* **78**, 314–322.

Cherif, M. and Benhamou, N. (1990) Cytochemical aspects of chitin breakdown during the parasitic action of a *Trichoderma* sp. on *Fusarium oxysporum* f.sp. *radicis-lycopersici*. *Phytopathology* **80**, 1406–1414.

Chet, I. (1990) Mycoparasitism-recognition, physiology and ecology. In: *New Directions in Biological Control: Alternatives for Suppressing Agricultural Pests and Diseases* (eds. R.R. Baker and P.E. Dunn), pp. 725–733, Alan R. Liss Inc., New York, USA.
Chung, Y.R. and Hoitink, H.A.J. (1990) Interactions between thermophilic fungi and *Trichoderma hamatum* in suppression of *Rhizoctonia* damping-off in a bark compost amended container medium. *Phytopathology* **80**, 73–77.
Chung, Y.R., Hoitink, H.A.J., Dick, W.A. and Herr, L.J. (1988a) Effects of organic matter decomposition level and cellulose amendment on the inoculum potential of *Rhizoctonia solani* in hardwood bark media. *Phytopathology* **78**, 836–840.
Chung, Y.R., Hoitink, H.A.J. and Lipps, P.E. (1988b) Interactions between organic-matter decomposition level and soilborne disease severity. *Agric. Ecosyst. Environ.* **24**, 183–193.
Clulow, S.A., Steward, H.E., Dashwood, E.P. and Wastie, R.L. (1995) Tuber surface microorganisms influence the susceptibility of potato tubers to late blight. *Ann. Appl. Biol.* **26**, 33–43.
Cook, R.J. and Baker, K.F. (1983) *The Nature and Practice of Biological Control of Plant Pathogens*. APS Press, St. Paul, Minnesota.
Craft, C.M. and Nelson, E.B. (1996) Microbial properties of composts that suppress damping-off and root rot of creeping bentgrass caused by *Pythium graminicola*. *Appl. Environ. Microbiol.* **62**(5), 1550–1557.
Daft, G.E., Poole, H.A. and Hoitink, H.A.J. (1979) Composted hardwood bark: a substitute for steam sterilisation and fungicide drenches for control of *Poinsettia* crown and root rot. *Hortscience* **14**, 185–187.
Dennis, C. and Webster, J. (1971) Antagonistic properties of species-groups of *Trichoderma* III Hyphal interaction. *Trans. Brit. Mycol. Soc.* **57**, 363–369.
Dick, W.A. and McCoy, E.L. (1993) Enhancing soil fertility by addition of compost. In: *Science and Engineering of Composting: Design, Environmental, Microbiological and Utilization Aspects* (eds. H.A.J. Hoitink and H.M. Keener), pp. 622–644, Renaissance Publications, Ohio, USA.
Elad, Y., Chet, I., Boyle, P. and Henis, Y. (1983) Parasitism of *Trichoderma* spp. on *Rhizoctonia solani* and *Sclerotium rolfsii* – Scanning electron microscopy and fluorescence microscopy. *Phytopathology* **73**, 85–88.
Finstein, M.S. and Morris, M.L. (1975) Microbiology of municipal solid waste composting. *Adv. Appl. Microbiol.* **19**, 113–151.
Garrett, S.D. (1962) Decomposition of cellulose in soil by *Rhizoctonia solani Kuhn*. *Trans. Brit. Mycol. Soc.* **45**, 114–120.
Gorodecki, B. and Hadar, Y. (1990) Suppression of *Rhizoctonia solani* and *Sclerotium rolfsii* in container media containing composted separated cattle manure and composted grape marc. *Crop Prot.* **9**, 271–274.
Hadar, Y. and Mandelbaum, R. (1986) Suppression of *Pythium aphanidermatum* damping-off in container media containing composted liquorice roots. *Crop Prot.* **5**, 88–92.
Hammerschmidt, R., Métraux, J.-P. and van Loon, L.C. (2001) Inducing resistance: a summary of papers presented at the First International Symposium on Induced Resistance to Plant Diseases, Corfu, May 2000. *Eur. J. Plant Pathol.* **107**, 1–6.
Hardy, G.E.St.J. and Sivasithamparam, K. (1991) Suppression of *Phytophthora* root rot by a composted Eucalyptus bark mix. *Aust. J. Bot.* **39**, 153–159.
Harman, G.E., Chet, I. and Baker, R. (1980) *Trichoderma hamatum* effects on seed and seedling disease induced in radish and pea by *Pythium* spp. or *Rhizoctonia solani*. *Phytopathology* **70**, 1167–1172.
Hoitink, H.A.J. (1990) Production of disease suppressive compost and container media, and microorganism culture for use therein. U.S. Patent 4,900,348.
Hoitink, H.A.J. and Fahy, P.C. (1986) Basis for the control of soilborne plant pathogens with composts. *Annu. Rev. Phytopathol.* **24**, 93–114.
Hoitink, H.A.J. and Boehm, M.J. (1999) Biocontrol within the context of soil microbial communities: a substrate-dependent phenomenon. *Annu. Rev. Phytopathol.* **37**, 427–446.

Hoitink, H.A.J., VanDoren, D.M. Jr. and Schmitthenner, A.F. (1977) Suppression of *Phytophthora cinnamoni* in a Composted Hardwood Bark Potting Medium. *Phytopathology* **67**, 56–565.

Hoitink, H.A.J., Daughtery, M. and Tayama, H.K. (1987) Control of cyclamen *Fusarium* wilt – A preliminary report. *Ohio's Florist's Assn. Bul.* **693**, 1–3.

Hoitink, H.A.J., Inbar, Y. and Boehm, M.J. (1991) Status of compost-amended potting mixes naturally suppressive to soilborne diseases of floricultural crops. *Plant Dis.* **75**(9), 869–873.

Hoitink, H.A.J., Boehm, M.J. and Hadar, Y. (1993) Mechanisms of suppression of soilborne plant pathogens in compost-amended substrates. In: *Science and Engineering of Composting: Design, Environmental, Microbiological and Utilization Aspects* (eds. H.A.J. Hoitink and H.M. Keener), pp. 601–621, Renaissance Publications, Ohio, USA.

Hoitink, H.A.J., Madden, L.V. and Boehm, M.J. (1996a) Relationships among organic matter decomposition level, microbial species diversity, and soilborne disease severity. In: *Principles and Practice of Managing Soilborne Plant Pathogens* (ed. R. Hall), pp. 237–249, APS press, St. Paul, Minnesota.

Hoitink, H.A.J., Stone, A.G. and Grebus, M.E. (1996b) Suppression of plant diseases by composts. In: *The Science of Composting* – vol. 1 (eds. M. de Bertoldi, P. Sequi, B. Lemmes and T. Papi), pp. 373–381, Blackie Academic & Professional, Glasgow.

Hoitink, H.A.J., Stone, A.G. and Han, D.Y. (1997) Suppression of plant diseases by composts. *Hortscience* **32**, 184–187.

Iannoti, D.A., Grebus, M.E., Toth, B.L., Madden, L.V. and Hoitink H.A.J. (1994) Oxygen respirometry to assess stability and maturity of composted municipal solid waste. *J. Environ. Qual.* **23**, 1177–1183.

Inbar, Y., Chen, Y., Hadar, Y. and Hoitink, H.A.J. (1990) New approaches to compost maturity. *Biocycle* **31**, 64–69.

Inbar, Y., Boehm, M.J. and Hoitink, H.A.J. (1991) Hydrolysis of fluorescein diacetate in sphagnum peat container media for predicting suppressiveness to damping-off caused by *Pythium ultimum*. *Soil Biol. Biochem.* **23**(5), 479–483.

Kok, C.J., Hageman, P.E.J., Mass, P.W.T., Postma, J., Roozen, N.J.M. and van Vuurde, J.W.L. (1996) Processed manure as carrier to introduce *Trichoderma harzianum*: population dynamics and biocontrol effect on *Rhizoctonia solani*. *Biocontrol Sci. Technol.* **6**, 147–161.

Krause, M.S. (2001) Specific suppression of plant diseases provided by compost-amended substrates. Ph.D. thesis. The Ohio State University, Ohio, USA.

Krause, M.S., De Ceuster, T.J.J. and Hoitink, H.A.J. (2000) Natural compost-induced systemic resistance against bacterial spot of radish is based on the activities of several microorganisms. *Phytopathology* **90**(6), S43.

Kuc, J. (2002) Induced resistance in plants-molecular, environmental and practical implications. In: *Induced Resistance in Plants against Insects and diseases* (eds. A. Schmitt and B. Mauch-Mani), vol. 25(6), pp. 1–7 IOBC/wprs Bulletin.

Kuter, G.A., Nelson, E.B., Hoitink, H.A.J. and Madden, L.V. (1983) Fungal populations in container media amended with composted hardwood bark suppressive and conducive to *Rhizoctonia* damping-off. *Phytopathology* **73**(10), 1450–1456.

Kuter, G.A., Hoitink, H.A.J. and Chen, W. (1988) Effects of municipal sludge compost curing time on suppression of *Pythium* and *Rhizoctonia* diseases of ornamental plants. *Plant Dis.* **72**, 751–756.

Kwok, O.C.H., Fahy, P.C., Hoitink, H.A.J. and Kuter, G.A. (1987) Interactions between bacteria and *Trichoderma hamatum* in suppression of *Rhizoctonia* damping-off in bark compost media. *Phytopathology* **77**, 1206–1212.

Leong, S.A. and Zhiqiang (1997) Biosynthetic and regulatory aspects of siderophores. In: *Transition Metals in Microbial Metabolism* (eds. G. Winkelmann and C.J. Carrano), pp. 51–79, Hardwood academic publishers, USA.

Le Quéré, D., Gagné, S. and Caron, M. (1990) Ability of sphagnum peat moss to suppress damping-off caused by *Pythium ultimum*. *Can. J. Plant Pathol.* **12**, 335–336.

Lewis, J.A. and Papavizas, G.C. (1985) Effect of mycelial preparations of *Trichoderma* and *Gliocladium* on populations of *Rhizoctonia solani* and the incidence of damping-off. *Phytopathology* **75**, 812–817.

Lewis, J.A., Lumsden, R.D., Millner, P.D. and Keinath, A.P. (1992) Suppression of damping-off of peas and cotton in the field with composted sewage sludge. *Crop Prot.* **1**, 260–266.

Lievens, B., Vaes, K., Coosemans, J. and Ryckeboer, J. (2001) Systemic resistance induced in cucumber against *Pythium* root rot by vegetable, fruit and garden waste composts. *Comp. Sci. Util.* **9**(3), 221–229.

Liu, W.-T., Marsh, T.L., Cheng, H. and Forney, L.J. (1997) Characterization of microbial diversity by determining terminal restriction fragment length polymorphisms of genes encoding 16S rRNA. *Appl. Environ. Microbiol.* **63**(11), 4516–4522.

Lumsden, R.D., Lewis, J.A. and Millner, P.D. (1983) Effect of composted sewage sludge on several soilborne pathogens and diseases. *Phytopathology* **73**, 1543–1548.

Malek, R.B. and Gartner, J.B. (1975) Hardwood bark as a soil amendment for suppression of plant parasitic nematodes on container grown plants. *Hortscience* **10**, 33–35.

Mandelbaum, R., Hadar, Y. and Chen, Y. (1988) Composting of agricultural wastes for their use as container media: effect of heat treatments on suppression of *Pythium aphanidermatum* and microbial activities in substrates containing composts. *Biol. Wastes* **26**, 261–274.

Mandelbaum, R. and Hadar, Y. (1990) Effects of available carbon source on microbial activity and suppression of *Pythium aphanidermatum* in compost and peat container media. *Phytopathology* **80**, 794–804.

Meera, M.S., Shivanna, M.B., Kageyama, K. and Hyakumachi, M. (1994) Plant growth promoting fungi from zoysiagrass rhizosphere as potential inducers of systemic resistance in cucumbers. *Phytopathology* **84**(12), 1399–1406.

Michel Jr F.C., Marsh, T.J. and Reddy, C.A. (2002) Bacterial community structure during yard trimmings composting. In: *Microbiology of Composting*. (eds. H. Insam, N. Riddech and S. Klammer), pp. 25–42, Springer, Heidelberg.

Mihuta, L. and Rowe, R.C. (1986) *Trichoderma* spp. as biocontrol agents of *Rhizoctonia* damping-off of radish in organic soil and comparison of four delivery systems. *Phytopathology* **76**, 306–312.

Moustafa, A.M. and Hoitink, H.A.J. (1977) Suppression of *Pythium* damping-off of tomato in hardwood bark compost. *Proc. Am. Phytopathol. Soc.* **4**, 173.

Nakasaki, K., Kubo, M. and Kubota, H. (1996) Production of functional compost which can suppress phytopathogenic fungi of lawn grass by inoculating *Bacillus subtilis* into grass clippings. In: *The Science of Composting*, (eds. M. de Bertoldi, P. Sequi, B. Lemmes and T. Papi), vol. 1, pp. 87–95, Blackie Academic & Professional, Glasgow.

Nelson, E.B. and Hoitink, H.A.J. (1983) The role of microorganisms in the suppression of *Rhizoctonia solani* in container media amended with composted hardwood bark. *Phytopathology* **73**, 274–278.

Nelson, E.B., Kuter, G.A. and Hoitink, H.A.J. (1983) Effects of fungal antagonists and compost age on suppression of *Rhizoctonia* damping-off in container media amended with composted hardwood bark. *Phytopathology* **73**, 1457–1462.

Ownley, B.H. and Benson, D.M. (1991) Relationship of matric water potential and air-filled porosity of container media to development of *Phytophthora* root rot of rhododendron. *Phytopathology* **81**, 936–941.

Paulitz, T.C. (1990) Biochemical and ecological aspects of competition in biological control. In: *New Directions in Biological Control: Alternatives for Suppressing Agricultural Pests and Diseases*. (eds. R.R. Baker and P.E. Dunn), pp. 713–724, Alan R. Liss Inc., New York, USA.

Phae, C.G. and Shoda, M. (1990) Expression of the suppressive effect of *Bacillus subtilis* on phytopathogens in inoculated composts. *J. Ferment. Bioeng.* **70**(6), 409–414.

Pieterse, C.M.J., Van Pelt, J.A., Van Wees, S.C.M., Ton, J., Léon-Kloosterziel, K.M., Keurentjes, J.J.B., Verhagen, B.W.M., Knoester, M., Van der Sluis, I., Bakker, P.A.H.M.

and van Loon, L.C. (2001). Rhizobacteria-mediated induced systemic resistance: triggering, signalling, and expression. *Eur. J. Plant Pathol.* **107**, 51–61.

Punja, Z.K. (1997). Comparative efficacy of bacteria, fungi, and yeasts as biological control agents for diseases of vegetable crops. *Can. J. Plant Pathol.* **19**, 315–323.

Roberts, D.P. and Lumsden, R.D. (1990) Effect of extracellular metabolites from *Gliocladium virens* on germination of sporangia and mycelial growth of *Pythium ultimum*. *Phytopathology* **80**, 461–465.

Ryckeboer, J. (2001) *Biowaste and Yard Waste Composts: Microbiological and Hygienic Aspects – Suppressiveness to Plant Diseases*. Ph.D. thesis. Katholieke Universiteit Leuven, Leuven, Belgium.

Ryckeboer, J. (2002) The Fate of Plant Pathogens during Anaerobic Digestion and Composting. Biocycle **43**(4), 50–53.

Ryckeboer, J. and Coosemans, J. (1996a) The influence of GFT-compost extracts on the motility of juveniles of *Heterodera schachtii in vitro*. In: *The Science of Composting* (eds. M. de Bertoldi, P. Sequi, B. Lemmes and T. Papi), vol. 2., pp. 1294–1300, Blackie Academic & Professional, Glasgow.

Ryckeboer, J. and Coosemans, J. (1996b) The influence of biowaste composts on *Pythium* and rootknot nematodes (*Meloïdogyne incognita*) on tomato plants (*Lycopersicon esculentum* Mill.) *in vivo*. *Med. Fac. Landbouww. Univ. Gent* **61**(2b), 567–576.

Ryckeboer, J. and Coosemans, J. (1997a) Compost en plantenziekten: Compostveredeling biedt nieuwe toekomstperspectieven. *Vlacovaria* **3**, 10–11.

Ryckeboer, J. and Coosemans, J. (1997b) De invloed van compost op plantenziekten: het gebruik van GFT-compost draagt bij tot het onderdrukken van verwelkingsziekten. *Vlacovaria* **6**(1), 2–3.

Ryckeboer, J., Deprins, K. and Coosemans, J. (1998) Compost onderdrukt de kiemplantenschimmels *Pythium ultimum* en *Rhizoctonia solani*: veredelde compost doet beter! *Vlacovaria* **7**(3), 20–26.

Ryckeboer, J., Deprins, K. and Coosemans, J. (1999) The influence of biowaste and garden waste on diseases caused by *Pythium ultimum* and *Rhizoctonia solani* related to the antagonists *Trichoderma hamatum* and *Flavobacterium balustinum*. In: *Proceedings of the International Conference ORBIT 99 on biological treatment of waste and the environment* (eds. W. Bidlingmaier, M. de Bertoldi, L.F. Diaz and E.K. Papadimitriou), vol. 2, pp. 523–531, Rhombos Verlag, Berlin.

Ryckeboer, J., Vaes, K., Vanwynsberghe, L. and Coosemans, J. (2002) The suitability of mature biowaste and yard waste composts as carrier substrates for bacterial and fungal biocontrol agents. In: *Proceedings of the 2002 International Symposium "Composting and Compost Utilization", Columbus, Ohio,* (eds. F.C. Michel, R.F. Rynk and H.A.J. Hoitink), pp. 1353–1365, The JG Press Inc., Emmaus, Pennsylvania, USA.

Schneider, R.W. (1985) Suppression of *Fusarium* yellows of celery with potassium chloride and nitrate. *Phytopathology* **75**, 40–48.

Schnürer, J. and Rosswall, T. (1982) Fluorescein diacetate hydrolysis as a measure of total microbial activity in soil and litter. *Appl. Environ. Microbiol.* **43**(6), 1256–1261.

Schüler, C., Biala, J., Bruns, C., Gottschall, R., Ahlers, S. and Vogtmann, H. (1989) Suppression of root rot on peas, beans and beetroots caused by *Pythium ultimum* and *Rhizoctonia solani* through the amendment of growing media with composted organic household waste. *J. Phytopathol.* **127**, 227–238.

Sivan, A. and Chet, I. (1989) The possible role of competition between *Trichoderma harzianum* and *Fusarium oxysporum* on rhizosphere colonization. *Phytopathology* **79**, 198–203.

Tahvonen, R. (1982a) The suppressiveness of Finnish light coloured sphagnum peat. *J. Sci. Agric. Soc. Finl.* **54**, 345–356.

Tahvonen, R. (1982b) Preliminary experiments into the use of *Streptomyces* spp. isolated from peat in the biological control of soil and seed-borne diseases in peat culture. *J. Sci. Agric. Soc. Finl.* **54**, 357–359.

Thomashow, L.S. and Weller, D.M. (1988) Role of a phenazine antibiotic from *Pseudomonas fluorescens* in biological control of *Gaeumannomyces graminis var. tritici*. *J. Bacteriol.* **170**, 3499–3508.

Tränkner, A. (1992) Use of agricultural and municipal organic wastes to develop suppressiveness to plant pathogens. In: *Biological Control of Plant Diseases* (ed. E.C. Tjamos), pp. 35–42, Plenum, New York.

Tränkner, A. (1995) Improvement of plant health by the application of compost. In: *Proceedings of the International Symposium "Biological Waste Management – A Wasted Chance?"* (eds. W. Bidlingmaier and R. Stegmann), pp. TP041, Bochum, 4–6 April.

Trillas-Gay, M.I., Hoitink, H.A.J. and Madden, L.V. (1986) Nature of suppression of *Fusarium* wilt of radish in container medium amendment with composted hardwood bark. *Plant Dis.* **70**, 1023–1027.

Tronsmo, A. (1996) *Trichoderma harzianum* in biological control of fungal diseases. In: *Principles and Practice of Managing Soilborne Plant Pathogens*. (ed. R. Hall), pp. 213–236, The American Phytopathological Society, St. Paul, Minnesota, USA.

Tuitert, G., Szczech, M. and Bollen, G.J. (1998) Suppression of *Rhizoctonia solani* in potting mixes amended with compost made from organic household waste. *Phytopathology* **88**, 764–773.

Vaes, K. (2000) *Compostveredeling: Een Nieuw Perspectief in de Biologische Bestrijding van Pythium ultimum.* M.Sc. thesis. Katholieke Universiteit Leuven, Leuven, Belgium.

Van Loon, L.C., Bakker, P.A.H.M. and Pieterse, C.M.J. (1998) Systemic resistance induced by Rhizosphere bacteria. *Annu. Rev. Phytopathol.* **36**, 453–483.

Van Peer, R., Niemann, G.J. and Schippers, B. (1991) Induced resistance and phytoalexin accumulation in biological control of *Fusarium* wilt of carnation by *Pseudomonas* sp. strain WCS417r. *Phytopathology* **81**, 728–734.

Wei, G., Kloepper, J.W. and Tuzun, S. (1991) Induction of systemic resistance of cucumber to *Colletotrichum orbiculare* by select strains of plant growth-promoting rhizobacteria. *Phytopathology* **81**, 1508–1512.

Weller, D.M., Raaijmakers, J.M., McSpadden Gardener, B.B. and Thomashow, L.S. (2002) Microbial population responsible for specific soil suppressiveness to plant pathogens. *Annu. Rev. Phytopathol.* **40**, 309–348.

Wells, H.D., Bell, D.K. and Jaworski, C.A. (1972). Efficacy of *Trichoderma harzianum* as a biocontrol agent *Sclerotium rolfsii*. *Phytopathology* **62**, 442–447.

Wolffhechel, H. (1988) The suppressiveness of sphagnum peat to *Pythium* spp. *Acta Hortic*. **221**, 217–222.

Zhang, W., Dick, W.A. and Hoitink, H.A.J. (1996) Compost-induced systemic acquired resistance in cucumber to *Pythium* root rot and anthracnose. *Phytopathology* **86**, 1066–1070.

Zhang, W., Han, D.Y., Dick, W.A., Davis, K.R. and Hoitink, H.A.J. (1998) Compost and compost water extract-induced systemic acquired resistance in cucumber and Arabidopsis. *Phytopathology* **88**, 450–455.

Zucconi, F., Forte, M., Monaco, A. and de Bertoldi, M. (1981) Biological evaluation of compost maturity. Biocycle **22**, 27–29.

16

Indicators for determination of stability of composts and recycled organic wastes

A. de Guardia, A. Tremier and J. Martinez

16.1 INTRODUCTION

Spreading on land may constitute a method of organic wastes recycling through agricultural use. Nevertheless raw organic wastes like green wastes, wastewater treatment sludge, manure or biowaste may exhibit properties like material heterogeneity, moisture, pathogens concentrations and biodegradable organic matter (OM) content, which prevent their direct spreading on land. Indeed, even if raw biodegradable OM may, in some special cases of agricultural use, provide nutrients and then enhance soil microbiological activity, it may also be responsible for depressive effects on plant growth. Moreover, biodegradable OM also induces quick development of biochemical transformations, which may lead to harmful effects like odour and gaseous emissions during transport, storage and land application.

© 2004 IWA Publishing. *Resource Recovery and Reuse in Organic Solid Waste Management.* Edited by Piet Lens, Bert Hamelers, Harry Hoitink, Werner Bidlingmaier. ISBN: 1 84339 054 X. Published by IWA Publishing, London, UK

Controlling the environmental conditions which the biomass is exposed to can regulate these spontaneous transformations and their harmful effects. Composting and methanization based on the development of respectively aerobic and anaerobic biomass are both processes that allow this regulation and produce a residue that can be applied to land.

Founded on both aerobic microbiological activity and engineering processes, composting treatments lead to obtain stabilized, dryer, sanitized and more homogeneous materials, which are then ready for spreading. Firstly, during the active phase, also called fermentation period, stabilization occurs by consumption of the most easily biodegradable fraction of OM in the presence of oxygen. It involves a microbial heat production, which enables material drying and sanitization. Sufficient biodegradable OM is required to initiate composting and almost all biodegradable OM should have been consumed before compost application. So stability should not be due to the lack of a nutrient necessary for microbial growth (oxygen, water, etc.) in the medium, as this situation would recover depending on the utilization of the product. The depressive effects that one tries to avoid would then re-appear. The indicator of stability should therefore rely on the compost intrinsic characteristics and not on a poor management of the composting process. Secondly, stabilization also occurs by biosynthesis of more stable organic molecules, i.e. the humic acids, during the curing period which follows fermentation.

One often mixes up stability and maturity concepts (Epstein 1997). Stability may be described as a slow down of microbial activity within the product and the achievement of a very low biodegradability of OM (Iannotti *et al.* 1992; Bernal *et al.* 1998). From a theoretical point of view, the slow down of microbiological activity is often considered to be achieved at the end of the fermentation period. On the other hand, maturity is usually associated with plant growth potential and phytotoxicity (Epstein 1997; Bernal *et al.* 1998). These concepts are related to the content of compost humic matter or phytotoxic compounds but also to the capacity of compost to immobilize soil nitrogen. As all these characteristics depend more or less on microbial activity and compost biodegradable OM content, the maturity concept may be closely linked with the one of stability. In the following, only the concept of stability will be used. Hence, stability indicators will illustrate both microbiological consumption of the most easily biodegradable fraction and humification.

Numerous papers in composting literature proposed stability indicators. Stability may be directly estimated on the composting site. Yet, the methods for determination of stability often require the sampling of composting material or compost. Taking into account the numerous phenomena ruling the composting process, these methods are based on thermodynamic, physical, chemical and biological properties, which evolutions are supposed closely linked to material transformations towards stabilization. The aim of this paper is to summarize the existing indicators and to appreciate their reliability in term of the evaluation of OM stability.

16.2 PROCESS STABILITY INDICATORS

16.2.1 Odour

On site odour emissions are often caused by reduced compounds, produced during raw wastes storage. As an example, a too high moisture preventing aeration may cause odour emissions. Yet, odour emissions could also appear after a few days composting during the degradation of OM through the release of ammonia. Table 16.1 gives the chemicals involved in odour emissions on composting plants.

Most of the time odour is not measured but estimated by composting plant manager. A quantitative measure of odour is also possible on air sampled on site (see Chapter 24). A panel of people then evaluates the diluted sub-samples. For each dilution, the panel has to establish if the odour is detectable or not. At the end of the tests the odour limit is calculated as the concentration for which 50% of the panel can detect the odour. The concentration is expressed as odour unit (o.u.), which corresponds to the value of the dilution factor. For instance if the dilution factor is 10, the odour source will be 10 o.u. A detailed method for dynamic olfactometry is described by the European standard EN 13725 (2003). Iglésias Jiménez and Perez Garcia (1989) and Hue and Liu (1995) recommend practising odour estimation during or after material turning. Recently, electronic noses have been proposed for continuous measurement of odorous gases on site (Nicolas *et al.* 2000; Hamacher *et al.* 2003). These methods, which are based on electronic sensors detection of specific reference molecules, should enable compost facilities operators to detect when malodourous level exceeds the threshold and to identify the odour origin on site. Results remain rather qualitative and the methods still need more validation.

On a process point of view, reduced molecules should be removed if aeration is sufficient to allow aerobic microbiological activity to develop. Composting treatment should also last until ammonia odour disappears, thanks to ammonia stripping to the atmosphere or its transformation to nitrites and nitrates. Moreover, whereas harmful odours are removed, they are then replaced by an odour of wet soil commonly described as "humus odour" (due to geosmine, a molecule produced by mesophilic microorganisms prevailing during the curing phase (Iglésias Jiménez and Perez Garcia 1989). Hence, disappearance of harmful odours and

Table 16.1 Odorous molecules contributing to the odorous emissions of composting plant (Martin and Laffort 1991; Haug 1993).

Molecules type	Examples	Odour description
Fatty acids	Acetic acid and butyric acid	Vinegar and rancid butter
Ketones and aldehydes	Acetone and butyric aldehyde	Fruits smell for ketones Acrid and rancid smell for aldehydes
Nitrogenous compounds	Pyridine and derived amino acids	Irritant smell
Sulphurated compounds	Mercaptans and ethyl sulphide	Putrid smell
Inorganic gases	Ammonia	Irritant smell

development of humus odour, both traduce the evolution of the composting material stability. However, considering the high subjectivity of its measurement, odour should rather be used as an environmental estimation of the treatment progress rather than as a real indicator of the biodegradable OM stabilization.

16.2.2 Material temperature

Microbiological activity is responsible for heat production. Heat produced is partly lost through transfer to the outside and the rest accumulates in the composting material. When production is superior to losses, accumulation is positive and the temperature increases. On the contrary, if production gets lower than losses, the temperature decreases.

Temperature is often determined by "thermocouples" or thermic sensors. Considering temperature gradients within the composting material (De Nobili and Petrussi 1988), one may propose, in the case of a forced aeration process with gas collection, to measure temperature in the exhaust gaseous flow. Temperature of this flow is 10–15°C inferior to the material's one.

At the beginning of the composting treatment, provided the wastes are enough biodegradable and that aeration rate, temperature and moisture allow microorganisms' growth, material temperature increases up to 60–70°C and may remain constant for several days. Consumption of biodegradable OM means depletion of microbial nutrients stock, which then leads to the decrease of both microbiological activity and material temperature. Hence, material temperature can be interpreted as indicative of biodegradability of fresh organic wastes and stability of composts (De Nobili and Petrussi 1988). One has to notice that temperature will decrease slower in case of the natural convective aeration process than in case of the forced aeration process. Indeed, heat losses are greater in the second case.

Nevertheless, all along the composting treatment, aeration rate, temperature and moisture may impede microorganisms' growth and lead either to obstruct temperature augmentation or accelerate temperature decreases. Thus, the temperature decrease should be carefully interpreted to determine whether it is due to a lack of moisture or a too high temperature, or really to the removal of biodegradable OM (i.e. the real stabilization of OM). Taking into account the heterogeneity of material and the fact that aeration is not equally distributed within the composting material, Iglésias Jiménez and Perez Garcia (1989) suggested as a limit of stability a constant temperature around the ambient one and the lack of evolution of this parameter following mixing and/or rewetting of the composted material. Moreover, the temperature may not be compared between different composts, as the absolute value of the temperature depends on the type of substrate and the kind of process (a high aeration may lead to a decrease of temperature).

16.2.3 Oxygen consumption and carbon dioxide production

As already said, the composting treatments are based on aerobic metabolisms of biomass. Thus, microbiological activity is proportional to the oxygen consumption by microorganisms. An increase of the oxygen consumption rate corresponds to the development of biomass whereas its decrease is due to a lower microbiological activity. Biodegradable OM consumption by microorganisms at the beginning of the composting process is responsible for an increase of the oxygen consumption. The removal of biodegradable OM will cause the decrease of the oxygen demand as already explained for heat production (see Section 16.2.2).

Oxidation of OM leads to the production of carbon dioxide and biomass. The ratio of carbon dioxide production above oxygen consumption is indicative of biomass production. Thus, provided carbonates do not accumulate in the composting material, the variations of carbon dioxide production and oxygen consumption are quite similar.

In the case of forced aeration, the exhaust gas collection enables the measurement of oxygen or carbon dioxide concentration. Both oxygen consumption and carbon dioxide production are estimated by comparing the measured values to O_2 and CO_2 concentrations in air. Oxygen concentration will first decrease to raise back to 20.8% whereas carbon dioxide concentration will first increase to slow down back to 0.04%. If aeration is performed through natural convection, the lack of gaseous flow makes the rigorous determination of oxygen and carbon dioxide concentrations in the composting material more complicated. Indeed, heterogeneity of aeration imposes a number of local measurements, which could modify the original composting conditions.

Both oxygen and carbon dioxide concentrations can be followed online. The oxygen concentration measurement is either based on the electrochemical sensors or paramagnetic detectors. Carbon dioxide concentration can be measured either with infrared detectors or by trapping CO_2 in sodium hydroxide (NaOH) and then titrating NaOH back with acid.

Both the increase and decrease of oxygen consumption or carbon dioxide production are indicators of biodegradability and stability of, respectively, fresh organic wastes and composts. However, once more, oxygen consumption or carbon dioxide production decrease may be caused by environmental conditions as non-homogeneous aeration, too high temperatures or, too high or too low moisture contents. It should then be carefully interpreted.

16.3 MOST COMMON PHYSICAL AND CHEMICAL STABILITY INDICATORS

16.3.1 Odour, colour and granulometry

The characterization of composts through the estimation of their odour, colour and granulometry was used for ages. Though these methods remain empirical, odour, colour and granulometry may help identify if a composting treatment has to be carried on or not.

As described in Section 16.2.1, the well-known soil humus smell seems to be a clue of a compost with stable OM.

On a colour point of view, humic substances, synthesized during the curing period, are aromatic molecules with numerous double bonds, which are responsible for black colour of stabilized composts. Colour is often visually estimated and composts should be dark brown or black (Epstein 1997). Colour can be quantified by spectroscopy too (see Section 16.6). Iglésias Jiménez and Perez Garcia (1989) managed to correlate colour thanks to its light intensity value Y to the C/N ratio for household waste and wastewater treatment sludge: $Y = 0.388\,\mathrm{C/N} + 8.13$. Y may be used as an indicator of stability of the compost with set values comprised between 11 and 13.

Compost structure also evolutes during the process: the granulometry reduces with both mechanical turning and biodegradation of material.

Subjectivity of compost characterization in term of odour and colour leads to consider these parameters as not accurate stability indicators. Considering granulometry, its evolution during composting may be very different from one substrate to another. Moreover, it can be largely influenced by technical intervention on the compost (mixing and screening because of granulometry threshold for further spreading). Hence, granulometry cannot be used as stability indicator.

16.3.2 Moisture

The relative humidity (RH) of a material is the ratio between the mass of water contained in the material and the mass of the wet material. The common method to obtain this value is the following: a wet sample is placed in a cup of known weight (T), previously dried. The cup is weighted (m_1) and then placed in the oven at 105°C during 24 h. After this time, the cup is weighted (following cooling in a dessicator) again (m_2). The RH is calculated as follows: $\mathrm{RH} = (m_1 - m_2)/(m_1 - T)$. For details, one may refer to the European standard EN 13040 (1999).

Due to high temperatures, moisture usually decreases significantly during composting, 30–40% during fermentation (de Bertoldi *et al.* 1983; Riffaldi *et al.* 1986; Grebus *et al.* 1994) and much less during curing (Grebus *et al.* 1994). Engineering processes influence water removal too. For example, Grebus *et al.* (1994) observed that material moisture decreased from 60% to 38% in less than 12 days fermentation under forced aeration. However, as moisture depends mainly on composting process management strategies (forced aeration or not, etc.) it cannot be considered as a reliable stability indicator.

16.3.3 Chemical characterization of composts

The chemical indicators are mainly based on carbon and nitrogen changes within the OM during the composting process. Moreover, parameters such as pH, conductivity and cation exchange capacity (CEC) are also considered as they traduce the chemical evolution of OM.

We must keep in mind that the OM is mainly degraded within the aqueous phase called biofilm, which surrounds solids. Loss of solid material is due to transfers into the liquid phase. So chemical analysis will be applied on both the solid substrate and biofilm. Depending on the studied phase, recorded parameters will follow a different evolution.

In the following paragraphs parameters determined within the global transforming substrate (OM, chemical oxygen demand (COD), carbon, nitrogen, C/N ratio, CEC), and parameters determined within the aqueous extract (pH, conductivity, COD, biochemical oxygen demand (BOD), carbon, C/N ratio, ammoniacal nitrogen and nitrates) will be distinguished.

Parameters are usually expressed in g/g dry matter (DM). Taking into account the removal of DM by biodegradation, mass balances will be useful to quantify the real variation of the parameter.

16.3.3.1 Chemical indicators on global substrate

(a) Organic matter The OM and ash content (mineral matter, MM) of any substrate are usually measured by calcinations of the dried sample. The OM content is calculated by difference between DM and ash. It is usually performed by calcination at 550°C for 6 h. A standardized method is available on the following reference: EN 13039 (1999).

The content in OM is a key parameter measured in most of the composting studies. The OM content varies widely between different fresh materials. This parameter tends to decrease during composting, mainly during the fermentation phase (Garcia *et al.* 1992a; Inbar *et al.* 1993; Avnimelech *et al.* 1996). Considering ash content remains constant all along the composting treatment, OM concentration measurements allow calculating OM removal in mass which is more interesting from an engineering point of view:

$$m\mathrm{DM_t} = m\mathrm{OM_t} + m\mathrm{MM_t} \tag{16.1}$$

$$m\mathrm{MM_t} = \mathrm{K} \quad \text{and} \quad \%\mathrm{OM_t} = \frac{m\mathrm{OM_t}}{m\mathrm{DM_t}}$$

then

$$1 = \%\mathrm{OM_t} + \frac{\mathrm{K}}{m\mathrm{DM_t}} \quad \text{or} \quad m\mathrm{DM_t} = \frac{\mathrm{K}}{1 - \%\mathrm{OM_t}} \tag{16.2}$$

Replacing the expression of $m\mathrm{DM_t}$ in equation (16.2) by using equation (16.1), one obtains:

$$m\mathrm{OM_t} = K\left(\frac{\%\mathrm{OM_t}}{1 - \%\mathrm{OM_t}}\right) \tag{16.3}$$

Equation (16.3) enables to express the mass ratio between initial OM mass and final OM mass:

$$\frac{m\mathrm{OM_F}}{m\mathrm{OM_0}} = \frac{\%\mathrm{OM_F}/(1 - \%\mathrm{OM_F})}{\%\mathrm{OM_0}/(1 - \%\mathrm{OM_0})} \tag{16.4}$$

with mX, mass of X; $\%X$, X concentration in DM.

For example with an initial organic content of 70% and a final organic content of 55%, the final OM mass will represent 52% of the initial one. As a comparison, the loss expressed in concentration would only be of 21%.

If the loss of OM traduces the microbial activity during the composting process, the OM content cannot be considered as a stability indicator. Indeed, as raw materials, composts exhibit a wide range of values (Table 16.2), they cannot be compared by this parameter. Moreover, the OM content expresses not only the biodegradable part of the substrate.

(b) Total carbon and total organic carbon Total carbon (TC) can be determined through chemical oxidation (Hue and Liu 1995) or thermal oxidation (NF ISO 10694 1995). Chemical oxidation implies reaction of the substrate with potassium dichromate. TC is then measured out with iron sulphate. Thermal oxidation means dry combustion of the substrate and then the measurement of the released CO_2 by conductimetry, infrared or chromatographic detection. In both cases, measurement of mineral carbon on the substrate's ashes enables to obtain the total organic carbon (TOC) content.

Carbon is the main component of OM. The study of TC or TOC is often performed on composts with the main aim of establishing the C/N ratio. There are a number of literature data available, the results are expressed in mg C/g dry compost. TC and TOC of raw substrates vary widely, which in addition to differences in engineering processes, leads to wide variations within composts. Table 16.3 presents some values for various types of composts.

During the composting process, the carbon content decreases, mainly during the fermentation phase. Benitez *et al.* (1999) studied a mixture of sludge and

Table 16.2 OM content of various types of composts (from Charonnat *et al.* 2001).

Type of compost	OM (%)
Green wastes	20–45
Biowastes	36–38.7
Household wastes	37.6–42
Sewage sludge	52.5–68

Table 16.3 TOC content of various compost material (Charonnat *et al.* 2001).

Type of compost	TOC content (mg/g)
Green wastes	85–455
Biowastes	122–706
Household wastes	76–347
Sewage sludge	136–517

observed reduction of up to 35% of the initial TC during the fermentation phase, while subsequent reduction during the curing phase was only 6.5%. Similar observations were reported by Benistant (1978) from compost of municipal waste with figures of, respectively, 50% and 7% during the fermentation and curing phases. The same observations were assessed by Riffaldi *et al.* (1986), Bernal *et al.* (1998) and Paré *et al.* (1998).

For the same reasons, which have been exposed for OM (see Section 16.3.3.1(a)), TC and TOC cannot be used as stability indicators.

(c) Chemical oxygen demand By definition, the COD is the overall quantity of oxygen necessary to degrade all the oxidizable compounds in the material: OM but also mineral compounds like nitrites, sulphides, chlorides, etc. So the COD is a picture of the total oxidizability of the matter: it includes both oxidizability by biodegradation and chemicals. But during the composting process only oxidation by biodegradation occurs.

The COD on global substrate is usually measured using the Lossin method (1971) adapted from the wastewater protocol. The protocol consists in an oxidation of 0.2–0.3 g of sample using potassium dichromate ($K_2Cr_2O_7$, 1N) and concentrated sulphuric acid (H_2SO_4). The COD is expressed in mg of oxygen per gram of dry sample.

Lossin applied his method to household waste and observed that fresh wastes had a COD value of 900 mg O_2/g. This concentration decreases during the process of composting. Lossin proposed to assess stability when the COD was inferior to 700 mg O_2/g. Godden *et al.* (1986) applied the method to cattle manure and noticed that fresh manure exhibited COD values inferior to 700 mg O_2/g.

Once again the wide variations of COD contents and the lack of distinction between biodegradable and non-biodegradable fractions of OM prevent the use of the COD content as stability indicator. However, the COD mass balance should correlate to oxygen consumption and should be, as a consequence, linked to the microbial activity evolution.

(d) Total nitrogen Nitrogen is a key element of OM and of some mineral compounds. In its organic form it is present in proteins, amino acids, etc., while its mineral forms comprise nitrates (NO_3^-), nitrites (NO_2^-) and ammoniacal nitrogen forms (NH_4^+ and NH_3).

The most common method to determine the total nitrogen remains the Kjeldahl method, which consists in a mineralization of organic nitrogen into ammoniacal nitrogen (EN 13654-1 2001). Thermal oxidation of DM followed by gas chromatography quantification of nitrogen is also proposed by Paré *et al.* (1998) (NF ISO 13878 1998). Total nitrogen content is expressed as mg N/g dry compost.

As carbon, nitrogen is a parameter widely determined on composts as it helps to establish the C/N ratio. The nitrogen content of raw substrates is rather variable. Some typical data are given in Table 16.4. The substrates with high concentrations in total nitrogen are often named nitrogenous substrates. These are wastewater treatment sludge or some livestock effluents like poultry litter.

Table 16.4 Nitrogen content of various raw substrates (Haug 1993).

Type of substrate	Total N (mg/g)
Sewage sludge	40–70
Manures	12–63
Green wastes	20–60
Urban wastes	20–29

The organic nitrogen may be mineralized by ammonification during the first stages of composting. It results in a production of ammonium and an increase in pH. The presence of ammonium, associated with high temperature, pH above 7.5–8 and consequent aeration rate, may lead to nitrogen losses by ammonia volatilization, particularly during the thermophilic stages. Alternatively, ammonium may be immobilized by microorganisms that use it as nitrogen source and transform it back to organic nitrogen. Moreover, ammonium may be transformed into nitrate by nitrification. A lack of oxygen may result in the use of nitrate as an oxygen source by denitrifying bacteria, and then in N_2 emission. A part of nitrogen losses may be also explained by harmful gases such as N_2O and NO_x emitted under unfavourable aeration conditions during the nitrification and denitrification processes. N_2 may also be biologically fixed into the material during the mesophilic stages.

Nitrogen content was often described as increasing during fermentation and then remaining constant during curing (Riffaldi *et al.* 1986; Bernal *et al.* 1996, 1998), its increase being often interpreted as biological fixation. As an example, Fraser and Lau (2000) observed that the weight of total nitrogen in pile increases by 24% during fermentation. Yet, taking into account the low level of nitrogen content in OM, compared to carbon, the nitrogen concentration may increase, not because of a nitrogen mass increase but because of the DM removal (Riffaldi *et al.* 1986). From both scientific and engineering points of view, this observation should promote mass balance calculations following concentration measurements. In the case of high nitrogen contents such as in wastewater treatment sludge or livestock effluents, depending on engineering processes, losses of nitrogen may be significant too (Paré *et al.* 1998; Benitez *et al.* 1999).

Nitrogen content cannot be used as stability indicator. However, nitrogen content of the compost is a good indicator of the fertilizing value of the product, though it does not give any indication on the real availability of the N for crop utilization.

(e) Ratio C/N on the solid fraction The C/N ratio is usually obtained by determining the content of TOC and Kjeldahl nitrogen in the compost material.

Usually the C/N ratio decreases during composting mainly during the fermentation phase, although some authors studying the compost of sewage sludge have observed increase or stability of the C/N ratio (Garcia *et al.* 1992a; Benitez *et al.* 1999). This ratio is often proposed as an indicator of stability or maturity.

Table 16.5 C/N ratio for some type of compost (Haug 1993; Epstein 1997; Charonnat *et al.* 2001).

Type of substrate	C/N in raw material	C/N in composted material
Sewage sludge	6–11	9.2–28.7
Manures	15–25	Around 10
Green wastes	11–41	10–26
Urban wastes	20–30	14.4–22.9

Indeed, considering that microorganisms need 15–30 times more carbon than nitrogen to grow, C/N variations were supposed to reach C/N values measured in microorganisms cells or soils humus, i.e. 10 (Haug 1980; Mathur *et al.* 1993a). An "ideal" C/N value of 30 for the initial material is also described as the most favourable situation for a rapid composting.

The Swiss Institute for Standardization (EAWAG) suggested in 1975 a C/N value of 20–25 for describing a compost mid-mature, and a value below 20 for a mature compost (EAWAG 1976). Juste *et al.* (1980) proposed, as a limit of maturity for urban compost, a limit of 0.75 for the ratio between the C/N final and initial values $[(C/N)_f/(C/N)_i]$. C/N ratios before and after composting are given for different type of substrates in Table 16.5.

Though for substrates from the same origin (yard wastes, wastewater treatment sludge, manure, etc.), the C/N ratio seems to vary towards identical values, it cannot be seriously used as stability indicator. The main drawback of the C/N ratio is that it does not account for only biodegradable forms of carbon and nitrogen but also for non-biodegradable fractions like lignin.

(f) Cation exchange capacity The CEC is defined as the amount of exchangeable cations that a substrate can adsorb per unit of mass. Usually the CEC is expressed as milli-equivalent per 100 g of OM (meq/100 g OM), which corresponds to centimoles of H^+ per kg of OM.

Harada and Inoko (1980) and Lax *et al.* (1986) proposed methods to measure CEC. Both consist in extractions trough acid scrubbing or adsorption on activated carbon. Cations are first replaced by H^+ then substituted by Ba^{2+} through percolation on materials of barium acetate $(BaOac)_2$ 1 N or barium chloride $(BaCl_2)$ 0.1 N solutions.

During the composting process the CEC tends to increase all along the fermentation phase and then stabilize. Iglésias Jiménez and Perez Garcia (1989), Riffaldi *et al.* (1986) and Inbar *et al.* (1990) observed that CEC was twice to three times superior in composts than in initial substrates. Inbar *et al.* (1990) and Mathur *et al.* (1993a) correlated CEC increases to that of humic acids concentrations. Yet, Stentiford and Pereira Neto (cited by Riffaldi *et al.* 1986) mentioned that CEC concentration during composting was rather due to the decrease of OM content than to an absolute increase of the exchangeable cations quantity. Indeed, CEC expressed on the basis of MM remained constant.

Harada and Inoko (1980) proposed to assess stability when CEC was superior to 60 meq/g OM. Bernal *et al.* (1998) and Garcia *et al.* (1992a) do not agree with that threshold value because some raw substrates have a CEC greater than 60 meq/g OM. Garcia *et al.* (1992a) observed that at the end of fermentation the ratio CEC/TOC was similar for all substrates in spite of a wide range of CEC of raw substrates. Thus, the authors proposed to consider that the material was stable when the CEC/TOC became superior to 3.5. Forster *et al.* (1993) wrote that a same threshold cannot be used for all organic materials.

From an agronomic perspective, the determination of the CEC in a compost is interesting because a high CEC avoids cations (Na^+, Ca^{2+}, Mg^{2+}, K^+) leaching. These cations are then available as nutrients for plants. However, stabilization of CEC does not rely rigorously on the stabilization of OM. Moreover, as shown by Forster *et al.* (1993), the wide variations of CEC of initial substrates prevent to define threshold values and to use it as a stability indicator.

16.3.3.2 Chemical indicators on aqueous extracts

(a) Aqueous extracts preparation Aqueous extracts are obtained trough stirring of composting material in an aqueous solution followed by the filtration of the suspension. Ratio of compost in aqueous solution vary between 0.1 and 0.2 g dry compost per g water (Chanyasak and Kubota 1981; Baca *et al.* 1990; Hue and Liu 1995; Blanco and Almendros 1997). Bernal *et al.* (1998) used ratios around 0.01 g dry compost per g water. Chanyasak and Kubota (1981) and Baca *et al.* (1990) practised extraction with hot water (60°C). Stirring lasts from 30 min to 4 h. Blanco and Almendros (1997) maintained extraction for 24 h. Hue and Liu (1995) centrifugated suspensions for 10 min at 10,000 g. Actually most of the time filtration is carried on with 0.45 μm filters.

Indicators measured on aqueous extracts are usually expressed as g per g of compost DM.

(b) pH and conductivity pH is measured with the help of a pH probe which allows the measurement of the potential difference between a reference probe (calomel electrode in saturated KCl) and the probe in solution. The potential difference is due to the balance between H^+ and OH^- ions concentrations which are responsible for acidity or basicity of the solution. The method was normalized for composts by CEN (EN 13037 1999).

pH is often measured on composts. Usually pH decreases at the beginning of composting due to release in biofilm of carbonates and organic acids (de Nobili and Petrussi 1988; Iglésias Jiménez and Perez Garcia 1989; Inbar *et al.* 1990). Avnimelech *et al.* (1996) did not observe acidification. Then, enzymatic attacks of OM generates basic compounds and particularly ammonia, which are responsible for the increase of the pH. Finally emission of ammonia and drying effects may cause a low decrease of pH, which final value is around neutrality. Lasaridi and Stentiford (1998a) reported such evolutions for different organic wastes (Table 16.6).

Table 16.6 Examples of pH evolution during composting.

Type of substrate	pH evolution during composting		
	Initial value	Maximal value during the thermophilic phase	Final value
Sludge and straw	5.89	8.32	7.52
Urban wastes	5.01	8.22	7.62
Paper mill sludge	7.28	8.30	7.68

Considering that acids have phytotoxic effects and enhance the solubility of heavy metals, composts with low pH should be used carefully. A low value of final pH may signify organic acids remain in material and that composting time was insufficient. Yet, pH around neutrality is not sufficient to consider that the OM is stable.

Conductivity measurement is based on electrical current measurement between two platinum electrodes. In aqueous extracts, conductivity is closely linked to the concentration of ions and particularly of mineral salts. Conductivity is expressed as Siemens per metre (S/m) or deciSiemens per metre (dS/m). The measurement method was normalized by CEN (EN 13038 1999).

Avnimelech *et al.* (1996) and Lasaridi and Stentiford (1998a) observed that conductivity of household wastes decreased during fermentation. Avnimelech *et al.* (1996) interpreted that decrease by chemical complexation of cations with OM, which meant a better stability of OM. However, Inbar *et al.* (1993) and Lasaridi and Stentiford (1998a) reported that the conductivity did not significantly change during cattle manure and wastewater treatment sludge fermentations, whereas Inbar *et al.* (1993) mentioned it increased during curing.

The absence of a rigorous tendency of the evolution of conductivity during composting, the influence of the drying effect, which may be responsible for conductivity increases without any real increase of the quantity of ions and the complexity of reactions between ions and OM prevent to use conductivity as a stability indicator. Yet, a high conductivity may mean too high salts concentrations, which could result in phytotoxic effects.

(c) Total carbon and total organic carbon Measurement methods of TC and TOC rely on chemical or thermal oxidation of aqueous extract (EN 1484 1997). Measurement units are respectively mg TC/l and mg TOC/l. Taking into account extraction ratios, calculation allows expression as mg TC/g DM and mg TOC/g DM.

Dissolved carbon concentrations are much lower than the carbon content in global substrate (Table 16.7). Bernal *et al.* (1998) and Benitez *et al.* (1999) reported a continuous decrease during fermentation from 10% to 75% of the dissolved carbon from different organic wastes: wastewater treatment sludge mixed with green waste, livestock manure, household wastes. Both observed a decrease of TOC in the range of 5–20% during curing. Mathur *et al.* (1993b) mentioned that dissolved

Table 16.7 Examples of dissolved carbon content in different type of substrates (Garcia *et al.* 1992a; Mathur *et al.* 1993b; Benitez *et al.* 1998).

Type of substrate	Dissolved carbon content (mg/g)
Cattle manure	19.7
Urban wastes	47.6
Sewage sludge	15.9

TOC first increased and then decreased until the end of curing. Though Garcia *et al.* (1992a) and Bernal *et al.* (1998) agreed that composting material have similar dissolved TOC concentrations whatever their origin, Garcia *et al.* (1992a) proposed a stability threshold value of 5 mg/g DM whereas Bernal *et al.* (1998) increased the stability threshold to 17 mg/g DM.

Consumption of dissolved available carbonaceous molecules is responsible for a decrease of the amount of carbon in the biofilm. Yet, the transfer from solid OM to biofilm through enzymatic attack increases the amount of carbon into the biofilm. Moreover, removal of DM reduces the consumption effect as dissolved TOC; is expressed as g/g DM. These processes explain the difficulty to agree on a common stability threshold value and prevent to use dissolved TOC expressed in concentration as stability indicator.

(d) Chemical oxygen demand COD measurement in aqueous extracts is based on oxidation with potassium dichromate ($K_2Cr_2O_7$) followed by titration with ammonium sulphate $(NH_4)_2SO_4$ of remaining dichromate. COD is expressed as mg O_2/l and takes into account extraction ratios, calculation allows its expression as mg O_2/g DM composting material. The measurement method was normalized trough NF T90-101 (2001).

In a study of the household wastes composting, Benistant (1978) observed that dissolved COD first increased from 25.2 to 37.6 then decreased to 8.1 mg O_2/g DM during fermentation. During curing, COD slowly decreased from 8.1 to 7.2 mg O_2/g DM. Consumption of dissolved OM by microorganisms should be responsible for a decrease of the amount of COD in the biofilm at the first step of composting. Yet, as the COD concentration is expressed as mg O_2/g DM, the decrease of the amount of DM reduces the COD decrease tendency. Moreover, enzymatic attack of DM leads quickly to the release in biofilm of OM, which augments the amount of dissolved COD. When that release slows down, consumption of dissolved OM and lower removal of DM explain both the decrease of the COD concentration at the end of fermentation and during curing period.

(e) Biochemical oxygen demand The measurement method of the BOD_5 was normalized by CEN, EN 1899-1 and EN 1899-2 (1998). The method consists in addition of biomass to the solution then in measuring the dissolved oxygen concentration after 5 days incubation at constant temperature (20°C). BOD_5 is expressed as mg O_2/l.

Benistant (1978) applied BOD measurement on aqueous extracts from household wastes. He observed BOD variations were similar to those of COD. BOD first increased from 7.5 to 13.7 then decreased to 4.4 mg O_2/g dry compost. BOD still decreased down to 0 during curing period. Mathur *et al.* (1993b) observed the same variations on livestock effluents, biowaste and municipal waste. Both mentioned that BOD expressed as mg O_2/g dissolved TOC decreased all along fermentation.

Taking into account that OM oxidation occurs only by biodegradation along the composting treatment, BOD variations can be explained as they were for COD even if COD relies on the total oxidability of the OM. In return expression of BOD as mg O_2/g dissolved TOC describes a continuous decrease of the biodegradable fraction within TOC.

Though the evolution of the BOD shows the behaviour of the biodegradable fraction of the OM, as explained for COD, a single BOD measurement on a compost extract cannot be used as stability indicator.

(f) Nitrogen forms Nitrogen forms present in aqueous extracts are organic nitrogen (N_{org}), ammoniums (NH_4^+), nitrates (NO_3^-) and nitrites (NO_2^-).

Ammoniacal nitrogen is usually determined by steam distillation. Distillates are trapped in boric acid, followed by back titration of the boric acid (H_3BO_3) with sulphuric acid (H_2SO_4). Colorimetric methods with nitroprussiate or Nessler reagent followed by measurement of spectrophotometric absorption at, respectively, 630 and 420 nm may be used too.

Measurement method of Kjeldahl nitrogen (EN 25663 1993) consists in digestion in sulphuric acid in order to transform organic nitrogen in ammonium. The ammonium content is then titrated with the help of steam distillation: trapping in boric acid or sodium hydroxide (NaOH), followed by back titration of trapping solution. Organic nitrogen is obtained by substraction of ammoniacal nitrogen from Kjeldahl nitrogen.

Oxidation of total nitrogen into nitrates through digestion with a potassium persulphate solution ($K_2S_2O_8$), followed by back titration of nitrates by a colorimetric method may be used too.

Nitrates may be titrated through the use of a specific electrode, by ionic chromatography, through a colorimetric reaction with sodium salicylate and then spectrophotometric absorption measurements at 415 nm, or finally by digestion with Dewarda catalyst to reduce nitrates into ammonia followed by titration of ammonia. Most of those methods were normalized to characterize wastewater (EN ISO 10304-1 1995, 1996; ISO 7890-3 1988).

Papers in literature account mainly for the variations of ammonium and nitrate ions in biofilm. Aqueous extracts originated from numerous organic wastes: poultry livestocks, wastewater treatment sludge and household wastes mixed with different vegetal residues (cotton, maize and sorgho) (Bernal *et al.* 1998), wastewater treatment sludge (Benitez *et al.* 1999), paper sludge mixed with corn straw

(Riffaldi *et al.* 1986), green wastes (Grebus *et al.* 1994), cattle manure (Mathur *et al.* 1993b). Some authors observed that ammonium concentration continuously decreased during fermentation (Bernal *et al.* 1998; Benitez *et al.* 1999) and some others observed first an increase then a decrease of the ammonium concentration (Riffaldi *et al.* 1986; Mathur *et al.* 1993b; Grebus *et al.* 1994; Bernal *et al.* 1998). That decrease went on during curing. On the other hand, nitrates concentrations were first very low then increased until the end of the curing period. Very low concentrations in ammonium and higher concentrations in nitrates were proposed as indicative of stability. Bernal *et al.* (1998) considered compost was stable when both ammonium concentration and the ratio $N\text{-}NH_4^+/N\text{-}NO_3^-$ were inferior respectively to 0.4 mg/g and 0.16.

Dynamics of nitrogen were described earlier in Section 16.3.3.1(d). The content of different nitrogen forms do not only depend on biodegradation but also on emission of ammonia and nitrification/denitrification steps. Then as explained for TOC, COD and BOD, the complexity of mechanisms ruling nitrogen dynamics does not allow to consider OM is stable when ammoniums and nitrates concentrations get inferior or superior to defined values. Moreover, threshold values, which could have been determined for some specific organic materials could not be used for the others.

(g) C/N ratios Different C/N ratios have been defined. All imply dissolved carbon (C_{diss}) whereas different nitrogen forms are considered: total dissolved nitrogen (N_{diss}), dissolved organic nitrogen ($Norg_{diss}$), organic nitrogen measured on global substrate ($Norg_T$) and total nitrogen measured on global substrate (N_T).

Measurement methods are those used for the determination of carbon and nitrogen forms. Carbon and nitrogen contents are expressed in the same unit and C/N ratios are dimensionless.

Chanyasak and Kubota (1981) gave ratio $C_{diss}/Norg_{diss}$ in the range 2.95–29.5 for wastewater treatment sludge mixed with green wastes and household wastes whereas Garcia *et al.* (1992a) calculated ratios C_{diss}/N_{diss} in the range 6.9–13.0 for the same type of mixtures. All observed that the C/N ratios decrease during fermentation to remain almost stable during curing. Table 16.8 gives the different threshold values proposed as stability indicators for the C/N ratio.

Table 16.8 C/N ratio threshold used as stability indicators in aqueous extracts.

Authors	C/N ratio	Threshold value
Garcia *et al.* (1992a)	C_{diss}/N_{diss}	2
Chanyasak and Kubota (1981); Iglésias Jiménez and Perez Garcia (1992)	$C_{diss}/Norg_{diss}$	6
Hue and Liu (1995)	$C_{diss}/Norg_T$	0.7
Bernal *et al.* (1998)	$C_{diss}/Norg_T$	0.55
Garcia *et al.* (1992a)	C_{diss}/N_T	0.3

As explained for carbon and nitrogen, the C/N ratio cannot be used as stability indicator.

Taking into account that biofilm is the seat of both consumption and transfer of nutrients, from solid OM to biofilm, continuous monitoring on aqueous extract of pH, TOC, COD, BOD, etc. helps to identify the different steps in composting processes. Yet, considering the biofilm content is the result of numerous phenomena, which may have opposite effects, the definition of threshold values concerning stabilization processes indicators should not be envisaged.

16.4 MICROBIOLOGICAL INDICATORS

The microbiological indicators rely directly on microbiological activity through biomass or metabolic processes quantification. Oxygen consumption, carbon dioxide production, adenosine triphosphate (ATP), phospholipids and specific enzymes titration take part to this type of indicators.

16.4.1 Specific microorganisms counting

Methods consist in compost sample incubation, in conditions favouring growth of microorganisms with particular properties followed by counting of these microorganisms. So in order to quantify mesophilic and thermophilic microorganisms, growing preferentially around 20–40°C and 40–80°C, Davis *et al.* (1992) suggested incubation around 30°C for 3 days with soya gelose as nutrient for mesophilic microorganisms and incubation at 60°C for 18–20 h with peptone gelose for thermophilic ones. The results are given in units forming colonies per unit mass (log UFC/g).

These methods showed that mesophilic microorganisms concentrations were superior at the beginning of fermentation then became inferior when material temperature exceeded 40°C and finally were still predominant when temperature decreased and during curing period (Mathur *et al.* 1993a). Davis *et al.* (1992) mentioned thermophilic microorganisms concentrations increased when temperature increased but always remained inferior to mesophilic concentrations. Both observed that bacteria were always more numerous than fungi.

Riffaldi *et al.* (1986) studied variations of ammonifying, proteolytic, nitrifying and cellulolytic bacteria during the composting of wastewater treatment sludge mixed with straw. Ammonifying and proteolytic bacteria varied similarly. They increased until the end of thermophilic period then decreased towards very low concentrations. On the contrary, nitrifying bacteria concentrations first decreased then increased when ammonifying and proteolytic bacteria concentrations fell down. Nitrifying bacteria remained numerous after fermentation even if a slight slow down was observed. Cellulolytic bacteria concentrations also increased all along fermentation and then slightly decreased. The authors proposed to consider that OM was stable when nitrifying and cellulolytic bacteria were maximum.

These variations observed on composts originating from nitrogenous and cellulolytic materials are generally in agreement with the degradation of nitrogenous molecules implying ammonifying and proteolytic bacteria. In the same time cellulolytic bacteria develop. When release of nitrogenous molecules from solid OM slows down, nitrifying bacteria develop to transform ammonia into nitrates and remain active during maturation.

Those studies help understand biological processes during composting treatment. The methods should be improved in order to be used as stability indicators.

16.4.2 ATP and phospholipids quantification

The measurements of ATP and phospholipids have been performed on many compost substrates (green waste, animal manures, municipal wastes, etc.).

The ATP is an energy rich molecule synthesized during the metabolism of microorganisms. It is an energy source for many cellular reactions.

The ATP measurement methods are based on compost extraction with distilled water, then addition of luciferine–luciferase followed by photometric quantification (Horiuchi *et al.* 2003). Tiquia *et al.* (1996, 2002) observed during composting of pig manure that ATP increased at the beginning of fermentation to remain stable afterwards for several days and then decreased down to low concentrations during curing. The authors noted that ATP and material temperature had quite similar variations. Iglésias Jiménez and Perez Garcia (1989) accounted for a faster decrease of ATP during the thermophilic phase.

Phospholipids are cell constituents of microorganisms. Rajbanshi *et al.* (1998) proposed a method based on serial extractions followed by reactions with colorimetric indicators. Phosphates are finally quantified by spectrophotometric absorption at 830 nm.

Derikx *et al.* (1990) and Rajbanshi *et al.* (1998) followed phospholipids during composting of, respectively, green wastes and livestock wastes mixed with straw. Phospholipids concentrations first increased, remained constant for several days then fell down. Concentrations remained constant during curing. Final phospholipids concentrations were quite similar whatever the original substrate. Hence, Rajbanshi *et al.* (1998) assessed stability for concentrations inferior to 0.1 mmol PO_4^{3-}/kg DM.

Rajbanshi *et al.* (1998) considered phospholipids measurement was the most rigorous method to follow composting biomass. Yet, the method remains complicated. Additional results are necessary to evaluate if ATP and phospholipids can be used as stability indicators.

16.4.3 Enzymatic activities

Enzymes are proteins synthesized by microorganisms to catalyse biochemical transformations. Each enzyme is the catalyst of a specific reaction, which may be hydrolysis or oxido-reduction. Numerous enzymes were studied in research about composting. Table 16.9 gives those enzymes and the implied reactions.

Table 16.9 Examples of enzymes implied in the composting process.

Enzyme	Catalysed reaction
β-Glucosidase	Carbohydrates hydrolysis
Protease	Proteins hydrolysis
Urease	Urea hydrolysis
Phosphatase	Ester-phosphate bonds hydrolysis
Xylanase	Xylases (contained in hemicellulose) hydrolysis
Invertase	Sucrose hydrolysis
Deshydrogenase	Hydrogen transfer from an oxidizable substrate to oxygen molecules
Peroxydase	Oxidation of substrates thanks to H_2O_2 reduction in H_2O

Ayuso *et al.* (1996) and Fang *et al.* (1998) proposed methods to measure activity of a few enzymes. All are based on hydrolysis, reduction or oxidation of a specific substrate, previously in contact, at 37°C for a few hours, with compost sample or enzymatic extract issued from compost followed by titration of the catalysed reaction products. Enzymatic activity is expressed as mol or g reaction product per g material per unit time.

Benitez *et al.* (1999), Ayuso *et al.* (1996) and Fang *et al.* (1998) measured activities of β-glucosidase, phosphatase, protease, urease, deshydrogenase during wastewater treatment sludge composting. Some authors observed a slight increase at the beginning of fermentation. All, except Godden *et al.* (1986) studying cattle manure composting, accounted for the decrease of enzymatic activities whether after peak temperature or at the end of fermentation. Enzymatic activities were stable or even slightly increased during curing. Final composts exhibited different enzymatic activities. As an example, Forster *et al.* (1993) observed that the deshydrogenase activity ranged from 0.83 to 5.89 mg TPF/g DM/day in composts originating from wastewater treatment sludge and household wastes mixed with straw and bark.

Methods to measure enzymatic activities are difficult to set-up and remain specific to every research team. Enzymatic assays also depend on pH and temperature. Hence, it would be rather difficult to use it as stability indicator.

16.4.4 Respirometric activity

By definition respirometry consists in measuring the uptake of O_2 and the release of CO_2, both processes being the result of microorganisms respiration during the degradation of OM. Respirometric activity is generally expressed as mg O_2 consumed or mg CO_2 produced per g DM per time unit.

Respirometric activity is usually measured through the incubation of compost sample for a few days at temperature in the range of 20–37°C (Germon *et al.* 1980; Iannotti *et al.* 1994; Rajbanshi *et al.* 1998). Incubated compost amount vary from 5 g (Rajbanshi *et al.* 1998) to 1 kg (Germon *et al.* 1980). Moisture is adjusted from 50% (Iannotti *et al.* 1994) to 80% of the water retention capacity

(Germon *et al.* 1980). Respirometric activity measurements are based on quantification of whether oxygen is consumed or carbon dioxide is produced during incubation. Carbon dioxide production may be quantified through carbon dioxide trapping in an alkaline solution (NaOH), then back titration with acid of the trapping solution. Oxygen consumption is estimated through the measurement of pressure differential following carbon dioxide trapping or by oxygen partial pressure measurement at the end of incubation. The method proposed by Iannotti *et al.* (1994) is a bit different. First, Iannotti *et al.* (1992, 1994) maintains aeration with a continuous flow of air for 16 h then air flow is stopped and oxygen concentration in the gaseous phase is measured with a probe usually used to measure dissolved oxygen concentration. Lasaridi and Stentiford (1998b) and Stentiford (2002), proposed another method, which consists in measuring the dissolved oxygen concentration in aqueous solutions containing compost sample. Specific indicators were also proposed. Binner and Zach (1999) presented the AT4 and AT7 (German Index for respiration activity in 4 or 7 days) that measures the cumulative quantity of consumed oxygen during 96 or 168 h. Adani *et al.* (2001, 2002) proposed the Dynamic Respiration Index (DRI) that is the mean instantaneous oxygen consumption kinetics, measured on a sample during 24 h after that the maximum respirometric activity has been reached. Recently, a new respirometric test has been proposed by Tremier *et al.* (2002). This test enables to continuously measure the respirometric activity of substrates (about 2 kg), put in composting conditions with control of moisture and temperature. This test will lead to quantify biodegradable OM in the substrate.

Usually all the authors account for an increase at the beginning of fermentation then a decrease of both oxygen consumption and carbon dioxide production (Kaiser 1983a, b; Iglésias Jiménez and Perez Garcia 1989; Iannotti *et al.* 1994; Stentiford 2002). In some cases increase was not observed. Garcia *et al.* (1992a) and Hue and Liu (1995) observed that stabilization thresholds were quite the same whatever the composts originated from. Table 16.10 gives different threshold values proposed by the authors to assess stability of OM.

Providing measurement methods settle conditions quite similar to composting conditions, oxygen consumption and carbon dioxide production are directly

Table 16.10 Examples of respirometric threshold to assess OM stability.

Authors	Threshold value on O_2 consumption	Authors	Threshold value on CO_2 consumption
Germon *et al.* (1980) Nicolardot *et al.* (1986)	0.96 mg O_2/g DM/day	Garcia *et al.* (1992a)	18 mg C-CO_2/g TOC
AT7 (Binner and Zach 1999)	11 mg O_2/g DM	Hue and Liu (1995)	2.88 mg CO_2/g DM/day
DRI (Adani *et al.* 2001, 2002)	1000 mg O_2/kg OM/h		

proportional to metabolic activity. Respirometric methods then enable to assess whether composting treatment led to stabilization of OM or not. Yet, the respirometric test would have to be able to identify if the decrease in oxygen consumption or carbon dioxide production is really due to the deterioration of biodegradable OM storage or to a lack or an excess of moisture, too high temperature (Derikx *et al.* 1990), insufficient aeration, etc. Further research on the respirometric activity, its modelling and the standardization of a respirometric method should lead to obtain reliable threshold values assessing whether the OM is stable or not.

16.4.5 Practical tests

16.4.5.1 Self-heating test

When a fermentative material initially at ambient temperature is placed in favourable conditions of humidity and aeration, the breakdown of its OM provides an increase in temperature. The gap between the initial and the maximal temperature observed after a few days is called the self-heating capacity.

The Dewar method is used in the case of composts. The 50% humidified material is placed in a Dewar flask of 2 l volume and 10 cm internal diameter. The flask is closed and a thermocouple is inserted in the compost through the lid. The measure is continuous during 10–15 days. The self-heating capacity is the difference between ambient temperature and the inside maximum temperature, expressed in °C. This method has been proposed and described by Brinton *et al.* (1995) who observed that self-heating capacity increased at the beginning of composting and decreased afterwards. The authors also noticed self-heating duration could be different though the same substrate was tested. Mathur *et al.* (1993a) mentioned compaction could disturb the measure. Brinton *et al.* (1995) suggested to use such a measurement to classify a range of products (Table 16.11).

In fact, an increase of temperature during the self-heating test certainly indicates the presence of biodegradable OM in the substrate, i.e. the substrate is unstable. However, the lack of self-heating can be caused by environmental parameters such as moisture and therefore it should be interpreted with much caution if used as stability indicator.

Table 16.11 Stability level based on self-heating capacity of composts.

Self-heating capacity (°C)	Stability level	Description of the product
0–10	5	Very stable and mature compost
10–20	4	Mid stable in maturation compost
20–30	3	Still in fermentation active compost
30–40	2	Fresh compost very active and beginning of fermentation
40–50 and above	1	Raw substrates

16.4.5.2 Solvita test

The Solvita test is based on the evaluation of the quantity of CO_2 and ammonia emitted by an organic substrate placed in a closed flask during 4 h. Two hundred and fifty ml jars are half filled with the organic sample to be tested. Two paddles of colorimetric indicators (one for CO_2 and one for NH_3) are placed above the sample. The jars are kept, hermetically closed, at ambient temperature for 4 h. Paddles colours are then read and reported on a colorimetric scale (1–8 for decreasing concentration of CO_2 and 1–5 for decreasing concentration of NH_3). Crossing of the two scales gives the stability level of the product, ranged from 1 to 8 (8 for mature compost and 1 for raw material).

CO_2 and partly ammonia emissions are representative of the microbial activity within the substrate (see Sections 16.4.4 and 16.3.3.1(d)). In this way, the Solvita test is an interesting indicator coupling information about OM degradation. But, gases emissions are largely influenced by the sample preparation: moisture, compaction, etc. Moreover, colour reading on the paddles may be a large variation source between several manipulators. Lastly, depending on the initial carbon and nitrogen contents in the raw substrate, it would be very difficult to compare results obtained for substrates from different origins. Hence, Solvita is an easy test but not really reliable as stability indicator.

16.5 BIOCHEMICAL INDICATORS

An other type of indicators are biochemical indicators which measurement is based on a series of extractions of composting material with organic or aqueous acidic solvents supposed to extract particular fractions: lipids, carbohydrates, aqueous soluble fraction, cellulose, hemicellulose, lignin and finally humic substances. The extract or the solid residue containing the required fraction is then characterized through only weight or carbon content measurement. Some authors developed indexes calculated on the basis of biochemical composition and which values assess stabilization level.

16.5.1 Volatile fatty acids in aqueous extracts

Volatile fatty acids (VFA) are carboxylic acids (R—COOH). Those are often synthesized during fermentation processes and particularly under anaerobic conditions. Concentrations in formic and acetic acids may be high in raw substrates partly due to their production during storage.

VFA are usually analysed in aqueous extracts with the help of ionic or gaseous phase chromatography. Results are expressed in mg VFA/g DM.

VFA concentrations increase at the beginning of fermentation then continuously decrease (Saviozzi *et al.* 1987). At the end of curing, VFA concentrations should be very low (Chanyasak *et al.* 1980; Saviozzi *et al.* 1987; Baziramakenga and Simard 1998). VFA have toxic effects on plant growth. De Vleeshauwer *et al.*

(1981, cited in Anid 1982) proposed to consider that OM is stable when acetic acid cannot be detected in aqueous extracts anymore.

Even if a correlation between the VFA content and phytotoxicity was demonstrated, VFA content alone cannot be used as OM stability indicator.

16.5.2 Carbohydrates

16.5.2.1 Soluble carbohydrates

Soluble carbohydrates in the aqueous fraction are easily biodegradable by microorganisms. Morel *et al.* (1979, 1986) practised extraction of 3 g compost with 50 ml water for 24 h at 80°C. Addition of phenol and sulphuric acid was responsible for a specific reaction and spectrophotometric absorption at 490 nm allowed the quantification of carbohydrates. Garcia *et al.* (1992a) used Brink's method based on reaction with anthrone followed by absorption measurement at 625 nm. Results are expressed as g/per g DM or g/g TOC.

Morel *et al.* (1986) and Garcia *et al.* (1992a) mentioned that soluble carbohydrates decreased all along the composting treatment and particularly during the fermentation phase. That result is in agreement with the fact that soluble carbohydrates are the most easily biodegradable fractions, which are quickly consumed at the beginning of fermentation. These authors also observed biodegradation was closely correlated to compost age, TOC concentration and carbohydrate concentration. Morel *et al.* (1979) proposed a biodegradation index:

$$\mathrm{BI} = 3.166 - 0.011\,\mathrm{A} + 0.059\,\mathrm{TOC} + 0.832\,\ln \mathrm{Glu}_{sol}$$

with:

BI = biodegradation index, expressed as mineralization rate of carbon following incubation for 21 days of compost mixed with soil,
A = compost age (day),
TOC = total organic carbon concentration (g/100 g DM),
Glu_{sol} = soluble carbohydrates concentration (g/100 g DM).

Table 16.12 gives threshold values of BI proposed by Nicolardot *et al.* (1986) and Garcia *et al.* (1992a).

Taking into account that the test considers only soluble carbohydrates and does not care about hydrolysable fractions, it cannot be proposed as a stabilization indicator. Indeed, hydrolysable fractions of carbohydrates are still a biodegradable part of OM. Then by considering only the soluble part, the OM stability is overestimated.

Table 16.12 Threshold values for the BI based on soluble carbohydrates measurement.

Authors	Threshold value on soluble carbohydrates	Biodegradation index
Garcia *et al.* (1992)	<0.1%	<2
Nicolardot *et al.* (1986)	–	<2.4

16.5.2.2 Hydrolysable carbohydrates

As defined, hydrolysable carbohydrates are carbohydrates in which consumption by microorganisms occurs after their hydrolysis.

After extraction in aqueous solution (see Section 16.5.2.1), Morel *et al.* (1979) practised extraction of carbohydrates contained in solid residue. The extraction was achieved with sulphuric acid (H_2SO_4, 3N) at 80°C for 24 h. After centrifugation of the suspension, addition of phenol (80%) is responsible for a specific reaction allowing the quantification of carbohydrates by measurement of absorption at 490 nm.

The hydrolysable carbohydrates content decreases during composting treatment (Morel *et al.* 1979, 1986). There was not any threshold value proposed for that parameter.

16.5.3 Lipidic fraction

Lipidic molecules are insoluble in water but soluble in organic solvent. Dinel *et al.* (1996a, b) practised extraction with first diethylether (DEE) then chloroform ($CHCl_3$). Indeed, DEE enabled extraction of short molecules supposed to be biodegradable whereas $CHCl_3$ was used to extract longer molecules supposed non-biodegradable. So concentrations of lipids extracted with DEE decreased quickly whereas lipids extracted with $CHCl_3$ decreased much more slowly. Dinel *et al.* (1996a, b) proposed to consider that OM was stable when [lipids_{CHCl_3}/total lipids] > 2.5 and [$\text{lipids}_{DEE}/\text{lipids}_{CHCl_3}$] < 2.5.

This approach should be interesting considering materials with a high lipid content such as some agro-food industry wastes or wastewater treatment sludge.

16.5.4 Van Soest's fractionation and BSI and BCM indexes

Van Soest and Wine's (1967) fractionation method were applied by Linères and Djakovitch (1993) and Robin (1997) to characterize fresh organic wastes and composts biodegradability.

Robin's method (1997) consists in extraction in series with inorganic and organic solvents to obtain aqueous soluble fraction (SOL), hemicellulose (HEM), cellulose (CELL), lignine (LIG) and MM. The method developed by Linères and Djakovitch (1993) does not take into account MM and the cellulose fraction is quantified with Wende's method (NF V03-040 1993). In both cases the amount of each fraction is obtained by differential weighting between the different dried solid residues followed by their OM content measurement. Mineralization of numerous organic wastes was performed. Mineralization consisted in an incubation of raw organic wastes (wastewater treatment sludge, household waste, green waste and manure) or composts (organic amendments and commercial amendments) mixed with soils followed by measurement of the remaining carbon. Incubation was practised for 6 months by Linères and Djakovitch (1993) whereas it lasted 40 days by Robin (1997). Mineralization rates were correlated to biochemical composition

(SOL, HEM, CELL, LIG and MM). Then, the authors were able to propose relations to calculate stability level defined as Biochemical Stability Index, BSI (Linères and Djakovitch 1993) and Biochemical Characterization of Material, BCM (Robin 1997):

$$\mathrm{BSI} = a - b\,\mathrm{SOL} - c\,\mathrm{HEM} - d\,\mathrm{CELL} + e\,\mathrm{LIG}$$

$$\mathrm{BCM} = 0.3221\,\%\mathrm{SOL} - 0.7155\,\%\mathrm{HEM} + 0.617\,\%\mathrm{CELL} + 1.8919\,\%\mathrm{LIG} + 0.0271\,\%\mathrm{MM}.$$

Table 16.13 gives BSI and BCM values measured on different fresh organic wastes and composts. Robin (1997) considers that the stable OM content is low when BCM is inferior to 34 whereas the stable OM content is high when BCM is in the range 42–100. Composting treatment is responsible for the decrease of soluble, hemicellulolytic and cellulolytic fractions whereas lignin, remains stable in weight and increases when expressed in concentration (Linères and Djakovitch, 1993). On the contrary, Robin (1997) claims that cellulose contributes to increase the remaining stable OM content too.

Linères and Djakovitch (1993) and Robin (1997) showed BSI and BCM were not correlated to the C/N ratio and claim that the BSI and CBM allowed to compare the behaviour in soils of different organic materials. Thus, these indicators enable to select composted solid-waste amendment as a function of a specific agronomic use.

These methods are contested because they are based on the hypothesis that (i) a solvent allows extraction of a single molecular family, (ii) biochemical families do not interact to each other in soils and (iii) lipidic fraction is not considered (Chaussod 1999a, b). The published results are still insufficient to assess whether that type of approach could lead to obtain an indicator allowing prediction of OM amendment behaviour in soils.

Table 16.13 Example of BSI and BCM values of different organic substrates.

Organic substrate	BCM value	Organic substrate	BSI value
Wastewater treatment sludge	0–40	Droppings	5–10
Commercial amendments	0–100	Straw	10–15
Commercial fertilizer	20–50	Green wastes	#15
Composts	20–80	Fresh wood	#20
Manure	25–80	Fresh manure	#25
Crops residues	25–30	Household composts	30–40
Household waste compost	35	Composted wood	50–60
Organic wastes mixtures compost	65–75	Green wastes compost	#70
		Pine bark	80–90
		Peat	#90

16.5.5 Adani's fractionation

Adani *et al.* (1995) proposed a fractionation leading to the extraction of the lipidic fraction, aqueous soluble fraction, hemicellulose, cellulose and lignin. Blanco and Almendros (1997) proposed some modifications to extract lignin.

Adani *et al.* (1995) applying this method observed that the lignin content increased during composting whereas the other fractions decreased. These results were confirmed by Genevini *et al.* (1997).

16.5.6 Humic substances

Many researchers performed humic substances extraction. Humic substances are extracted from the solid fraction by means of an alkaline solvent: NaOH 0.1 M (Morel *et al.* 1986; Inbar *et al.* 1990; Hue and Liu 1995; Bernal *et al.* 1998), $Na_2P_2O_7$ 0.5 M (Roletto *et al.* 1985), $K_4P_2O_7$ 0.1 M (Prudent *et al.* 1995), NaOH 0.1 M + $Na_2P_2O_7$ 0.1 M (De Nobili and Petrussi 1988; Ciavatta *et al.* 1993; Adani *et al.* 1995). Considering an alkaline solvent could lead to overestimate humic substances by solubilizing non-humic molecules, Adani *et al.* (1995) practised humic substances extraction after extraction of the aqueous soluble, lipidic, hemicellulose and cellulose fractions. Humic and fulvic acids are then separated by acid precipitation, usually with H_2SO_4, of humic acids. The filtrate contains fulvic acids and non-humic substances. Adsorption of fulvic acids on polyvinylpyrrollidone (PVD) column followed by elution (NaOH 0.5 N) enables collection of fulvic acids (Sequi *et al.* 1986). The carbon content of each extracted fraction is then quantified: C_{ALC}, carbon in alkaline solvent, C_{HA}, carbon in humic acids, C_{FA}, carbon in fulvic acids, C_{NH}, carbon in non-humic substances. The results are expressed as g/g TOC or g/g DM.

The increase, during the fermentation step of composting, of carbon contained in alkaline solvent (g C/g TOC) was interpreted by an increase in humic acids to the detriment of fulvic acids (Morel *et al.* 1986; Riffaldi *et al.* 1986; Inbar *et al.* 1990). However, the amount of carbon in alkaline solvent expressed as g/g DM or g/g MM was observed as decreasing (Riffaldi *et al.* 1986; Ciavatta *et al.* 1993; Hue and Liu 1995). The following interpretation was then proposed: the amount of humic acids remains stable whereas the amount of fulvic acids decreases during the first step of the composting treatment. During the curing phase, the humic and fulvic acids contents become stable.

Table 16.14 gives indicators defined by the authors to get threshold values assessing the stability of OM.

Humic substances are stable molecules partly responsible for amendment properties of soils. Their presence in compost is actually a clue of the evolution stage of the OM. Nevertheless, humification processes during composting treatment remain still badly known. As a consequence, one cannot conclude that easily biodegradable OM is totally consumed because of the presence of humic substances. Moreover, methods to evaluate the humic fractions contents are not reliable. Simple alkaline extraction methods are not selective enough (Adani *et al.*

Table 16.14 Threshold value for several stability indexes based on the humic substances content.

Authors	Indicator	Threshold value
Roletto *et al.* (1985)	C_{ALC} (%TOC)	>7% (manure and green wastes)
Iglésias Jiménez and Perez Garcia (1989)	C_{ALC} (%TOC)	>5% (household wastes)
Roletto *et al.* (1985)	C_{ALC} (%DM)	>3% (manure and green wastes)
Roletto *et al.* (1985)	C_{HA} (%TOC)	>3.5% (manure and green wastes)
Ciavatta *et al.* (1993)	$HI = C_{NH}/(C_{HA} + C_{FA})$	>1 (non-mature household composts) <0.5 (mature household composts)
Adani *et al.* (1995)	$SI = C_{HA\ Adani}{}^{*}/C_{HA}$	>0.6 to assess stability >0.8 to assess maturity

* $C_{HA\ Adani}$: measured by Adani method.

1995; Veeken *et al.* 2000). Functional characterization of humic substances by spectroscopic methods (see Section 16.6.1) are too sophisticated and expensive to be currently used (Epstein 1997). As a consequence humification indexes should not be considered as really adequate stability indicators.

16.6 SPECTROSCOPIC INDICATORS

Spectroscopic methods are applied either on the global substrate or on aqueous extract. These techniques linked with several fractionation methods of the OM do not supply stability indicators. However, they translate the OM stabilization by showing the degradation of small organic molecules and the development of aromatic structures.

16.6.1 ^{13}C NMR on the global solid and of humic substances

Nuclear magnetic resonance spectroscopy of carbon (^{13}C NMR) is devoted to describe the structural environment of carbon atoms in terms of specific regions of the spectrum where several chemical groups can be distinguished (Table 16.15).

Inbar *et al.* (1989, 1991) used this method to follow the evolution of the OM of cattle manure and compost from grape pulp. Their observations help to describe two peaks representatives, one at 75 ppm of the carbohydrates and the other at 108 ppm of the polysaccharides. Moreover, the area of the peak at 58 ppm representative of lignin increases during the process. These authors calculated an aromaticity index, which used to increase with increasing age of the compost. ^{13}C NMR was also applied to humic substances extracted from these composts (Inbar *et al.* 1990b,

Table 16.15 NMR regions of compost and humic substances spectrum.

Limit values of chemical displacement in ppm	Structure of the compounds in this region of the spectrum	Specific peaks of this region of spectrum	
0–112	Aliphatic	58	Methoxy groups and lignin
		75	Carbohydrates or aliphatic alcohols
		107	Polysaccharides
112–163	Aromatic	133	Aromatic
		155	Phenols and aromatic amines
163–190	Esters and carboxyl groups	175	Carboxyl groups
190–215	Aldehydes and ketones structures		This region specific of aldehydes and ketones has not specific identified peaks

1992). Humic acids spectra exhibit the same peaks as composts: at 58 ppm for cattle manure and at 75 ppm for grape pulp.

Actually NMR is mostly adapted to laboratory studies interested in the evolution of composts. It should be developed to offer more quantitative information.

16.6.2 Infrared spectrum on global solid and humic substances

Infrared analyses is a spectrometric method, which enables to obtain the functional pattern of a compound. The spectra are taken from the raw substrates or from the composted material. Several methods were proposed by Inbar *et al.* (1989, 1990b, 1991, 1992), Chefetz *et al.* (1996) and Tseng *et al.* (1996).

The major peaks observed on compost samples and humic substances are presented in Table 16.16.

The spectra from raw materials are characterized by numerous peaks within the region from 1720 to 1000 cm^{-1}. Among these peaks, the polysaccharides are the most intense. Such peaks decrease during the composting process. Concurrently, the aromatic peak (1655 cm^{-1}) increase with the age of compost. Briefly during the composting, an increase in the carboxylic and aromatic groups is observed concurrently to the decrease of aliphatic and polysaccharides groups. The spectra of composts are also characterized by a small number of peaks in comparison to the raw material's spectra.

Infrared determination offers very accurate information on the nature of the compounds present in the compost. It is a complementary quantitative information relatively to NMR and chemical analysis. Nevertheless, infrared determination is not accurate enough to be used as stability indicator *per se*.

Table 16.16 Functional groups determined on infrared spectrum of raw and composted materials.

Wave lengths (cm^{-1})	Link of functional group
3300–3400	Hydrogen link of the HO groups
3090	C—H aromatic
2930	C—H aliphatic
1720	C═O with COOH
1655	C═C aromatic; COO–; H link of C═O groups
1520	C═C of aromatic cycles
1450	CH_2; CH_3
1425	COO^-; CH_2
1385	COO^-; CH_3
1240–1200	Aromatic C; C—O
1150–1160	Aliphatic CH_2
1100–1040	Polysaccharides C—O

16.6.3 Ultraviolet–visible absorption of aqueous extracts

Ultraviolet (UV)–visible spectroscopy is practised in the range 200–800 nm. Aromatic molecules concentrations increase when absorption lengths increase. Usually tested absorption lengths were 280, 465 and 665 nm (Mathur *et al.* 1993b; Blanco and Almendros 1997; Lasaridi and Stentiford 1998b). Whatever the absorption length, optical density (OD) usually increases at the beginning of composting, then slows down after the thermophilic phase. Mathur *et al.* (1993) and Blanco and Almendros (1997) defined the absorption ratio between 465 and 665 nm, E4/E6 = OD at 465/OD at 665 nm. For Blanco and Almendros (1997), the unexpected increase of E4/E6 means degradation occurs in the biofilm whereas humification does not.

UV spectroscopy was also practised on alkaline extracts. Morel (1982) applied extraction with $Na_4P_2O_7$ 0.1 M. Morel (1982), Anid (1982) and Riffaldi *et al.* (1986) measured absorption at 350, 450, 550 and 625 nm. Whatever the absorption length, the OD increased at the beginning of the fermentation and then stabilized. Prudent *et al.* (1995) used $K_4P_2O_7$ (0.1 M) as alkaline solvent. Then they proposed to interpret the UV spectrum of the analysed compost by comparing it to different spectra obtained with characteristic composts (with ages from a few days to a few months) or acid humic fractions originating from these composts.

16.6.4 Gel permeation fractionation of aqueous extracts

Gel permeation chromatography was used to fractionate dissolved OM as a function of molecular sizes. The method leads to a chromatogram on which each signal corresponds to a liquid fraction collected at the exit of the permeation column.

Yoshida and Kubota (1979) used Sephadex columns (G10, G15, G25). Elution was performed with water (Yoshida and Kubota 1979) or NH_4OH 0.1 M (Chanyasak *et al.* 1980). The smaller are the molecules, the longer is the elution time. Measurement of absorption at 280 nm is practised to characterize the collected fractions.

The best results were obtained with G15 Sephadex column getting fractionation in the molecular weight range 0–1500 Da. The authors showed low molecular weight compounds like peptides, amino acids and glucides decreased during composting treatment whereas polymers as polysaccharides and proteins increased. These evolutions were responsible for a lower resolution of spectra originating from raw wastes compared to an easier resolution of spectra from composts.

Prudent *et al.* (1995) fractionated extracts obtained with $K_4P_2O_7$ 0.1 M solvent. Sephadex columns G75 and G25 allowing fractionation in the ranges 3000–70,000 and 1000–6000 Da were used. Absorption of the collected fractions were measured between 200 and 350 nm.

Prudent *et al.* (1995) applied this method to household wastes submitted to a fermentation in a rotative reactor for 3 days followed by curing in windrows for 12 months. Fractionation of alkaline extracts was practised to understand the synthesis of humic substances. The study showed that polymerization occurred during the first step of curing and was followed by mineralization of OM.

16.6.5 Isoelectric focusing on aqueous extracts

Ciavatta *et al.* (1993) and Govi *et al.* (1995) applied isoelectric focusing (IEF) on alkaline extracts supposed to contain humic substances. Extracts originated from household waste composts and mixtures of straw with solid residue of pig slurry treatment.

The method is based on the measurement of molecules' affinity with H^+ or HO^- ions. Alkaline extract is placed in a gel (containing an ampholyte) submitted for a few hours to an electric tension (1200 V). Molecules displace as function of their affinity with pH. Coomasie blue reagent is added to the gel in order to colour displacements followed by measurement of absorption at 633 nm. Govi *et al.* (1995) dialysed alkaline extract before applying IEF. The method remains qualitative.

16.6.6 Circular chromatographic test

The method was proposed by Hertelendy (1974). Alkaline extract is laid down at the centre of a circular filter moistened with barium nitrate ($Ba(NO_3)_2$). Elution of the alkaline extract depends on the molecules' size. Hertelendy (1974) observed that chromatograms of fresh organic wastes or sampled at the beginning of the composting treatment were dark coloured at the periphery whereas the centre remained almost white. All along the composting treatment, the centre gets a dark colouration whereas some peaks appear around the centre. The chromatogram centre of a mature compost is almost black and is surrounded with "clear" peaks.

16.7 THERMAL FRACTIONATION METHODS

16.7.1 Thermogravimetry

The thermogravimetric method proposed by Blanco and Almendros (1994) consists in heating 10 mg powder (200 μm) of dried compost from 20 to 580°C (20°C/mn) and monitoring continuously mass losses. Compost was previously extracted with water to avoid MM interferences. Mass losses as function of time lead to a chromatogram which peaks are associated to the degradation of specific molecules. For example, peaks around 300°C were interpreted to correspond to the degradation of carbohydrates and low condensed lignin whereas high condensed lignin and aromatic polymers degraded from 350 to 550°C (Blanco and Almendros 1994, 1997). Only few results obtained with that method have been published so far.

16.7.2 Pyrolysis coupled with chromatography or mass spectrometry

Pyrolysis is a thermal fractionation method. Organic fragments or molecules released by this method are then analysed through gas chromatography or mass spectrometry.

Garcia *et al.* (1992b) and Van Bochove *et al.* (1996) used pyrolysis followed by chromatography and mass spectrometry, respectively, to study the OM structure evolution during composting. They observed an increase of fragments number coming from aromatic compounds. They achieved to distinguish three steps during the process: cellulose depolymerization during the thermophilic phase, then monomers consumption and lastly increase in lignin content. These methods are purely qualitative and rather difficult to interpret.

16.8 PHYTOTOXIC INDICATORS

This type of indicators relies on tests, which underline the effects of a compost sample or of an extract of compost on crop's growth. These tests allow measuring the immediate toxicity of the compost thus inhibiting the germination, or the latent toxicity by inhibiting the root growth. A recent study also proposes to study the inhibition of specific carriers in the chloroplasts extracted from plants. Such indicators are devoted to assess the agronomic relevance of the wastes being transformed.

16.8.1 Seed growing tests

Seed growing tests consist in measuring germination and roots lengths following 24 h incubation at 27–28°C of seeds placed on a paper filter previously moistened with aqueous extract of compost. Cress seeds were used by Zucconi *et al.* (1981)

and Anid (1982) whereas Rajbanshi *et al.* (1998) used sorgho seeds. Differences in methods also exist concerning seeds number, aqueous extract volume and incubation time. Germination and roots length are compared to a blank obtained with seeds in distilled water.

Zucconni *et al.* (1981) defined germination indice, GI:

$$\text{GI} = (\text{number of germinated seeds/number of seeds}) \times (\text{total roots length/blank roots length})$$

Germination tests were applied on aqueous extracts from green wastes, household wastes and wastewater treatment sludge. Germination rate usually increases during fermentation to get a maximal value in the curing phase (Baca *et al.* 1990; Van de Kerkhove 1990). On the contrary, roots length may decrease during curing (Phillip 1996). The salts concentrations limit length growth. Results obtained by Zucconi *et al.* (1981) and Rajbanshi *et al.* (1998) assess that OM can be considered stable when the GI gets superior to 50%.

Considering the method is rather easy to use and the results are obtained after 1 or 2 days incubation, seed growing tests are particularly interesting for compost producers or users. Yet, the influence of dilution rates of aqueous extracts and the prediction of culture hazards require further studies.

16.8.2 Growing tests

Growing tests are mostly performed on mixtures of compost and soil and more seldom on compost alone. Table 16.17 gives some examples of existing protocols applied to different crops. Growing tests consist in determining the DM production on the leaves, shoots, roots and global DM.

Usually, it is observed that young composts inhibit crops' growth, while this inhibition decreases with increasing age of composts (Benistant 1978; Inbar *et al.* 1993; Chefetz *et al.* 1996). Baca *et al.* (1990) and Blanco and Almendros (1997) observed a decrease in the global DM if compost is too old. Depressive effects of young composts may be explained by consumption of oxygen and nitrogen by still biodegradable OM. Excess of salts could be responsible for the decrease of growth in the case of old compost.

Benistant (1978) also studied the dilution rate of compost in soil. Plant tests led Iannotti *et al.* (1994) to remark cucumber was poorly sensitive for growing tests and radish growth was inhibited. Juste *et al.* (1980) explained radish growth inhibition by a too high sensitivity to poor lighting.

Research on growing tests in order to describe the agronomic quality of composts should be carried on.

16.8.3 Mushrooms growing tests

Growing tests of *Verticilium cinnabarium and Chaetomium gracilis* inoculated in a medium containing nutrients and compost were applied by Kaiser (1983a and 1983b), Mathur *et al.* (1993a), Anid (1982) and Mustin (1987).

Table 16.17 Examples of protocols for growing tests for composts.

Plant test	Culture medium	Sowing	Protocol of growing	Authors
Maize	Pure compost or mix of compost (50% humidity) and peat with 90 and 75% of compost	5 seeds	28 days, 22°C (day), 16°C (night)	Juste *et al.* (1980)
Bean	Pure compost or mix of compost (50% humidity) and peat with 90 and 75% of compost	5 seeds	28 days, 22°C (day), 16°C (night)	Juste *et al.* (1980)
Sunflower (*Helianthus annus*)	Mix of compost and calcareous	5 seeds pre-germinated	21 days glasshouse	Baca *et al.* (1990)
Cucumber	Mix of compost and perlit/peat (15/30/50, V/V)	8 seeds	After 10 days reduce the number of plants to 4. 21 days in glasshouse 20–30°C.	Iannotti *et al.* (1994)
	Mix of compost and perlit (1/1)	10 seeds	After 7 days reduce the number of plants to 3. One month in glasshouse.	Chefetz *et al.* (1996)
Ryegrass	Mix of compost and perlit and peat (15/30/55, V/V)	19 seeds	21 days glasshouse 23–28°C.	Iannoti *et al.* (1994)
	Mix of compost and soil (20 to 40 g/kg	29 seeds	Irrigation twice a day. 25°C, 65% humidity.	Blanco and Almendros (1997)

16.8.4 Inhibition of photosynthesis processes

The method was adjusted by Helfrich *et al.* (1998). Thylakoïdes membranes are, in chloroplasts, the seat of some photosynthesis processes. Mixing thylakoïdes membranes with compost aqueous extracts may inhibit photosynthesis and could be responsible for phytotoxic effects. Inhibition observed when composts were young was interpreted as the presence of VFA in aqueous extracts.

16.9 CONCLUSION

Many papers in the composting literature deal with the determination of stability indicators for recycled organic wastes and composts. These indicators are measured directly on composting site or require compost sampling.

Material temperature and oxygen consumption enable real controls of composting processes. Providing aeration, moisture and temperature do not limit microbiological activity, the decrease of both material temperature and oxygen consumption are really due to the degradation of biodegradable OM amount, which means stabilization of compost.

The OM, total nitrogen and CEC measurements also help predict compost amendment effect.

Some other indicators are measured on aqueous extracts: pH, TC, COD, BOD, nitrogen forms, C/N ratios. Yet, considering that biofilm composition is the result of numerous phenomena which may have opposite effects, the definition of threshold values for stabilization indicators should not be envisaged.

Providing the controlled conditions of aeration, temperature, moisture, incubation do not limit microbiological activity, the respirometric activity measurement can rigorously account for biodegradability or stability of fresh organic wastes or composts. Yet, only a relevant choice of the respirometric indicator will lead to the rigorous assessment of stability.

The self-heating test appears useful for composting sites managers. However, the temperature increase should be correlated to biodegradable OM content to improve the reliability of this test.

Researches on (1) the understanding of humification processes during composting treatment; (2) amendment effects of humic substances and (3) on the improvement of humic substances extraction methods are necessary to go on with them as stability indicators.

Finally results from phytotoxic tests should be correlated to biodegradable OM content to envisage their use as stability indicators. However, phytotoxic test may provide interesting information on crops growth following compost spreading. That approach should be further studied in order to obtain a good prediction of composts impacts on soil and crops following the spreading.

REFERENCES

Adani, F., Genevini, P.L. and Tambone, F. (1995) A new index of organic matter stability. *Compost Sci. Util.* **3**(2), 25–37.

Adani, F., Lozzi, P. and Genevini, P. (2001) Determination of biological stability by oxygen uptake on municipal solid waste and derived products. *Compost Sci. Util.* **9**(2), 163–178.

Adani, F., Ubbiali, C., Tambone, F., Centemero, M. and Genevini, P.L. (2002) Static and dynamic respirometric indexes – Italian research and studies. In: *The Biological Treatment of Biodegradable Waste – Technical Aspects.* The Environment DG and the Joint Research Centre of the European Commission (eds.), Workshop Conference, 8–10 April 2002, Brussels.

Anid, P.J. (1982) Caractérisation de l'état de maturation du compost. *Ann. Gembloux* **88**, 119–131.

Avnimelech, Y., Bruner, M., Ezrony, I., Sela, R. and Kochba, M. (1996) Stability indexes for municipal solid waste compost. *Compost Sci. Util.* **4**(2), 13–20.

Ayuso, M., Hernandez, T., Garcia, C. and Pascual, J.A. (1996) Biochemical and chemical-structural characterization of different organic materials used as manures. *Biores. Technol.* **57**, 201–207.

Baca, M.T., Delgado, I.C., Sanchez-Raya, A.J. and Gallardo-Lara, F. (1990) Comparative use of cress seed germination and physiological parameters of *Helianthus annuus* L. to assess compost maturation. *Biol. Wastes* **33**, 251–261.

Baziramakenga, R. and Simard, R.R. (1998) Low molecular weight aliphatic acid contents of composted manures. *J. Environ. Qual.* **27**, 557–561.

Benistant, D. (1978) Caractérisation de la maturité des composts d'ordures ménagères. Chaire d'agronomie, Ministère de l'agriculture, Ecole Nationale d'Ingénieurs des Travaux Agricoles – Dijon, Laboratoire de microbiologie des sols, INRA, Dijon.

Benitez, E., Nogales, R., Elvira, C., Masciandaro, G. and Ceccanti, B. (1999) Enzyme activities as indicators of the stabilization of sewage sludges composting with *Eisenia Foetida*. *Biores. Technol.* **67**, 297–303.

Bernal, M.P., Navarro, A.F., Roig, A., Segarra, J. and Garcia, D. (1996) Carbon and nitrogen transformation during composting of sweet sorghum bagasse. *Biol. Fertil. Soil.* **22**(1–2), 141–148.

Bernal, M.P., Paredes, C., Sanchez-Monedero, M.A. and Cegarra, J. (1998) Maturity and stability parameters of composts prepared with a wide range of organic wastes. *Biores. Technol.* **63**, 91–99.

Binner, E. and Zach, A. (1999) Laboratory tests describing the biological reactivity of pretreated residual wastes. In: *Organic Recovery and Biological Treatment – ORBIT 99* (eds. W. Bidlingmaier, M. deBertoldi, L. Diaz and E. Papadimitriou), pp. 255–261. Rhombos-Verlag, Weimar.

Blanco, M.J. and Almendros, G. (1994) Maturity assessment of wheat straw composts by thermogravimetric analysis. *J. Agric. Food Chem.* **42**, 2454–2459.

Blanco, M.J. and Almendros, G. (1997) Chemical tranformation, phytotoxicity and nutrient availability in progressive composting stages of wheat straw. *Plant Soil* **196**, 15–25.

Brinton Jr. W.F., Evans, E., Droffner, M.L. and Brinton, R.B. (1995) Standardized test for evaluation of compost self-heating. *BioCycle.* November, 64–69.

Chanyasak, V., Yoshida, T. and Kubota, H. (1980) Chemical components in gel chromatographic fractionation of water extract from sewage sludge compost. *J. Ferment. Technol.* **58**(6), 533–539.

Chanyasak, V. and Kubota, H. (1981) Carbon/nitrogen ratio in water extract as measure of composting degradation. *J. Ferment. Technol.* **59**(3), 215–219.

Charonnat, C., Deportes, I., Feix, I. and Merillot, J.M. (2001) *Approche de la qualité des composts de déchets en France*. ADEME Editions, Paris.

Chaussod, R. (1999a) "Caractériser" la matière organique: pourquoi et comment? Commentaires sur I.S.B., C.B.M. et K1, première partie. *Echo MO* **15**, 3–4.

Chaussod, R. (1999b) "Caractériser" la matière organique: pourquoi et comment? Commentaires sur I.S.B., C.B.M. et K1, seconde partie. *Echo MO* **16**, 3–4.

Chefetz, B., Hatcher, P.G., Hadar, Y. and Chen, Y. (1996) Chemical and biological characterization of organic matter during composting of municipal solid waste. *J. Environ. Qual.* **25**, 776–785.

Ciavatta, C., Govi, M., Pasotti, L. and Sequi, P. (1993) Changes in organic matter during stabilization of compost from municipal solid wastes. *Biores. Technol.* **43**, 141–145.

Davis, C.L., Hinch, S.A., Donkin, C.J. and Germishuizen, P.J. (1992) Changes in microbial population numbers during the composting of pine bark. *Biores. Technol.* **39**, 85–92.

De Bertoldi, M., Vallini, G. and Pera, A. (1983) The biology of composting: a review. *Waste Manage. Res.* **1**, 157–176.

De Nobili, M. and Petrussi, F. (1988) Humification Index (HI) as evaluation of the stabilization degree during composting. *J. Ferment. Technol.* **66**(5), 577–588.

Derikx, P.J.L., Op Den Camp, H.J.M., Van der Drift, C., Van Griesven, L.J.L.D. and Vogels, G.D. (1990) Biomass and biological activity during the production of compost used as a substrate in Mushroom cultivation. *Appl. Environ. Microbiol.* **56**(10), 3029–3034.

Dinel, H., Schnitzer, M. and Dumontet, S. (1996a) Compost maturity: chemical characteristics of extractable lipids. *Compost Sci. Util.* **4**(1), 16–25.

Dinel, H., Schnitzer, M. and Dumontet, S. (1996b) Compost maturity: extractable lipids as indicators of organic matter stability. *Compost Sci. Util.* **4**(2), 6–12.

Epstein, E. (1997) *The Science of Composting.* CRC Press LLC, Baca Raton, FL.

Fang, M., Wong, J.W.C., Li, G.X. and Wong, M.H. (1998) Changes in biological parameters during co-composting of sewage sludge and coal ash residues. *Biores. Technol.* **64**, 55–61.

Forster, J.C., Zech, W. and Wûrdinger, E. (1993) Comparison of chemical and microbiological methods for the characterization of the maturity of composts from contrasting sources. *Biol. Fertil. Soil.* **16**, 93–99.

Fraser, B.S. and Lau, A.K. (2000) The effect of process control strategies on composting rate and odor emission. *Compost Sci. Util.* **8**(4), 274–292.

Garcia, C., Hernandez, T., Costa, F. and Ayuso, M. (1992a) Evaluation of the maturity of municipal waste compost using simple chemical parameters. *Commun. Soil. Sci. Plant Anal.* **23**(13–14), 1501–1512.

Garcia, C., Hernandez, T., Costa, F., Ceccanti, B. and Calcinai, M. (1992b) A chemical-structural study of organic wastes and their humic acids during the composting by means of pyrolysis-gas chromatography. *Sci. Total Environ.* **119**, 157–168.

Genevini, P., Adani, F. and Villa, C. (1997) Rice hull degradation by co-composting with dairy cattle slurry. *Soil Sci. Plant Nutr.* **43**(1), 135–147.

Germon, J.C., Nicolardot, B. and Catroux, G. (1980) *Mise au point d'un test rapide de détermination de la maturité des compost*, Convention d'étude n° 79–509, Ministère de l'environnement et du Cadre de Vie.

Godden, B., Penninckx, M.J. and Castille, C. (1986) On the use of biological and chemical indexes for determining agricultural compost maturity: extension to the field scale. *Agricul. Wastes* **15**, 169–178.

Govi, M., Ciavatta, C., Sitti, L. and Gessa, C. (1995) Evaluation of the stabilization level of pig organic waste: influence of humic like compounds. *Commun. Soil Sci. Plant Anal.* **26**(3–4), 425–439.

Grebus, M.E., Watson, M.E. and Hoitink, H.A.J. (1994) Biological, chemical and physical properties of composted yard trimming as indicators of maturity and plant disease suppression. *Compost Sci. Util.* **2**, 57–71.

Hamacher, T., Niess, J., Schulze Lammers, P., Diekmann, B. and Boeker, P. (2003) Online measurement of odorous gases close to the odour threshold with a QMB sensor system with an integrated preconcentration unit. *Sensor. Actuator. B.* **95**, 39–45.

Harada, Y. and Inoko, A. (1980) The measurement of the cation exchange capacity of composts for the estimation of the degree of maturity. *Soil Sci. Plant Nutr.* **26**(1), 127–134.

Haug, R.T. (1980) *Compost Engineering: Principles and Practices.* Ann Arbor Science Publishers Inc., MI.

Haug, R.T. (1993) *The Practical Handbook of Compost Engineering.* Lewis Publishers, Boca Raton, FL.

Helfrich, P., Chefetz, B., Hadar, Y., Chen, Y. and Schnabl, H. (1998) A novel method for determining phytotoxicity in composts. *Compost Sci. Util.* **6**(3), 6–13.

Hertelendy, K. (1974) Paper chromatography – a quick method to determine the degree of humification of refuse compost. *IRCWD News* **7**, 1–3.

Horiuchi, J.I., Ebie, K., Tada, K., Kobayashi, M. and Kanno, T. (2003) Simplified method for estimation of microbial activity in compost by ATP analysis. *Biores. Technol.* **86**, 95–98.

Hue, N.V. and Liu, J. (1995) Predicting compost stability. *Compost Sci. Util.* **3**(2), 8–15.

Iannotti Frost, D., Toth, B.L. and Hoitink, H.A.J. (1992) Compost stability. *Biocycle* November, 62–66.
Iannotti, D.A., Grebus, M.E., Toth, B.L., Madden, L.V. and Hoitink, H.A.J. (1994) Oxygen respirometry to assess stability and maturity of composted municipal solid waste. *J. Environ. Qual.* **23**, 1177–1183.
Iglésias Jiménez, E. and Perez Garcia, V. (1989) Evaluation of city refuse compost maturity: a review. *Biol. Wastes* **27**, 115–142.
Inbar, Y., Chen, Y. and Hadar, Y. (1989) Solid-state carbon-13 nuclear magnetic resonance and infrared spectroscopy of composted organic matter. *Soil Sci. Soc. Am. J.* **53**, 1695–1701.
Inbar, Y., Chen, Y., Hadar, Y. and Hoitink, H.A.J. (1990a) New approaches to compost maturity. *Biocycle* **31**(12), 64–69.
Inbar, Y., Chen, Y. and Hadar, Y. (1990b) Humic substances formed during the composting of organic matter. *Soil Sci. Soc. Am. J.* **54**, 1316–1323.
Inbar, Y., Chen, Y. and Hadar, Y. (1991) Carbon-13 CPMAS NMR and FTIR spectroscopic analysis of organic matter transformations during composting of solid wastes from wineries. *Soil Sci.* **152**(4), 272–282.
Inbar, Y., Hadar, Y. and Chen, Y. (1992) Characterization of humic substances formed during the composting of soil wastes from wineries. *Sci. Total Environ.* **113**(4), 35–48.
Inbar, Y., Hadar, Y. and Chen, Y. (1993) Recycling of cattle manure: The composting process and characterization of maturity. *J. Environ. Qual.* **22**, 857–863.
Juste, C., Soldà, P. and Dureau, P. (1980) *Mise au point de tests agronomiques légers permettant de déterminer simultanément la phytotoxicité globale des ordures ménagères et leur degré de maturation,* Convention d'étude n° 77–147, Ministère de l'environnement et du cadre de vie, INRA Bordeaux.
Kaiser, M. (1983a) L'analyse microbiologique des composts (1ère partie). *Compost Informat.* **10** (1er trimestre), 9–13.
Kaiser, M. (1983b) L'analyse microbiologique des composts (2éme partie). *Compost Informat.* **13** (3ème trimestre), 1–5.
Lasaridi, K.E. and Stentiford, E.I. (1998a) Biological parameters for compost stability assessment and process evaluation. *Acta Hort.* **469**, 119–128.
Lasaridi, K.E. and Stentiford, E.I. (1998b) A simple respirometric technique for assessing compost stability. *Water Res.* **32**(12), 3717–3723.
Lax, A., Roig, A. and Costa, F. (1986) A method for determining the cation-exchange capacity of organic materials. *Plant Soil* **94**, 349–355.
Linères, M. and Djakovich, J.L. (1993) *Caractérisation de la stabilité biologique des apports organiques par l'analyse biochimique*. Matières organiques et agriculture, Quatrièmes journées de l'analyse de terre (Gemas). Cinquième forum de la fertilisation raisonnée (Comifer). 16–18 novembre 1993 (eds. Decroux et Ignazi), 159–168.
Lossin, R. (1971) Measurement of the chemical oxygen demand of compost. *Compost Sci.* March–April, 31–32.
Martin, G. and Laffort, P. (1991) *Odeurs et désodorisation dans l'environnement*, Ed. TEC & DOC Lavoisier, Paris.
Mathur, S.P., Owen, G., Dinel, H. and Schnitzer, M. (1993a) Determination of compost biomaturity: I. Literature review. *Biol. Agricul. Horticul.* **10**, 65–85.
Mathur, S.P., Dinel, H., Owen, G., Schnitzer, M. and Dugan, G. (1993b) Determination of compost biomaturity: II. Optical density of water extracts of composts as a reflection of their maturity. *Biol. Agricul. Horticul.* **10**, 87–108.
Morel, J.L., Jacquin, F. and Guckert, A. (1979) *Tests de détermination de la maturité des composts urbains*, Compte rendu de fin de contrat, Conventions n° 75124–77137, Ministère de l'environnement et du Cadre de vie, Ministère des universités, ENSIA Nancy.
Morel, J.L. (1982) L'évolution de la maturité des composts urbains par une méthode colorimétrique. *Compost Informat.* **10**, 4–9.

Morel, J.L., Guckert, A., Nicolardot, B., Benistant, D., Catroux, G. and Germon, J.C. (1986) Etude de l'évolution des carctéristiques physico-chimiques et de la stabilité biologique des ordures ménagères au cours du compostage. *Agronomie* **6**(8), 693–701.

Mustin, M. (1987) *Le compost: gestion de la matière organique.* Ed. François Dubusc, Paris.

Nicolardot, B., Chaussod, R., Morel, J.L., Guckert, A., Benistant, D., Catroux, G. and Germon, J.C. (1986) Appréciation de la maturité des composts urbains en relation avec leurs effets sur la production végétale. *Agronomie* **6**(9), 819–827.

Nicolas, J., Romain, A.C., Wiertz, V., Maternova, J. and André, Ph. (2000) Using the classification model of an electronic nose to assign unknown malodours to environmental sources and to monitor them continuously. *Sensor. Actuator. B.* **69**, 366–371.

Paré, T., Dinel, H., Schnitzer, M. and Dumontet, S. (1998) Transformations of carbon and nitrogen during composting of animal manure and shredded paper. *Biol. Fertil. Soil.* **26**, 173–178.

Philip, J.M. (1996) *Recherche d'indices d'évaluation de la maturité de composts d'ordures ménagères et estimation de leur valeur agronomique*, Mémoire d'obtention du diplôme d'ingénieur agronome de l'INA-PG, Institut national agronomique Paris Grignon, INRA – Centre de Grignon – Unité de Sciences des sols.

Prudent, P., Domeizel, M., Massiani, C. and Thomas, O. (1995) Gel chromatography separation and U.V. spectroscopic characterization of humic-like substances in urban composts. *Sci. Total Environ.* **172**, 229–235.

Rajbanshi, S.S., Hidekazu, E., Sakamoto, K. and Inubushi, K. (1998) Stabilization of chemical and biochemical characteristics of grass straw and leaf mix during in-vessel composting with and without seeding materials. *Soil Sci. Plant Nutr.* **44**(4), 485–495.

Riffaldi, R., Levi-Minzi, R., Pera, A. and De Bertoldi, M. (1986) Evaluation of compost maturity by means of chemical and microbial analyses. *Waste Manage. Res.* **4**, 387–396.

Robin, D. (1997) Intérêt de la caractérisation biochimique pour l'évaluation de la proportion de matière organique stable après décomposition dans le sol et la classification des produits organo-minéraux. *Agronomie* **17**, 157–171.

Roletto, E., Barberis, R., Consiglio, M. and Jodice, R. (1985) Chemical parameters for evaluating compost maturity. *Biocycle*. March, 46–47.

Saviozzi, A., Riffaldi, R. and Levi Minzi, R. (1987) Compost maturity by extract analysis. In: *Compost Quality and Use* (eds. M. De Bertoldi, M.P. Ferranti, P. L'Hermite and F. Zucconi), pp. 359–367. Elsevier, London.

Sequi, P., De Nobili, M., Leita, L. and Cercignani, G. (1986) A new index of humification. *Agrochimica* **XXX**(1–2), 175–179.

Stentiford, E., (2002) The specific oxygen uptake rate (SOUR). In: *The Biological Treatment of Biodegradable Waste – Technical Aspects,* The Environment DG and the Joint Research Centre of the European Commission (eds.), Workshop Conference, 8–10 April, Brussels.

Tiquia, S.M., Yam, N.F.Y. and Hodgkiss, I.J. (1996) Microbial activities during composting of spent pig manures sawust litter at different moisture contents. *Biores. Technol.* **55**, 201–206.

Tiquia, S.M., Wan, J.H.C. and Tam, N.F.Y. (2002) Dynamics of yard trimmings composting as determined by dehydrogenase activity, ATP content, arginine ammonification, and nitrification potential. *Process Biochem.* **37**, 1057–1065.

Tremier, A., De Guardia, A. and Massiani, C. (2002) A respirometric method to study biodegradation kinetics of mixed sludge and woody bulking agent. *Proceedings RAMIRAN 2002, Strbske Pleso, Slovak Republic.* May 14–18 (FAO eds.), 323–328.

Tseng, D.Y., Vir, R., Traina, S.J. and Chalmers, J.J. (1996) A Fourier–Transform infrared spectroscopic analysis of organic matter degradation in a bench-scale solid substrate fermentation (composting) system. *Biotechnol. Bioeng.* **52**, 661–671.

Van Bochove, E., Couillard, D., Schnitzer, M. and Schulten, H.R. (1996) Pyrolysis-field ionization mass spectrometry of the four phases of cow manure composting. *Soil Sci. Soc. Am. J.* **60**, 1781–1786.

Van de Kerkove, J.M. (1990) Evolution de la maturité de trois déchets urbains en cours de compostage, Thèse de Doctorat en Sciences agronomiques, soutenue le 14 septembre 1990 à l'Institut National Polytechnique de Lorraine.

Van Soest, P.J. and Wine, R.H. (1967) Use of detergents in the analysis of fibrous feeds. IV. Determination of plant cell wall constituents. *J. Association Anal. Chem.* **50**, 50–55.

Veeken, A., Nierop, K., de Wilde, V. and Hamelers, B. (2000) Characterisation of NaOH-extracted humic acids during composting of a biowaste. *Biores. Technol.* **72**, 33–41.

Yoshida, T. and Kubota, H. (1979) Gel chromatography of water extract from compost. *J. Ferment. Technol.* **57**(6), 582–584.

Zucconi, F., Pera, A., Forte, M. and De Bertoldi, M. (1981) Evaluating toxicity of immature compost. *Biocycle.* March–April, 54–57.

EN 1484 (1997) Water analysis – Guidelines for the determination of total organic carbon (TOC) and dissolved organic carbon (DOC). CEN Publication date 1997-05-21.

EN 1899-1 (1998) Water quality – Determination of biochemical oxygen demand after *n* days (BOD*n*) – Part 1: Dilution and seeding method with allylthiourea addition. CEN Publication date 1998-03-18.

EN 1899-2 (1998) Water quality – Determination of biochemical oxygen demand after *n* days (BOD*n*) – Part 2: Method for undiluted sample. CEN Publication date 1998-03-18.

EN 13037 (1999) Soil improvers and growing media – Determination of pH. CEN Publication date 1999-12-08.

EN 13038 (1999) Soil improvers and growing media – Determination of electrical conductivity. CEN Publication date 1999-12-08.

EN 13039 (1999) Soil improvers and growing media – Determination of organic matter content and ash. CEN Publication date 1999-12-08.

EN 13040 (1999) Soil improvers and growing media – Sample preparation for chemical and physical tests, determination of dry matter content, moisture content and laboratory bulk density. CEN Publication date 1999-12-08.

EN 13654-1 (2001) Soil improvers and growing media – Determination of nitrogen – Part 1: Modified Kjeldahl method. CEN Publication date 2001-09-19.

EN 13725 (2003) Air quality – Determination of odour concentration by dynamic olfactometry. CEN Publication date 2003-04-16.

EN 25663 (1993) Water quality – Determination of Kjeldahl nitrogen – Method after mineralization with selenium. CEN Publication date 1993-09-30.

EN ISO 10304 – 1 (1995) Water quality – Determination of dissolved fluoride, chloride, nitrite, orthophosphate, bromide, nitrate and sulfate ions, using liquid chromatography of ions – Part 1: Method for water with low contamination. CEN Publication date 1994-11-03.

EN ISO 13395 (1996) Water quality – Determination of nitrite nitrogen and nitrate nitrogen and the sun of both by flow analysis (CFA and FIA) and spectrometric detection. CEN Publication date 1996-06-09.

ISO 7890-3 (1988) Qualité de l'eau. Dosage des nitrates. Partie 3: méthode spectrométrique avec l'acide solfosalicylique.

NF ISO 10694 (Juin 1995) Qualité du sol – Dosage du carbone organique et du carbone total après combustion sèche (analyse élémentaire).

NF ISO 13878 (Juillet 1998) Qualité du sol – Détermination de la teneur totale en azote par combustion sèche (analyse élémentaire).

NF V03-040 (Octobre 1993) Produits agricoles et alimentaires – Détermination de la cellulose brute – Méthode générale.

NF T90-101 (February 2001) water quality – Determination of chemical oxygen demand (COD).

17

Manure-based biogas systems – Danish experience

J.B. Holm-Nielsen and T. Al Seadi

17.1 THE PLACE OF BIOGAS IN THE DANISH ENERGY STRATEGY

17.1.1 Introduction

The long-term strategy plan for energy of the Danish Government – Energy 21 – emphasises the fact that the main concern of the Danish energy planning is the environment. The main objective of Energy 21 is therefore a reduction of the CO_2 emissions from the entire society of 50% by year 2030, compared with the average level of 1988–1990. This can be achieved by out-phasing fossil energy and in-phasing renewable energy sources.

The biogas sector has therefore received growing attention and recognition in Denmark during the last decade and is together with other bioenergy systems part

 Resource Recovery and Reuse in Organic Solid Waste Management. Edited by Piet Lens, Bert Hamelers, Harry Hoitink, Werner Bidlingmaier. ISBN: 1 84339 054 X. Published by IWA Publishing, London, UK

of the CO_2 emission reduction strategy. Biogas shall provide 20 PJ (Peta-Joule) to the national energy production by year 2030, an eightfold increase, compared with year 1990 level and by this the highest growth compared with other renewable energy sources. The experience of the last years proved that the strategic objectives of Energy 21 in the area of biomass for energy are rather difficult to reach due to relative high costs of technology for solid biomass conversion for combined heat and power (CHP) generation, but this does not include anaerobic digestion technologies.

17.1.2 Biogas state of the art in Denmark

In 2002, 3.35 PJ biogas was produced in Denmark (Table 17.1). Most of it originates from the large-scale, manure-based co-digestion plants, wastewater treatment plants and landfill plants. Commercial farm-scale biogas plants are increasing in numbers and production. Farm-scale biogas production in 1999 was 0.070 PJ, 0.129 PJ in 2000, 0.179 PJ in 2001, and 0.380 PJ in 2002. The theoretical biogas potential in Denmark is estimated at 37 PJ of which 26 PJ (70%) is represented by animal manure (Table 17.2). Energy 21 forecasts 20 PJ by year 2030 that shall mainly emerge from manure-based biogas plants (Nielsen *et al.* 1998).

Table 17.1 Biogas plants and production in Denmark, 2002 (Danish Energy Agency 2002).

Type of biogas plant	Amount of plants	Production 2002 (PJ)
Wastewater treatment plants	64	0.875
Landfill plants	26	0.604
Industrial waste treatment plants	5	0.150
Manure-based plants		
Centralised, co-digestion	20	1.342
Farm-scale plants	45	0.380
Total	160	3.351

Table 17.2 Potential, actual production and targets for biogas in Denmark (Danish Energy Agency 2002).

	Potential (PJ)	Production 1999 (PJ)	Target production 2012 (PJ)	Target production 2030 (PJ)
Animal manure	26.0	0.50	5.00	14.0
Sewage sludge	4.0	0.79	1.15	1.5
Organic waste from industries	2.5	0.81	0.75	1.5
Household waste (organic)	2.5	0.01	0.50	2.0
Green waste (parks and gardens)	1.0	0.00	0.20	0.8
Landfill gas	1.0	0.55	0.60	0.2
Total	37.0	2.66	8.20	20.0

17.1.3 The overall framework for the development of manure-based biogas systems in Denmark

The development of manure-based biogas systems in Denmark takes place within a framework of favourable preconditions (Al Seadi 2000), which can be resumed as follows.

General environmental and energy targets

- Danish Government target of 20% reduction of the CO_2 emissions by 2005, compared with the level of 1988–1990 in average.
- Danish Environment Protection Agency target of recycling 30% of domestic waste by 2004 and 50% by long term (Danish Ministry of Environment and Energy 1998).
- Danish Energy Agency target of a fourfold increasing of the biogas production in 2005, compared with 1995 level (Danish Ministry of Environment and Energy 1996).

Government research, development, and demonstration programmes

- Energy research programmes: funding of research, development, and demonstration projects (R, D&D).
- Renewable energy development programme: grants for reviews and demonstration (pilot) projects.
- Follow-up programmes: collection, systematisation, and dissemination of the gathered experience among farmers, plant operators, consulting companies, plant constructors and authorities.

The legislative push

- Requirement of 9 months slurry storage capacity of animal manure and a concomitant restriction of the season for slurry application in spring and early summer.
- The harmony rules, restricting the amount of manure applied per hectare.
- Control and regulation of the application of waste products for agricultural purposes.
- Interdiction to landfill any organic wastes.
- Heavy taxation of waste incineration and exemption of recycling.
- Power companies obliged to purchase electricity produced on biogas at prices according to the law.

Basic economic preconditions

- Government investment grants of 20–40% of the investment costs to build a biogas plant, lowered to 0–20% in year 2000.
- Biogas and heat from biogas are exempted from energy taxation.
- State production grant of DKK 0.27 per kWh electricity produced.
- Building a biogas plant is financed by long-term (20 years) low-interest rate loans.

Other important favourable issues

- Up to 48 million tonnes animal manure per year are produced in Denmark.
- Food and fodder processing industries concentrated in the high-manure density areas, making co-digestion of organic wastes convenient and economically advantageous.
- Positive perception of the advantages of co-digestion of a variety of biomasses, supported therefore by the veterinary and environmental authorities (Bendixen 1999).
- Possibility of selling the heat through the district heating systems, which are widespread in Denmark. In the future the heat will be used for industrial as well as processing purposes.
- Heat demand is widespread in the society. House heating is necessary 8–10 months per year in Denmark.

17.2 BIOGAS TECHNOLOGIES

17.2.1 Centralised co-digestion plants

The first generation of centralised co-digestion plants in Denmark were built in the early 1980s with the only aim of producing renewable energy (Danish Energy Agency 1995). The acknowledgement of the environmental consequences of the intensive animal production, strengthening of the legislation regarding manure storage and application, regulations of waste production and treatment increased the interest for biogas plants which proved to play a new role as providers of manure storage, manure distributors, and organic waste treatment facilities.

In recognition to this, the Danish Government financed successive R, D&D programmes and follow-up programmes which proved that the co-digestion of manure and organic industrial waste concepts for producing biogas offer integrated solutions to a range of environmental problems in relation to agriculture, organic waste treatment, and renewable energy production. (Danish Energy Agency 1995).

Twenty manure-based centralised co-digestion plants are in operation in Denmark, processing approximately 1.5 million tonnes animal manure and 325,000 tonnes industrial biomass per year, producing 50.1 million m^3 biogas (1999 figures).

The centralised co-digestion concept was continuously developed and improved and represents today an integrated system of renewable energy production, manure and organic waste treatment, and nutrient recycling, generating intertwined agricultural, and environmental benefits. The main streams of the integrated co-digestion concept are shown in Figure 17.1.

Animal manure and slurry are collected from the pre-storage tanks at the farms, transported to the biogas plant, mixed and co-digested with maximum 15–25% digestible organic wastes (also called alternative biomass) from food processing industries and municipalities, and submitted to a controlled sanitation process ensuring effective pathogen reduction. New EU legal standards have been adopted in

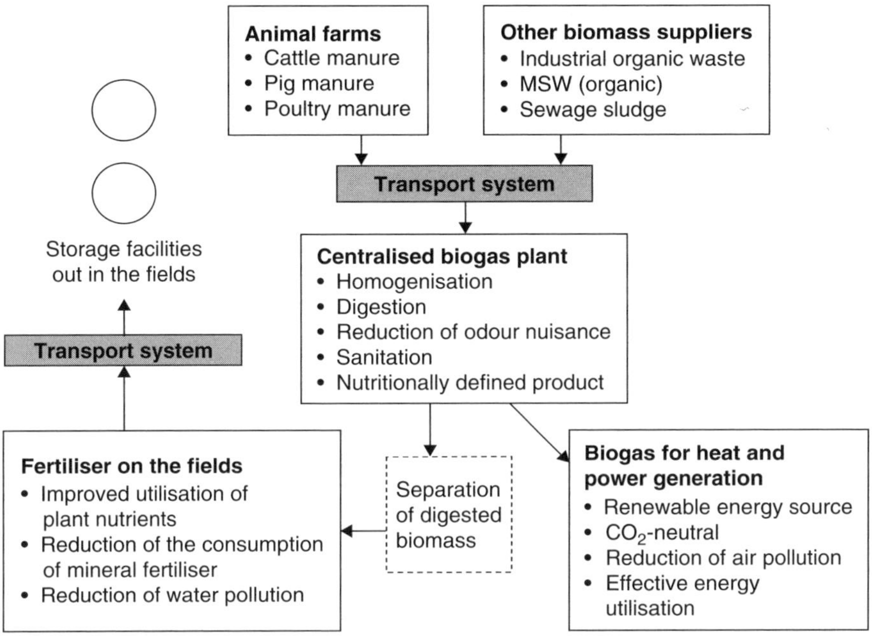

Figure 17.1 The main streams of the integrated concept of centralised co-digestion plant.

October 2002 by the regulation (EC) No. 1774, laying down health rules concerning animal by-products not intended for human consumption.

The digestion process takes place at mesophilic (30–40°C) or thermophilic (50–55°C) temperatures. The retention time in the reactor is of 2–25 days. A controlled sanitation process takes place as well, where pathogens are effectively reduced, and the disease/pathogen cycles are broken. The digested biomass is transferred to the storage tanks, covered with a gas proof membrane for the recovery of the remaining biogas production (up to 15% of the total biogas volume). Some plants are equipped with installations for fibre and phosphorous separation of the digested biomass, these technologies are under strong development from year 2000 onwards (Al Seadi and Møller 2003).

The digested biomass is transported back to the farmers at their storage tanks, placed out in the fields, as a pathogen free, nutritionally defined bio-fertiliser, to be integrated in the fertilisation plan of each farm and applied to crops during the growing season. The slurry suppliers got up to 40% investment grants to build their slurry storage tanks back in the 1990s. The farmers receive back only that amount of digested biomass containing the amount of plant nutrients – nitrogen (N), phosphorus (P), and potassium (K) – they are allowed by law to apply on their crops. The excess of digested biomass is sold to arable farms in or out of the region.

The biogas plant effectuates and supports the cost of the biomass transport. The biogas produced is used for CHP generation. The power is sold to the grid and the heat is distributed through the district heating to heat consumers and some of it, in some cases, used by the biogas plant for process heating.

The biogas production cycle represents an integrated system (Figure 17.1) of renewable energy production, resources utilisation, organic wastes treatment, and nutrient recycling and redistribution, generating intertwined agricultural, energetic, and environmental benefits, as listed below:

- renewable energy production,
- cheap and environmentally healthy organic waste recycling,
- less greenhouse gas emission,
- pathogen reduction through sanitation,
- improved fertilisation efficiency,
- less nuisance from odours and flies,
- economical advantages for the farmers.

The Danish livestock counts approximately 2.4 million livestock units, producing approximately 48 million tonnes of manure per year, of which approximately 1.5 million tonnes is supplied to the biogas plants (2000 figures).

The co-digestion of animal manure with 15–25% organic industrial wastes is done in order to increase the biogas yield (Danish Energy Agency 1995). It is estimated that 75% of the suitable and accessible organic wastes are already utilised for biogas production in Denmark. The paucity of organic waste can become a barrier for the further development of centralised co-digestion systems, thus one of the actual issues is to find ways to improve the biogas yield from animal manure, with less dependence on organic waste addition.

17.2.2 Farm-scale biogas plants in Denmark

Forty-five farm-scale biogas plants are in operation in Denmark in 2003. These plants were built from 1980 up to now, the largest numbers in the last 5–6 years. Most of the biogas plants are co-digesting animal manure and small amounts of organic waste, primarily from food industries.

The new actions and initiatives concerning farm-scale biogas plant development, ongoing from 1995, targeted primarily the large-scale pig farms in Denmark, with a high consumption of heat and power. This resulted in a range of technical and conceptual improvements and some promising economical results for the new built plants. The analyses carried out as part of the follow-up programme for farm-scale biogas plants showed that the main cause for poor economy for most of the plants is related to high investment and operational costs and low stability of the anaerobic digestion process and operational stability of the gas engines (Hjort-Gregersen 1998). These problems can be overcomed by better education, training, and management.

The farmer's interest in farm-scale plants is growing nowadays, as they are interested in the business opportunities and the environmental assets that farm-scale biogas production offers.

Three basic concepts of farm-scale biogas plants have been built and operated in Denmark until now. The main characteristics of the three basic concepts are briefly presented below.

17.2.2.1 The "Black Smith" plant concept

The main parts of this type of plant are given in Figure 17.2: a horizontal, rust free steel digester (150–200 m^3), equipped with a slowly rotating stirring system, a pre-storage tank for manure, a storage tank for digested biomass, a storage capacity for biogas, and a CHP unit (Hjort-Gregersen 1998). The plant co-digests animal manure and fish oil waste. The process is mesophilic up to semi-thermophilic (35–48°C) and the hydraulic retention time is 15–25 days. The gas yield is normally of 40–50 m^3 biogas per m^3-digested biomass. Bigger plants of this concept are now equipped with vertical black steel biogas digesters (800–1000 m^3), which are equipped with slowly rotating stirring propellers.

17.2.2.2 The "Soft Top" plant concept

The schematic representation of the concept is shown in Figure 17.3. Typical for this type of plant is the "two in one" (Hjort-Gregersen 1999) slurry storage and digester tank, covered by a gas tight "Soft Top" membrane for the collection of

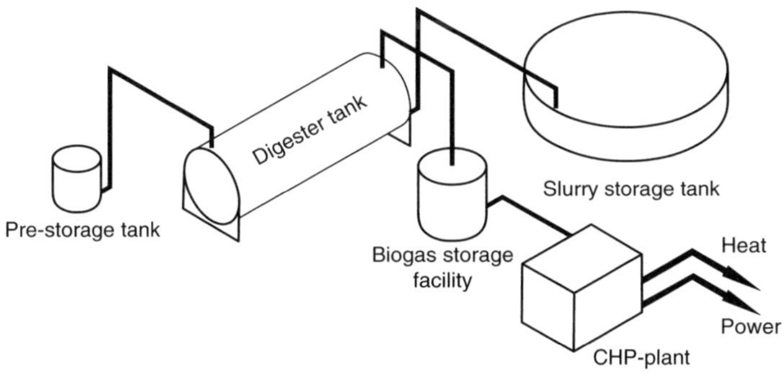

Figure 17.2 Schematic representation of the Black Smith plant concept (Hjort-Gregersen 1999).

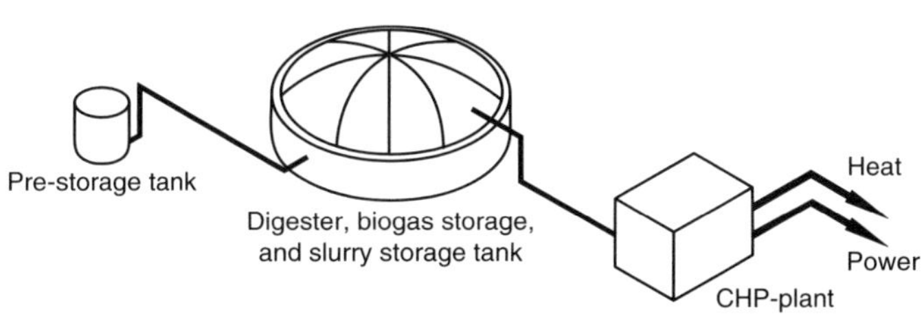

Figure 17.3 Schematic representation of the Soft Top plant concept (Hjort-Gregersen 1999).

biogas. The Soft Top membrane floats on the surface of the slurry by means of a ring, pressing its margins by the edge of the tank and is inflated by the emerging gas production. The tank is stirred by an electric propel. Furthermore, there is a pre-storage tank for slurry and a CHP-unit. The process temperature is 22–25°C and the hydraulic retention time is more than 50 days.

17.2.2.3 The "Soft Cover" plant concept

The main components of the Soft Cover are presented in Figure 17.4. The membrane in the top of the tank is fastened on the edges of the tank and supported by a central mast and functions as a gas storage facility.

The digester is slimmer and is built inside the storage tank, placed tangentially to it. The 200 m^3 digester is insulated and stirred with an electrical propel stirrer. The process temperature is 35–37°C.

The new trend for the farm-scale biogas plants is to build integrated plants for upgrading and processing digested manure into two or several fractions: solid (fibres) and liquid (nutrient concentrates and process water (H_2O)).

17.2.2.4 Investment and economy of farm-scale biogas plants

Table 17.3 compares the characteristics of the three concepts and their initial investment costs.

Economic results from a set of improved concepts indicate that Danish farm-scale biogas plants are at the point of commercial breakthrough (Dansk BioEnergy 1999).

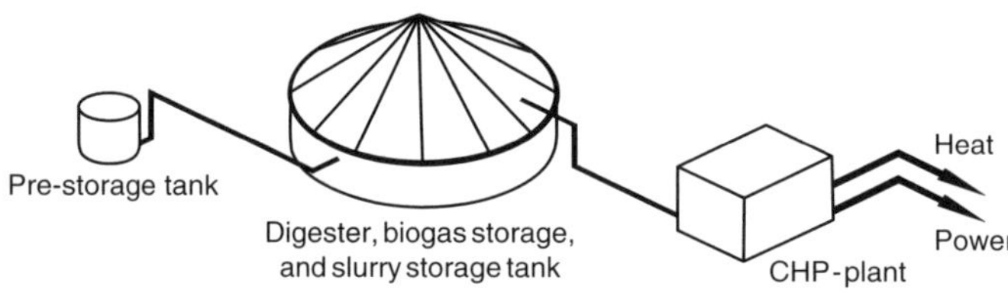

Figure 17.4 The schematic representation of the Soft Cover concept (Hjort-Gregersen 1999).

Table 17.3 Initial investment costs of the basic farm-scale plant concepts (Hjort-Gregersen 1999).

	Black Smith plant	Soft Top plant	Soft Cover plant
Capacity (m^3/day)	12	6	14
Electricity production capacity (kW)	85	15	37
Investment (1000 DKK)	2.150	620	1.113
Investment grants (1000 DKK)	650 (30%)	186 (30%)	723 (65%)
Net investment (1000 DKK)	1.500 DKK, 200.000 Euro	434 DKK, 58.000 Euro	390 DKK, 52.000 Euro

It is a clear precondition that they must be technically well functioning, and sufficient biogas can be derived from the available biomass. Heat utilisation is important to the economic results, though electricity sales are the primary source of revenues. Increased value of energy production may be achieved in some cases by using the most efficient CHP-systems, electric output of up to 38% + 50% heat or more.

The ongoing concentration of livestock production increases farm-scale biogas opportunities in Danish agriculture. The economic incentives seem to be strong enough to initiate a major enlargement of the number of farm-scale biogas plants. Consequently, economic results have to be further improved. Future objectives in research and development are: refined and cheaper plant concepts, improved operation stability, search for applicable supplementary biomass types, development and highly efficient small CHP-systems, or alternatives in the future as fuel cells and microturbines.

17.3 TECHNICAL EXPERIENCE WITH CENTRALISED BIOGAS PRODUCTION IN DENMARK

A range of technical and non-technical improvements affected the concept of centralised co-digestion, so that there is a good foundation today for establishing commercial centralised biogas plants. The economy of the plants can still be vulnerable in some aspects. The premises for establishing a new plant must therefore be carefully evaluated in each particular case ensuring maximum quality standards in order to avoid problems during the start up and during operation.

The basic premises for establishing a biogas plant are:

- The existence and accessibility of the biomass resources, both animal slurries and organic wastes.
- A group of co-interested farmers, as suppliers of fresh slurries and receivers of the digested biomass.
- Sales market for the produced biogas, either for heat or for heat and power production. Sales of heat imply the existence of a district heating net, to ensure the utilisation of the produced heat.
- An established form of organisation, able to plan, build, and operate the plant.

All the plants attempt to achieve a high and stabile gas production and by this an acceptable economic situation. The economic situation depends not only on the level of production, but on a multitude of other factors such as the quality of the initial planning and organising work, operational stability, low maintenance costs, interactive management of the plant, etc.

17.3.1 Concepts and technologies

In the 1990s Danish efforts have focused on developing large centralised co-digestion biogas plants and during the last 3 years also to improve the farm biogas plants by utilising the entire biogas sector experience at all scales of biogas plants.

From the outset, the centralised plant concept was solely based on the utilisation of animal manure delivered from a group of suppliers within a radius of up to about 10 km from the plant. The addition of various categories of organic waste from industry soon proved to be beneficial, and in recent years some of the plants have been using wastewater sludge and source-separated household wastes, the organic fraction. The supplementary waste, for which a receiver fee is often paid to the biogas plant, usually yields more gas per m^3 than animal manure.

Transport and storage systems are also connected to the centralised plants. In some cases, a post-treatment of the digested slurry occurs (separation). The plants are well suited for recycling and redistributing the slurry and nutrients, partly to the group of suppliers, partly to other farmers like arable farmers who can use the surplus slurry and crop nutrients. By doing so, the plant solves an important assignment for agriculture and the environment. The processing of slurry in the biogas system reduces obnoxious smells during spreading, and the slurry is also more homogenous and its nutrient content easier to declare and apply on cropland. Moreover, the numbers of disease germs are significantly reduced.

The size of the 20 centralised co-digestion plants in operation varies from 25 m^3 to more than 500 m^3 of feedstock per day. Half of the plants are based on mesophilic operation and the other half on thermophilic operation. Considerable technological progress has been made in recent years concerning the centralised plants:

- Higher and more reliable production through the addition of organic waste as well as through technical improvements, gas extraction from storage tanks, and improved control and regulation.
- Development of pumps, mixers, and heat exchangers resulting in savings in power consumption and fewer interruptions of operation.
- Improved and less expensive gas purification and gas transport using low-pressure systems.
- Criteria for neutralising infectious matter.
- Techniques for preventing obnoxious smell around the plants.
- Cheaper slurry transport and improved logistics concerning redistribution and utilisation of slurry.

As far as the farm plants are concerned, corresponding progress has occurred in the fields of biogas production and technology. Development efforts are being stated partly on high-technology plants and partly on simpler, cheaper plants with a covered slurry tank and the reactor as an integrated section.

The primary conclusion is that the construction of technically well-functioning plants is now possible. Moreover, the plants can be adopted to meet various requirements and preferences regarding the production and utilisation of biogas, which can unite agricultural, environmental, and energy-related interest.

Due to better organisation, technical and technological improvements, the operational stability has considerably improved and is considered to be one of the explanations of the increased biogas production. It has been proven that operational stability has also a beneficial effect on the anaerobic digestion process.

Important steps forward have been made in order to better understand and control the anaerobic process and the parameters that are influencing it. A combination of practical experience and research results has improved the know-how about anaerobic digestion process among plant operators and suppliers.

17.3.2 Biogas purification

17.3.2.1 Hydrogen sulphide removal

Simple and cheap methods of biogas purifying have been developed. Draining of H_2O and removing hydrogen sulphide (H_2S) are the main problems concerning biogas purification (Ellegaard and Kristensen 1997).

Removing H_2S from biogas is necessary for environmental considerations regarding emissions to the atmosphere if the biogas is burned in a boiler. Furthermore, if the biogas is used in a gas engine for CHP generation, like in most cases in Denmark, the level of H_2S must not be higher than 500–700 ppm on conventional gas engines. This is necessary in order to avoid increased corrosion and rapid and costly deterioration of the lubrication oil and engine spare parts.

The levels of H_2S in biogas are various and depend on the chemical composition of the supplied biomass. If operating on manure alone, the levels of H_2S can reach up to 3000 ppm. Co-digesting manure with other biomass can increase or reduce the concentration of H_2S in the produced biogas, but wastes can contain ferrous compounds like $FeCl_3$ that will keep the H_2S content in biogas at a suitable low level for running the gas engine.

The earlier method to reduce the H_2S content in the biogas, based on adding an iron–chloride to the supplied biomass, proved to be expensive if used continuously and aggressive in corrosion (Ellegaard and Kristensen 1997; Al Seadi 2000). A method largely used in Denmark is the biological oxidation, obtained by injecting 2–5% v/v air into the raw biogas. The oxidation of H_2S will take place resulting in formation of sulphur and/or sulphurous acid:

$$2H_2S + O_2 \rightarrow H_2O + 2S^0$$
$$2H_2S + 3O_2 \rightarrow 2H_2SO_3$$

Addition of air can be done in the top space of the digester, in a separate purification tank, with better control, or in a post-digestion and storage tank at the biogas plant.

17.3.2.2 Low-pressure gas transmission

Low-pressure systems proved to be more suitable for the transmission of biogas compared to high-pressure systems and became widespread in the 1990s. As a consequence, operational stability increased and gas transmission costs were considerably reduced. All biogas transmission systems are equipped this way today.

17.3.2.3 Odour control

A significant odour reduction occurs when digesting the manure, making it easier to handle and to spread on the fields. Nevertheless, the centralised co-digestion plant can be a source of odour nuisance if some important measures considering odours prevention are not taken when building the plant. During the last years, one of the barriers for increasing the number of centralised plants in Denmark has been the odour problem around the plants (Hjort-Gregersen 1999). Special efforts of finding solutions to the problem ended in some important improvements (Hjort-Gregersen 1999).

The causes of odour around the plant can be multiple (Hjort-Gregersen 1999). Some of the most frequent are the territorial emplacement of the plant in the immediate neighbourhood of the settlement areas, uncovered or not tight storage tanks, handling feed stocks outdoors, insufficient ventilation, insufficient suction from storage tanks, deficient treatment of the air obtained from the suction system, supply of heavily smelling feedstock, etc.

Special attention must be given to the treatment of the air resulting from the suction systems (Danish Energy Agency 1995). Passing the air through a bark filter or a compost filter (bio-filter) will result in biological cleaning of the air. Compost filters are a relatively new method, cheap to establish, but there is not sufficient operational experience with it. The biogas plant in V. Hjermitslev uses the controlled compost filter with good results (Al Seadi 2000).

Active coal filters are also used, but the need of regular replacement of the saturated filters can be relatively costly (Danish Energy Agency 1995). The method is used at Hashøj biogas plant. Cleaning the air via combustion in a boiler (catalytic combustion) with generation of process heating for the biogas plant is used at Lintrup and Nysted biogas plants. Combustion in engines, together with biogas (catalytic combustion with regenerative heat exchanging) is currently used at Blaabjerg biogas plant. The method is rather expensive, but the results are fairly positive (Al Seadi 2000).

17.3.3 Biomass handling

17.3.3.1 Pumps and stirring systems

The pumps and the stirring systems are the key elements of handling the biomass (Danish Energy Agency 1995). Great efforts have been made in order to solve a range of technical and economical problems related to it and to find pumping systems and mixing techniques suitable for this kind of biomasses.

Most of the plants are currently using eccentric snail pumps, mono pumps for continuous feeding and extraction of biomass from the digesters, demanding high pressure and low flow. Centrifugal pumps with open impeller and cutters carry out the transfer of biomass within the system, demanding low pressure and high flow.

Regarding biomass stirring, the submerged electrically driven medium speed stirrers, largely utilised in the beginning, proved to be relatively expensive in operation and difficult to access for servicing and inspection, especially when mounted inside a digester. Continuously slow rotating stirrers, centrally mounted in the top of the digesters are considered a better alternative. It requires that the level of biomass in the digester should be correctly adjusted in order to avoid floating layers.

17.3.3.2 Sand removal

Animal manure but also other types of biomass contain various amounts of sand and minerals (Ellegard and Kristensen 1997). The presence of few percentage of sand in the biomass flow causes problems, as it is heavily loading the stirring systems, the pumps and the heat exchangers, causing fouling, obstructions, and heavy wear. When accumulated inside the tanks and digesters, the active volume of these will be reduced. The general principles in order to avoid sand problems at the Danish biogas plants are as follows (Ellegard and Kristensen 1997):

- Regularly emptying of pre-storage and storage tanks.
- Establishing of a sufficiently large pre-storage capacity.
- Adequate stirring strategy.
- Adequate placement of the pumping pipe stubs, in order to avoid sand circulation.
- Avoiding highly sand containing biomass.
- Utilisation of specially developed methods of sand evacuation from the digesters.

17.3.3.3 Biomass transport

The biogas plant is responsible for the transport of animal slurries from the suppliers to the plant and the digested biomass from the plant to the farmers. This task is related to relatively high costs for the biogas plant that can represent up to 1/3 of the operating costs. At the same time it does not involve any costs for the slurry suppliers. The digested biomass is delivered free of charge back to the storage tanks, close to the fields to be spread on, bringing economical benefits for the involved farmers.

In order to minimise the transportation costs, it is vital for the plant to be placed optimal, in relation with the slurry density in the area. To make slurry transportation more efficient and lowering the transportation costs means transportation capacities of trucks with 32 m^3 volume tankers and getting into developing pipeline pumping systems as well.

The transport task is well planned, avoiding empty drive and completed with a sustained awareness campaign among farmers about the importance of H_2O sawing systems at the stables, and of reducing the volume of the transported biomass, by supplying slurry with appropriate dry matter content. Introduction of a bonus-penalty system regarding the dry matter content in the supplied biomass gave positive results at many centralised biogas plants in Denmark.

Fibre separation and concentration of the digested biomass can also contribute to the reduction of transport expenses. Six centralised plants are equipped with installations for fibre separation of the digested biomass, but the results are modest so far. It is estimated that the fibre fraction still contains up to 1/3 of the biogas potential as slowly biodegradable lignocelluloses organic matter. Selling the separated fibre fraction as compost product was not successful and the separation and concentration technologies are still under development, but since 2000 with rapid increasing interest. New innovative technologies now occur in conceptual systems for separating liquid and solid fractions and nutrients in purified fractions, the documented value and treatments costs are not yet verified in all details (Al Seadi and Møller 2003).

17.3.3.4 Gas recovery from storage tanks

Experience showed that the digestion process continues inside the storage tanks after the digested biomass is pumped out of the digesters. Up to 15% of the biogas production can be recovered from the storage tanks if covered with a gas proof membrane. Gas recovery systems are now established at both the new and the old plants in Denmark. The initial one step process becomes a two steps process, where the second step is post-digestion at lower temperature.

17.4 ORGANISATION AND FINANCING

17.4.1 Ownership and organisation

The organisation and the ownership of the Danish centralised co-digestion plants differ from plant to plant according to the local conditions and interest. One of the frequent organisation forms is the co-operative company, where the farmers supplying manure to the plant are the owners (nine plants), and in some cases the owners are the farmers and the heat consumers (five plants). Furthermore, three plants are owned and operated by municipalities, two are private foundations, and one plant is a limited company.

Most of the plants are organised as co-operative companies with the farmers who deliver manure to the plants as share members and owners (Ellegaard and Kristensen 1997). The company's board of directors is responsible for managing the plant by a manager and employing the necessary staff. Furthermore, the board is responsible for all economically and legally binding agreements, regarding construction of the plant, feed stock supply, effluent fertiliser distribution, and eventually redistribution, sale of energy, financing, etc. The co-operative company has proved to be an organisational structure that can function with suitable built-in economic and professional incentives to efficiently pursue the aims of the biogas plant.

17.4.2 Economy – funding and financing

Biogas projects demand important investments. Financing is therefore one of the key elements in order to increase the utilisation of biogas. The financing schemes must be adapted to this type of investments, with long payback periods, as biogas projects are fitting into the basic supply chain of energy as heat and continuously produced electricity feed into the electricity grid.

The Danish biogas projects are mainly financed by means of index-linked annuity loans, guaranteed by the municipalities, supplemented with subsidies. Ordinary mortgage loans are used as well, and in increasing extent, due to low bank interest.

The index-linked annuity loans are low-interest loans, which secure the investor against inflation through a revaluation of the unpaid debts with the inflation rate. The maturity is 20½ years. This type of loan proved to be the most suitable for the biogas plants meeting the demands for long maturity, low interest, and low-initial instalments. The disadvantages of the loans are that they are raised by ordinary sales of bonds, at the stock exchange at market price, implying a depreciation risk that may induce some uncertainty in the planning phase.

The establishment of all the plants was financed from 20% to 40% of the total investment costs by government investments grants. The share of the government grants was higher for the earlier plants, but it has been lowered as the operational stability and the economic results of the plants improved in the last years and from 2003 no grants will be supplied anymore.

The remaining investment costs are financed by indexed mortgage loans, guaranteed by municipalities. Some plants raised bank loans or loans guaranteed by county councils. A number of the early plants were partly financed by traditional mortgage loans, but this type of loans is no longer used in the 1990s.

Only one centralised plant, Ribe Biogas A/S, the only plant organised as a limited company, controlled its own capital as part of the original plant financing. This financial concept seems to be the strongest way of financing, where the basic financing and responsibility of the boards of directors are the same group of people, supplemented with low rate band or financial institution loans, and securing of relative high prices of green produced electricity. This pathway of financing will be the major way in the future for new plants.

Considering the Danish case story, the conclusion from the assessment of the results from centralised biogas plants is that it is possible, under favourable conditions, to achieve a balance in the company economy, without grants for initial investments (Hjort-Gregersen 1999). It is mandatory to have favourable local conditions for selling the biogas and for supplying biomass. A gas yield of at least 35 m^3 per m^3 biomass must be achieved, which means that easy digestible biomass must be added to the slurry, which biogas potential is only of 20–25 m^3 per m^3 biomass (Hjort-Gregersen 1999). The achievement of easy digestible biomass often provides gate fees, contributing to the improvement of plants economy (Hjort-Gregersen 1999).

The biogas plants are still very vulnerable from a commercial point of view, and the actual breakthrough is still uncertain. Relapse is still a very real possibility (Hjort-Gregersen 1999). A major increase in the number of biogas plants implies that the systems must be able to survive economically, mainly on the basis of the animal manure, which is by far their greatest resource. The environmental pressure for sustainable solutions of handling wet organic wastes means that there are many interesting projects in the pipelines in the years to come including co-digestion of manure and organic wastes (Hjort-Gregersen 1999).

A European financing scheme could be established using the positive Danish experience, with long-term index-linked loans, guaranteed by the local or regional authorities, supplemented with national or EU subsidies. The price paid for biogas must correspond to current rules on tax exemption and electricity production subsidies (Holm-Nielsen and Al Seadi 1997).

17.5 DRIVING FORCES FOR FURTHER ANAEROBIC DIGESTION DEVELOPMENT

The experience gathered up to now shows that the premises for establishing new biogas plants are generally favourable and the economic risks are considerably reduced compared with the past. There is a potential to establish at least 20–30 co-digestion centralised biogas plants and several farm-scale plants in Denmark in the first decade of the new millennium, and even a much larger number to fulfil the environmental and energy tasks and goals before 2025.

The strengthening of the environmental legislation regarding handling and recycling of organic waste will contribute to enlarge the fundament of organic waste supply to the biogas plants. Some centralised biogas plants are also capable of treating wastewater sewage sludge and household waste, which represent supplementary biomass resources.

Co-digestion of animal manure and approximately 25% easy digestible biomass (i.e. organic industrial waste) have boosted the biogas production, but have made the centralised co-digestion plants dependent on this type of alternative biomass. This can limit the future expansion of this type of plants as it is estimated that approximately 70% of the existing potential of the digestible organic waste is currently utilised. Reaching the biogas development goals of Energy 21 raises the question of finding other sources of alternative biomass within the agricultural sector (DEA, Energy 21 1996). Some steps forward have been made in co-digesting straw, grass, crops waste, agricultural sub-products, etc. This innovative route is led by the German and Austrian biogas sectors, with which the Danish biogas sector has close relationship.

The ambitious eightfolding of the overall biogas production until year 2025, as underlined in the Danish energy plan "Energy 21", is expected to happen mainly by the contribution of manure-based biogas plants. In order to reach this objective, straightened efforts are to be made in all the biogas-producing sectors to achieve the following:

- Improved economy of the centralised co-digestion plants and independence of supplying highly digestible organic waste.
- Improved technical solutions concerning co-digestion of the organic fraction of source-separated municipal solid waste.

- Further improvement of nutrient efficiency in the digested biomass and reduction of the loss of nutrients.
- Further development and improvement of low-cost farm-scale biogas plants.

One of the important elements of the future development of manure-based biogas production in Denmark is the active involvement and motivation of the farmers. The EU-nitrogen directive and the Danish Water Environment Plan II (1998) and new coming EU directives like the animal by-product directive from 10/2002 and the Danish Water Environment Plan III (2003/2004) prescribe a further reduction of the input of nitrogen and phosphorus from animal manure per ha and demand higher utilisation rates of nutrients in animal manure. From the farmer's point of view this means an increased need for additional storage capacity and for redistribution of nutrients as new concentrated fertilisers. Separation and volume reduction technologies for digested biomass are at the point of commercial breakthrough and will become an integrated part of the manure-based systems in the near future.

The impact of organic farming on the development of manure-based biogas is still an open issue, as well as it is a task for the research sector, the environmentalists to work together, and to find a common ground.

The liberalisation of the Danish and European energy market will have an impact on the overall production of renewables. In spite of this, the Danish Government is committed to apply a progressive policy of reaching the political goals in the area of renewable energy systems following the agreements from Rio, Kyoto, Buenos Aires, and other UN agreements as well as EU commitments.

The public perception and the awareness about the environmental benefits of anaerobic digestion of animal manure and organic wastes are generally positive in Denmark. There is a need for further research work to be carried out, to analyse and quantify the agricultural and environmental effects of biogas production, and the need to include all of them in the national energy policy.

To achieve a satisfactory evaluation of centralised biogas plants, a thorough socio-economic analysis has to be carried out. If biogas plants are established on manure-based solely, the deployment of biogas cannot be justified. However, based on long-term economic results, where agricultural benefits and benefits in industry concerning treatment of organic waste are included in the socio-economic analysis, this picture changes, and specifically larger biogas plants are favourable for the society at large (Nielsen *et al.* 2002). If furthermore the benefits environmental externalities are taken into account, the utilisation of biogas plants in the larger scales considered becomes very attractive from the socio-economic point of view.

17.6 CONCLUSION

The two main concepts of manure-based biogas plants in Denmark – the large-scale centralised co-digestion plants and the farm-scale plants – represent integrated systems of renewable energy production, manure and organic waste treatment, and nutrient recycling, emphasising the environmental, energy, and agricultural benefits of anaerobic digestion.

Recognising the significant contribution of biogas systems for solving a number of environmental problems in the field of agriculture, emission of green house gasses, and organic waste treatment, the Danish Government has supported biogas by creating a favourable framework for its development and implementation such as: appropriate legislative framework, RD&D programmes, investment grants, and other subsidies. The future development of manure-based biogas systems in Denmark will be influenced by many factors, among which the strengthening of the environmental regulations, the liberalisation of the energy market, the organic trends in the farming sector as well as the previously gathered technical, and organisational experience.

REFERENCES

Al Seadi, T. (2000) *Danish Centralised Biogas Plants – Plant Descriptions.* Bioenergy Department, University of Southern Denmark, Esbjerg, Denmark.

Al Seadi, T., Hjort-Gregersen, K. and Holm-Nielsen, J.B. (2000) The impact of the legislative framework in the implementation of manure based centralised co-digestion systems in Denmark. In: *Proceedings of the 1st World Conference on Biomass for Energy and Industry Purposes*. Seville, 5–9 June 2000.

Al Seadi, T. and Møller, H.B. (2003) Separation of slurry – a potential option for the animal production sector. In: *Proceedings of the European Biogas Workshop "The Future of Biogas in Europe II"*. Esbjerg, 2–4 October.

Bendixen, H.J. (1999) Hygienic safety – results of scientific investigations in Denmark (Sanitation requirements in Danish biogas plants). In: *Proceedings of IEA Bioenergy Workshop on Hygienic and Environmental Aspects of Anaerobic Digestion: Legislation and Experiences in Europe.* Stuttgart, Hohenheim, 29–31 Marts.

Danish Energy Agency (1995) *Biogasfællesanlæg fra idé til realitet.* Copenhagen, Denmark (only in Danish).

Danish Ministry of Environment and Energy (1996) Energy 21, Copenhagen, Denmark.

Danish Ministry of Environment and Energy (1998) Waste 21, *Danish Government's Waste Management Plan 1998–2004*. Copenhagen, Denmark.

Dansk BioEnergy (1999). *Særnummer af Juni.* BioPress, 8240 Risskov, Denmark (only in Danish).

Ellegard, L. and Kristensen, E.B. (1997) *Manure Based Biogas Plants in Denmark. Configuration, Dimensioning and Operational Technologies*. BWSC, Denmark.

Hjort-Gregersen, K. (1998) Danish Farm Scale Biogas Concepts – at the point of commercial breakthrough. In: *Proceedings of the International Conference on Biomass for Energy and Industry*. Würzburg, Germany, 8–11 June.

Hjort-Gregersen, K. (1999) Centralised biogas plants. Integrated energy production, waste treatment and nutrient redistribution facilities. *Conclusions and Economic Perspectives*. Danish Institute of Agricultural and Fisheries Economics, Copenhagen, Denmark.

Holm-Nielsen, J.B. and Al Seadi, T. (1997) The future of biogas in Europe. *In the Final Report of Waste for Energy Net II.* The Altener Programme, EC/DG XVII Energy.

Nielsen, P.S., Salomonsen, K. and Holm-Nielsen, J.B. (1998) *Biogas Policies and Biogas Utilisation in Denmark.*

Nielsen, L.H. *et al.* (2002) *Socio-economic analysis of centralised biogas plants*. Danish Research Institute of Food Economics, report no. 136.

18

Electricity production from agricultural wastes through valorisation of biogas

T. Ahrens and P. Weiland

18.1 INTRODUCTION

The times for energy production are about to change in the near future. New energetical pathways have to be found to support the increasing energy demand of the growing world population with renewable resources (Figure 18.1).

Figure 18.1 shows informations, which are based on an actual internet investigation. Agricultural biogas plants located on farms are able to give their input to a new energy infrastructure. After implementation of the "Renewable Energy Sources Act" (EEG) in April 2000 in Germany, which grants a certain allowance for producing electrical energy from renewable resources (for example biomass), the number of agricultural biogas plants in Germany rose significantly, as Schattauer (2003) presented (Figure 18.2).

 Resource Recovery and Reuse in Organic Solid Waste Management. Edited by Piet Lens, Bert Hamelers, Harry Hoitink, Werner Bidlingmaier. ISBN: 1 84339 054 X. Published by IWA Publishing, London, UK

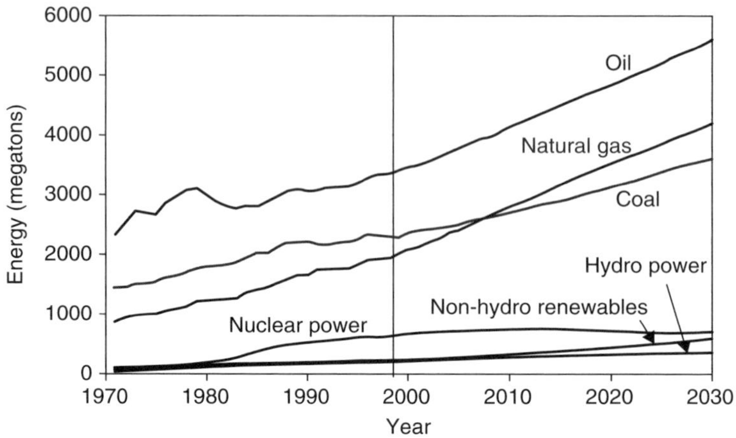

Figure 18.1 Development of world primary energy demand.

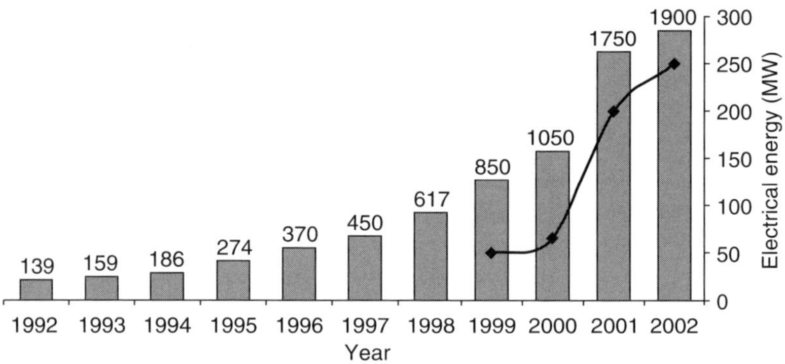

Figure 18.2 Number of biogas plants in Germany.

An effective and positive biogas plant line operation is adhered to some rules, for example stable substrate conditions. Ensiled energy crops and agricultural residues like cattle manure are able to ensure this. Running a biogas plant with an agricultural scale of 800 m^3 fermenter volume (and below) based just on a substrate basis from waste material markets is not commendable, because these markets are mostly not able to guarantee a continuous substrate supply for the plant. Especially in little fermenters, the anaerobic bacteria population might react very sensitively on a non-continuous substrate input, causing deep trouble for the biogas process. Therefore, there is a strong interest in getting information about digestability and possible methane yields from renewable resources like ensiled energy crops (maize, wheat, oat, rye, rape, sunflower, etc.) out of mono-fermentation and co-fermentation processes with manure.

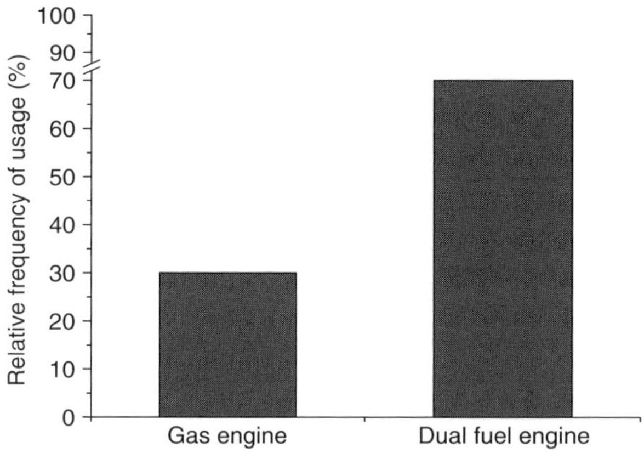

Figure 18.3 Relative frequency of different biogas utilisation techniques.

Another important aspect is energy conversion. Up to now most of the biogas plants located in Germany use combined heat-and-power (CHP) units for combusting biogas and producing electrical and thermal energy (Figure 18.3). Rieger and Ehrmann (2003) acquired that plant owners are normally interested in electrical energy production but the degree of electrical efficiency of CHP units is not very high (related to the installed power between 25% and 35%). Therefore, most of the energy content of biogas (up to 55%) becomes heat, which normally entails no financial revenue for the plant owner.

In contrast, utilisation of the rest heat (after heating the fermentation process) in an advanced heating system (for example for household and stables) causes high-investment costs. Therefore, normally the surplus of the produced heat is disposed into the environment.

For getting out of this problem, modernising of actual energy converting machines is necessary and, furthermore, new and more efficient energy converting systems have to be developed. One alternative for this is the fuel cell technology, which brings up interesting aspects for the future utilisation of biogas (biogas upgrading to natural gas quality, biogas reforming into hydrogen, electrical utilisation of hydrogen from biogas in low-temperature fuel cells and direct electrical utilisation of biogas in high-temperature fuel cells).

18.2 THE BIOGAS PROCESS

18.2.1 Basic aspects for the biogas process

The production of biogas is a natural process, which is for example located in swamps, soils or cattle paunches. Biogas plants are technical solutions for simulation of the degradation processes that occur in nature. In principle, most kinds of

Table 18.1 Substrate frequency in agricultural biogas plants in Germany.

	Ensiled maize	Ensiled grass	Cereal strip waste	Fat	Fodder residues	Grass cut	Bread	Glycerine	Further agricultural residues
Frequency of input (% of plants)	66	47	28	22	16	16	13	9.4	100
Percent by weight of total substrate input (%)	11	3.6	5.3	4.7	1.7	1.1	1.3	1.9	69.4

organic materials are suitable for the biogas process. Rieger *et al.* (2003) found out that common substrates in agricultural biogas plants are:

- liquid and solid wastes from livestock husbandries (like manure);
- ensiled energy crops;
- harvesting residues;
- residues from waste material market.

Table 18.1 shows the relative frequency of different substrates used in Germany.

Braun (1982) described that the organic load of a fermenter is catabolised by anaerobic bacteria (hydrolysis, acidogenesis, acetogenesis and methanogenesis) over carbonic acids as intermediate products to methane within a retention time of 20–30 days (Figure 18.4).

Braun (1982) also said that the methane concentration in biogas mixtures varies between 55% and 65% depending on the substrate and fermentation conditions. The biogas is normally combusted in energy conversion systems for producing electricity and heat. Figure 18.5 shows the principal flow circle of organic material and energy around the biogas process as it was presented by Ehrmann and Ahrens (2003).

Important aspects for implementing biogas technology on farms are:

- electrical and thermal efficiency of the process;
- economical aspects;
- technical aspects concerning the fermentation process;
- process requirements (gas quality/gas production);
- substrate aspects.

In Germany, the energy production out of renewable ressources is subsidized by government. The "Erneuerbare-Energien-Gesetz" (EEG), which became operative in April 2000, grants certain allowances depending on the kind of utilised renewable energy. The following links lead to the wording of the law:

www.verbraucherministerium.de/landwirtschaft/nachw-rohstoffe/eeg-gesetz.pdf and *www.umweltministerium.de/sachthemen/erneuerbar/eeg.php*

The two main aims of this law are stimulating renewable energy for multiplying the shares of renewable energy in the whole energy infrastructure in Germany

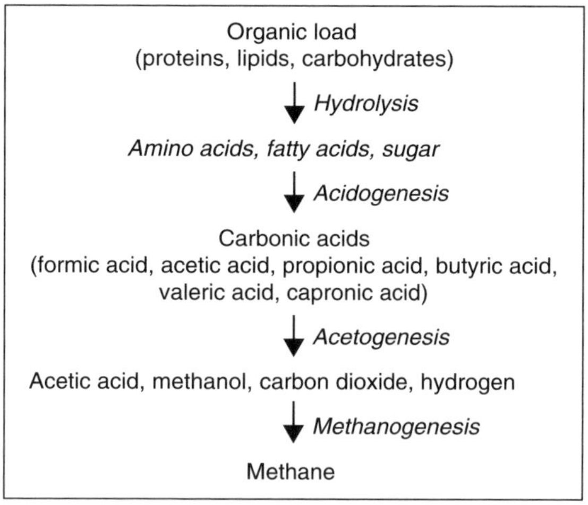

Figure 18.4 Microbiological pathway for biogas production.

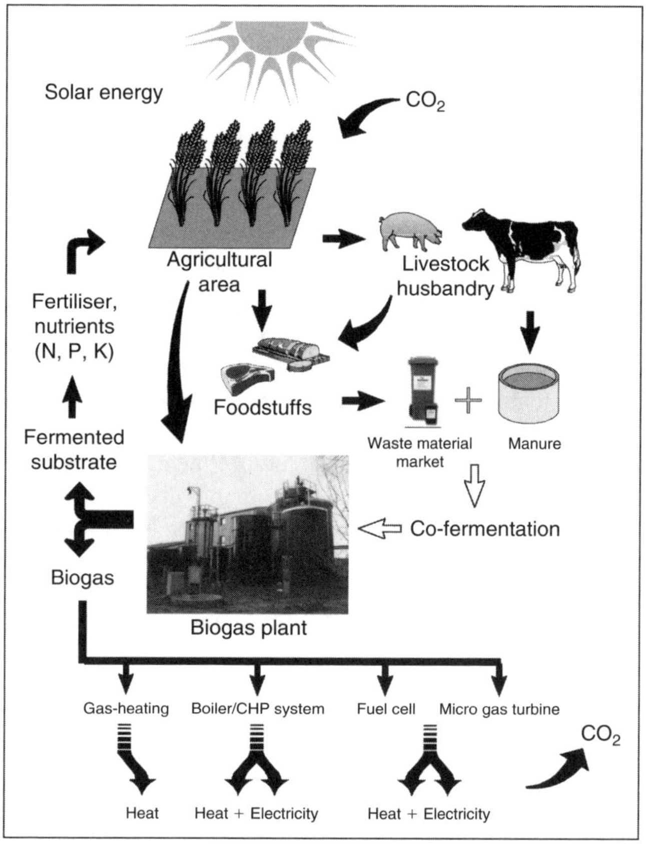

Figure 18.5 Material and energy flow within the biogas process.

and doubling the contingent of renewable energy referred to the whole energy consumption in Germany until year 2010.

In §5 of the EEG an allowance of up to 10.23 €-Cent per supplied kW of electricity out of biomass is ensured to plant operators. This is the most important aspect for biogas production and establishment of agricultural biogas plants on farms in Germany.

18.2.2 Substrate aspects for the biogas process

Rieger *et al.* (2003) also published that important substrates for agricultural biogas plants are (next to animal dung and manure) energy crops and plant residues. Therefore, informations about these substrates are very helpful for biogas plant operators. Targets of research work are therefore examining reachable degradation rates and methane yields in mono-fermentation processes and co-fermentation processes with manure.

In addition, the following aspects are also important:

- energy efficient use of set-aside agricultural land;
- efficient substrate base for agricultural biogas plants;
- climate protection in terms of CO_2 saving;
- reachable level of gas quality;
- influence of mechanical pre-disintegration on aerobic degradability.

18.2.2.1 Experimental equipment

Examinations are normally accomplished in laboratory scale. Figure 18.6 shows experimental fermenters. In batch experiments, the essential retention time for the highest possible degradation rate of the examined substrate is determined. With semi-continuous experiments the whole process circle of a biogas plant is simulated at different retention times and with different substrate mixtures (in mono- and co-fermentation).

For appraisal of the used substrates the following parameters should be acquired: Chemical oxygen demand (COD), organic dry matter (oDM) and content of volatile fatty acids as important measures for input and degradation of organic load in the fermentation process. Further dry matter (DM) of substrate as an important handling parameter and ortho-phosphate (PO_4-P), total nitrogen content (Kjeldahl-N) and ammonium nitrogen content (NH_4-N) as measures for nutritive value of the used substrates and the residues of the fermentation process. The assessment of the fermentation process is done with the degradation rate of liquid phase (concerning COD, oDM, DM) (%) and degradation rate of gas phase (%), which describes the amount of formed methane related to the COD input. Furthermore, methane productivity (m^3 methane/m^3 fermenter volume), methane yield (m^3 methane/kg reduced COD), methane yield (m^3 methane/kg COD_{input}) and reached biogas quality have to be examined.

(a) (b)

Figure 18.6 Semi-continuous stirred tank reactor (a) and batch reactor (b).

The theoretical maximum methane yield was defined as the result of a stoichiometric calculation by Weiland (2001):

$$CH_4 + 2O_2 \rightarrow CO_2 + 2H_2O$$

1 mol of CH_4 equals a volume of 22.4 l
2 mol of O_2 equal a weight of 64 g
1 g COD = 22.4 l CH_4/64 g O_2 = 0.35 l CH_4 g/O_2

For the aerobic oxidation of 22.4 l methane 64 g oxygen are needed, this equals a relation between methane and oxygen of 0.35 : 1. From this analytic relationship results a maximum reachable anaerobic methane yield of 0.35 l methane per g COD. This coherency is very helpful for the appraisement of an anaerobic fermentation process.

18.2.2.2 Experimental results

The following energy crops were examined: maize (reference material), rye, Jerusalem Artichoke, oat, sunflower, triticale, rape and wheat. As substrate mixtures manure/ensiled maize, manure/ensiled maize/ensiled rye and manure/ensiled maize/ensiled rape were also tested.

Depending on the input material and the retention time, significant differences concerning anaerobic degradability and produced amount of biogas were obtained, but based on the results of Ahrens and Weiland (2002) the rate of mechanical predisintegration had only a minor influence (Table 18.2).

Advanced mechanical pre-disintegration heats up the ensiled material and might cause a loss of volatile compounds. This may end in a decrease of the methane

Table 18.2 Effect of mechanical pre-disintegration (ensiled maize).

	Chopped	Squashed	Ripped
Methane yield (m^3 CH_4/kg COD_{input})	0.32	0.27	0.28
Methane yield (m^3 CH_4/kg oDM_{input})	0.43	0.38	0.41
Methane yield (m^3 CH_4/kg LM_{input})	0.13	0.11	0.11
Degradation rate (gas phase) (%)	90	75	75

LM: liquid matter.

Table 18.3 Methane yield of various substrates (mono-fermentation) at an organic load of 2 g oDM/l fermenter volume × d.

	Methane yield		
	l CH_4/g COD_{input}	l CH_4/g oDM_{input}	l CH_4/g LM_{input}
Triticale chopped/squashed	0.21	0.31/0.32	0.10
Rape chopped/squashed	0.18/0.19	0.30/0.35	0.08/0.09
Oat chopped/squashed	0.17/0.18	0.24/0.28	0.07/0.08
Jerusalem Artichoke squashed	0.17	0.24	0.06
Sunflower squashed	0.13	0.20	0.04
Wheat squashed	0.20	0.29	0.11
Rye squashed	0.18	0.29	0.09
Maize chopped/squashed/ripped	0.22/0.23/0.20	0.30/0.33/0.30	0.09

Table. 18.4 Methane yield of various substrates (co-fermentation with manure) at an organic load of 2 g oDM/l fermenter volume × d

	Methane yield		
	l CH_4/g COD_{input}	l CH_4/g oDM_{input}	l CH_4/g LM_{input}
Manure/maize	0.25	0.33	0.04
Manure/maize/rye	0.23	0.35	0.03
Manure/maize/rape	0.24	0.36	0.03

production and is not recommendable. A pre-disintegration before the ensiling process might lead to better results.

Experimental results of Ahrens and Weiland (2002), which are presented in Tables 18.3 and 18.4, show that the methane yield per input of organic matter rises in co-fermentation with manure, because of the positive influence of manure as a source of essential trace elements and buffering substances for the biogas process. The methane yield per liquid matter input decreases, because the water content of manure is up to 90% and higher.

18.2.3 Indirect effects of biogas production

18.2.3.1 *Utilisation of set-aside agricultural area*

For a calculation example the following aspects were decided:

- Set-aside agricultural area: 15 ha (average value for Germany).
- Harvest of maize per area and year: 55 tons fresh matter per ha and year.

This results in a crop yield of 825 tons liquid matter of ensiled maize material per ha and year. With

- a COD of ~390 g/kg liquid matter,
- an experimental degradation rate of ~66% and
- referring to Weiland (2001) a maximum methane yield of 0.35 m^3 CH_4/kg COD_{input}

An amount of 75.000 m^3 methane per year is reachable.

18.2.3.2 Economical aspects

With the heating value of methane of 9.97 kW/m^3, defined by Beitz and Grote (2001), a financial benefit of ~67.300 € per year (granted by EEG) is theoretically reachable. Therefore a highly efficient energy converting system is desirable.

18.3 ENERGY CONVERTING ASPECTS

18.3.1 Energy converting processes

Nowadays best available technologies CHP-units are installed at biogas plants for energetical utilisation of biogas. Although this technology is well developed, there are several disadvantages:

- an intensive need of maintenance,
- a relatively low degree of electrical efficiency,
- a high emission of acoustic noise and exhaust gas,
- a high level of lost heat,
- an increased requirement of operational supplements (like motor oil).

Figure 18.7 shows a 160 kW CHP unit, installed at a biogas plant in Germany.

Utilisation of biogas with fuel cell technology might compensate these disadvantages, but one general conclusion of the "Deutscher Wasserstoff-Energietag" which took place in Essen, Germany (2002), was, that this is a new field of technology with only little experience. Furthermore, there is the need for cleaning biogas and reforming it to hydrogen. Last but not least, referring to Höhlein (2003) the investment costs for a complete fuel cell system are up to now much higher than for a comparable CHP system (roundabout a factor of 10, see Table 18.5).

Figure 18.7 160 kW CHP unit, installed at a biogas plant in Germany.

18.3.2 Fuel cell technology

The main focus has to be the choice of an appropriate fuel cell system. High-temperature fuel cells have a lower sensitivity to disturbing compounds in the process gas, but are very expensive. Low-temperature fuel cell systems need an effective gas cleaning step, but the investment costs in direct comparison to high-temperature systems are lower (Table 18.5 and Figure 18.8).

Therefore, there are several reasons for establishing a low-temperature fuel cell system instead of a high-temperature system at a biogas plant:

- lower investment costs,
- easier maintenance in case of technical problems,
- clear system structure and a flexible range of performance,
- modular structure allows upgrading,
- biogas is upgraded to natural gas quality,
- external reforming process for producing regenerative hydrogen,
- separated energy converting unit,
- system upgrade from low-temperature fuel cells to high-temperature fuel cells is much easier than vice versa.

On the other hand several technical and economical parameters have to be acquired before a proton exchange membrane fuel cell (PEM-FC) system can be established at a biogas plant, these are for example:

- gas quality of biogas depending on different substrates,
- reforming process for generating hydrogen out of biogas,
- rate of efficiency of the whole process chain,

Table 18.5 Electrical efficiency/specific costs of energy converting systems published by Höhlein (2003).

		PEM-FC	SO-FC[a]	MC-FC[b]	CHP unit
Electrical power	$kW_{el.}$	250	100	280	100–1000
Electrical efficiency	%	~34	~47	~48	~38
Specific costs	€/$kW_{el.}$	~10.000	~20.000	~8.000	~500–1000

[a] SO-FC: solid oxide fuel cell.
[b] MC-FC: molten carbonate fuel cell.

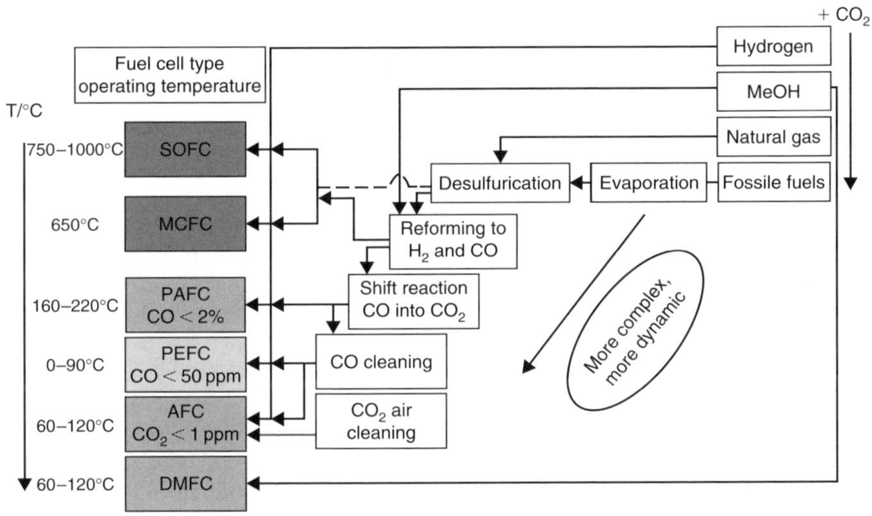

Figure 18.8 Boundary conditions for different fuel cell systems, published by Höhlein (2003).

- dynamical and electrical behaviour,
- life cycle, investment and operation costs, etc.

18.3.3 Trace gas analytics

One main aspect in connecting low-temperature fuel cells to biogas plants is the strong demand in a high-quality process gas (clean hydrogen with a concentration of disturbing trace gas compounds below 10 ppm). Therefore, the spectrum of disturbing trace gas compounds in biogas mixtures depending on different substrates for the fermentation process was measured with a combination of gas chromatography with mass spectrometer technology (GC-MS) by Ahrens and Weiland (2003). The detected mass spectra of pentane as one important trace gas compound in the biogas of processing plants is shown in Figures 18.9 (a).

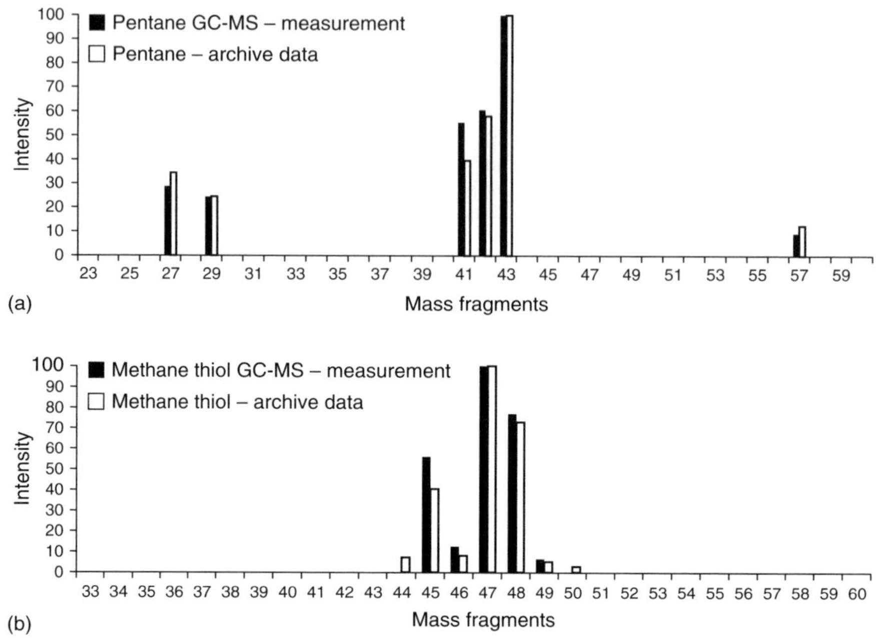

Figure 18.9 GC-MS analysis of (a) pentane and (b) methane thiol.

Figure 18.9 (b) shows the detected mass spectrum of methane thiole as another important trace gas compound in biogas mixtures:

Depending on variation of substrates several other trace gas compounds were detected. Actually a matrix is under construction, which will show the variety of trace gas compounds in biogas depending on the substrate input of the examined biogas plant. This matrix will be published within a scientific report.

Therefore the most important result of the examinations was a significant relationship between different substrates and certain trace gas compounds in biogas and a lower concentration of trace gas compounds in biogas mixtures from renewable resources in comparison to biogas mixtures out of substrates from waste material markets.

A further general result of Ahrens and Weiland (2003) was that the concentration rate of disturbing trace gas compounds in most biogas mixtures was significantly lower than 10 ppm. Therefore most biogas mixtures from agricultural biogas plants are suitable for high-efficient energy-converting processes based on catalysed reaction steps (like fuel cell technology for example).

18.4 PEM-FC PILOT PLANT

Figure 18.10 shows the whole process flow sheet of a PEM-FC pilot plant developed by the farmatic biotech energy AG, Nortorf (Germany), which is connected to an

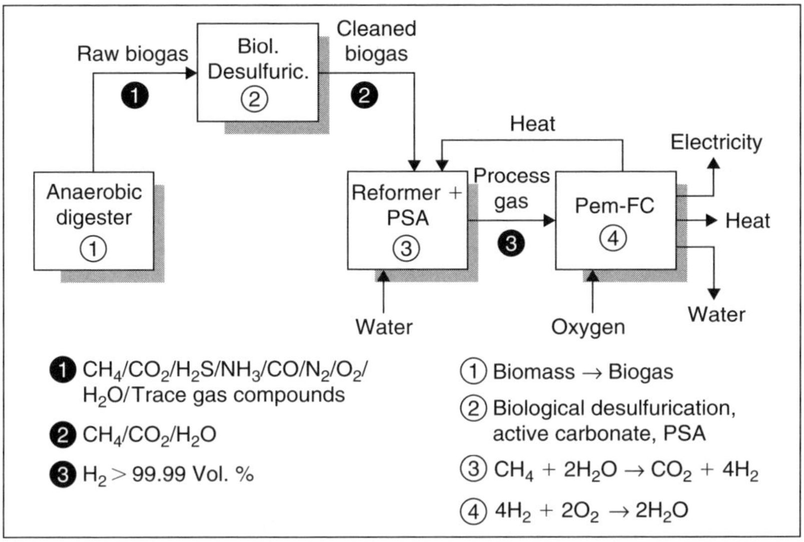

Figure 18.10 Process Flow Sheet of a PEM-FC pilot plant.

agricultural biogas plant located at the Federal Agricultural Research Centre (FAL) in Braunschweig (Germany).

This scheme was chosen because process steps 1 and 2 are well established at a high number of agricultural biogas plants. Therefore a representative comparison between a CHP unit and process steps 3 and 4 becomes possible. Furthermore, this scheme contains three variations for trend-setting utilisation of biogas in the future. The first is producing a clean gas mixture with a very low concentration of disturbing trace gas compounds (after step 2), the second is producing a hydrogen-rich process gas (after step 3) and finally the third is the electrical utilisation of biogenous hydrogen with fuel cell technology (with step 4).

Figure 18.11 gives a view of the PEM-FC pilot plant.

The following Figures 18.12–18.14 present a zoom into the single process steps.

The desulfurication unit works at a temperature of 37°C at aerobic conditions and reaches desulfurication rates of 99% and higher (referred to H_2S).

With the hydrogen reforming system (HCG) a volume flow of up to $10\,m^3$ desulphurised biogas is transferred to a hydrogen-rich process gas with hot steam reforming at a temperature of 950°C and a pressure of 8 bar. In addition this process gas is cleaned with a pressure swing adsorption unit and becomes hydrogen with a pureness of at least 99.99%. Subsequently, the biogenous hydrogen enters the PEM-FC (Figure 18.14) and is converted to electricity, water and heat.

Figure 18.14 shows a prototype test stack with an electrical power of 25 W. A system upgrade up to an electrical power of 1000 W is already done. Main target of the research work is a direct comparison of energy recovery and economical aspects of a fuel cell power unit with a CHP motor unit (as it was shown in Figure 18.7).

Figure 18.11 PEM-FC pilot plant (farmatic biotech energy ag) at FAL.

Figure 18.12 Biological desulfurication unit (farmatic biotech energy ag).

Figure 18.13 Hydrogen Reforming System, HCG (CarboTech Anlagenbau).

Figure. 18.14 PEM-FC test stack (farmatic biotech energy ag).

Figure 18.15 Biogas plant, FAL Braunschweig, Germany.

Further on all important parameters concerning the plant operation (like for example efficiency and profitability of each single process step and the whole process chain) are acquired and will be published in a scientific report.

Figure 18.15 shows the biogas plant located at the FAL Braunschweig.

The biogas plant has a fermenter volume of 600 m^3, it runs with separated cattle manure in mono-fermentation within a retention time of ~40 days at mesophilic conditions (37°C). The biogas production rate is ~10 m^3 per hour with a methane concentration of ~63%.

18.5 CONCLUSION

With the electrical utilisation of biogas with fuel cell technology a new field of technology is entered. Within this, trace gas analytics of biogas mixtures, biogas upgrading, hydrogen reforming and further technological aspects are matters of particular interest.

Furthermore there is the need to find substrates for the biogas process, which are able to give a high-specific biogas yield with a low amount of disturbing trace gas compounds in the produced biogas at the same time.

The presented research work allows insights to important technological aspects like biogas upgrading to natural gas quality, cleaning, upgrading, reforming of complex gas mixtures and last but not least demonstration of effectiveness of fuel cell technology under practical conditions. Against this background a clear stimulus of research and development in this field of technology may be expected.

REFERENCES

Ahrens, T. and Weiland, P. (2002) Technologische Bewertung der Gasaufbereitung und Verstromung von Biogas in einer Brennstoffzelle einschließlich der Beurteilung der Biogasqualität bei Einsatz unterschiedlicher nachwachsender Rohstoffe, 3. Zwischenbericht, FNR – Project FKZ 22009000, Federal Agricultural Research Centre (FAL), Braunschweig, Germany.

Ahrens, T. and Weiland, P. (2003) Technologische Bewertung der Gasaufbereitung und Verstromung von Biogas in einer Brennstoffzelle einschließlich der Beurteilung der Biogasqualität bei Einsatz unterschiedlicher nachwachsender Rohstoffe, 4. Zwischenbericht, FNR – Project FKZ 22009000, Federal Agricultural Research Centre (FAL), Braunschweig, Germany.

Beitz, W. and Grote, K.-H. (2001) Dubbel – Taschenbuch für den Maschinenbau, 20. Auflage, Springer Verlag, Germany.

Braun, R. (1982) Biogas – Methangärung organischer Abfallstoffe: Grundlagen u. Anwendungsbeispiele/R. Braun. – Wien; New York: Springer 1982, ISBN 3-211-81705-0 (Wien, New York).

Ehrmann, T. and Ahrens, T. (2003) Poster Presentation within the exhibition “Tage der Chemie”, Braunschweig, Germany.

farmatic biotech energy ag (2003), Personal information, Nortorf, Germany.

Federal Ministry for the Environment, Nature Conservation and Nuclear Safety (2003) Renewable Energy Sources Act – EEG, Germany.

General Information (2002) Deutscher Wasserstoff-Energietag “Technik, Infrastruktur, Märkte”, Essen, Germany.

Höhlein, B. (2003) Brennstoffzellen für stationäre Anwendungen – Optionen und Herausforderungen, “Life Needs Power – Perspektiven für Energie-nutzung”, VDEW ZVEI, Hannover, Germany.

Internet presentation (2003) http://www.worldenergyoutlook.org/weo/pubs/weo2002/WEO21.pdf.

Rieger, C., Ehrmann, T. and Weiland, P. (2003) Wissenschaftliches Mess-programm zur Bewertung von Biogasanlagen im landwirtschaftlichen Bereich, FNR-Project FKZ 00NR179, personal information, Federal Agricultural Research Centre (FAL), Braunschweig, Germany.

Schattauer, A. (2003) Personal information, FAL Braunschweig, Germany.

Weiland, P. (2001) Erneuerbare Energie in der Landwirtschaft, Braunschweig, Germany.

19

Trendsetter: biogas in European vehicles

G. Landahl and C. Plombin

19.1 INTRODUCTION

19.1.1 Mobility, the environment and our energy future

Most people will agree that the Western world's car growth and congestion already have become unsustainable. Since the 1970s, the use of private cars has doubled, and the flow of traffic is continuously decreasing. Car ownership is further expected to increase with 50% in the next 15 years. This also has an environmental aspect. Road traffic has replaced industry as the main global source of air-borne pollution, noise, and heavy metals. Simultaneously, it significantly contributes to global warming.

Also the development of the world's energy demand and resources makes it a necessity to find ways out of this situation. Already now, it is known that the production of crude oil will reach its maximum in about a generation. Some predict in 2010 others in 2030. When the availability of crude oil decreases and demand continues to grow at approximately 2% per year, fuel prices will increase. Therefore,

 Resource Recovery and Reuse in Organic Solid Waste Management. Edited by Piet Lens, Bert Hamelers, Harry Hoitink, Werner Bidlingmaier. ISBN: 1 84339 054 X. Published by IWA Publishing, London, UK

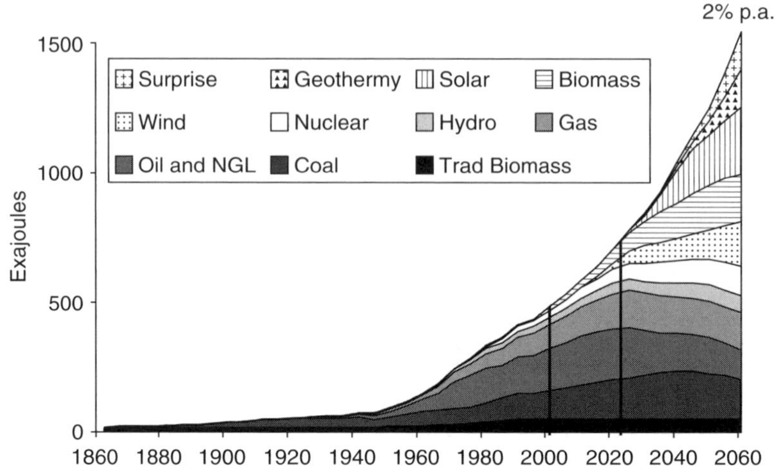

Figure 19.1 A long-term energy scenario showing how demand may develop and how it may be met up to with different energy sources.

new energy sources must be found. In 50 years from now, alternative energies may dominate (Figure 19.1).

When the prices for conventional fuels reach certain levels, alternative energy sources will become more interesting. This transition from a fossil fuel to alternative fuels-based economy will surely have a lot of impact on the European society.

19.1.2 Combining advanced mobility management with clean fleets

Currently, there is almost no diversity in transport propulsion fuels. The fossil fuels dominating current transportation systems are subject to future price increases when supplies diminish. Using renewable sources and realising higher energy efficiency contributes to the diversity and security of energy supply in Europe. In the long term the complete replacement of fossil energy with renewable sources such as biomaterial, hydroelectric power, and waste is necessary.

Sustainable transport can be achieved by combining advanced mobility management schemes with clean vehicle fleets. The *Trendsetter* project (Figure 19.2) is a large demonstration project focusing on both heavy vehicles such as lorries, busses and vans, and on private cars. The project comprises of some 50 different demonstration projects in five cities with 17 partners. The demonstration projects are organised in different work packages covering all aspects of urban transport. The measures intend to show how changes in goods and passenger transport can be accomplished but also how the transportation can be done with clean, alternatively fuelled vehicles.

The short-term implementation of renewable energy systems in *Trendsetter* supports the European Union (EU) objective of protecting the environment and

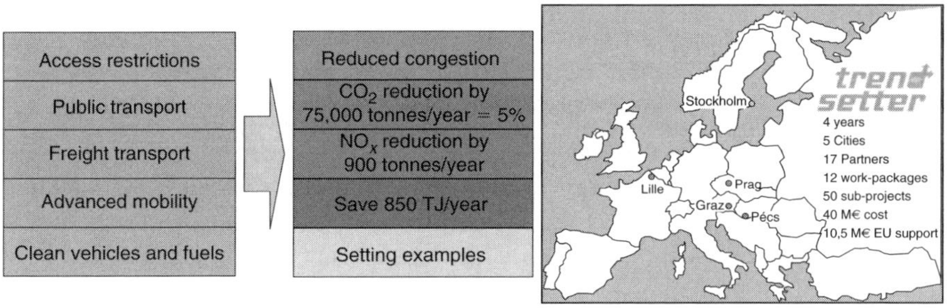

Figure 19.2 The *Trendsetter* project combines a wide range of different measures to help reduce congestion and negative impact on our environment as well as saving energy. The project started in 2002 and will go on to 2006 when the 50 different measures have been evaluated and the results disseminated.

reducing CO_2 emissions from the transport sector. This contributes also to the conservation of resources and reduces the overall CO_2 emissions (energy production and vehicle operation) to a minor level.

The results will prove that cities can achieve Kyoto and EU goals, e.g. by reducing fossil CO_2 emission by 5% through combinations of measures such as the use of biogas from sewerage sludge and the use of clean and efficient alternatively fuelled vehicles. The results will be used in the cities but will be disseminated to other cities who can take part of these examples and be inspired to follow suite. The results also are intended to help the EU form new policies in the field of energy, transport, and the environment.

19.2 BIOGAS PRODUCTION AND USE IN EUROPE

19.2.1 A growing production

For the last decade, the use of biogas coming from sewage collection, farms, and industrial treatment has risen constantly. Nowadays biogas plants are easily available on the market, and biogas constructions have been installed all over Europe. UK and Germany are the two leading countries in producing biogas (Euroberv'ER, Annual balance sheet, April 2002). They have mostly developed the farms and landfill biogas plants. France, Italy, Netherlands, Sweden, Spain, and Denmark form the second group, which have not only developed farms plants, but have also municipal sewage and dumps plants producing biogas.

19.2.2 Inefficient biogas utilisation

19.2.2.1 Biogas production compared to the other energies

The renewable energies are a small part (9%) of the total energy production of Europe (Figure 19.3). Biogas is a renewable energy considered as biomass (4% of the total energy production). In that small figure, biogas represents 0.7% of the total energy production in 2002.

Biogas consumption Only 50% of the biogas production is used. Indeed, Europe still burns off in flare stacks a big part of its produced biogas. The first use of biogas is producing heat for internal consumption of the plant (57%), injecting electricity/gas into the grid (37%); 4.5% of this upgraded biogas production is used as co-generation fuel (for both heat and electricity production) and at least 1.5% is used for producing biogas-fuel for vehicles (Figure 19.4).

Forecasts The European Commission is targeting 1000 MW of installations at the conclusion of the Campaign for Take Off (end of 2003) and 15 million tonnes of oil equivalents of biogas production in 2010 (*1 Ton Equivalent Oil* that means the energy produced compared to a tonne of oil). As of present, the first threshold and benchmark has already been passed (mid 2003). On the other hand, the 2010 incentive is going to be much more difficult to attain. The difference between the EU incentives and effective production can be attributed to the general lack of

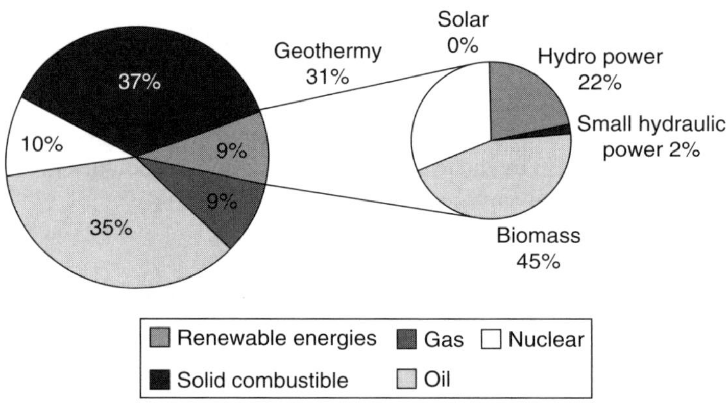

Figure 19.3 Part of the biomass in the whole energy production of Europe 1977 (*source*: www.eubionet.vtt.fi).

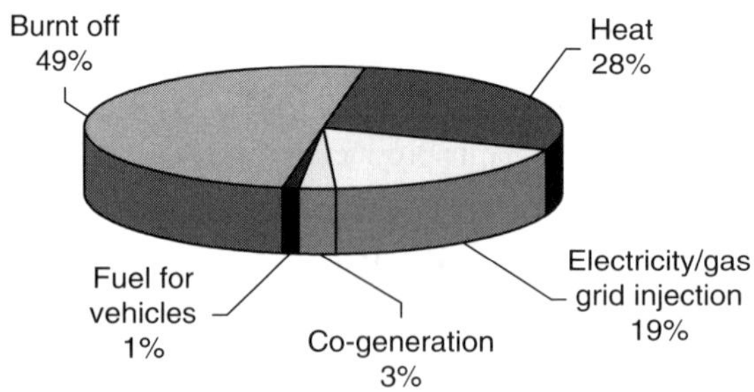

Figure 19.4 Biogas use in Europe (*source*: Biogas Barometer December 2002).

favourable installation and generating incentives put forward by governments across Europe.

19.3 BIOGAS AS FUEL FOR VEHICLES

19.3.1 Why developing biogas as vehicle fuel?

The main reasons for developing biogas as a fuel for vehicles (Figures 19.5 and 19.6) are:

- avoiding polluting emissions, does not add to global warming, low local emissions;
- avoiding noise pollution;
- reducing dependency on imported fuels;
- stopping wasting the produced biogas into flare stacks;
- saving money by selling biogas-fuel;
- saving money by purchasing biogas-fuel instead of petrol or diesel fuel.

19.3.2 Cities who developed biogas for vehicles

Nowadays, there are about 12 European cities that run biogas vehicles. The policies and the environmental commitment of the state support those projects. That

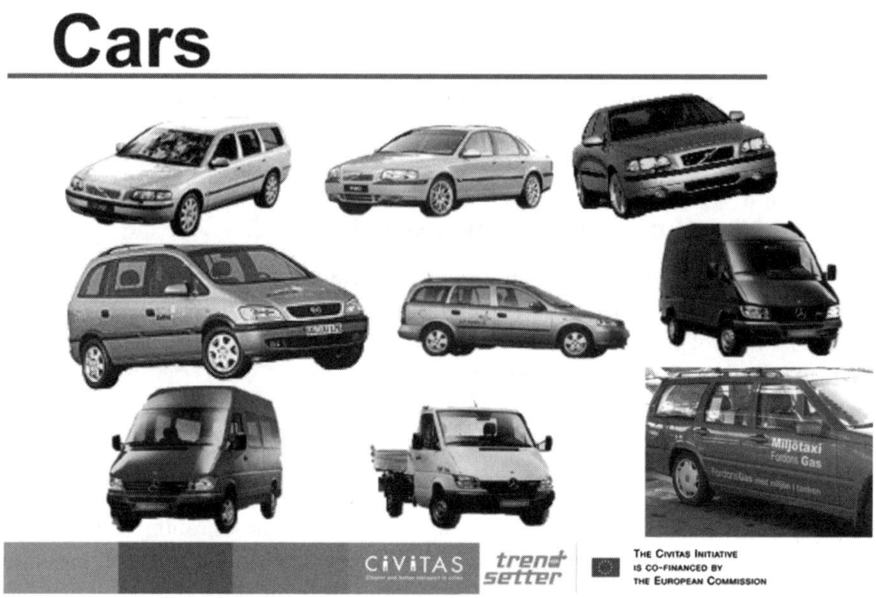

Figure 19.5 There are many car models available in the market for either dedicated methane use or for dual-fuel solutions (providing two tanks one for gas and one for petrol).

Figure 19.6 There are many models of busses and lorries available in the market for methane use. They are often appreciated in urban areas where their lower emissions and quiet engines help improve the cities environment compared to diesel engines.

is why the most advanced countries in the use of biogas-fuel are Sweden and Switzerland (Table 19.1).

19.3.3 Distribution of the biogas-fuel

It is very important to develop an efficient distribution infrastructure when distributing the biogas. Indeed, the consumer needs several refuelling sites in order to avoid waste of time and fuel to reach a fuelling station. The distribution solutions to refuel the fuelling sites are as follows:

- Direct from production to fuelling station in pipeline (Figure 19.7b).
- By natural gas pipeline. In that case the biogas is upgraded to the same quality as natural gas (reduction of unwanted chemical components – like sulphur, hydrocarbons, and oxygen – high compression, odorisation) and pumped into the existing natural gas grid in order to be distributed as a “green gas”.
- By special trailers (mobile refuelling trucks). Here, the biogas produced at the plant is pumped and stocked into special trailers that can victual the remote refuelling stations (Figure 19.7a).

Table 19.1 Cities that today have biogas available for vehicles to use. Name, size (number of inhabitants), the source of the biogas produced and the vehicles type using the gas. In the full report more information about the produced quantities etc are available.

City	Size (inhabitants)	Source	Annual production	Vehicles park	Production cost	Sale
Stockholm	600,000	Sewage	1.5 MN m^3 (+3 MN m^3)	425 cars, 3 lorries	0.7 €/m^3	0.9 €/m^3 1SEK less than petrol
Lille	1.2 millions	Sewage	1 MN m^3	124 buses	0.75 €/m^3	
Uppsala	200,000	Food industries manure	2 MN m^3	46 buses (+2 hybrid buses: electricity/biogas)		0.85 €/m^3
Zurich	1.2 millions		Natural gas + biogas	9 trucks only biogas, 600 light vehicles		
Roma	2.8 millions	Sewage dump	3.5 MN m^3	12 lorries		
Reykjavik	170,000	Landfill		42 municipal cars		1.1 €/m^3
Linköping	132,000	Food industries organic waste	3 MN m^3	68 buses, 200 cars		0.83 €/7 m^3
Göteborg	800,000		Natural gas i + biogas	1490 vehicles		0.81 €/m^3
Trollhättan	52,000	Sewage food industries	0.6 MN m^3/year	100 cars, 10 buses, 5 lorries	0.2 €/m^3	0.61 €/m^3
Helsingborg	300,000	Landfills		60 vehicles		0.88 €/m^3
Kristianstad	70,000	Sewage manure	1.5 MN m^3	22 buses, 2 lorries, 50 cars		0.66 €/m^3
Kalmar	57,000	Sewage, manure	2.5 MN m^3	35 cars, 5 lorries		0.77 €/m^3
Jönköping	118,500	Sewage	1.5 MN m^3	95 cars		0.75 €/m^3
Eslöv	29,000	Sewage	0.2 MN m^3	50 cars, 2 buses, 2 lorries	0.33 €/m^3	0.66 €/m^3

Source: *Trendsetter*.

Figure 19.7 Distributions of biogas by (a) special biogas-freighters or by (b) direct pipe to refuelling facilities.

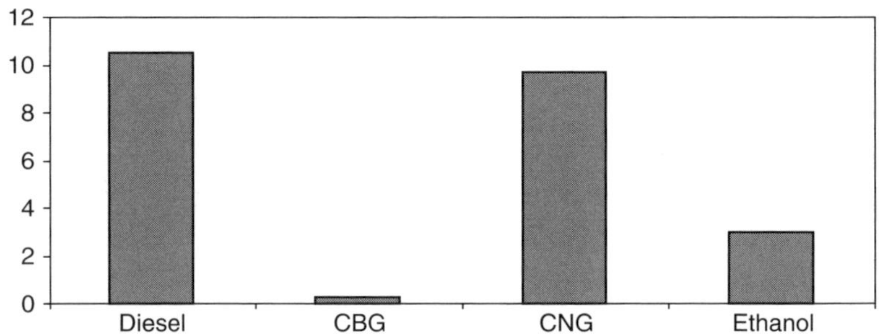

Figure 19.8 Emissions of fossil CO_2 from heavy-duty vehicles (kg/10 km). CBG is compressed biogas, CNG is compressed natural gas (*source*: SWECO).

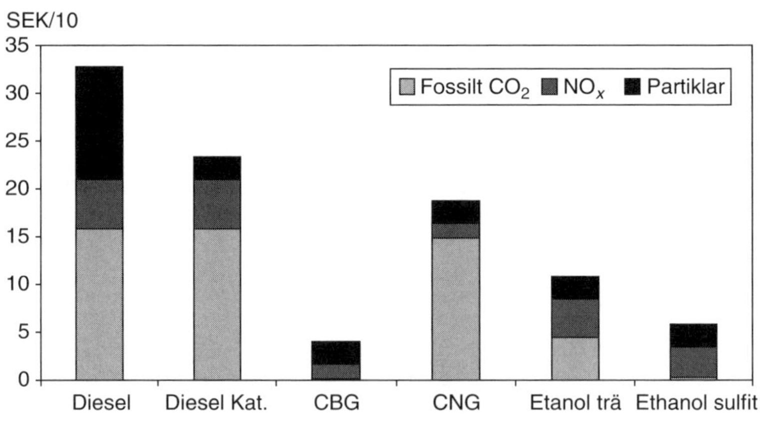

Figure 19.9 Environmental costs for different pollutants from heavy-duty vehicles. CBG is compressed biogas; CNG is compressed natural gas. Etanol trä is ethanol produced from the wood industry; ethanol sulfit is ethanol produced from the pulp industry. The costs are calculated by the Swedish method developed by SIKA and represent the complete life cycle of the fuel (http://www.sika-institute.se/english_fr.html) 9SEK = 1€. As shown, biogas has the most favourable emission characteristics (*source*: SWECO).

19.3.4 Environmental and costs benefits

As shown in Figure 19.8, biogas used as vehicle fuel gives lower CO_2 emissions compared to other fuels.

The low level of pollutant emissions as NO_x, carbon mono-oxide (CO), and particles permits to evaluate the costs saved and thanks to the use of biogas. Figure 19.9 shows that biogas is clearly the best alternative seen from the point of environmental costs.

19.4 *TRENDSETTER* BIOGAS PROJECTS IN STOCKHOLM AND LILLE

19.4.1 The *Trendsetter* projects in Lille and Stockholm

The implementation of *Trendsetter* will make Stockholm and Lille probably the largest users of biogas for transport in Europe. Producing biogas from domestic sewage and landfill waste makes sense for cities, since environmental concerns about methane emission from these sources already necessitates after treatment of sewage sludge digestion. In effect, using biogas closes the energy loop by redefining waste as a fuel feedstock. Cities gain energy security and utilise an existing resource. While biogas cannot replace all fossil fuels, it can potentially reduce their use and is a complement to other clean vehicle technologies.

In addition, biogas pilot production facilities will be opened and techniques refined. This will encourage new biogas producers to enter the market and ultimately reduce the production cost of biogas. Vehicle producers will also be encouraged to produce dedicated biogas or bi-fuelled vehicles. Common procurement has the potential to decrease the purchase price for biogas vehicles, based on electric vehicle procurement strategies developed in the THERMIE project ZEUS.

The distribution of alternative fuels will be developed in these cities as well as the building of new fuelling facilities.

New vehicles will be procured and demonstrated. Demonstration will involve both busses, lorries, vans, and cars. A total of some 1400 vehicles are expected by 2006. In some cases common procurement processes will be used to help obtain better market penetration. Special emphasis will be put on persuading the business society to start using clean vehicles. Demonstrating clean vehicles in taxi fleets will be one part of this, since taxies are very visible in the cities traffic and this may inspire others to follow suite.

Supporting measures to help facilitate this shift to clean vehicles will be demonstrated. Incentives to clean vehicle use will be provided through the implementation reduced parking fees for clean vehicles.

19.4.2 New (bio)gas buses in Lille

Lille's local bus authority, The Syndicat Mixte des Transports, will triple the number of gas buses in the municipal bus fleet, raising the share of gas buses in the fleet to 50% by 2005.

The authority will gradually increase the use of biogas, utilising biogas produced in the planned plant for fermenting organic waste and a new bus depot adjacent to the waste treatment plant.

Currently, 13% of Lille's buses are gas powered, most using natural gas and a few using biogas produced in the pilot facility at a sewage sludge treatment plant.

The large-scale introduction of biogas buses is a result of the positive experience from the buses currently in operation. Lille will also build a high-speed refuelling station where buses can be re-fuelled with natural gas overnight. A similar facility will be introduced at the planned biogas bus depot.

19.4.3 Clean vehicles in Stockholm's commercial fleets and biogas-fuelling stations

It is now time to concentrate public efforts on encouraging private companies to choose clean vehicles. Private companies are very important to developing a market for these vehicles, as they buy about 70% of the cars sold in Stockholm.

In Stockholm, city efforts to promote cleaner vehicles have primarily focused on using clean vehicles in the municipal fleet. This was a strategic decision; these vehicles were city owned, and the city could therefore use its own purchasing decisions to develop a market for cleaner fuels and associated infrastructure. A network of ethanol and biogas-fuelling stations has been established and is continuously expanding. Now the second generation of clean vehicles is available on the market, with operating economy and performance competitive with their petrol-fuelled counterparts. At the same time, many companies and organisations are concentrating their environmental efforts on reducing the impact of their own transport and travel.

This *Trendsetter* measure will:

- conduct a market study identifying the determinants of company car purchases and necessary specifications for clean vehicles;
- produce an information campaign promoting clean vehicles and fuels aimed at private companies;
- provide a fleet of clean test vehicles in close co-operation with Stockholm car dealers;
- identify companies with large company fleets that may be interested in receiving information about clean vehicles;
- measure the number of clean vehicles sold and the volume of renewable fuels purchased in Stockholm as a result this measure, and assess the environmental impact of this type of incentive programme;
- provide subsidies covering the extra-costs for private companies to invest in clean vehicles.

Fuelling possibilities are necessary to spread the use of biogas. Thus, two new biogas-fuel stations will be built in Stockholm in important business districts where there are no possibilities to fuel biogas today.

Easy access to a fuel station is an important factor for choosing a clean vehicle. Around 8 million Nm^3 biogas per year can be available to the local biogas service stations in Stockholm, but the net of fuel stations is insufficient. Within the time frame of CIVITAS, the City of Stockholm plans to establish two new biogas stations. These new stations are vital to the CIVITAS project as it is currently only possible to fuel biogas at four places in Stockholm.

The new fuel stations will be placed in important business districts of Stockholm. Special studies will be carried out to find out if it is possible to read biogas pressure remotely, if there are less expensive refuelling technique, and finally also what the infrastructure looks like in the region close to Stockholm. This project goes hand in hand with the work on focusing on getting private companies to choose clean vehicles, e.g. gas vehicles.

19.4.4 Heavy biogas vehicles in Stockholm

The environmental benefits of using biogas are most profound when biogas is used in heavy vehicles in city traffic. Therefore the City of Stockholm plans to introduce more heavy biogas vehicles.

Within this *Trendsetter* measure, investment grants are being offered for private companies, municipal companies, and city administrations that invest in biogas distribution lorries refuse trucks and buses. The grants will cover 30% of the extra-cost of choosing a biogas alternative to a conventional diesel vehicle.

The investment grants are administered by the Stockholm Environment and Health Administration, and are open to all the city's companies and administrations. Subsidised vehicles will be evaluated with respect to technical performance and user acceptance.

For the past 2 years, two biogas-fuelled waste collection lorries have been operating in Stockholm's Old Town. Now the City of Stockholm Waste Management Administration will replace another eight to ten diesel waste collection lorries with biogas vehicles.

The two biogas-fuelled waste collection lorries have been greatly appreciated both by drivers and by residents, especially because they are so much quieter than their conventional counterparts. Residents are less disturbed by noise from the waste collection lorries, which operate from early morning throughout the day. For drivers, the reduced noise means improved working conditions.

Based on these positive experiences, the City of Stockholm Waste Management Administration will replace another eight to ten diesel waste collection lorries with biogas vehicles. Contractors own all waste collection lorries. Thus, this replacement will be achieved by requiring a certain percentage of biogas vehicles when tendering for contracts for the collection of domestic solid waste in the City of Stockholm (Figure 19.10).

Within *Trendsetter*, the Waste Management Administration will:

- require city contractors to operate one or two biogas-fuelled waste collection lorries;
- offer *Trendsetter* co-financing to contractors for extra-costs incurred investing in biogas collection lorries, up to a total of eight to ten vehicles in the project;

Figure 19.10 Practical examples of biogas vehicles. (a) Biogas refuse freighter in Stockholm. These freighters, so far only used in the medieval old town, have become very popular because of their low noise and air emissions. (b) Fuelling station selling biogas to vehicles in the inner city.

- use the biogas vehicles in daily operation;
- evaluate operational use as well as citizens' attitudes regarding biogas-fuelled waste collection lorries.

19.5 CONCLUSION

Biogas is getting more and more developed across Europe, and thanks to its renewable sources and current EU incentives, it has a real potential of fossil energy replacement. However, still 50% of the biogas production is just burnt, and the rest is used as heat fuel. It has been proven that the best environmental way to use this biogas is the production as vehicle fuel (wastewater upgrading, diesel/petrol economy, replacement and no global warming effects). This upgrading method is the less employed because of its high investment costs and few-known possibilities. But, thanks to European grants, projects, and associations, biogas as vehicle fuel is being developed.

A common legislation still does not exist about the production, quality, and distribution of biogas. That is why biogas as vehicle fuel is still depending on the country policies and on the government sensitivity and concerns.

As it is a very new development, the financial output of such a production cannot be demonstrated yet (even if they are economical viability). However, the positive environmental impact has already been demonstrated, and the customers who tried biogas vehicles presented a very good opinion of this friendly environmental fuel.

20

Bioplastics from waste materials

Paulo C. Lemos, Luísa S. Serafim, Ana M. Ramos and Maria A.M. Reis

20.1 INTRODUCTION

Synthetic plastics have a wide range of applications in different industry sectors. It is estimated that their production exceeds 100 million of tonnes per year and that in European countries the consumption of plastics is about 60 kg per capita. Worldwide, every year, about 15 million tonnes of plastics are discharged into landfills. Furthermore, several hundred thousand tonnes of plastics are discarded to marine environments and accumulate in certain regions of the ocean each year (Reddy *et al.* 2003).

Since petroleum-based plastics are not biodegradable, the accumulation of such high amounts of recalcitrant compounds causes a waste disposal problem. Recycling is not a cost-effective solution for many types of plastics, especially those used in packaging. Incineration is also expensive and dangerous due to the harmful chemical compounds released (Reddy *et al.* 2003). The introduction of biodegradable plastics with similar material properties to replace synthetic plastics is a possible solution for waste disposal problems. Bioplastics are the possible

 Resource Recovery and Reuse in Organic Solid Waste Management. Edited by Piet Lens, Bert Hamelers, Harry Hoitink, Werner Bidlingmaier. ISBN: 1 84339 054 X. Published by IWA Publishing, London, UK

competitors of synthetic plastics because they are biodegradable, thereby being capable of overcoming the problem of pollution and waste accumulation.

Several types of biodegradable plastics are available in the market, e.g. starch-based, polyhydroxyalkanoates (PHA), polylactates (PLA) and polyglycolides (PGA). PHA are among the most promising bioplastics because they are thermoplastics, their strength and toughness are similar to polypropylene and these properties can be changed by altering their composition. They are completely resistant to moisture and have very low oxygen permeability.

Until now the main obstacle to the replacement of synthetic plastics by biopolymers is the great cost difference (€9/kg for PHA against €1/kg for synthetic plastics) (Biby 2002). The price of PHA depends on the substrate cost, PHA yield on substrate and on the downstream process used for polymer recovery (Lee 1996).

A great effort has been devoted to reduce the production costs of PHA. The substrate is one of the main cost factors on PHA production. Indeed, raw materials used in PHA production may account for 40–50% of the total operating costs (Choi and Lee 1997; Meesters 1998; Van Wegen *et al.* 1998). Since the cost of the carbon sources may account for 70–80% of the raw materials, the price of PHA can be substantially reduced if cheap organic substrates are used. Also development of more productive cultures and more efficient fermentation/recovery processes can contribute to the reduction of PHA production costs.

The use of renewable sources obtained from waste organic carbon and mixed microbial cultures (e.g. activated sludge) can significantly decrease the price of PHA. An economic evaluation showed that the price of PHA can be reduced to €4/kg if low cost substrates and mixed cultures are used (Meesters 1998). This value is still higher than the price of polypropylene, but this disadvantage is offset by its complete biodegradability. However, this price is lower than those of other biodegradable polymers, such as PLA, diol-diacid-based aliphatic polyesters and starch-based polymers, which are sold in the range of €5–12/kg (Choi and Lee 1997).

In this work an overview of PHA production using a wide range of substrates and organisms is presented.

20.2 PHA

PHA produced by bacteria are stored in the cell cytoplasm as granules of 0.2–0.5 μm surrounded by a membrane (Sudesh *et al.* 2000). PHA storage inclusions can be visualized by microscopic observation in phase contrast appearing as refractive granules inside the cells. A proper differentiation of this intracellular polymer is usually obtained with lipophilic dyes: Sudan Black and Nile blue. Figure 20.1 shows a Nile blue staining of PHA fluorescence granules in activated sludge.

PHA have a general chemical formula as presented in Figure 20.2. More than 100 different monomer units have been identified as constituents of PHA in various bacteria including 3-hydroxyalkanoates of 3–12 carbon atoms with a large variety of R-pendant groups, 4-hydroxyalkanoates of 4–8 carbon atoms, 5-hydroxypentanoate, 5-hydroxyhexanoate and 6-hydroxydodecanoate (Lee 1996). Polyhydroxybutyrate

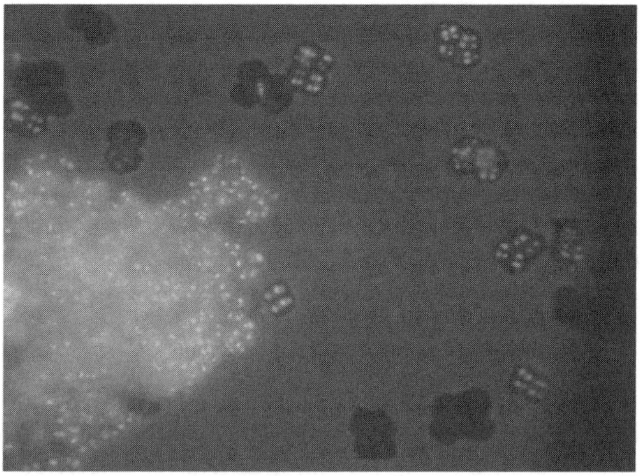

Figure 20.1 Microscopic (epifluorescence 1000×) observation of PHA granules (bright spots) in activated sludge by Nile blue staining.

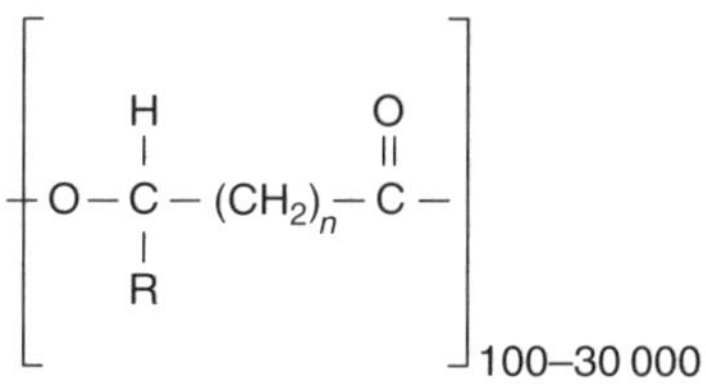

PHA

		Polymer
$n = 1$	R = hydrogen	Poly(3-hydroxypropionate)
	R = methyl	Poly(3-hydroxybutyrate)
	R = ethyl	Poly(3-hydroxyvalerate)
	R = propyl	Poly(3-hydroxyhexanoate)
	R = pentyl	Poly(3-hydroxyoctanoate)
	R = nonyl	Poly(3-hydroxydodecanoate)
$n = 2$	R = hydrogen	Poly(4-hydroxybutyrate)
$n = 3$	R = hydrogen	Poly(5-hydroxyvalerate)

Figure 20.2 General chemical formula of PHA and some representative members (Lee 1996).

(PHB), a lipid-like polymer of 3-hydroxybutyrate, is the most common and best-characterized PHA stored by bacteria. Other frequently PHA stored by bacteria include polyhydroxyvalerate (PHV), polyhydroxymethylvalerate (PHMV) and polyhydroxymethylbutyrate (PHMB). Their presence and relative proportion are dependent upon the type of carbon substrate used by the microorganism.

PHA can be divided into two classes of polymers, namely short chain length (scl-PHA) containing 3-hydroxybutyric and 3-hydroxyvaleric acid units, and medium chain length (mcl-PHA) which contain monomers ranging from C_6 (3-hydroxyhexanoate) to C_{14} (3-hydroxytetradecanoate).

Copolymers, composed by different monomeric units, can be synthesized by feeding the microbial culture with a mixture of substrates. A copolymer of 3-hydroxybutyrate and 3-hydroxyvalerate, P (3HB-co-3HV), is industrially produced by a pure culture of *Ralstonia eutropha* from propionic acid and glucose. As referred below (see Section 20.4.2), the copolymer has improved material properties relatively to the homopolymer.

20.3 PHA PRODUCTION

20.3.1 Organisms

PHA are synthesized by many Gram-negative and Gram-positive bacteria from at least 75 different genera. Low molecular weight PHA are also detected as components of the cytoplasm and cellular membrane in organisms such as *Escherichia coli*. It is also a membrane constituent of yeasts, plants and animals (Braunegg *et al.* 1998).

20.3.1.1 Bacteria

Pure cultures Some examples of pure bacteria industrially used for PHA production comprise *R. eutropha, Alcaligenes latus* and several *Pseudomonas* sp. (*P. oleovorans, P. denitrificans, P. putida*). Several other microorganisms are referred in the literature as able to accumulate PHA, although with a lower yield (Table 20.1, see Section 20.3.2). In some cases these polymers are accumulated intracellularly up to a content of 90% of cell dry weight, under conditions of nutrient limitation, acting as carbon and energy reserves. Some pure cultures of natural-PHA producing bacteria have a long generation time and a relatively low optimal growth temperature. Usually they are hard to lyse and can contain enzymes for PHA degradation. These factors can influence the extraction and recovery of PHA granules (Reddy *et al.* 2003).

Metabolic engineering has been used to introduce genes in non-producing strains, to modify the biochemical pathways in order to expand the substrate utilization range, to enhance PHA synthesis and to produce new types of PHA. An example is the recombinant *E. coli*, in which the *R. eutropha* PHA biosynthesis genes in a high-copy-number plasmid was inserted in order to increase productivity. The introduction of plasmid-containing genes from *R. eutropha* in *P. oleovorans,* which was only able to synthesize mcl-PHA, allowed for the production of a mixture of short and mcl-PHA (Sudesh *et al.* 2000).

The PHA biosynthesis by microorganisms can be explained by two different metabolic pathways, one represented by *R. eutropha* and the other one by *Pseudomonas* sp. This division was based on the monomers composition of PHA produced by these organisms (Figure 20.3).

For *R. eutropha* two acetyl-coenzymeA (acetyl-CoA) are drifted to the synthesis of PHB by condensation to acetoacetyl-CoA and reduction to hydroxybutyryl-CoA (consuming NADPH), which is finally incorporated in the growing chain of PHB. Apart from glucose, other compounds like lactic acid, *n*-alkanoates with even number of carbons or carbon dioxide give rise to PHB. The *n*-alkanoates with odd number of carbons are also utilized forming a copolymer of HB and HV. The incorporated monomers usually contain between 3–5 carbons.

Pseudomonas belonging to the rRNA-homology group I generally synthesizes mcl-PHA from several alkanes, alkanols or alkanoates, and not scl-PHA. However, some can accumulate HB and mcl-PHA. The hydroxyacyl-CoA substrates

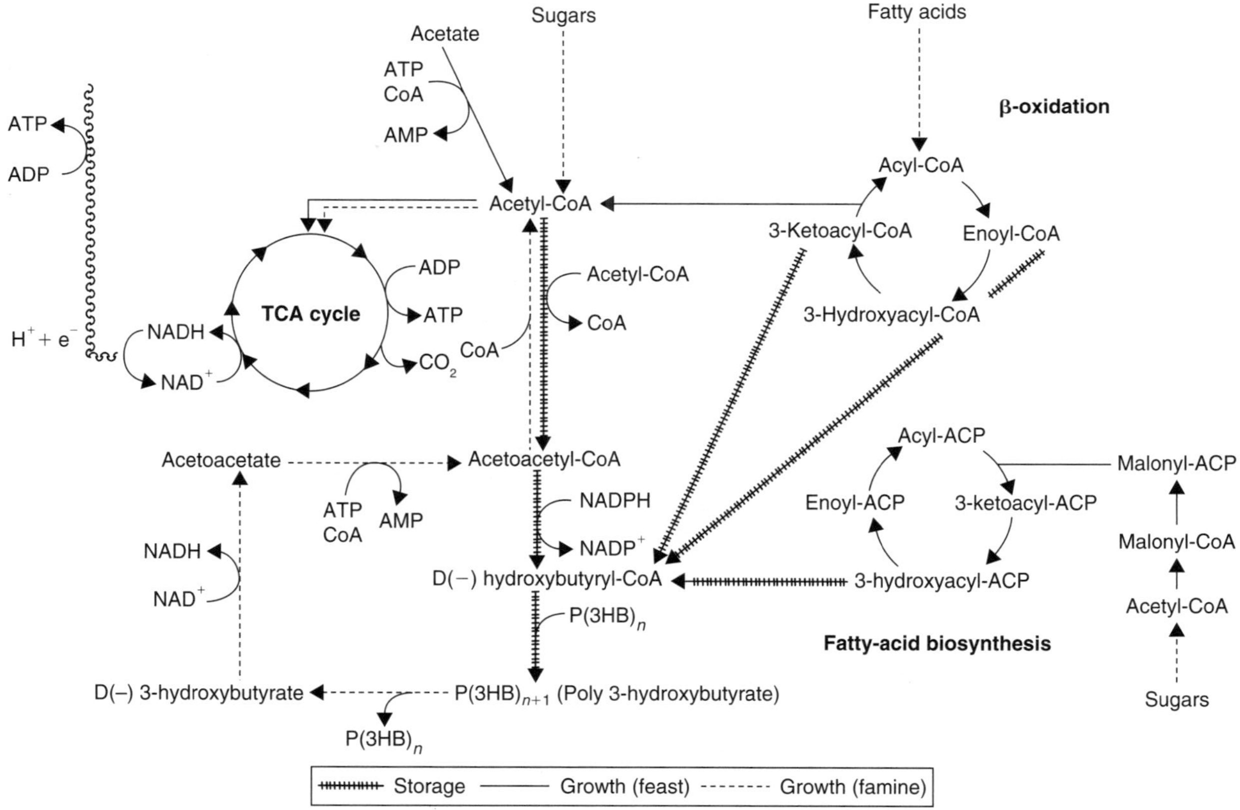

Figure 20.3 Metabolic pathway for synthesis and degradation of PHA by diverse organisms (modified from Sudesh *et al.* 2000).

for mcl-PHA containing between 6 and 14 carbons are formed from the intermediates of the fatty acid ß-oxidation pathway (Sudesh *et al.* 2000).

The majority of *Pseudomonas* belonging to rRNA-homology group I, like *P. aureofaciens*, *P. citronellolis*, *P. mendocina*, *P. putida*, but not *P. oleovorans*, can also synthesize mcl-PHA derived from unrelated carbon sources suchlike carbohydrates. These organisms produce primarily PHA containing 3-hydroxydecanoate as the main monomer from substrates like gluconate, fructose, acetate, glycerol or lactate. The 3-hydroxyacyl-CoA precursors are formed via the biosynthetic fatty acid pathway. As the fatty acid biosynthetic intermediate is a 3-hydroxyacyl-ACP (acyl-carrier protein), an additional step is needed to convert it to the corresponding 3-hydroxyacyl-CoA (Sudesh *et al.* 2000).

Mixed cultures Accumulation of PHA is also observed in mixed microbial cultures present in wastewater treatment processes (e.g. activated sludge). PHA storage by mixed cultures occurs in systems where electron donor and acceptor availability are separated (e.g. anaerobic/aerobic dynamics) or when the substrate is not continuously available for the microorganisms (aerobic dynamic feeding). In the first case, PHA plays a specific role in the ecophysiology of a certain group of bacteria, the most well known are the polyphosphate accumulating organisms (PAOs) and glycogen accumulating organisms (GAOs) (Levantesi *et al.* 2002). The second case concerns a more general aspect of microbial growth under non-steady-state conditions. Production of PHA by PAOs and GAOs is relatively lower compared with the one obtained by activated sludge submitted to aerobic dynamic feeding, commonly known as "feast and famine" process (Beccari *et al.* 1998). In this case no specific group of microorganisms related with PHA production was so far identified. The ability to store internal reserves gives these microorganisms a competitive advantage over those without this possibility, when facing transient substrate supply.

Even though no experimental evidence for mixed cultures exists, it can be anticipated that the metabolic pathways involved in carbon consumption and polymer storage, in this case, is the same as reported for pure culture (Figure 20.3). For activated sludge submitted to aerobic dynamic feeding, considering acetate as the carbon substrate, the produced acetyl-CoA is partially channelled to the tricarboxylic acid (TCA) cycle for growth and NAD(P)H production, and partially deviated to PHA production. The accumulated intracellular reserve is metabolized when no external substrate is available.

20.3.1.2 Plants

Crop plants can also synthesize PHA. Some plants presently used for PHA synthesis are *Gossypium hirsutum* and *Zea mays*. Cheap resources (water and sun) can be used for the production of PHA on an agronomic scale when compared with fermentative processes involving bacteria. The yield of PHA from plants is relatively low (4% of plant dry weight) (Poirier *et al.* 1995). If this yield could be improved to 20–40% of plant dry weight, the price of PHA could be potentially reduced to €0.20–0.50/kg (Reddy *et al.* 2003). However, these values of PHA yields can only be reached by creating transgenic plants with high productivity and stability over generations. *Arabidopsis thaliana* was transfected from *R. eutropha* and able to produce PHB granules in the cytoplasm, vacuole and nucleus but resulted in growth defects on the plant. As plant cells are highly compartmentalized, the insertion of PHA bacterial genes should be located in places where high acetyl-CoA concentrations are present. The placement of the plasmids on plastids of *A. thaliana* improved the accumulation of PHB on the leaves to 14% of dry weight (Braunegg *et al.* 1998).

As acetyl-CoA is the precursor of both PHB and oils, oilseeds crops can be a good agent for PHA production, if the acetyl-CoA is drifted towards PHA (Reddy *et al.* 2003). Cultures such as rapeseed, sunflower and soya bean can potentially represent a good investment in research on genetic engineering of plants for PHA production.

In the present work only bacterial production of PHA will be considered.

20.3.2 Substrates

The majority of the published works concerning PHA production by pure microbial cultures use simple substrates like sugars (glucose, fructose or sucrose) or carboxylic acids (propionate or butyrate). *Burkholderia cepacia* can use sucrose and gluconate. *R. eutropha* can store PHA from carbon dioxide and from more complex compounds, like 1,5-pentanediol, 1,7-heptanediol, 4-hydroxyvaleric acid or 4-valerolactone (Braunegg *et al.* 1998). Alcohols, like methanol or pentanol, are stored by some strains of *Pseudomonas* (Kim *et al.* 1994). Furthermore, some species of the *Pseudomonas* genus are able to convert more complex molecules into PHA, like *n*-alkanes, *n*-alkanoates, *n*-alkenes with long or branched chains (Braunegg *et al.* 1998). Storage of PHA by activated sludge is frequently studied using short chain organic acids like acetate, propionate and butyrate as carbon sources. Acetate, which is effectively stored by activated sludge, is not usually utilized by pure cultures to produce PHA. On the contrary, glucose is not a good substrate to produce PHA by mixed cultures since it is converted mainly into another polymer, glycogen.

Table 20.1 summarizes the production of PHA from different substrates by microbial cultures. A large variety of bacteria can store high amounts of PHA from different substrates with similar productivities and yields. Furthermore, in some cases, the amount of PHA stored by activated sludge is in the range of that synthesized by pure cultures.

As pointed out above, the price of PHA can be substantially reduced if cheap substrates are used. This goal can be achieved if complex substrates, like agricultural or food industrial residues and wastewaters, are used instead of expensive pure substrates. Table 20.2 lists examples of PHA yields and productivities obtained from complex substrates and pure cultures. The amount of PHB obtained from those substrates is, in many cases, similar to that obtained from simple substrates.

Table 20.1 PHA produced by pure and mixed cultures from simple substrates.

Conditions	Substrate	PHA	$Y_{P/S}$	Cell concentration	q_p	%PHA	References
Mixed cultures							
Aerobic	Acetate	PHB	0.52	0.16	0.086	31.0	Beccari *et al.* (1998)
Dynamic	VFA[a]	P(HB/HV)	–[b]	0.16	0.058	62.0	
Feeding	Acetate	PHB	0.36–0.54	1.02–2.80	0.32–0.74	12.0	Beun *et al.* (2000)
	Acetate[c]	PHB	0.52–0.61	1.10–1.30	0.69–1.01	–	Beun *et al.* (2002)
	Acetate	PHB	0.56	2.05	0.082	–[b]	Dircks *et al.* (2001)
	Acetate	PHB	0.35	2.20	0.026	–[b]	Beccari *et al.* (2002)
	Ethanol	PHB	0.37	2.00	0.0056	–[b]	
	Acetate	PHB	0.36	0.39	0.77	–[b]	Dionisi *et al.* (2001)
	Acetate	PHB	0.42–0.62	1.50–2.60	0.70–1.20	54.5	Reis *et al.* (2003)
	Propionate	–[b]	0.027	2.80	0.14	14.0	
Pure cultures							
R. eutropha	Glucose	PHB	0.35	93.0	1.90	76.0	Kim *et al.* (1994)
R. eutropha	CO_2/H_2	PHB	–[b]	85.0	1.54	72.0	Kim *et al.* (1994)
E. coli Rec	Glucose + HPr	P(HB/HV)	0.29	110	2.18	77.3	Choi and Lee (1997)
E. coli Rec	Glucose + LB[d]	PHB	–[b]	117	2.11	76.0	Kim *et al.* (1994)
P. extorquens	Methanol	PHB	–[b]	223	1.12	61.0	Kim *et al.* (1994)
P. oleovorans	Octanoate	PHB	–[b]	–[b]	–[b]	68.0	Reddy *et al.* (2003)
P. denitrificans	Pentanol	PHV	–[b]	–[b]	–[b]	55.0	Reddy *et al.* (2003)
B. cepacia	Glucose + HPr	P(HB/HV)	–[b]	35.0	0.29	45.7	Silva *et al.* (2000)
A. vinelandii	Glucose	PHB	0.34	40.1	0.68	79.8	Lee (1996)
A. latus	Sucrose	PHB	0.17	143	2.50	50.0	Choi and Lee (1997)
M. organophilum	Methanol	PHB	0.19	250	1.86	52.0	Choi and Lee (1997)

Units: storage yield, $Y_{P/S}$, in g PHA/g substrate; cell concentration in g/l; production rate, q_p, in g PHA/l h; %PHA in g PHA/g biomass.
[a]Mixture of acetate, propionate (HPr), butyrate and valerate.
[b]Not reported.
[c]Acetate in a very high concentration (5 times the previous one).
[d]Luria broth.

Table 20.2 PHA produced by pure cultures from wastes, by-products or renewable sources.

Microorganism	Substrate	PHA	Productivity production[a]	%PHA	References
A. beijerinckii	Corn-steep liquor and molasses	PHB	–[b]	–[b]	Purushothaman *et al.* (2001)
A. chroococcum	Starch	PHB	0.0535	74.2	Kim (2000)
	Alpechin[c]	P(HB/HV)	–[b]	80.0	Pozo *et al.* (2002)
A. vinelandii UWD	Molasses	PHB	0.61	66.0	Kim (2000)
	Swine waste liquor	P(HB/HV)	–[b]	34.0	Suk *et al.* (1997)
	Tapioca hydrolysis	PHB	1.03	58.0	Kim (2000)
Klebsiella spp.	Malt waste	P(HB/HV)	–[b]	–[b]	Wong *et al.* (2002)
	Soya waste	P(HB/HV)	–[b]	–[b]	Wong *et al.* (2002)
Pseudomonas cepacia	Xylose	PHB	0.0259	60.0	Kim (2000)
E. coli Rec	Whey	PHB	0.48	80.0	Kim (2000)

[a]Productivity in g PHA/l h.
[b]Not reported.
[c]Wastewater from olive oil mills.

The PHA price can be further reduced by simultaneously using complex substrates and mixed cultures. Indeed, utilization of open mixed microbial cultures facilitates the use of complex substrates since microbial populations can adapt continuously to changes in substrate. Selection of microorganisms occurs on the basis of the high capacity for PHA storage. Therefore, the selection is based on ecological principles rather than making use of sterile conditions as in the case of pure cultures. Consequently, there is no need for sterilization and sterile fermentation systems (Reis *et al.* 2003).

Furthermore, in some cases, complex wastes should be subjected to a preliminary step of fermentation to convert the carbon compounds to short chain carboxylic acids, like acetate or lactate. This fermentation step can be preformed by mixed cultures of acidogenic and acetogenic organisms that produce low volatile fatty acids (VFA) under anaerobic conditions at low pH (5–7).

20.3.3 Process technology

20.3.3.1 Pure cultures

Industrial PHA production is currently carried out using pure cultures as referred (Table 20.1). Some of the cultures (*R. eutropha*, *P. oleoverans* and *M. organophilum*) synthesize PHA under nutrient limitation such as nitrogen, phosphorus, magnesium, potassium, oxygen or sulphur in the presence of excess of carbon source

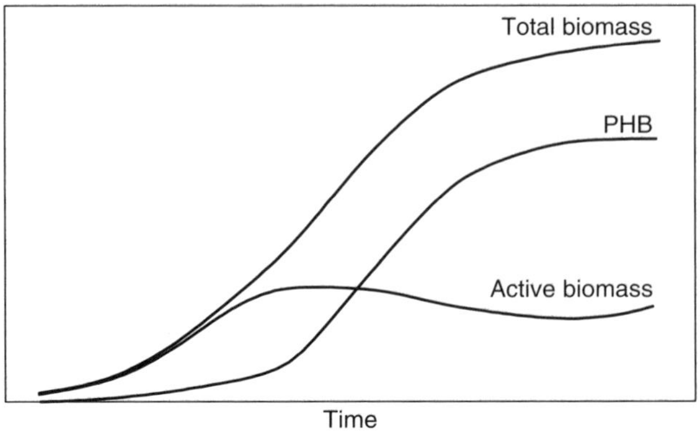

Figure 20.4 PHA production by *R. eutropha* (modified from Lee 1996).

(Sudesh *et al.* 2000). The process is performed in two steps: in the first one cells are grown without nutrient limitation and in the second, cells are supplied with the medium containing the carbon source and the limiting nutrient. PHA accumulation occurs in the second step. Figure 20.4 shows the evolution of PHA produced in a batch process using *R. eutropha*. Other pure cultures such as *A. latus*, a mutant strain of *A. vinelandii* and recombinant *E. coli* do not require nutrient limitation for the synthesis of PHA and can accumulate the polymer during growth.

20.3.3.2 Mixed cultures

As mentioned above mixed microbial cultures are able to synthesize PHA. Production of PHA is particularly effective when activated sludge is submitted to consecutive periods of external substrate accessibility (feast) and unavailability (famine). Under these dynamic feeding conditions, during excess of external carbon substrate, the uptake is driven to simultaneous growth of biomass and polymer storage. After substrate exhaustion, the stored polymer can be used as energy and carbon source. In this process, denominated by "feast and famine", polymers are synthesized under conditions that are not limiting for growth.

Figure 20.5 shows storage of PHB by a mixed culture submitted to a spike of acetate after a 10 h period of famine. Ammonia is consumed while PHB is being stored, indicating that growth and PHB accumulation occur simultaneously.

Table 20.1 depicts values of productivities and PHA content for different substrates and processes using pure cultures and mixed cultures under "feast and famine" conditions. It is worth noting that values of the amount of polymer stored per amount of substrate consumed ($Y_{P/S}$) for pure cultures are in the range of those obtained for the "feast and famine" process using mixed cultures. The volumetric productivities are much lower for mixed culture than for pure cultures, which can be explained by the higher cell concentration used in the later case. However, specific productivities (g PHA/g cell h) are about one order of magnitude higher for

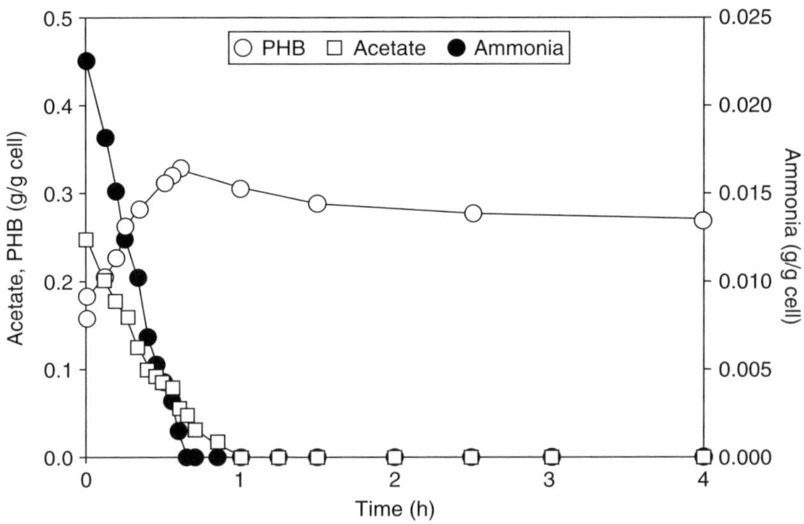

Figure 20.5 Production of PHA by mixed aerobic cultures using the "feast and famine" process.

mixed cultures than for pure cultures. Furthermore, results presented for mixed cultures were obtained in studies where the main objective was to understand the mechanisms of PHA storage rather than the optimization of PHA production. Higher cell concentrations can be obtained if a cell growth phase is included before the PHA accumulation step.

One important parameter to be considered in the optimization of a process is the cell PHA content (% of PHA/dry weight). Values reported for pure cultures are in the range of 70–80%. Using mixed cultures the maximum sludge PHA content so far published was 66.8%, using acetate as substrate (Table 20.1). The PHA content of mixed cultures submitted to feast and famine conditions can reach about 80% if reactor operational parameters are optimized (Serafim *et al.* 2004).

Analysis of data presented in Table 20.1 shows that activated sludge submitted to "feast and famine" conditions can reach similar yields and higher specific productivities than pure cultures. The "feast and famine" process is commonly carried out in a sequencing batch reactor (SBR). The SBR reactor is usually operated with cycles of about 12 h (Serafim *et al.* 2002). The feast period, where the substrate is available, takes about 2 h while the famine phase is extended for almost 10 h. Part of the biomass containing PHA is harvested at the end of the feast period. The extension of feast and famine phases depends on the culture used and on the substrate concentration.

In the case of waste materials being composed by fermentable carbon (e.g. polysaccharides), an upstream fermentation step is required in order to produce VFA, which are the precursors for PHA synthesis. As shown in Figure 20.6 fermentation precedes the PHA accumulation step where cultures are selected on the basis of its storage capacity. Since the polymer quality (ratio of HB/HV) as well as the quantity (PHA content in the cell) is determined by the relative concentration of the VFA, these two aspects can be defined by adjusting the pre-fermentation conditions. For the same substrate, the composition of VFA produced is a function of the reactor operational conditions namely organic load, pH and hydraulic retention time (Reis *et al.* 1991; van Munch and Greenfield 1998).

20.4 PHA UTILIZATION

20.4.1 PHA recovery

The process recovery of the polymer from the microbial biomass can affect the molecular weight. PHB and other PHA are readily extracted from microorganisms by refluxing with chlorinated hydrocarbons (dichloromethane or chloroform) without significant polymer fractionation (Figure 20.6). The polymer is precipitated with methanol or ethanol, leaving low molecular weight lipids in solution. However, the large-scale use of solvents is neither economic nor environmentally friendly, and other methods have been adopted. One of the most common techniques involves the extraction with an alkaline hypochlorite solution. The occurrence of some degradation by using this method, due to polymer hydrolysis is likely to occur (Lafferty *et al.* 1988; Anderson and Dawes 1990). Treatment of the bacterial cells with enzymes and surfactants that solubilize all of the components except PHA can also be used to isolate the biopolyesthers (Doi 1990).

More recently other extraction methods have been reported (Choi and Lee 1999). They include the use of several acids (HCl and H_2SO_4), alkalis (NaOH, KOH and NH_4OH), and surfactants (dioctylsulphosuccinate sodium salt, hexadecyltrimethylammonium bromide, sodium dodecylsulphate, polyoxyethylene-p-ter-octylphenol and polyoxyethylene (20) sorbitan monolaurate). NaOH and KOH proved to be very efficient and economical for the recovery of P(3HB), due to the mild conditions (30°C and 1 h) involved in the procedure. With these bases, polymers bearing high purities and yields, without fractionation, were obtained.

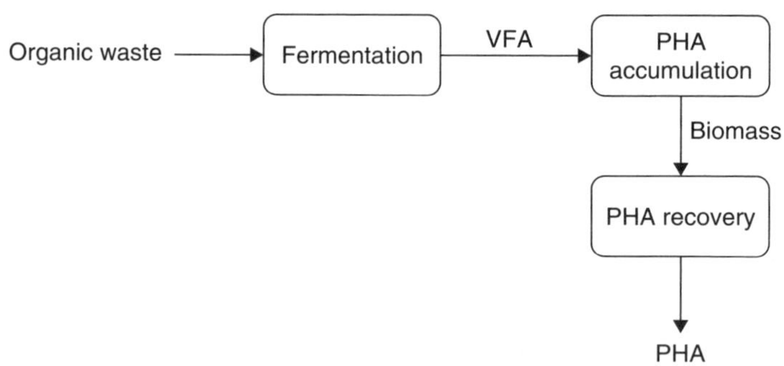

Figure 20.6 PHA production scheme from organic wastes.

20.4.2 PHA characteristics

In Table 20.3 are listed some examples of characteristics of bacterial polyesters obtained from mixed and pure cultures (weight average molecular weight, M_w; polidispersity, M_w/M_n, being M_n the number average molecular weight; T_m, melting point; T_g, glass transition temperature and degree of crystallinity). The information presented is not exhaustive because for PHA obtained from mixed cultures only a study reported by the authors was found in literature (Reis *et al.* 2003). In what concerns the characterization of PHA produced by pure cultures much information is available. Nevertheless, complete studies correlating the most important properties (polymer composition, average molecular weights, polydispersity, crystallinity and transition temperatures) with the producing microorganisms and carbon sources used, are very scarce. As a consequence, the following considerations about PHA characteristics are based on the general trends of these properties reported on different origins of the classical literature about bacterial PHA.

According to literature, even for PHB obtained from pure cultures of different bacteria, the value of M_w can vary in the range 1.7×10^5–4.5×10^6, as reviewed by Serafim *et al.* (2001). An overall evaluation of the presented data shows that the homo and copolymers obtained by mixed cultures bear average molecular weights, polydispersities and crystallinity in the same range of variation than those obtained by the use of pure cultures.

The molecular mass of the PHA is a very important feature, as it has an impact on their physical properties and the applications to be considered (Inoue and Yoshie 1992). While PHA used as drug release carriers must have M_w values in the range $6–10 \times 10^4$ its application in commodity products needs at least M_w equal to 4×10^5 to guarantee the mechanical properties (Lafferty *et al.* 1988). The molecular weight depends on the producing microorganism, the concentrations of the carbon sources and the environmental conditions used (Anderson and

Table 20.3 Comparison between characteristics of PHA obtained by mixed cultures in different processes and PHA obtained by pure cultures.

HB : HV mol ratio (%)	M_w (10^{-5})	M_w/M_n	T_g (°C)	T_m (°C)	Cryst.[a] (%)	Reference
Mixed cultures						
90 : 10	4.0	3.3	1.6	168	34	Reis *et al.* (2003)
75 : 25	17	2.5	56	139	4	Reis *et al.* (2003)
100	3.5	1.2	−19	145	30	Reis *et al.* (2003)
Pure cultures						
3 : 97[b]	3.7	2.3	−44	42	8	Matsusaki *et al.* (2000)
88 : 12[b]	4.9	1.4	−13	106	29	Matsusaki *et al.* (2000)
100	12	1.8	4	178	69	Matsusaki *et al.* (2000)

[a] Determined assuming a melting enthalpy value of 132 J/g for the 100% crystalline PHB (Miguel *et al.* 2001).
[b] P(3HB-co-3HA) with HA = C6–C12 (mcl 3-hydroxyalkanoates units).

Dawes 1990; Asenjo *et al.* 1995). The polydispersity, which is a measure of the molecular weight distribution, is very narrow usually, varying in the range 1–3, depending on the microorganism (Anderson and Dawes 1990; Lemos *et al.* 1998). The mechanical properties are also deeply correlated with the polymer average molecular weight: the longer the polymer chain length the higher its resistance.

The more common PHA, scl-PHA, are semi-crystalline polymers with a wide range of crystallinity values, depending on the chemical structure of the monomer unit or the molar percentage of the two monomer units, in PHA copolymers such as P(HB/HV) (Lafferty *et al.* 1988). PHB commonly exhibit a degree of crystallinity in the range 60–80%, which decreases to 30–40% as the 3HV content of P(3HB-co-3HV) increases to 30 mol% (Anderson *et al.* 1990). Mcl-PHA are elastomers, presenting very low crystallinity and values of melting temperatures in the range 39–61°C (de Koning *et al.* 1995).

The crystallinity degree, T_g and some mechanical properties of P(3HB) are similar to polypropylene (Young's modulus and tensile strength) being the elongation to break lower. Nevertheless, PHB is stiffer and more brittle than polypropylene (Lafferty *et al.* 1988; Anderson and Dawes 1990; Lee 1996). However, the chemical characteristics are quite different from polypropylene, and PHB possesses far inferior solvent resistance but better natural resistance to UV weathering (Anderson and Dawes 1990).

The polymer density, melting point, glass transition temperature and mechanical properties are strongly dependent on the molecular weight, polymer composition and crystallinity (Lafferty *et al.* 1988; Anderson and Dawes 1990). The glass transition temperature and melting point of P(3HB) are approximately 5°C and 175°C, respectively. These values decrease slightly for the copolymers, and as much as the 3HV units content increases. For copolymer containing 30 mol% 3HV the T_m is reduced to 143°C (Lafferty *et al.* 1988; Anderson and Dawes 1990; Lee 1996).

The incorporation of different monomeric units, other than HB, as in the P(3HB-co-3HV) copolymer developed by Metabolix Inc. presents improved mechanical properties when compared with PHB homopolymer, due to a decrease in crystallinity and T_m. The copolymer becomes tougher (increase in impact strength), more flexible (decrease in Young's modulus), and the elongation to break also increases as the content in 3HV units in the copolymer increases (Lafferty *et al.* 1988; Anderson and Dawes 1990; Lee 1996).

Thus, the final material properties can be tailored by controlling the fraction of 3-hydroxyvalerate incorporated into the copolymer during the fermentation process.

20.4.3 Applications

The more common PHA applications include packing films mainly used in bags, containers and paper coatings or disposable items namely, feminine hygienic products, cosmetic containers, etc.

As PHA are biocompatible and have no cytotoxic responses, they can be used in medical and pharmaceutical applications including slow drug release, surgical sutures and bone plates (Lafferty *et al.* 1988). Indeed, D(–)–3-hydroxybutyric acid is a normal metabolite in human blood and does not cause any significant toxicity. Also, human tissue does not contain bacteria or fungi able to degrade PHA. Hydrolytic degradation *in vivo* is almost exclusively caused by the simple hydrolytic degradation due to water, which occurs at very slow rates. This last aspect has hampered the wide applications of PHA in medicine and pharmacy.

In agriculture PHA can be used as biodegradable matrix for the sustained release of plant growth regulators or pesticides (Lafferty *et al.* 1988). These biopolymers can also be used in the synthesis of chiral compounds (enantiometrically pure chemicals) and are raw materials for the production of paints.

Furthermore, PHA can be blended with synthetic polymers forming semi-biodegradable plastics. PHA are miscible to some degree with a number of functional polymers, such as Nylon, polyethylene oxide, polyvinyl acetate, polyvinyl chloride, and polysaccharides, because of hydrogen bonding (Doi 1990). Blends with polysaccharides can be obtained with amylose, dextran, dextrin and sodium alginate. However, rate of PHA degradation is negatively affected by the presence of polysaccharides.

20.5 BIODEGRADABILITY OF PHA

One of the important characteristics of PHA is that they are completely biodegradable in soil, in activated sludge or in seawater. Some microorganisms, such as bacteria and fungi, excrete extracellular depolymerases that degrade PHA and utilize the decomposed compounds as carbon source for growth. Microorganisms where PHA depolymerase was found include: *A. faecalis*, *Pseudomonas limoignei*, *Rhodospirillum rubrum*, *Baccilus megateruim* and *Azotobacter beijerinkii* (Anderson and Dawes 1990). Degradation can be performed in aerobic conditions, resulting in carbon dioxide and water formation or under anaerobic conditions, with carbon dioxide and methane as the main end products.

The rate of degradation depends on the polymer composition, crystallinity and crystal size perfection, exposed surface area, microbial activity and environmental conditions, such as pH and temperature (Spyros *et al.* 1997). It is usually assumed that PHB depolymerases attack preferentially the amorphous regions of the semi-crystalline PHB. The degradability decreases as the overall crystallinity increases. The molecular weight of the PHA plays an important role on the degradation, as well: low molecular weight polymers are generally degraded more rapidly than high molecular weight ones (Spyros *et al.* 1997). A complete degradation of films of PHA was observed between 2 and 8 weeks (Doi 1990). Comparing with PLA and PGA, PHA are faster degraded. Indeed, degradation time for PLA and PGA vary from few months to years.

20.6 CONCLUSIONS

The main features of PHA, besides its complete biodegradability, are their similar properties to polypropylene and that they can be tailored by changing the organic substrate composition. PHA might replace, in a near future, synthetic plastics in a wide range of applications. However, the price of PHA is still very high compared with that of petrochemical-derived plastics. To be commercially competitive with the later, the price of PHA has to be reduced. This can be accomplished by using more productive cultures, cheaper substrates and more cost-effective recovery processes.

The use of mixed microbial cultures and waste material as substrate seems to be a possible solution to reduce the price of PHA to values close to the other biodegradable plastics such as polylactide and aliphatic polyesters. Also current advances in recovery processes based on the use of environmental friendly solvents or enzymatic techniques can contribute to the development of a more sustainable production process and to reduce the price of PHA to a competitive level.

Acknowledgements

We acknowledge the financial support of Fundação para a Ciência e Tecnologia (FCT) under the project POCTI/35675/BIO/2000. Luísa S. Serafim and Paulo C. Lemos acknowledge FCT for grants PRAXIS XXI BD/18287/98 and BPD/20197/99, respectively.

REFERENCES

Anderson, A.J. and Dawes, E.A. (1990) Occurrence, metabolism, metabolic role, and industrial uses of bacterial polyhydroxyalcanoates. *Microbiol. Rev.* **54**, 450–472.

Asenjo J., Schmidt A., Andersen, P. and Andrews, B. (1995) Effect of single nutrient limitation on poly-β-hydroxybutyrate molecular weight distribution in *Alcaligenes eutrophus*. *Biotechnol. Bioeng.* **46**, 497–502.

Beccari, M., Majone, M., Massanisso, P. and Ramadori, R. (1998) A bulking sludge with high storage response selected under intermittent feeding. *Water Res.* **32**(11), 3403–3413.

Beccari, M., Dionisi, D., Giuliani, A., Majone, M. and Ramadori, R. (2002) Effect of different carbon sources on aerobic storage by activated sludge. *Water Sci. Technol.* **43**(6), 157–168.

Beun, J.J., Paletta, F., Van Loosdrecht, M.C.M. and Heijnen, J.J. (2000) Stoichiometry and kinetics of poly-β-hydroxybutyrate in aerobic, slow growing, activated sludge cultures. *Biotechnol. Bioeng.* **67**(4), 379–389.

Beun, J.J., Dircks, K., van Loosdrecht, M.C.M. and Heijnen, J.J. (2002) Poly-β-hydroxybutyrate metabolism in dynamically fed mixed microbial cultures. *Water Res.* **36**, 1167–1180.

Biby, G.D. (2002) Degradable plastics: http://www.icma.com/inf/polymers.htm

Braunegg, G., Lefebvre, G. and Genser, K.F. (1998) Polyhydroxyalkanoates, biopolyesters from renewable resources: physiological and engineering aspects. *J. Biotechnol.* **65**, 127–161.

Choi, J. and Lee, S.Y. (1997) Process analysis and economic evaluation for poly(3-hydroxybutyrate) production by fermentation. *Bioprocess Eng.* **17**, 335–342.

Choi, J. and Lee, S.Y. (1999) Efficient and economical recovery of poly(3-hydroxybutyrate) from recombinant *Escherichia coli* by simple digestion with chemicals. *Biotechnol. Bioeng.* **62**(5), 546–553.

de Koning, G. (1995) Physical properties of bacterial poly((R)-3-hydroxyalkanoates). *Can. J. Microbiol.* **41**, 303–309.

Dionisi, D., Majone, M., Tandoi, V. and Beccari, M. (2001) Sequencing batch reactor: influence of periodic operation on performance of activated sludges in biological wastewater treatment. *Ind. Eng. Chem. Res.* **40**, 5110–5119.

Dircks, K., Henze, M., van Loosdrecht, M., Mosbæk, H. and Aspegren, H. (2001) Storage and degradation of poly-β-hydroxybutyrate in activated sludge under aerobic conditions. *Water Res.* **35**(9), 2277–2285.

Doi, Y. (1990) *Microbial Polyesters*. VCH Publishers, Inc., New York.

Inoue, Y. and Yoshie, N. (1992) Structure and physical properties of bacterially synthesized polyesters. *Prog. Polym. Sci.* **17**, 571–610.

Kim, B.S. (2000) Production of poly(3-hydroxybutyrate) from inexpensive substrates. *Enz. Microb. Technol.* **27**(12), 774–777.

Kim, B.S., Lee, S.C., Lee, S.Y., Chang, H.N., Chang, Y.K. and Woo, S.I. (1994) Production of poly(3-hydroxybutyric acid) by fed-batch culture of *Alcaligenes eutrophus* with glucose concentration control. *Biotechnol. Bioeng.* **43**, 892–898.

Lafferty, M.R., Korsato, B. and Korsato, W. (1988) Microbial production of poly-β-hydroxybutyric acid. In: *Biotechnology* (eds. H.J. Rehm and G. Reed), Vol. 6B, *Special Microbial Processes*, pp. 136–176. UCH, Weinheim.

Lee, S.Y. (1996) Plastic bacteria? Progress and prospects for polyhydroxyalkanoate production in bacteria. *Trends Biotechnol.* **14**, 431–438.

Lemos, P.C., Viana, C., Salgueiro, E.N., Ramos, A.M., Crespo, J.P.S.G. and Reis, M.A.M. (1998) Effect of carbon source on the formation of polyhydroxyalkanoates (PHA) by a phosphate-accumulating mixed *culture. Enz. Microb. Technol.* **22**, 662–671.

Levantesi, C., Serafim, L.S., Crocetti, G.R., Lemos, P.C., Rossetti, S., Blackall L.L., Reis M.A.M. and Tandoi, V. (2002) Analysis of the microbial community structure and function of a laboratory scale enhanced biological phosphate removal reactor. *Environ. Microbiol.* **4**(10), 559–569.

Matsusaki, H., Abe, H. and Doi, Y. (2000) Biosynthesis and properties of poly(3-hydroxybutyrate-co-3-hydroxyalkanoates) by recombinant strains of *Pseudomonas* sp. *61-3*. *Biomacromolecules* **1**, 17–22.

Meesters, K. (1998) *Production of Poly-3-Hydroxyalkanoates from Waste Streams*. Design Report, Delft University of Technology, Delft, NL.

Miguel O., Egibur, J.L. and Iruin, J.J. (2001) Blends of bacterial poly(3-hydroxybutyrate) with synthetic poly(3-hydroxybutyrate) and poly(epichloridrin): transport properties of carbon dioxide and water vapour. *Polymer* **42**, 953–962.

Poirier, Y., Nawrath, C. and Somerville, C. (1995) Production of polyhydroxyalkanoates, a family of biodegradable plastics and elastomers, in bacteria and plants. *Bio/Technology* **13**, 142–149.

Pozo, C., Martinez-Toledo, M.V., Rodelas, B. and Gonzalez-Lopes, J. (2002) Effects of culture conditions on the production of polyhydroxyalkanoates by *Azotobacter chroococcum H23* in media containing a high concentration of alpechin (wastewater from olive oil mills) as primary carbon source. *J. Biotechnol.* **97**(2), 125–131.

Purushothaman, M., Anderson, R.K.I., Narayana, S. and Jayaraman, V.K. (2001) Industrial byproducts as cheaper medium components influencing the production of polyhydroxyalkanoates (PHA): biodegradable plastics. *Bioprocess Biosys. Eng.* **24**(3), 131–136.

Reddy, C.S.K., Ghai, R., Rashmi and Kalia, V.C. (2003) Polyhydroxyalkanoates: an overview. *Biores. Technol.* **87**, 137–146.

Reis, M.A.M., Lemos, P.C., Martins, M.J., Costa, P.C., Gonçalves, L.M.D. and Carrondo, M.J.T. (1991) Influence of sulfates and operational parameters on volatile fatty acids concentration profile in acidogenic phase. *Bioproc. Eng.* **6**, 145–151.

Reis, M.A.M., Serafim, L.S., Lemos, P.C., Ramos, A.M., Aguiar, F.R. and van Loosdrecht, M. (2003) Production of polyhydroxyalkanoates by mixed microbial culture. *Bioprocess Biosyst. Eng.* **23**(6), 377–385.

Serafim, L.S., Lemos, P.C., Ramos, A.M., Crespo, J.P.S.G. and Reis, M.A.M. (2001) Polyhydroxyalkanoates production by activated sludge. In: *Biorelated Polymers: Sustainable Polymer Science and Technology* (eds. E. Chiellini, M.H. Mendes Gil, G. Braunegg, J. Buchert, P. Gatenholm and M. van der Zee), pp. 147–177. Kluwer Academic Plenum Publishing, New York.

Serafim, L.S., Lemos, P.C. and Reis, M.A.M. (2002) Change in metabolism of PHA accumulation by activated sludge modifying operating conditions. *Water Sci. Technol.* **46**(1-2), 353–356.

Serafim, L.S., Lemos, P.C., Oliveira, R. and Reis, M.A.M. (2004) Optimization of Polyhydroxybutyrale Production by Mixed Cultures Submitted to Aerobic Dynamic Feeding Conditions. *Biotech. Bioeng.* **87**(2), 145–160.

Silva, L.F., Gomez, J.G.C., Oliveira, M.S. and Torres, B.B. (2000) Propionic acid metabolism and poly-3-hydroxybutyrate-co-3-hydroxyvalerate (P3HB-co-3HV) production by *Burkholderia* sp. *J. Biotechnol.* **76**, 165–174.

Spyros, A., Kimmich, R., Briese, B.H. and Jendrossek, D. (1997) ^{1}H NMR imaging study of enzymatic degradation in poly(3-hydroxybutyrate) and poly(3-hydroxybutyrate-co-hydroxyvalerate). Evidence for preferential degradation of the amorphous phase by PHB depolymerase B from *Pseudomonas lemoignei*. *Macromolecules* **30**, 8218–8225.

Sudesh, K., Abe, H. and Doi, Y. (2000) Synthesis, structure and properties of polyhydroxyalkanoates: biological polyesters. *Prog. Polym. Sci.* **25**, 1503–1555.

Suk, C.K, Wook, R.H., Ho, P.C. and Goodrich, P.R. (1997) Poly(hydroxybutyrate-co-hydroxyvalerate) from swine waste liquor by *Azotobacter vinelandii* UWD. *Biotechnol. Lett.* **19**(1), 7–10.

Van Munch, E. and Greenfield, P.F. (1998) Estimating VFA concentrations in prefermenters by measuring pH. *Water Res.* **32**(8), 2431–2441.

Van Wegen, R.J., Ling, Y. and Milddelberg, A.P.J. (1998) Industrial production of polyhydroxyalkanoates using *Escherichia coli*: an economic analysis. *Trans. Inst. Chem. Eng.* **76**(Part A), 417–426.

Wong, P.A.L., Chua, H., Lo, W., Lawford, H.G. and Yu, P.H. (2002) Production of specific copolymers of polyhydroxyalkanoates from industrial waste. *Appl. Biochem. Biotechnol.* **98–100**, 655–662.

21

Biological production of hydrogen from waste and biomass

Gerasimos Lyberatos, Hariklia N. Gavala, Ioannis V. Skiadas, Michael Kornaros, Katerina Stamatelatou and Kostas Angelopoulos

21.1 INTRODUCTION

Nowadays, there is an increasing interest of the international community in renewable energy sources. Microorganisms have the potential of converting biomass (wastes/wastewater or plants) to valuable liquid or gaseous substances that could be used as energy sources and especially as fuels. Moreover, European Community encourages the research to increase the microbial production of fuels (Rose 1994; Atlas and Bartha 1998) in order to solve two problems: the agricultural overproduction and the high carbon dioxide emissions, associated with the greenhouse effect.

Ethanol and methane produced through anaerobic processes are among the best known microbial products and so far much research has been conducted in

Edited by Piet Lens, Bert Hamelers, Harry Hoitink, Werner Bidlingmaier. ISBN: 1 84339 054 X. Published by IWA Publishing, London, UK

order to elucidate the mechanisms and optimize their generation (Dalianis *et al.* 1996; Gavala *et al.* 1999). During the last decade, much attention has been paid to hydrogen gas and its potential use as a fuel for transportation purposes (automobiles) and electricity generation. The main advantage is that hydrogen is a clean and renewable energy source, possessing a high-energy yield (122 kJ/kg), not contributing to the greenhouse effect (Benemann 1996; Lay *et al.* 1999). For comparison, note that the energy yield of ethanol and methane are 26,500 and 50,120 kJ/kg, respectively.

The main drawback of hydrogen use, i.e. the difficulty in securing a safe storage, has been overcome by intense research on metal hydrides, which reversibly adsorb hydrogen at room temperature and low pressures (Ramachandran and Menon 1998; Noike and Mizuno 2000). Moreover, extensive research on the use of hydrogen gas in electrochemical power generation devices (fuel cells) is nowadays in progress in Japan and Europe with very promising developments (Ramachandran and Menon 1998).

Hydrogen can be generated in a number of ways, i.e. through fossil fuels processing or by electrolysis using solar power; processes that could be prohibitively expensive.

The biological production of hydrogen is much more attractive if wastes/wastewater could be used as the raw material (Mizuno *et al.* 2000a). However, the technology of the production of hydrogen through microbial processes still remains at an infant stage. Research on it started mainly in Japan in the early 1990s.

Hydrogen production by microorganisms from organic compounds is accomplished either through the direct use of the chemical energy contained in these substrates (heterotrophic fermentation) or through the additional use of light energy (photoheterotrophic fermentation, Benemann 1996). In both cases, hydrogen production is directly linked with energy metabolism. Hydrogen is produced as the ultimate product of the reducing activity of the enzymes hydrogenase and nitrogenase (Figures 21.1). Feredoxin is the main electron donor for both enzymes. This compound is reduced by NADH or NADPH (Glass *et al.* 1977; Bryant and Adams 1989) which in turn may originate from glycolysis. Both photoheterotrophic and heterotrophic processes have their advantages and disadvantages, but the hydrogen yield that has been achieved so far does not exceed 20% in either of them (Benemann 1996). In addition, so far no commercial processes have been developed based on microbial hydrogen production. Wang *et al.* (2003) recently compared yields (Table 21.1) while Levin *et al.* (2003) recently compared the rates obtained from the different biohydrogen producing processes (Table 21.2).

21.2 FERMENTATION

21.2.1 Photoheterotrophic fermentation

The majority of studies carried out on the biological production of hydrogen in the last 20 years have focused on the use of photosynthetic bacteria. Photoautotrophs

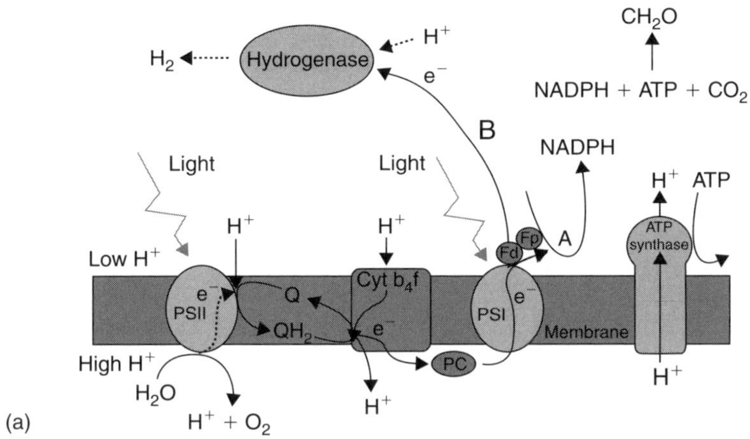

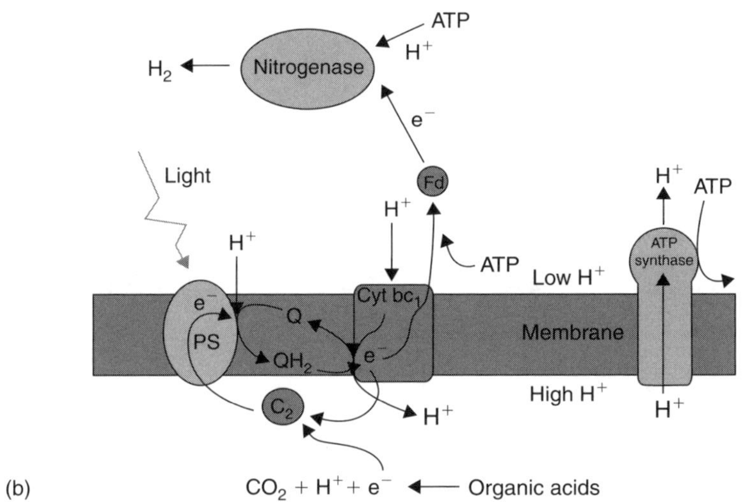

Figure 21.1 (a) Hydrogenase-mediated and (b) nitrogenase-mediated hydrogen production (Akkerman *et al.* 2002).

(green algae and cyanobacteria) are responsible for photolysis (Figure 21.1). They simply convert water into hydrogen and oxygen using light energy. Photoheterotrophic microorganisms produce hydrogen from organic matter, and thus are the only ones presenting an interest from a waste treatment point of view. They behave as follows depending on the conditions:

- Aerobic conditions:

$$nCO_2 + nH_2O + h\nu \rightarrow C_n(H_2O)_n + n/2O_2 \quad (21.1)$$

$$N_2 + 8H^+ + 8e- + 16ATP \rightarrow 2NH_3 + H_2 + 16ADP + 16Pi \text{ (nitrogenase)} \quad (21.2)$$

Table 21.1 Examples of yields obtained depending on the reactor, the substrate and the seed (from Wang *et al.* 2003).

Author	Reactor type	Substrate	Seed	Max. yield (g/kg VSS)
Tanisho and Ishiwata (1994)	Continuous reactor	Molasses	*Enterobacter aeroenes*	20.8[a]
Tanisho and Ishiwata (1995)	Continuous reactor with urethane foam to support the flocs	Molasses	*Enterobacter aeroenes*	15.6[a]
Kataoka *et al.* (1997)	Continuous reactor	Glucose and polypeptone	*Clostridium butylicum*	24.44[a]
Lay *et al.* (1999)	Serum Bottles	Organic municipal solid waste	Sludge of anaerobic digester	14 (37°C)[a]
Lin and Chang (1999)	Continuous reactor	Glucose	Consortium from sewage digester	19.0[a]
Zhu *et al.* (1999)	Batch reactor with agar gel immobilization	Tofu wastewater	Immobilized *Rhodobacter sphaeroides*	19.3 (30°C)[a,b]
Zhu *et al.* (1999)	Batch reactor with agar gel immobilization	Glucose	Immobilized *R. sphaeroides*	29.8 (30°C)[a,b]
Cheng *et al.* (2000)	Batch reactor	Concentrate of boiling treated sludge	Consortiums in boiled treated sludge	1.44[c]
Lay (2000)	Serum bottles	Starch	Sludge from anaerobic digester	95.6[a]
Mizuno *et al.* (2000a)	Continuous reactor without nitrogen sparging	Glucose	*Clostridium* sp.	9.3[a]
Mizuno *et al.* (2000a)	Continuous reactor with nitrogen sparging	Glucose	*Clostridium* sp.	15.8[a]
Mizuno *et al.* (2000b)	Batch reactor	Bean curd manufacturing waste	Anaerobic microflora	6.3
Bai *et al.* (2001)	Serum bottles	Peptone	Boiling treated sludge from UASB	1.98 (35°C)[a,b]
Lay (2001)	Serum bottles	Crystalline cellulose	Sludge from anaerobic digester	4.36[a]
Liang *et al.* (2001)	Serum bottles	Peptone	UASB sludge inhibited by chloroform	2.96[a]
Liang *et al.* (2001)	Serum bottles	Glucose	UASB sludge inhibited by chloroform	9.92[a]

[a] All substrates are assumed volatile and ash-free.
[b] The hydrogen was assumed to be collected at 1 atm.
[c] These values were reported in the original work.

Table 21.2 Comparison of rates of biohydrogen production (Levin *et al.* 2003).

$BioH_2$ system	H_2 synthesis rate (mmol $H_2/(l \times h)$)
Direct photolysis	0.07
Indirect photolysis	0.355
Photo-fermentation	0.16
CO-oxidation by *R. gelatinosus*	96.0
Dark-fermentations	
Mesophilic, pure strain[a]	21.0
Mesophilic, undefined[b]	64.5
Mesophilic, undefined	121.0
Thermophilic, undefined	8.2
Extreme thermophilic, pure strain[c]	8.4

[a] *Clostridium* sp. no. 2.
[b] A consortium of unknown microorganisms cultured from a natural substance and selected by the bioreactor culture conditions.
[c] *Caldicellulosiruptor saccharolyticus.*

- Anaerobic conditions:

$$C_2H_4O_2 + 2H_2O + h\nu \rightarrow 2CO_2 + 4H_2 \ (\Delta G_o = 75.2\,\text{kJ}) \quad (21.3)$$

The most promising results have been based on the genera *Rhodopseudomonas* and *Rhodobacter*. Successful hydrogen production based on organic wastes was first demonstrated by Zurer and Bachofen (1979), based on the photosynthetic bacterium *Rhodospirillum rubrum*. The maximum production rate was 20 ml H_2/h per g volatile suspended solids (VSS). Since most of the energy generated during photoheterotrophic fermentation originates from the organic substrate, there is a relatively low requirement for light energy. Consequently, the required size of a photobioreactor is relatively small compared with the corresponding requirements for photoautotrophic growth. However, the fermentation efficiency remains low, due to the fact that the cellular photosystem becomes saturated for small light intensities (6.5 $\pm$ 20 Klux, Sasikala *et al.* 1993), and this limits the maximum hydrogen production rate.

The optimal bioreactor conditions (pH, temperature, light intensity and frequency, cell density, etc.) for hydrogen production may differ substantially from the optimal growth conditions (Sasikala *et al.* 1993). Consequently, decoupling of microbial growth from hydrogen production could lead to higher overall hydrogen yields (Hillmer and Gest 1977; Sasikala *et al.* 1993). For example, Mao *et al.* (1986) demonstrated the advantages of first growing the microbes under dark aerobic conditions, before exposing them to anaerobic light conditions.

Another unique type of hydrogen production is through the attainment of the water gas shift reaction, CO and H_2O to H_2 and CO_2, by photosynthetic bacteria under dark conditions (Uffen 1976). CO is present in synthesis gas or in thermally gasified biomass, e.g. wood chips (Weaver *et al.* 1998). It seems that hydrogenase, when associated with CO is most suitable for a practical application (Weaver *et al.* 1998). The enzyme allows the production of hydrogen from CO at rates as high as 3 mmol/min per g VSS.

21.2.2 Heterotrophic fermentation

The heterotrophic process has the advantage that hydrogen can be continuously produced in a reactor without the need for light energy (Mizuno *et al.* 2000b). Heterotrophic fermentations aiming at hydrogen production may be carried out by a large variety of species, such as strict anaerobes (*clostridia, ruminococci* and *archaea*), facultative anaerobes (*Escherichia coli* and *Enterobacter aerogenes*) and aerobes (e.g. *Alcaligenes eutrophus* and *Bacillus licheniformis*) under anoxic conditions (Nandi and Sengupta 1998). Generation rates as high as 11 ml H_2/h per g VSS have been obtained using *Clostridium* grown on cellobiose (Taguchi *et al.* 1996). Among the various microbes, clostridia are the most promising fermentative bacteria since they are characterized by a rapid metabolism with a high potential for hydrogen yield (Noike and Mizuno 2000).

21.2.2.1 Substrates

Many organic compound types, carbohydrates, lipids and aminoacids can serve as hydrogen sources. There are many studies on the hydrogen production by pure cultures and from specific, sugar-based substrates reviewed by Nandi and Sengupta (1998). Only recently, attention has been paid to the production of hydrogen from wastes/wastewater by using a non-defined, mixed anaerobic culture (Ueno *et al.* 1996; Lay *et al.* 1999; Okamoto *et al.* 2000; Mizuno *et al.* 2000b; Noike and Mizuno 2000; Nielsen *et al.* 2001). It is worthy to note that fermentation of the organic fraction of municipal solid wastes with an adjusted moisture content of 85% resulted in a hydrogen composition of the biogas higher than 60% and a hydrogen production rate of 45 ml/g VSS/h (Lay *et al.* 1999).

From all the aforementioned studies it is well founded that the main source of hydrogen in fermentation processes are the carbohydrates. Consequently, wastes/wastewater rich in carbohydrates could be considered as potential resources of hydrogen if they undergo anaerobic fermentation under the required conditions for hydrogen recovery. The production of hydrogen increases with increasing carbohydrates concentration, but there is an upper threshold where the bacterial metabolism shifts to the production of alcohols that is not favourable for the production of hydrogen.

Due to positive Gibbs free energy change (ΔG) under standard conditions ($\Delta G_0 = +3.2$ kJ/mol), there is practically no energy generation during the (theoretical) complete bioconversion of glucose to hydrogen and CO_2 (Thauer 1976):

$$C_6H_{12}O_6 + 12H_2O \rightarrow 12H_2 + 6HCO_3^- + 6H^+ \tag{21.4}$$

It has been assessed that the highest yield per mole of glucose is 4 mol of hydrogen, based on the reaction (Solomon *et al.* 1995; Kengen *et al.* 1996):

$$C_6H_{12}O_6 + 4H_2O \rightarrow 2CH_3COO^- + 4H_2 + 2HCO_3^- + 4H^+ \quad (21.5)$$

The ΔG_0 of this reaction is -206 kJ/mol, and suffices for microbial growth. The rest of the hydrogen potential is contained in intermediates such as acetate, and ethanol and lactate to a lesser extent. The conversion of acetate to hydrogen through the reaction:

$$CH_3COOH + 2H_2O \rightarrow 4H_2 + 2CO_2 \quad (21.6)$$

is thermodynamically unfavourable under moderate temperatures ($\Delta G_0 = +104.6$ kJ/mol) and is directly dependent on the hydrogen partial pressure. For hydrogen partial pressures of 10–100 Pa, the oxidation of acetate becomes thermodynamically feasible at temperatures above 40°C (Lee and Zinder 1988b).

21.2.2.2 Process conditions

The operating conditions are vital for a hydrogen producing fermentation. Critical factors for the optimal production of hydrogen through fermentative processes are stirring conditions, pH, nutrients and organic carbon concentration, hydrogen partial pressure and the complete absence or the deliberate inhibition of hydrogenotrophic methanogens (Sparling *et al.* 1997; Noike and Mizuno 2000; Okamoto *et al.* 2000). In principle, it is possible to achieve metabolism shifts towards hydrogen, by optimizing process conditions, especially by reducing nutrients, partial pressure under 50 Pa (Lee and Zinder 1988b) or through temperature increase.

As far as the overall thermophilic heterotrophic fermentation is concerned, the hydrogen gas partial pressure must be kept low (<2 kPa), otherwise the growth rate will be reduced (Kelly 1988) and lactate may be produced instead of acetate (Janssen and Morgan 1992) or even alanine (Kengen and Stams 1994), thus reducing the overall efficiency of hydrogen production from glucose. Above 2 kPa, hydrogen may inhibit growth (Schaefer and Schoenheit 1991; Schroeder *et al.* 1994). Therefore, high hydrogen partial pressures limit growth and hydrogen production, as well as influence the selectivity of hydrogen.

For heterotrophic fermentation, hydrogen transfer from the liquid to the gaseous phase needs to be improved, through the development of suitable separation processes. The limited solubility of hydrogen in water does not prevent concentrations as high as 80 times the solubility limit under dynamic conditions (Pauss *et al.* 1990). The continuous stirring of the anaerobic culture and the removal of the hydrogen by another gas supply, i.e. nitrogen, helps to keep a low hydrogen partial pressure, resulting thus in enhanced hydrogen production rate and yield (Mizuno *et al.* 2000a). A number of palladium-based membranes have been developed in order to separate the hydrogen from the gas stream and are already used in industry as hydrogen purifier (Nielsen *et al.* 2001).

Higher temperatures favour the hydrogenase efficiency of thermophiles, since the affinity of hydrogen to hydrogenase is reduced (Adams 1990), while thermodynamic equilibrium favours hydrogen production (Lee and Zinder 1988b). Indeed, a bacterium has been isolated that is capable of converting anaerobically, acetate into hydrogen and CO_2 at 60°C (Lee and Zinder 1988a). Since then, similar organisms have been isolated from activated sludge (Ahring *et al.* 1995) and from thermal Springs in Iceland at 70°C (Ahring 1992), proving thus that ΔG at these temperatures is sufficiently negative to sustain growth.

The pH and the hydraulic retention time (HRT) are two other critical factors determining the extent of the hydrogen production. In the study of Lay (2000), working with starch and household waste as the raw material and a heat-shocked digested sludge as an inoculum, it has been concluded that a pH of 4.7–5.7 and an HRT of 15–20 h are the most favourable.

The complete absence or the inhibition of hydrogenotrophic methanogens is also a prerequisite for achieving maximum hydrogen production. Maintaining a low enough HRT and/or pH can assure the absence of these organisms.

21.2.2.3 Effect of reactor type

Most research studies on biological fermentative hydrogen production from sugars/carbohydrates using mixed cultures have been conducted in conventional continuous stirred tank reactors (CSTR) while investigations on hydrogen production in high rate systems allowing attached microbial growth are scarce. Gavala *et al.* (2003, 2004) studied the mesophilic (35°C) biological production of hydrogen from glucose in a continuous stirred tank type bioreactor (CSTR) in comparison with an upflow anaerobic sludge blanket bioreactor (UASB) containing sterilized granules, both inoculated with digester sludge at various HRT.

Some indicative results from this study are shown in Figure 21.2. One can see that the rate of hydrogen production at the UASB increased sharply at the HRT of 2 h while the transition of the CSTR was smooth. It is noticeable that the hydrogen production rate in the UASB exceeded more than two times that of the CSTR at the HRT of 2 h. This suggests that the UASB is more productive for the biological production of hydrogen compared to the CSTR technology at low HRT. In addition, the UASB configuration is more stable than the CSTR at the HRT tested regarding pH, glucose uptake and microbial by-products (Gavala *et al.* 2004).

21.2.2.4 Coupling biohydrogen production to anaerobic digestion

From the waste/wastewater treatment point of view, fermentative hydrogen production mainly includes the transformation of sugars into volatile fatty acids (VFA) without much effect on the organic content of the wastewater. Thus, a complementary stage would be necessary for chemical oxygen demand (COD) elimination. Anaerobic digestion has been used in the past for the stabilization of sludges as

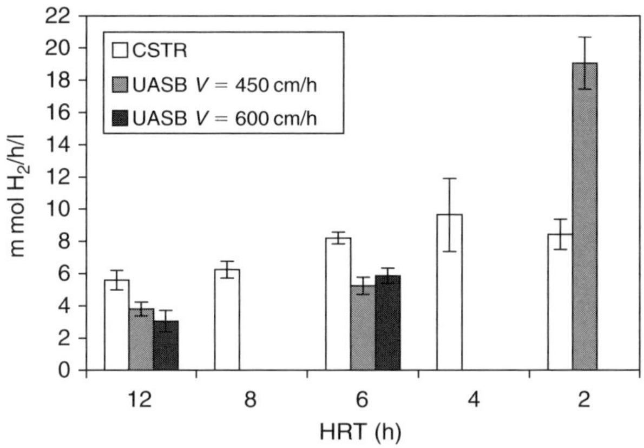

Figure 21.2 Hydrogen production rate presented along with the standard deviation during the CSTR and UASB experiment at different HRT and upflow velocities (V) of 450 and 600 cm/h (Gavala *et al.* 2004).

well as for the treatment of high organic content industrial and agro-industrial wastewaters.

Anaerobic digestion has many advantages compared to the conventional (aerobic) treatment technology such as the limited sludge production and nutrients requirement combined with the production of methane that can be used as an energy source. Since 1970 anaerobic digestion technology has been significantly developed with many industrial applications. It is well known that the formation of VFA during the acidogenesis of the organic matter is actually the precursor to methanogenesis. Thus, anaerobic digestion could be very efficiently coupled with a preliminary step for hydrogen production during which extended production of VFA takes place.

21.3 EXAMPLE APPLICATIONS OF BIOHYDROGEN PRODUCTION

Two example projects are outlined in the following as examples of the potential application of biological hydrogen production. The first, BIOTROLL, has to do with production of hydrogen from waste, while the second, PENNED, deals with production of biohydrogen from an energy crop.

21.3.1 BIOTROLL: integrated biological treatment and agricultural reuse of olive mill effluents with the concurrent recovery of energy sources

One major environmental concern in the Mediterranean countries (i.e. Spain, Italy, Greece, Portugal, Lebanon, Turkey, Tunisia and Syria) is the generation of olive mill wastewater from the traditional, three-phase, olive-oil producing industries. The three-phase processing of the olive fruits includes the milling of the olive fruit, followed by addition of warm (35°C) water and the subsequent centrifugation of this mixture using three-phase decanters. Thus, an oily phase that contains mostly the olive oil, a solid residue phase and an aqueous phase are produced. The solid residue phase (water content 40%) is further treated via drying and subsequent extraction of the remaining olive oil and the oil content of the stones with organic solvents. The remaining dry and solid product is used as a fuel. Finally, the aqueous phase is disposed as wastewater and it is characterized by a high organic content (COD up to 150 g/l, i.e. two orders of magnitude higher than domestic sewage).

Almost 30 millions m^3 of olive mill wastewater are generated every year around the Mediterranean Sea. Many studies exist on the physicochemical and/or biological treatment of olive mill wastewater with efficiencies of organic matter removal that lie between 65% and 95% (Tsonis and Grigoropoulos 1988; Chakchouk *et al.* 1994; Gavala *et al.* 1996; Hamdi 1996; Benitez *et al.* 1997; Marques *et al.* 1997; Andreozzi *et al.* 1998; Marques *et al.* 1998; Gavala *et al.* 1999; Benitez *et al.* 1999; Aktas *et al.* 2001) . However, none of the proposed management methods is practically applied due to the fact that effective wastewater treatment in the small-scale olive mills is prohibitively expensive.

Recently, the extraction of olive oil without addition of water and without production of wastewater has become possible, through the employment of two-phase decanters (Angerosa *et al.* 2000). The two-phase centrifugation of the milled olives, with no previous water addition, produces the olive oil containing phase and a semi-solid residue (60–70% water content) called olive pulp. So far, the treatment of the raw olive pulp includes its transportation to large-scale central units and the use of specially designed two-phase decanters for the recovery of the remaining olive oil (approximately 1–2%) and the separation of the olive stones (which are an excellent fuel) from the pulp material. This procedure results in increasing the water content of the olive pulp up to 80%. Most research efforts concerning the ultimate fate of the olive pulp have been focused on the composting as a treatment method (Madejon *et al.* 1998a, b; Nogales *et al.* 1998). The high water content, however, and the slurry nature of the olive pulp demands the use of large amounts of bulking material which leads to prohibitively large volumes of compost to be handled.

The olive pulp generated during the two-phase processing of the olive oil contains almost 80% (per dry weight) carbohydrates (Guillén *et al.* 1992) and thus it could be an ideal substrate for the production of energy in the form of hydrogen and biogas (methane). No research has been done so far on this aspect. The BIOTROLL project (EU Framework V Quality of Life Progamme: QLK5-CT-2002-02344) aims at studying the anaerobic bio-production of hydrogen from the olive pulp and the subsequent anaerobic treatment of the effluent with the simultaneous production of methane.

The efficient production of hydrogen and methane from the olive pulp presupposes an in-depth investigation and optimization of the factors that affect it as well as the development of an appropriate mathematical model. Specifically, BIOTROLL's objectives are:

(1) Determination of the optimal operational parameters for hydrogen production from the olive pulp.
(2) Investigation of the anaerobic digestibility and methane potential of the pretreated olive pulp. The operational parameters will be determined for optimum methane production and organic matter removal.
(3) Development of a kinetic model for both hydrogen and methane production from the olive pulp.
(4) Assessment of the feasibility of a two-step anaerobic process aiming at the production of hydrogen and methane.

The production of hydrogen from wastes/wastewater has received attention during the last decade. It has been proven that wastes/wastewater rich in carbohydrates could be used for hydrogen production. However, no study exists so far combining the hydrogen production with a waste purification technology. Furthermore, the olive pulp generated from the recently employed two-phase processing of olive oil could be regarded as a useful by-product that can be used for the production of energy in the form of hydrogen and methane (Figure 21.3). Last but not least, the proposed process could constitute a solution to the problem of olive pulp stabilization. The further implementation of this process will result in the reduction

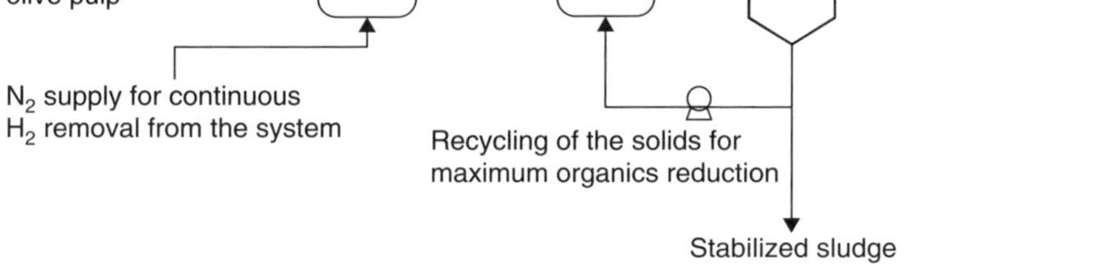

A: Anaerobic CSTR for hydrogen production
Hydraulic retention time: 0.5–1 days
pH: 4.7–5.7

B: Anaerobic CSTR for methane production and organic matter elimination
Hydraulic retention time: 15–30 days
pH: 6.7–7.5

C: Settling tank

Figure 21.3 The process proposed by BIOTROLL for the effective production of fuel from olive mill pulp.

of the heavy pollution from the generated olive mill wastewater during the traditional three-phase, olive oil processing in Mediterranean countries.

21.3.2 PENNED: biological hydrogen production as a fuel from organic farming of sweet sorghum

This program, funded by the Greek government, concerns the development of a novel process for the production of hydrogen-based electrical energy. The overall process is characterized by (a) almost complete energy utilization of biomass for the production of hydrogen and biogas, (b) increased performance based on fuel cell technology and (c) zero emissions.

The schematic diagram of the process is given in Figure 21.4. It consists of the following sub-processes:

(1) Biomass production based on organic farming of the energy plant *Sorghum bicolor* (L.) Moench Keller variety.
(2) Pretreatment and hydrolysis of the plant biomass.
(3) Valorization of the hydrolysis products for the fermentative production of H_2.
(4) Anaerobic digestion of the wastes from the fermentation process for biogas production.
(5) Usage of hydrogen in a fuel cell for the generation of electrical energy.

Individual targets of the PENNED project include:

(1) Development of techniques and methods of organic farming of sweet sorghum. In particular, it will be attempted to cover the culture needs on nutrients,

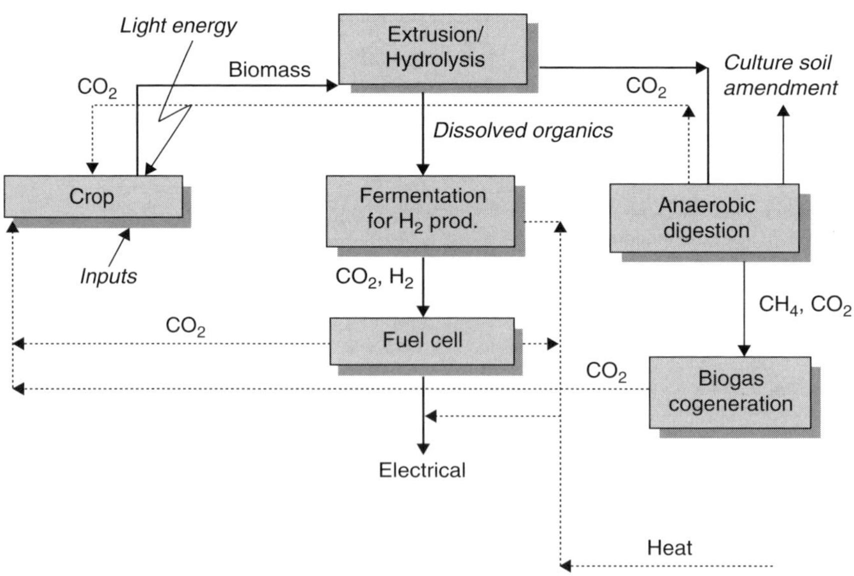

Figure 21.4 The energy producing process proposed in PENNED.

mainly nitrogen and phosphorus through the use of renewable sources such as atmospheric N and soil P reserves, as well as to determine the daily and seasonal water needs under conditions of organic farming.

(2) Development of lab procedures for the production of hydrogen from sweet sorghum biomass. This involves optimizing biological and/or physicochemical pretreatment methods, as well as development of optimized microbial process for the production of biomass from carbohydrates.

(3) Development of the know-how for the optimal production of biogas based on anaerobic digestion of the fermentation process wastes, and valorization of the digester waste as a soil amendment.

(4) Design and techno economic assessment of the overall process.

21.4 CONCLUSIONS

Biological hydrogen production from solid waste is still at an infant stage. So far yields of 10–20% have been reported but economic feasibility requires 60–80% (Benemann 1996). In addition, especially in the case of photoheterotrophic fermentation, overall production rates are still quite low and need to be increased substantially before commercial applications would be possible.

The development of high-yield hydrogen producing fermentation processes still requires substantial research. As far as photoheterotrophic hydrogen production is concerned, a significant improvement could result through an increase of the photosystem saturation level, so that the efficiency of light energy conversion might be increased. Optimization requires selection of the most efficient species, and subsequent optimization of the operating conditions.

Hyperthermophiles have the advantages that: (a) at their optimal temperatures competition against other heterotrophs is favoured and (b) the generated exoenzymes assist in the hydrolysis of biopolymers, such as starch, cellulose and hemicellulose, xylan, peptides, etc. As their potential has not been fully explored, a significant improvement in the yield and rate of hydrogen production is possible through their use. In both cases, i.e. heterotrophic as well as photoheterotrophic fermentations genetic improvements are possible (Benemann 1996).

Bioreactor design could improve yields, while combination of different processes (e.g. heterotrophic and photoheterotrophic fermentations) can significantly increase the yield, with a maximum theoretical possibility of 12 mol of hydrogen per mole of glucose.

Finally, in order to fully exploit the energy potential of organic solid waste, the residues from the hydrogen producing process should be further utilized for biogas production, as exemplified through the two research projects presented in this chapter.

REFERENCES

Adams, M.W.W. (1990) The metabolism of hydrogen by extremely thermophilic, sulfur-dependent bacteria. *FEMS Microbiol. Rev.* **75**, 219–238.

Ahring, B.K. (1992) Turn-over of acetate in hot springs at 70°C. In: *Thermophiles: Science and Technology International Conference*, p. 130. IceTec Publishers, Reykjavik, Iceland.
Ahring, B.K., Rintala, J., Nozhevnikova, A.N. and Mathrani, I.M. (1995) Metabolism of acetate in thermophilic (55°C) and extreme thermophilic (70°C) UASB granules. In: *Proceedings of the International Meeting on Anaerobic Process Bioenergy Environment*, p. 130, Copenhagen.
Akkerman, I., Janssen, M., Rocha, J. and Wijfels, R.H. (2002) Photobiological hydrogen production: photochemical efficiency and bioreactor design. *Int. J. Hydrogen Ener.* **27**, 1195–1208.
Aktas, E.S., Imre, S. and Ersoy, L. (2001) Characterization and lime treatment of olive mill wastewater. *Water Res.* **35**(9), 2336–2340.
Allen, S.E. (1989) *Chemical Analysis of Ecological Materials*. Blackwell Scientific Publication, London.
Andreozzi, R., Longo, G., Majone, M. and Modesti, G. (1998) Integrated treatment of olive oil mill effluents (OME): study of ozonation coupled with anaerobic digestion. *Water Res.* **32**(8), 2357–2364.
Angerosa, F., Mostallino, R., Basti, C., Vito, R. and Serraiocco, A. (2000) Virgin olive oil differentiation in relation to extraction methodologies. *J. Sci. Food Agricul.* **80**, 2190–2195.
Atlas, R.M. and Bartha, R. (1998) Microorganisms in mineral and energy recovery and fuel and biomass production. In: *Microbial Ecology, Fundamentals and Applications*, pp. 599–623. Benjamin/Cummings Publishing Company, Inc.
Bai, M.D., Cheng, S.S. and Tseng, I.C. (2001) Biohydrogen produced due to peptone degradation by pretreated seed sludge. *The IWA Asia-Pacific Regional Conference (WaterQual 2001)*, Fukuoka, Japan.
Benemann, J. (1996) Hydrogen biotechnology: progress and prospects. *Nature Biotechnol.* **14**, 1101–1103.
Benitez, F.J., Beltran-Heredia, J., Torregrosa, J. and Acero, J.L. (1997) Improvement of the anaerobic biodegradation of olive mill wastewaters by prior ozonation pre-treatment. *Bioprocess Eng.* **17**(3), 169–175.
Benitez, F.J., Beltran-Heredia, J., Torregrosa, J. and Acero, J.L. (1999) Treatment of olive mill wastewaters by ozonation, aerobic degradation and the combination of both treatments. *J. Chem. Technol. Biotechnol.* **74**(7), 639–646.
Bryant, F.O. and Adams, M.W.W. (1989) Characterization of hydrogenase from the hyperthermophilic archaebacterium, *Pyrococcus furiosus*. *J. Biol. Chem.* **264**, 5070–5079.
Chakchouk, M., Hamdi, M., Foussard, J.N. and Debellefontaine, H. (1994) Complete treatment of olive mill wastewaters by a wet air oxidation process coupled with a biological step. *Environ. Technol.* **15**(4), 323–332.
Cheng, S.S., Bai, M.D., Chang, S.M., Wu, K.L., Chen, W.C. and Chen, W.C. (2000) Studies on the feasibility of hydrogen production hydrolyzed sludge by anaerobic microorganisms. *The Twenty-fifth Wastewater Technology Conference*, Yunlin, Taiwan (in Chinese).
Claassen, P.A.M., van Lier, J.B., Lopez Contreras, A.M., van Niel, E.W.J., Sijtsma, L., Stams, A.J.M., de Vries, S.S. and Weusthuis, R.A. (1999) Utilisation of biomass for the supply of energy carriers. *Appl. Microbiol. Biotechnol.* **52**, 741–755.
Curt, M.D., Fernandez, J. and Martinez, M. (1996) Effect of nitrogen fertilization on biomass production and macronutrients extraction of sweet sorghum cv Keller. In: *Biomass for Energy and the Enviroment. Procceedings of the 9th European Bioenergy Conference*, Copenhagen 24–27 June, 1996.
Dalianis, C., Panoutsou, C. and Dercas, N. (1996) Sweet and fiber sorghum, two promising biomass crops. In: *First European Seminar on Sorghum for Energy and Industry*, pp. 173–176, Toulouse, France – 1–3 April, 1996.
Derkas, N. and Liakatas, A. (1999) Sorghum water loss in relation to irrigation treatment. *Water Res. Manage.* **13**, 39–57.
Gavala, H.N., Skiadas, I.V., Bozinis, N.A. and Lyberatos, G. (1996) Anaerobic codigestion of agricultural industries wastewaters. *Water Sci. Technol.* **34**(11), 67–75.

Gavala, H.N., Skiadas, I.V. and Lyberatos, G. (1999) On the performance of a centralized digestion facility receiving seasonal agroindustrial wastewaters. *Water Sci. Technol.* **40**(1), 339–346.

Gavala, H.N., Skiadas, I.V. and Ahring, B.K. (2003) Biological hydrogen production in conventional and high-rate systems. *Proceedings of the 4th Panhellenic Scientific Conference on Chemical Engineering Science*, Patras, Greece, May 29–31.

Gavala, H.N., Skiadas, I.V. and Ahring, B.K. (2004) Biological hydrogen production in conventional and high-rate systems, *submitted.*

Glass, T.L., Bryant, M.P. and Wolin, M.J. (1977) Partial purification of ferredoxin from Ruminococcus albus and its role in pyruvate metabolism and reduction of nicotinamide adenine dinucleotide by hydrogen. *J. Bacteriol.* **131**, 463–467.

Guillén, R., Heredia, A., Felizón, B., Jiménez, A., Montaño, A. and Fernández-Bolaños, J. (1992) Fibre fraction carbohydrates in *Olea europaea* (Gordal and Manzanilla var.). *Food Chem.* **44**, 173–178.

Hall, D.O., Scurlock, J.M.O., Bolhar-Nordenkamph, H.R., Leegood, R.C. and Long, S.P. (eds.) (1993) *Photosynthesis and Production in a Changing Enviroment. A Field and Laboratory Manual.* Chapman & Hall, London.

Hamdi, M. (1996) Anaerobic digestion of olive mill wastewaters. *Process Biochem.* **31**(2), 105–110.

Hillmer, P. and Gest, H. (1977) H_2 metabolism in the photosynthetic bacterium *Rhodopseudomonas capsulata*: production and utilization of H_2 by resting cells. *J. Bacteriol.* **129**, 732–739.

Janssen, P.H. and Morgan, H.W. (1992) Glucose catabolism by *Spirochaeta-Thermophila* RI 19.B1. *J. Bacteriol.* **174**(8), 2449–2453.

Kataoka, N., Miya, A. and Kiriyama, K. (1997) Studies on hydrogen production by continuous culture system of hydrogen-producing anaerobic bacteria. *Water Sci. Technol.* **36** (6–7), 41–47.

Kelly, R.M. (1988) Growth and gas production by the hyperthermophilic archaebactrium, *Pyrococcus furiosus. Biotech. Bioeng.* **32**, 438–444.

Kengen, S.W.M. and Stams, A.J.M. (1994) Formation of L-alanine as a reduced end product in carbohydrate fermentation by the hyperthermophilic archaeon *Pyrococcus furiosus. Arch. Microbiol.* **161**, 168–175.

Kengen, S.W.M., Stams, A.J.M. and Vos, W.M. de (1996) Sugar metabolism of hyperthermophiles. *FEMS Microbiol. Rev.* **18**, 119–137.

Kordesch, K. and Simader, G. (1996) *Fuel Cells and Their Application.* VCH, Weinheim.

Lay, J.J. (2000) Modeling and optimization of anaerobic digested sludge converting starch to hydrogen. *Biotechnol. Bioeng.* **68**, 269–278.

Lay, J.J., Lee, Y.J. and Noike, T. (1999) Feasibility of biological hydrogen production from organic fraction of municipal solid waste. *Water Res.* **33**(11), 2579–2586.

Lay, J.J. (2001) Biohydrogen generation by mesophilic anaerobic fermentation of microcrystalline cellulose. *Biotechnol. Bioeng.* **74**, 280–287.

Ledgard, S.F. and Steele, K.W. (1992) Biological nitrogen fixation in mixed legume/grass pastures. *Plant soil* **141**, 137–153.

Lee, M.J. and Zinder, S.H. (1988a) Isolation and characterization of a thermophilic bacterium which oxidizes acetate in syntrophic association with a methanogen and which grows acetogenically on H_2–CO_2. *Appl. Environ. Microbiol.* **54**, 124–129.

Lee, M.J. and Zinder, S.H. (1988b) Hydrogen partial pressures in a thermophilic acetate-oxidizing methanogenic coculture. *Appl. Environ. Microbiol.* **54**, 1457–1461.

Levin, D.B., Pitt, L. and Love, M. (2003) Biohydrogen production: prospects and limitations to practical application. *Int. J. Hydrogen Ener.* **29**(2), 173–185.

Liang, T.M., Wu, K.L. and Cheng, S.S. (2001) Hydrogen production of chloroform inhibited granular sludge. *The IWA Asia-Pacific Regional Conference (WaterQual 2001)*, Fukuoka, Japan.

Madejon, E., Galli, E. and Tomati, U. (1998a) Bioremediation of olive mill pomaces for agricultural purposes. *Fresenius Environ. Bull.* **7**(12A), 873–879.
Madejon, E., Galli, E. and Tomati, U. (1998b) Composting of wastes produced by low water consuming olive mill technology. *Agrochimica* **42**(3–4), 135–146.
Mao, X.Y., Miyake, J. and Kawamura, S. (1986) Screening photosynthetic bacteria for hydrogen production from organic acids. *J. Ferment. Technol.* **64**, 245–249.
Marques, I.P., Teixeira, A., Rodrigues, L., Dias, S.M. and Novais, J.M. (1998) Anaerobic treatment of olive mill wastewater with digested piggery effluent. *Water Environ. Res.* **70**(5), 1056–1061.
Mizuno, O., Dinsdale, R., Hawkes, F.R., Hawkes, D.L. and Noike, T. (2000a) Enhancement of hydrogen production from glucose by nitrogen gas sparging. *Biores. Technol.* **73**, 59–65.
Mizuno, O., Ohara, T., Shinya, M. and Noike, T. (2000b) Characteristics of hydrogen production from bean curd manufacturing waste by anaerobic microflora. *Water Sci. Technol.* **42**(3–4), 345–350.
Nandi, R. and Sengupta, S. (1998) Microbial production of hydrogen: an overview. *Crit. Rev. Microbiol.* **24**, 61–84.
Nielsen, A.T., Amandusson, H., Bjorklund, R., Dannetun, H., Ejlertsson, J., Ekedahl, L.G., Lundström, I. and Svensson, B.H. (2001) Hydrogen production from organic waste. *Int. J. Hydrogen Ener.* **26**, 547–550.
Nogales, R., Thompson, R., Calmet, A., Benitez, E., Gomez, M. and Elvira, C. (1998) Feasibility of vermicomposting residues from olive oil production obtained using two stage centrifugation. *J. Environ. Sci. Health, Part A – Toxic/Hazardous Subst. Environ. Eng.* **33**(7), 1491–1506.
Noike, T. and Mizuno, O. (2000) Hydrogen fermentation of organic municipal wastes. *Water Sci. Technol.* **42**(12), 155–162.
Okamoto, M., Miyahara, T., Mizuno, O. and Noike, T. (2000) Biological hydrogen potential of materials characteristic of the organic fraction of municipal solid wastes. *Water Sci. Technol.* **41**(3), 25–32.
Pauss, A., Andre, G., Perrier, M. and Guiot, S.R. (1990) Liquid-to-gas mass transfer in anaerobic processes: inevitable transfer limitations of methane and hydrogen in the biomethanation processes. *Appl. Environ. Microbiol.* **56**, 1636–1644.
Ramachandran, R. and Menon, R.K. (1998) An overview of industrial uses of hydrogen. *Int. J. Hydrogen Ener.* **23**, 593–598.
Rose, J. (1994) Biofuel benefits questioned. *Environ. Sci. Technol.* **28**, 63A.
Sasikala, K., Ramana, C.H.V., Rao, P.R. and Kovacs, K.L. (1993) Anoxygenic phototrophic bacteria: physiology and advances in hydrogen technology. *Adv. Appl. Microbiol.* **38**, 211–295.
Schaefer, T. and Schoenheit, P. (1991) Pyruvate metabolism of the hyperthermophilic archaebacterium *Pyrococcus furiosus*. Acetate formation from acetyl-CoA and ATP synthesis are catalysed by an acetyl-CoA synthetase (ADP-forming). *Arch. Microbiol.* **155**, 366–377.
Schroeder, C., Selig, M. and SchoÈnheit, P. (1994) Glucose fermentation to acetate, CO_2 and H_2 in the anaerobic hyperthermophilic eubacterium Thermotoga maritima: involvement of the Embden–Meyerhof Pathway. *Arch. Microbiol.* **161**, 460–470.
Sipma, J., Meulepas, R.J.W., Parshina, S.N., Stams, A.J.M., Lettinga, G. and Lens, P.N.L. (2003) Effect of CO, H_2 and sulfate on thermophilic (55°C) carbon monoxide conversion routes in two anaerobic bioreactor sludges. *Appl. Microbiol. Biotechnol.* **64**, 421–428.
Solomon, B.O., Zeng, A.P., Biebl, H., Schlieker, H., Posten, C. and Deckwer, W.D. (1995) Comparison of the energetic efficiencies of hydrogen and oxychemicals formation in Klebsiella pneumoniae and *Clostridium butyricum* during anaerobic growth on glycerol. *J. Biotechnol.* **39**, 107–117.

Sparling, R., Risbey, D. and Poggi-Varaldo, H.M. (1997) Hydrogen production from inhibited anaerobic composters. *Int. J. Hydrogen Ener.* **22**(6), 563–566.

Taguchi, F., Yamada, K., Hasegawa, K., Taki-Saito, T. and Hara, K. (1996) Continuous hydrogen production by *Clostridium* sp. strain no. 2 from cellulose hydrolysate in an aqueous two-phase system. *J. Ferm. Bioeng.* **82**, 80–83.

Tanisho, S. and Ishiwata, Y. (1994) Continuous hydrogen production from molasses by the bacterium *Enterobacter aerogenes*. *Int. J. Hydro. Ener.* **19**, 807–812.

Tanisho, S. and Ishiwata, Y. (1995) Continuous hydrogen production from molasses by fermentation using urethane foam as a support of flocks. *Int. J. Hydro. Ener.* **20**, 541–545.

Thauer, R.K. (1976) Limitation of microbial hydrogen formation via fermentation. In: *Microbial Energy Conversion* (eds. H.G. Schlegel and J. Barnea), pp. 201–294. Erich Goltze, GoÈttingen.

Tsonis, S.P. and Grigoropoulos, S. (1988) High-rate anaerobic treatment of olive oil mill wastewater. *J. Adv. Water Pollut. Contr.* **5**, 115–124.

Ueno, Y., Otsuka, S. and Morimoto, M. (1996) Hydrogen production from industrial wastewater by anaerobic microflora in chemostat culture. *J. Ferment. Bioeng.* **82**(2), 194–197.

Uffen, R.L. (1976) Anaerobic growth of a Rhodopseudomonas species in the dark with carbon dioxide as sole carbon and energy substrate. *Proc. Natl. Acad. Sci.,* USA **73**, 3298–3302.

Wang, C.C., Chang, C.W., Chu, C.P., Lee, D.J., Chang, B.V. and Liao, C.S. (2003) Producing hydrogen from wastewater sludge by *Clostridium bifermentans*. *J. Biotechnol.* **102**, 83–92.

Weaver, P., Maness, P.C., Rodes, C., Scahill, J., Dundorf, S. and Martin, S. (1998) Biological H_2 from fuel gases and from H_2O. *Proceeding of the 1998,US DOE Hydrogen Program Review*, pp. 1–8.

Zhu, H., Suzuki, T., Tsygankov, A.A., Asada, Y. and Miyake, J. (1999) Hydrogen production from tofu wastewater by *Rhodobacter sphaerodies* immobilized in agar gels. *Int. J. Hydro. Ener.* **24**, 305–310.

Zurer, H. and Bachofen, R. (1979) Hydrogen production by the photosynthetic bacterium *Rhodospirillum rubrum*. *Appl. Environ. Microbiol.* **37**, 789–793.

22

Hygienic safety in organic waste management

R. Böhm

22.1 INTRODUCTION

Products derived from biological wastes may be either recycled to agriculture and horticulture or sold as organic fertilizers or culture substrates to households. The agricultural utilization of organic wastes, whether they are from the farm itself or if they are recycled from other sources, require precautions in order to minimize risks for the involved animals, persons, plants and the environment. The risks may be caused by biotic factors like pathogens or abiotic factors like heavy metals. In this chapter, only the biotic factors will be regarded. In principle, stabilization of products from organic wastes may be achieved by physical, chemical or biological treatment alone or in combination. In this chapter, mainly the biological or more specific biotechnological treatment will be dealt with. Such a treatment may be done either in large-scale plants run in urban industrial or rural areas or in small-scale plants mostly run on farms in order to improve the farmers income. Both situations will be regarded here because special risks are connected with

 Resource Recovery and Reuse in Organic Solid Waste Management. Edited by Piet Lens, Bert Hamelers, Harry Hoitink, Werner Bidlingmaier. ISBN: 1 84339 054 X. Published by IWA Publishing, London, UK

handling materials, which may contain pathogens for man, animals and plants in an environment where the susceptible hosts are present. This is especially the case in rural composting or co-digestion plants, where municipal or industrial organic wastes are processed together with farm manure.

Mainly three hygienic risks do generally exist in handling biological wastes:

- Occupational health risks
- Environmental risk
- Risk concerning product safety.

Occupational health risks do exist in small-scale plants as well as in large-scale ones, but they are not the main subject of this contribution and will only be addressed in a short survey. More informations may be taken from other sources (Grüner 1996; Böhm *et al.* 1998). Environmental risks especially concerning spreading of pathogens close to animal houses and risks concerning epidemiological pathways closed via contaminated products as well as measures against such risks will be regarded here.

Recently European Union (EU) – regulations have been fixed which are dealing with risks due to animal by-products including those of faecal origin. Those regulations have to be regarded very carefully if they are fulfilling the intended purpose of avoiding risks due to other causative agents of notifiable animal diseases and other zoonotic agents than those causing transmissible spongiform encephalopathies (TSE agents). The hygienic parameters given there are originating from experiences with products of rendering plants, fixed 20 years ago. Therefore, some comments will be given here concerning those regulations which are up to a certain extend conflicting with actual scientific experiences and facts.

22.2 OCCUPATIONAL HEALTH ASPECTS

The collecting, handling and processing of source separated biological wastes generated a public discussion concerning occupational health risks due to bioaerosols. The involved personnel are either exposed to the microorganisms and related materials by direct contact with the wastes or via aerosol.

Results of recent investigations showed that there are, in principle, no differences between aerosols generated from biowastes, residual wastes and the traditional unseparated household wastes, which is especially the fact in collection, delivery and sorting of such material. This means, if ever there are existing health risks for workers in such facilities, they could not be new, but must have existed since decades (Martens *et al.* 1999).

First it has to be differentiated between primary and secondary sources concerning the emission of microorganisms, their components and/or their metabolites. Collection and handling of household wastes, especially of source separated biowastes, represents a sequence of different steps from which each one has a special importance with respect to propagation and the emission of microorganisms. Simplified this means that every step in which the collected material comes into contact with the environment and sets free the microorganisms is followed by a phase in which biological and physical influences are causing a microbiological process in the material itself, which will influence the hygienic state of the material in a positive or negative way (Böhm *et al.* 1996). It must be kept in mind, that not only the waste but also the persons handling the material may be a primary source of microorganisms too, because they are biotopes themselves. Moreover, the involved personnel represents a secondary source of germs by dirty skin, clothes and tools. This leads especially in small rooms, like driver cabins, to an extremely high microbial load of the air. Another secondary source may be rotating parts of machines, fans and similar equipment. If the waste air from a composting plant is treated in biowashers or biofilters parts of the indigenous microflora of such equipment may be also set free as a secondary aerosol.

The following risk factors must be considered:

- Bacteria: endotoxins, exotoxins, enzymes, metabolites.
- Fungi: glucanes, mycotoxins, metabolites (volatile).
- Viruses.

From the large variety of germs and biological products that may be present in the collected and treated wastes, only a very limited number of parameters may be detected due to principal and technical reasons. The most frequently measured parameters are the total germ count (bacteria) and the total fungal count. Both parameters are relatively unspecific and it must be questioned if any causal relationship may exist to occupational health effects. Even if more specific parameters like the amount of airborne *Aspergillus fumigatus* spores or thermophilic *Actinomyces* spp. are determined no dose/effect relationship can be found. This explains why negative health effects can be found in workers in one case and not in another. The microbial load in waste treatment plants mainly depends on the kind of material and the way of handling. In thermal utilization of waste, the microbial load is mainly caused by the germs present in the waste itself. In biological waste treatment, additional sources as the involved biotechnological process and the biological waste air treatment have to be considered.

Besides a broad variety of saprophytic microorganisms, wastes also contain pathogenic agents which are affecting human health. Those germs are present in unseparated household wastes and probably in residual wastes resulting from separate collection of biowastes too. With regard to pathogenic microorganisms, the untreated wastes are the main source. Therefore, its handling is mainly related to specific risks caused by pathogens transmitted via contact or stab wounds (e.g. *Clostridium tetani*).

The aerogenic distribution of microorganisms, their components or metabolites may happen at every stage of treatment up to the final thermal or biological processing and in the latter case handling of the resulting product too. Only in rare cases obligatory pathogenic germs are transmitted via aerosol. Generally, widely distributed microorganisms propagating in the collected material or involved in the biological degradation of the organic matter are released into the air. Until now one did not succeed in isolating *Salmonellas* from the air of a biocompost plant even if they may be present in the collected material up to 10^6 cfu/g moist weight (Scherer 1992). This means that the risk for a grown up healthy person to catch an aerogenic infection is neglectable against the risks of developing an allergic or a more complex multifactorial clinical picture like the organic dust toxic syndrome, short ODTS (Malmros 1994).

The following pathophysiological reactions and picture of illness may result from exposure to bioaerosols and volatile components from waste treatment:

- Exogenic allergic alveolitis (EAA).
- ODTS.
- Allergies caused by inhaled fungal spores.
- Allergic symptoms caused by ingestion of organic allergens.
- Allergies caused by contact to organic allergens.
- Intoxication by volatile components.

22.3 RISKS DUE TO PROCESSING AND UTILIZATION OF ORGANIC WASTES

A compilation of bacterial, fungal, parasitic and viral pathogens for man, animals and plants which may be present in organic wastes are given in Tables 22.1–22.4. Plant pathogenic organisms and viruses are mainly distributed directly by the product used as fertilizer or growth substrate to the susceptible plant. The general epidemiological risks due to handling and the utilization of organic wastes as fertilizers in agriculture with respect to animal pathogens and zoonotic agents are listed in Table 22.5. Those risks are not the same for every group of used wastes. There are quantitative and qualitative differences between manure, source separated biowaste, biogenic wastes of industrial origin and sludges from municipal wastewater purification.

Out of the variety of bacterial pathogens *Salmonella* spp. are the most relevant since they can infect or contaminate nearly all living vectors from insects to mammals. Multiresistant bacteria are coming more and more into focus since their transmission via the environment as well as the introduction of resistance genes into other bacteria may cause tremendous problems in human – and veterinary medicine. The roll of Clostridial toxins, which may be also detected in several organic wastes is not fully understood until now, since epidemiological proofs of

Table 22.1 A survey on obligatory and facultative pathogens for man and animals which had been isolated from biowastes and household wastes.

Bacteria	Citrobacter, *Clostridia*, Enterobacter, *E. coli*, *Klebsiella*, *Proteus*, Pseudomonads, *Salmonella*, *Serratia*, Staphylococci, Streptococci, *Yersinia*
Fungi	*Aspergillus* spp., e.g. *A. fumigatus*
Viruses	Adenovirus, Coxsackievirus, ECHO virus, Enterovirus, Hepatitis A virus, Herpesvirus suis, Paramyxovirus, Parvovirus, Pestivirus, Poliomyelitisvirus, Reovirus.

causing clinical cases are missing. Regarding the viral pathogens, enteroviruses and rotavirus are the most relevant ones from the point of view of environmental risks (Metzler *et al.* 1996). Special attention must be paid to the parasitic pathogens, not only to eggs of round- and tapeworms but to *Giardia lamblia* and especially *Cryptosporidium parvum* too, the later sometimes occurs in slurry from calves and cattle used in co-digestion. Nearly all gut related pathogens could be found in slaughterhouse effluents. If sludges are of plant origin or had been processed by using plant material, they may contain plant pathogenic viruses, fungi, bacteria, parasites and undesired weeds. Concerning the epidemiological importance of organic wastes of municipal origin for the transmission of pathogens several authors had given evidence in the past (Enigk 1980; Breer 1981; Ottolenghi and

Table 22.2 Hygienic relevance of different biological wastes originating from the households.

Type of waste	A	B
Meat leftovers (raw or insufficiently heated)		
– Meat cuttings, tendons, rinds, etc.	+	–
Food of animal origin		
– Egg shells	+	–
– Several meat and dairy products	+	–
– Raw milk products	+	–
– Leftovers from fish and shellfish	+	–
Other wastes (animal and man)		
– Dirty packing material for meat and products of animal origin	+	–
– Used litter and wastes from pets	+	+
– Used paper handkerchiefs and sanitary pads	+	–
– Diapers	+	–
Household wastes from		
– Potatoes	–	+
– Carrots	–	+
– Onions	–	+
– Tomatoes	–	+
– Cucumbers	–	+
– Salad	–	+
– Cabbage	–	+
– Beans	–	+
– Cut flowers	–	+
– Balcony and indoor plants	–	+
Garden wastes		
– Boughs and plant material	–	+
– Fruits	–	+
– Dead leaves and lawn trimmings (faecal contamination)	(+)	+
Other wastes (plant origin)		
– Paper	–	–
– Paperboard	–	–
– Organic packing material (e.g. wood wool)	–	–

A: may contain pathogens of man and animals; B: may contain plant pathogens and/or weed seeds.

Table 22.3 A selection of plant pathogenic bacteria and viruses (Menke 1992; shortened*).

Pathogen	Susceptible plant	Type
Xanthomonas campestris	White cabbage, turnip cabbage, swede cauliflower	B
Pseudomonas marginalis	Salad, endive	B
Pseudomonas phaseolicola	Beans	B
Pseudomonas lacrimans	Cubumber	B
Pseudomonas tabaci	Tabacco	B
Corynebacterium michiganense	Tomato	B
Corynebacterium sepedonicum	Potato	B
Erwinia phytophthora	Potato, carrot	B
Erwinia amylovora	Pomaceus fruits, flowers	B
Agrobacterium tumefaciens	Various hosts	B
Potato virus Y	Potato, tabacco, tomato	V
Potato virus X	Potato, tomato, tabacco, paprica, eggplants	V
Aucuba virus	Potato, tabacco, tomato	V
Tobacco-ring spot virus	Potato, tabacco, beans, cucumber	V
Rattle virus	Potato, tabacco	V
Tobacco-mosaic virus	Tobacco, tomato, paprica	V
Tobacco-necroses virus	Tobacco, beans	V
Horsebean-mosaic virus	Horsebeans, pea	V
Pea-mosaic virus	Peas, horsebeas	V
Bean-mosaic virus	Beans, runner beans	V
Yellow bean-mosaic virus	Beans, peas	V
Cauliflower-mosaic virus	Several cabbage species	V
Cucumber-mosaic virus	Cucumber, melon, pumpkin, spinach, peas, beans, salad, tomato, celery	V
Aucuba-mosaic virus	Cucumber, melon	V
Cabbage-ring spot virus	Cauliflower, white cabbage, horseradish, spinach, tobacco, rhubarb, flowers	V
Lettuce-mosaic virus	Salad, endive	V
Beet-mosaic virus	Spinach, root beet, leaf beet, peas	V
Onion-mosaic virus	Onion, leek	V

B: bacteria; V: viruses

* More details about transmission and resistance may be taken from the original paper, which also contains information concerning parasitic nematodes and weed seeds.

Hamparian 1987; Strauch 1998; Bendixen 1999; Böhm 1999a, b; Pesaro *et al.* 1999).

Special risks are connected with the reuse of catering wastes. Those risks are connected with the occurrence and survival of certain viral pathogens causing notifiable diseases in animals like classical swine fewer or foot and mouth disease (FMD). Special care has to be taken therefore if such materials, which may contain high amounts of infective viruses are treated in biotechnical aerobic or anaerobic processes. One important factor of special epidemiological importance is the long time survival as demonstrated in Table 22.6. In composting of such materials special risks are connected with bones. Bones are often found on the surface of windrows and they are attracting living vectors.

Table 22.4 A selection of plant pathogenic fungi (Menke 1992; shortened*).

Pathogen	Susceptible plant	Type
P. brassicae	Cabbage	F
Phoma apiicola	Cabbage	F
Peronospora brassicae	Celery	F
Peronospora spinaciae	Spinach	F
Peronospora destructor	Onion	F
Marssonina panattoniana	Salad	F
Sclerotinia minor	Salad	F
Botrytis cinerea	Salad	F
Bremia lactucae	Salad, endive	F
Cercospora beticola	Turnip	F
Aphanomyces raphani	Radish	F
Alternaria porri	Carrot	F
Septoria apii	Celery	F
Albugo tragoponis	Scorconera	F
Albugo candida	Horseradish	F
Turburcinia cepulae	Onion	F
Sclerotium cepivorum	Onion	F
Botrytis allii	Onion	F
Uromyces appendiculatus	Bean	F
Mycosphaerella pinodes	Peas	F
Ascochyta pinodella	Peas	F
Erysiphe polygoni	Peas	F
Cladosporium cucumernum	Cucumber	F
Sclerotinia sclerotiorum	Cucumber	F
Septoria lycopersici	Tomato	F
Alternaria solani	Tomato	F
Didymella lycopersici	Tomato	F
Rhizoctonia solani	Potato	F
Phytophthora infestans	Potato, tomato	F
Synchytrium endobioticum	Potato	F
Verticillium albo-atrum	Potato	F

F: fungi.
* More details concerning transmission and resistance may be taken from the original paper, which also contains information concerning parasitic nematodes and weed seeds.

Moreover, the heat penetration into bones is difficult, the delay times for temperature equilibration is 3 days at 55°C and 1 day at 61°C (Breitenfeld 2000). Material containing a relatively high amount of bones like catering wastes have to be regarded, therefore, very critically from the point view of veterinary epidemiology.

Animal manure from healthy herds generally represents a low epidemiological risk for farm animals. In certain situations a specific risk may result for humans especially if apparently healthy cattle is shedding enterohaemorrhagic *Escherichia coli* (EHEC) bacteria, but the relevant ways of transmission are not via slurry and environment. Moreover, the percentage of herds with those bacteria present in the manure is overestimated in continental Europe (Karuniawati *et al.* 2000). The same applies for *Salmonella* spp. in manure (Philipp *et al.* 1990). Nevertheless, if a herd is already infected or pathogens are shed during the incubation time a high

Table 22.5 Epidemiological importance of processed wastes and residuals as well as of the resulting products.

A	*Direct transmission to farm animals* Contamination of meadows Introduction of pathogens by storage and processing close to susceptible animals Aerogenic transmission by spreading the materials onto farmland
B	*Direct transmission to humans* Handling of contaminated products in the household Occupational exposure to contaminated products Accidental transmission to immunocompromised persons
C	*Indirect transmission to farm animals* Via feed from contaminated sites Via living vectors
D	*Indirect transmission to humans* Via introduction of zoonotic agents into the food chain Via food contaminated by living vectors
E	*Introduction into the environment* Generation of carriers in the fauna Introduction of undesired properties into the microflora

Table 22.6 Survival of a selection of viruses in meat and meat products.

Foot and mouth disease virus	
Frozen meat	About 240 days
Meat products	Up to 4 month
Classical swine fever virus	
Frozen meat	About 150 days

risk for transmission to other susceptible animal populations does exist and veterinary authorities have to take adequate measures.

A special epidemiological situation is given if central storage units for liquid manure are used by several different farmers. Under such conditions the closed inner farm circle is no longer existing and the epidemiological risks are increasing by the number of farms involved as well as the number of animals kept. Therefore, special hygienic precautions must be taken especially in co-digestion plants.

22.4 PREVENTION OF OCCUPATIONAL RISKS

European legislation requires several precautions with respect to biological working materials. This includes classification of the work into one of the defined risk groups, giving safety instructions to the workers, preventing infections by offering vaccination and providing protective measures, offering occupational health care and reporting to the authorities if works are carried out at higher risk classes. This concerns mainly the personnel exposed to the microorganisms and related materials by direct contact or via aerosol in organic waste treatment facilities, especially those working at the sorting belt. For prevention of infections by certain pathogens it is obligatory in Germany according LASI-LV 1 (1995) and LASI-LV 13 (1997) to vaccinate the personnel in waste treatment plants against tetanus and recommended to vaccinate against polyomyelitis and diphtheria. Vaccination against hepatitis B is strictly recommended for all workers handling untreated wastes. In general, protective clothes and – gloves have to be worn, smoking, eating and drinking have to be prohibited at such working places. If possible, hand-sorting should be generally avoided. It seems to be better to use electronic detectors, which will indicate the presence of undesired materials during the collection of wastes. They have shown to be very effective in order to reduce the heavy metal contents of compost from biowastes and thus can substitute hand-sorting (Krauss *et al.* 1996).

This applies for both, aerobic and anaerobic treatment. Regarding the two different biotechnological processes themselves, there is a basic difference concerning the emission of microorganisms and microbial products between aerobic and anaerobic biotechnological treatment. In composting, the setting free of aerosols is unavoidable while biogas production takes place in a closed vessel. This means that anaerobic treatment is more advantageous from this point of view than composting. Microbial aerosols may be avoided or reduced by composting in closed systems and by modern waste air treatment too, but generally composting in windrows leads to significantly high emissions of bioaerosols as shown in Table 22.7. If the anaerobically treated material is used in a liquid state no further processing which may generate aerosols will happen. On the other hand, if the solids

Table 22.7 Airborne microorganisms collected with a Sartorius MD 8 sampler in six composting plants during turning of windrows and screening.

Sampled area	Microbiological parameter	n	Minimum	Median	Maximum
Turning	Total bacterial count	48	9.2×10^3	2.4×10^5	5.6×10^6
Screening	Total bacterial count	36	3.7×10^3	1.0×10^6	1.7×10^7
Background	Total bacterial count	9	1.3×10^3	1.3×10^4	2.8×10^4
Turning	*E. coli*	48	n.n.	n.n.	1.6×10^2
Screening	*E. coli*	36	n.n.	n.n.	7.4×10^1
Background	*E. coli*	9	n.n.	n.n.	n.n.
Turning	Enterobacteria	48	n.n.	6.4×10^1	1.4×10^3
Screening	Enterobacteria	36	n.n.	1.6×10^2	1.1×10^4
Background	Enterobacteria	9	n.n.	n.n.	1.1×10^1
Turning	Aerobic sporeformers	48	9.2×10^2	1.9×10^5	2.8×10^6
Screening	Aerobic sporeformers	36	1.3×10^3	5.4×10^5	1.1×10^7
Background	Aerobic sporeformers	9	1.3×10^3	4.5×10^3	1.3×10^4
Turning	Total fungal count	48	1.4×10^3	3.7×10^5	1.8×10^7
Screening	Total fungal count	36	9.2×10^2	2.2×10^3	5.2×10^5
Background	Total fungal count	9	1.3×10^3	4.5×10^3	2.2×10^4
Turning	*A. fumigatus*	48	1.4×10^3	3.7×10^5	1.8×10^7
Screening	*A. fumigatus*	36	9.2×10^2	2.2×10^3	5.2×10^5
Background	*A. fumigatus*	9	2.7×10^2	2.8×10^3	1.5×10^4

n: number of measurements; n.n.: below detection level.

Table 22.8 Distribution of particle size of bioaerosols collected with the Anderson sampler* in six composting plants during turning and screening of compost.

Area in the plant	Microbiological parameter	Stage 1 >8 μm (%)	Stage 2 4.7–8 μm (%)	Stage 3 3.3–4.7 μm (%)	Stage 4 2.1–3.3 μm (%)	Stage 5 1.1–2.1 μm (%)	Stage 6 0.6–1.1 μm (%)
Turning	Total bacterial count	24.3	13.1	14.6	13.1	20	14.4
Screening	Total bacterial count	47.5	11.2	11.4	9.8	9.5	10.2
Turning	Total bacterial count	18.9	10.7	10.4	10.7	17.9	31.5
Screening	Total bacterial count	50.4	16.8	7.6	8.4	8.4	8.4

* Deposition in the respiratory tract: Stage 1: nose; Stage 2: pharynx; Stage 3: trachea and primary bronchi; Stage 4: secondary bronchi, Stage 5: terminal bronchi; Stage 6: alveoli.

Table 22.9 Total bacterial and total fungal count in cfu/m^3 air in driver cabins with and without fine dust filter in the ventilation system.

	Total bacterial count			Total fungal count		
Location	n	Without filter	With filter	n	Without filter	With filter
Plant 1	9	1.8×10^5	1.1×10^4	9	3.3×10^3	6.0×10^3
Plant 2	8	6.2×10^5	4.8×10^4	8	3.3×10^5	4.7×10^4

n: number of measurements.

from the treated product are separated and composted, the same situation as in genuine composting of biowastes can be expected, but no experimental data are available in this field until now.

In composting plants, it was found that during turning of windrows the aerosol fraction with particles which may be inhaled into the lung is bigger than at other places of the plants (Table 22.8). It must be considered that at this place the number of airborne *A. fumigatus* spores is relatively high too. Therefore, no steady working places should be in such areas and turning of windrows should be done by automatic devices. In general, all technical measures should be taken to keep the concentration of aerosols as low as possible in order to minimize the unspecific risk. Drivers' cabins should be equipped with air conditions in combination with an appropriate dust filter system. Table 22.9 demonstrates that this measure is effective in reducing the microbial aerosol in the cabin for at least 1 log step. This technical measure has to be accompanied by keeping the cabin strictly clean and by wearing always clean protection clothes.

22.5 HYGIENIC SAFETY OF PRODUCTS

Biological, chemical and physical methods may be applied in order to inactivate pathogens. All methods may be successful in order to reach this purpose but some have special limitations. Nature and number of pathogens basically determines the requirements concerning the applied process, as well as the epidemiology, e.g. the occurrence and distribution of the relevant pathogens in the target region. It seems, therefore, not reasonable to define the requirements to the process, e.g. as sporocidal, because anthrax caused by sporeforming bacteria is an extremely rare infection in northern, central and western Europe. In addition, due to other legal requirements the possibility that those agents will be introduced into wastes and sludges is nearly zero. But *Salmonella* may be found regularly in organic wastes, wastewater as well as sludge and the number of enteric infection caused by this pathogen is steadily rising. Moreover, it may affect a broad spectrum of species as hosts and/or carriers. Thus, treatment must inactivate those types of pathogens with sufficient safety, e.g. by reducing them at least for 5 log steps.

Since it is impossible to supervise the treated material for each of the pathogenic agents which may occur, other strategies have to be used in order to assure the hygienic safety of the processed material. The first step in such a strategy is to find out a representative indicator organism, which may be used for checking the product for hygienic safety as well as for evaluating the treatment process for its capability to inactivate pathogens which are of epidemiological relevance. The second step which is necessary in this connection is to define hygienic requirements for the treatment itself, since due to the high volume of the product to be controlled as well as to the inhomogeneity of distribution of pathogens in the material, only products processed in a validated process should be distributed to the consumer or user. This means that the following strategies may be combined with each other in order to assure hygienic safe utilization of the processed materials:

- Validation of treatment (disinfection by chemical, physical or biological means).
- Continuous registration of the relevant process parameters (e.g. temperature, pH-value, exposure time).
- Microbiological supervision of the final product (indicators).
- Restrictions for the utilization of the final product.

All three ways may be gone alone or in combination as had been discussed in the frame work of the European Sludge Directive 86–278–EWG, but the most important step forward is the strategy that treatment of such materials should only be done in a validated process.

22.6 STRATEGIES FOR VALIDATION OF THE ORGANIC SOLID WASTE TREATMENT PROCESS

Process safety concerning the inactivation of relevant transmissible agents for plants, man and animals as well as weed seeds and reproducible parts of plants may be validated best in two steps. The first step is the validation of the process as designed by the producer of the technical equipment in a basic procedure, the second step is a putting into service validation of a treatment process at the plant with the input material under practical conditions.

22.6.1 The German Regulations

The German LAGA M10 (1995), a precursor of the present legal regulations in Germany, offered a relatively broad approach to solve this problem with respect to composting. It was based on the ATV (German Association of Wastewater Experts) recommendations of 1988 for treatment of sewage sludge. Due to economical reasons this approach had been changed into a procedure given in Table 22.10, using *Salmonella senftenberg* W 775 (H_2S negative), Tobacco-mosaic virus, *Plasmodiophora brassicae* and tomato seeds as test organisms. More details concerning the methods to be used in validating the different processes are described elsewhere (ATV 1992).

Other test organisms than *Salmonella* are in discussion in the framework of the projected EU-compost directive and in other national approaches in Europe and the USA for the same purpose. The most important are:

- Enterococci
- *E. coli*
- *Campylobacter*
- Enteric cytopathogenic bovine orphan (ECBO) virus
- Bovine parvovirus (BPV).

Table 22.10 Example of a validation and supervision strategy for biogas and composting plants and the resulting products according to the German biowaste ordinance (1998).

I. Process validation

Investigated parameter	Direct validation of the process
Hygienic safety concerning risks for man, animals and plants	• New constructed compost plants (within 12 months after opening of the plant) • Already validated plants of new technologies have been invented or of the process has been significantly modified (within 12 months after invention of modification) • Existing plants without validation within the last 5 years before this validation strategy was invented (within 18 months)
Number of the test trials/recording of data	2 test trials at open air composting plants, at least one in wintertime
Number of test organisms	
Human and veterinary hygiene	1 test organism (*S. senftenberg* W 775, H_2S negative)
Photo-hygiene	3 test organisms (*P. brassicae*, tobacco-mosaic virus, tomato seeds)
Number of samples	
Sample per test trial	
Human and veterinary hygiene	24*
Phytohygiene	36*

*At small plans half the number of samples ($\leq$3000 t/a).

Table 22.10 (*Continued*).

II. Process supervision	
Investigated parameter	Indirect process supervision
Hygienic safety concerning risks for man, animals and plants	• Continuous registration of temperature at three representative locations in the process, responsible for the inactivation of the microorganisms and seeds • Recording of process data (e.g. turning of windrows, moisture of material, starting and finishing data)
Number of the test trials/recording of data	Continuous data recording to be filed for at least 5 years
Number of test organisms	
Human and veterinary hygiene	–
Photo-hygiene	–
Number of samples	
Sample per test trial	–
Human and veterinary hygiene	
Phytohygiene	

Table 22.10.3 (*Continued*).

III. Supervision of the final product	
Investigated parameter	Supervision of the final product
Hygienic safety concerning risks for man, animals and plants	Regular investigation of the final product for hygienic safety*,**
Number of the test trials Recording of data	Continuously all over the year at least • semi-annual (plants with ⩽3000 t/a throughput) • quarterly (plants ⩾3000 t/a throughput)
Number of test organisms	
Human and veterinary hygiene	No *Salmonella* in 50 g product detectable
Photo-hygiene	<2 seed capable of germinating and/or reproducible parts of plants on 1 l of product
Number of samples Sample per test trial Human and veterinary hygiene Phytohygiene	Throughput of the plants in t/a 1. ⩽3000 (six samples per year) 2. >3000–6500 (six samples per year plus one more sample for every 1000 t throughput) 3. >6.500 (12 samples per year plus one more sample for every 3000 t)

* Every statement concerning the hygienic safety of the product is always based on the result of the supervision of the final product together with the result of the validation of the process.
** Every sample is a "mixed sample" (about 3 kg) based on five single samples of the final product.

Enterococci in general may be used, but if the carriers used in the survival studies are exposed to a composting process, it may happen that the test material is contaminated with indigenous Enterococci. The differentiation between indigenous flora and the test strain is not possible under the given circumstances, other the proposed *S. senftenberg* which is rarely present in the material and can be easily identified by a natural marker (H_2S negative). Recent investigations with *Enterococcus faecalis* DSM 257 are indicating that this species may be better suitable for the intended purpose, since it can be quantitatively detected and easily differentiated on species level by online polymerase chain reaction (PCR) (Beyer *et al.* 2002). *E. coli* and *Campylobacter jejuni* are not as resistant as the above-mentioned *Salmonella* strains, the same applies for ECBO virus. The BPV is much more resistant than *Salmonella* and has still to be taken into account, but

Table 22.11 Inactivation times during semi-technical in vessel composting for selected viral pathogens in a temperature range between 25°C and 65°C (Moss and Haas 2000).

Pathogen	Initial titre*	Detection level*	Inactivation time (h)
Swine vesicular disease (SVD) virus	6–8 log 10 KID_{50}	1–4 log KID_{50}	27–72**
Foot and mouth disease (FMD) virus	6–7 log 10 PFU	1–2 log 10 PFU	12–144
Classical swine fever (CSF) virus	5–6 log 10 KID_{50}	1 log 10 KID_{50}	12–144
Aujeszky disease (AD) virus	5–6 log 10 KID_{50}	1 log 10 KID_{50}	20–192
African swine fever (ASF) virus	5–6 log 10 KID_{50}	1–2 log 10 KID_{50}	27–168**

KID: culture infective dose; PFU: plaque forming units.
* Reduced to full log steps.
** No sufficient inactivation within 27 h if not at least 55°C had been reached.

Table 22.12 Time (h) necessary for the inactivation of 1 log 10 (T_{90}-values) of a choice of viral pathogens determined in a semi-technical co-digestions plant (catering waste and slurry) at several temperatures. The longest survival times are given without taking the exposure technique (virus suspension or virus adsorbed to a membrane) into account (Moos and Haas 2000).

Pathogen	Type of slurry							
	CS	PS	CS	PS	CS	PS	CS	PS
	30°C		35°C		50°C		55°C	
SVD virus	30.33	28.5	17.83	13.7	0.13	0.06	0.04	0.02
FMD virus	13.0	11.5	11.16	5.0	0.43	0.40	0.13	0.20
CFS virus	0.9	0.73	0.46	0.43	0.15	0.12	0.05	0.06
AD virus	1.51	1.43	1.56	1.31	0.13	0.18	0.09	0.05
ASF virus	1.45	2.63	1.21	4.55	0.86	1.11	0.16	0.21

CS: cattle slurry; PS: pig slurry.

re-isolation and propagation are very labour and cost intensive. Virology requires specialized laboratories, this is for a general purpose a strong limitation but some indications are given for using BPV. It could be demonstrated that in composting *S. senftenberg* W 775 will cover the most relevant viral pathogens causing notifiable diseases in farm animals because the time necessary for their inactivation is generally shorter or in the same range as for *S. senftenberg* W 775, H_2S negative (Braumiller *et al.* 2000; Moss and Haas 2000). The latter results are summarized in Table 22.11 and they are demonstrating the range of time necessary for reducing the infectivity of the tested viral pathogens for 3–4 log 10 steps.

From Tables 22.12 and 22.13 it could be taken that in anaerobic treatment the situation is a little bit different. The viruses of FMDs and African swine fever have a higher logarithmic inactivation rates (T_{90}-values) than the test strain *S. senftenberg* at 55°C, all the other behave similar or are less resistant. Since the T_{90}-value is far below 1 h there is enough safety in the normal case in which low concentration of such non-ubiquitous pathogens could be attended. If there is a high risk, especially if catering wastes are processed, it seems to be better to use BPV as test virus as mentioned above.

The supervision with pathogens and seeds may be regarded as "direct process validation". It must be accompanied by continuous recording of measurable process data like temperature, pH-value, humidity, etc., in order to detect deviations and disturbances of the process over the whole year, which may result into an insufficient microbicidal effect.

The so-called "indirect process supervision" may be accompanied by process recommendations. An example is given according to the German legislation. Composting plants should operate with a moisture content of 45–50% in the material at a pH-value about 7. The recommended exposure time at 55°C should be at least two weeks in the whole material in windrows and 65°C for 1 week in the whole material in vessels. Mesophilic biogas plants (mean hydraulic retention time at least 20 days) should only process materials after heating it at 70°C for 60 min or they have to pasteurize the final liquid product at 70°C for 60 min. Alternatively, composting of the solid phase as described is allowed. Heating of the material in thermal biogas plants could be shorter (70°C/60 min) presuming that they are operated in a continuous mode or semi-continuous mode with intervals in filling and harvesting of <22 h and a mean hydraulic retention time of at least 20 days. If such a reactor is run batch wise, a real exposure time of 24 h is recommended.

The system of process validation and control has to be completed by a continuous supervision of the final product, at least twice a year.

22.6.2 European animal by-product regulation

The European animal by-product regulation is fixing general and special requirements for the processing of materials depending on the risk for transmission of prion diseases. It defines category 1, 2 and 3 materials, category 1 materials have the highest and category 3 materials the lowest risk. In this connection, only the processing of class 3 materials in composting and biogas plants will be stressed here. The specific requirements for the approval of biogas and composting plants

Table 22.13 Decimal destruction rates (T_{90}-values) found for several bacteria and viruses in co-digestion (catering waste and slurry) at different temperatures according to Hoferer *et al.* (2001).

Type of carrier		Type of slurry							
		CS	PS	CS*	PS	CS	PS	CS	PS
		30°C		35°C		50°C		55°C	
Bacteria									
EHEC	S	78.48	52.08	n	27.60	0.43	0.40	0.08	0.03
								0.03	0.02
S. senftenberg	S	56.40	41.76	34.08	25.20	0.60	0.53	0.11	0.06
							0.07	0.06	
Strept. faec.	S	186.24	74.40	143.04	93.36	7.48	11.26	1.70	1.64
Viruses									
ECBO	S	16.13	34.60	23.28	10.58	0.93	0.12	0.10	0.07
ECBO	A	43.44	24.72	25.20	17.36	0.61	0.12	0.24	0.07
ERV	S	13.53	23.98	n	n	1.38	0.89	0.13	0.28
ERV	A	34.08	25.92	n	n	0.96	0.72	0.54	0.20
Polio	S	n	22.32	n	n	1.05	0.71	0.09	0.07
Polio	A	n	32.16	n	n	0.63	0.18	0.07	0.03
BPV	S	n	n	317.53	n	23.26–28.41	11.99–31.06	7.87	6.17
BPV	A	n	n	180.24	n	20.41–10.48	14.27	4.67	5.47

n: not investigated; A: test organisms absorbed to membranes; S: suspended test organisms; CS: cattle slurry; PS: pig slurry.
*Long time stored pig slurry without catering wastes.

are mainly concerning the equipment and general hygienic measures as storage, vehicles, pest control, etc., no direct process validation is required.

According to EU law, the plants shall be equipped with:

- a pasteurization/hygienization unit, which cannot be bypassed in biogas plants;
- a closed composting reactor which cannot bypassed in composting plants;
- this unit/reactor must have installations for monitoring temperature against time;
- this unit/reactor must have recording devices for continuous recording of the measured data;
- this unit/reactor must have an adequate safety system to prevent insufficient heating;
- adequate facilities for the cleaning and disinfecting of vehicles and containers leaving the plant.

The regulations are also giving processing standards for biogas and composting plants. The requirements for the biogas plants are limited to the pasteurization/hygienization unit and do not even mention a thermophilic anaerobic process. The maximum particle size before entering the unit is fixed to 12 mm, the minimum temperature in all material to 70°C and the minimum exposure time without interruption to 60 min. The requirements for the composting process are similar, but no pasteurization/hygienization prior to the process is necessary. The minimum temperature in all material shall be 70°C for 60 min.

This is not a realistic approach, therefore, it is recommended from the practical and the scientific point of view, to validate at least the composting process by BPV as test organism (reduction for at least three log) if critical materials as total gut packages, slaughterhouse offal or catering wastes are being processed.

22.7 SUPERVISION OF THE FINAL PRODUCT

As mentioned above, the investigation of the final product in order to detect every pathogen which may be present in the material is impossible. Therefore representative indicator organisms have to be determined from the point of view of human and animal health as well as for the purpose of safe plant breeding and production. Those indicator organisms must fulfil several requirements:

- They have to be present with a high probability in the raw materials.
- The transmission via the final product must be a factor in epidemiology.
- The indicator should not be involved in the biotechnological process itself.
- The indicator should not be an organism which is generally present in soil and soil related materials.
- The method for isolation and identification must be simple, definitely and reliable if applied to a substrate with a complex microbiological matrix as compost or digested material.

With respect to public health and veterinary requirements several indicators and parameters are in discussion:

- *Salmonella*
- Enterococci (Streptococci of group E)

- *Staphylococcus aureus*
- Enterobacteriaceae
- *E. coli*
- *Campylobacter*
- *Yersinia* sp.
- *Listeria* sp.
- *Clostridium perfringens*
- Sulphite reducing Clostridia
- Entero viruses
- Rotavirus
- Eggs of nematodes
- Larves of nematodes

Since compost or the anaerobic digested material are products of a microbial degradation process and the knowledge about the microbiological ecology of such materials is still limited, it must be warned to use isolation and identification techniques common in clinical microbiology or in drinking water examination without careful validation in combination with the involved sample materials.

The variety of species to be present in environmental and such complex samples by far exceeds the limited number of species to be taken into account in se- and excreta as well as in body fluids and the variability in species is high and not yet fully understood. Moreover, microbial parameters which are used in the field of water hygiene and food inspection are not applicable to substrates like compost or sludges from anaerobic digestion, because most of those indicators belong to the indigenous flora of agricultural soils (Böhm 1995). If the limited reliability and applicability of methods adopted from clinical microbiology and water inspection for the intended field of use is taken into account as well as the fact that the exclusion of organisms which generally may be found in normal soils makes no sense for a substrate and fertilizer as, e.g. compost, the following microbial parameters are (with some exception in certain situations) inappropriate: *S. aureus*, Enterobacteriaceae, *C. perfringens*, sulphite reducing *Clostridia* and *Listeria*. For hygienic monitoring of the final product the European animal by-product regulations have fixed the parameter <300 Enterobacteriaceae per g. This is not to keep in biotechnological treatment, actual unpublished data of own investigations showed, that compost coming out of a validated process may contain up to 10^5 cfu Enterobacteriaceae per g, while *Salmonella* are reliable inactivated.

One parameter which seems to be very useful and reliable in this context is the absence or presence of *Salmonella*. They are generally found in fresh biowastes or untreated sewage sludge with a high probability in various concentrations. Since it is known that the probability of identifying a positive sample is basically related to the amount of investigated material a compromise between feasibility and reliability has to be found. It is proposed to investigate 50 g or 100 g (2×50 g) of material for the presence or absence of *Salmonella* with the method described in principle in the German biowaste ordinance using a pre-enrichment in buffered peptone water and an enrichment step (Rappaport *et al.* 1956; Edel and Kampelmacher 1969; Vassiliadis, 1983). The approach of the European animal by-product regulations is to investigate only 25 g of material as was the state of science 15 years ago (ATV 1988). This gives less good performance data as where 50 g of materials are used.

Some other parameters are still in discussion with respect to sewage sludge treatment and composting in the framework of EU directives. For example, Enterococci cannot be used as indicator in the examination of compost and compost related products but for the thermophilic anaerobic treatment in biogas plants as well as for pure thermal treatment, they are very valuable (Bendixen 1999). For *E. coli*, *Campylobacter* and *Yersinia*, beside the lack of reliable re-isolation techniques, it must be stated that their thermoresistance and with minor exceptions chemoresistance is lower than that of *Salmonella.* This means it will give no sense if they are used as additional microbial parameters for describing a hygienically safe product.

Enteroviruses are generally present in sludge of faecal origin but not regularly in sludges coming from other sources. In principle Enteroviruses may be used as additional indicator but the re-isolation procedure are as for all viruses from environmental samples labour- and costintensive. Their resistance in the involved treatment processes is not higher than that of *Salmonella*, this means, that the additional information resulting from using this indicator organisms are also low. The same applies for rotavirus, even it is of special environmental importance according to Metzler *et al.* (1996) and Pesaro *et al.* (1999).

The question if nematodes or nematode eggs are an useful indicator in this connection is not easy to answer. With respect to nematodes pathogenic for men and/or animals the experience shows, that even eggs of *Ascaris suum* are less thermoresistant than *Salmonella*, but behave different in chemical treatment, this means that if *Salmonella* would have not survived, e.g. the composting process *Ascaris* eggs and with them all other nematodes eggs would not have done either. This does not apply for treatment with slaked lime or long-term storage. This means that *Ascaris* eggs will not be a necessary indicator in all processes in which the thermal effect is the predominant one, but they will give valuable additional information if used in the supervision of all other treatment processes. But nematodes may be an indicator for insufficient storage conditions for the final product by which plant pathogenic nematodes may have invaded the compost. In order to identify this situation eggs or larvae of such species have to be properly identified. This requires special expertise which is generally not available in most of the involved laboratories, which means that a general parameter "free of nematode eggs and/or larvae" seems not to be easy to realize in this context. Another situation in which the investigation of the final products for the presence of nematode eggs makes sense is in co-digestion with liquid manure or sewage sludge if this feeding material had not been heated before entering the reactor.

No plant pathogenic virus, fungus or bacterium has been found so far which is comparable in importance to *Salmonellae* for the above-mentioned purpose. The only indicator which is widely distributed in biological wastes from households and in wastewater is tomato seed. Even this indicator will not totally cover all requirements, additionally reproducible parts of plant materials from biowastes have also to be taken into account. Therefore, it seems reasonable and feasible to define the term "phytohygienic safety" of the product as follows, as done in the German biowaste ordinance: "the final product may not contain more than two tomato seeds per litre product that are capable of germinating and/or reproducible plants parts". A suitable test method is described by Bundesgütegemeinschaft Kompost (1994).

22.8 RESTRICTIONS IN THE USE OF THE FINAL PRODUCT

Restriction in the use of fertilizers and substrates resulting from biotechnological waste treatment should either prevent introduction of undesired chemical residuals by contaminated crops into the food chain and prevent direct transmission of pathogens to susceptible animals and humans via feed and food, respectively. This had been practised in the past especially with sewage sludge. Such a strategy alone does not prevent the environmental risks or transmission of pathogens to vector populations, e.g. birds, which will generate indirect transmission cycles.

Several authors had given examples how birds can become carriers of *Salmonella* (Hellmann 1977). One of the sources of infection in seagulls has been found to be sewage treatment plants. The further ways of introduction of a certain lysotype of *Salmonella enteritidis* could be demonstrated by Köhler (1993). He identified the waste delivered from West-Berlin to a waste disposal site in the former German Democratic Republic and followed the introduction of this pathogen via birds into the chicken populations and finally to humans via food containing eggs. Williams *et al.* (1977) as well as several other authors like Coulsen *et al.* (1983), Mayr (1983) described the importance of vectors in the transmission of *Salmonella* to farm animals and humans. Foster and Spector (1995) described specific molecular mechanisms responsible for the ability of *Salmonella* to survive the environmental stress.

This means even if the fertilizers containing pathogens are immediately ploughed into the soil or injected sub-soil by special devices, they may generate carriers by attraction of certain species (seagulls by ploughing) or prolonged survival in sub surface soil layers. Thus restrictions in use are a tool with limited effects form the point of view of epidemiology and should be avoided if possible and feasible. Moreover, concerning plant pathogens and seeds this strategy is ineffective if the products are reused in agriculture.

22.9 CONCLUSIVE SUMMARY

In order to assure hygienic safety of composted products, treated by a biological aerobic or anaerobic process, that are deemed to be applied in agriculture or

Figure 22.1 Hygienic requirements in processing organic wastes by aerobic or anaerobic treatment.

horticulture, a three step control system is recommended which includes approved methods that are designed to minimize the costs and labour on the one hand and allow optimal product safety by using additive effects on the other. Figure 22.1 summarizes these strategies. Concerning test organisms to be used in direct process validation, the discussion in European context is not finished until now, *S. senftenberg* W 775 and *E. faecalis* DSM 257 may be used for general validation procedures as well as the BPV strain Haden and eggs of *Ascaris suis* for special purposes. The most appropriate parameter for supervision of the final product is the absence of *Salmonella* in 50 g, but depending on the involved substrates and materials as well as the intended field of application other parameters may be useful as well.

REFERENCES

ATV-Abwassertechnische Vereinigung (1988) Entseuchung von Klärschlamm, 2. Arbeitsbericht der ATV (VKS-Arbeitsgruppe 3.2.2). *Korrespondenz Abwasser* **35**, 71–74.

ATV-Abwassertechnische Vereinigung (1992) ATV_Merkblatt Nr. 365 Hygiene der Abfallbehandlung – Hinweise zu baulichen und organisatorischen Maßnahmen, Abwassertechnische Vereinigung, D-53758 Hennef.

Bendixen, H.G. (1999) Hygienic safety – results of scientific investigations in Denmark (Sanitation requirements in Danish biogas plants). In: *IEA Bioenergy Workshop (Proceedings)*, pp. 27–47, Deutsche Veterinärmedizinische Gesellschaft e.V., Frankfurter Str. 89, D-35392 Giessen.

Beyer, Ch., Beyer, W., Philipp, W. and Böhm, R. (2002) Untersuchungen zur Eignung von Indikatororganismen zur Beurteilung der Hygienisierung in Anaerobanlagen – Evaluierung und Etablierung eines neuen PCR-Konzeptes zum quantitativen Nachweis phyto- und seuchenhygienisch relevanter Organismen. II. Seuchenhygienisch relevante Organismen. Abschlussbericht Forschungsvorhaben Az. 08965. Deutsche Bundesstiftung Umwelt, Postfach 1705, D-49007 Osnabrück.

Böhm, R. (1995) Keimemissionen bei der Kompostierung. In: *Biologische Abfallbehandlung*, (ed. K.J. Thomé-Kozmiensky), pp. 508–526, EF-Verlag für Energie und Umwelttechnik GmbH, Berlin.

Böhm, R., Fack, Th. and Philipp, W. (1996) Anforderungen an die biologische Abfallbehandlung aus der Sicht des Arbeitsschutzes. In: *Biologische Abfallbehandlung* III. (eds. K. Wiemer, and M. Kern), pp. 335–365, M.C. Baeza-Verlag: Witzenhausen.

Böhm, R., Martens, W. and Bittighofer, P.M. (1998) Aktuelle Bewertung der Luftkeimbelastung in Abfallbehandlungsanlagen. Baeza-Verlag: Witzenhausen.

Böhm, R. (1999a) Hygienic requirements in aerobic and anaerobic treatment of biological wastes. In: *Proceedings of the 8th International Conference on the FAO ESCORENA Network on Recycling of Agricultural, Municipal and Industrial Residues in Agriculture (formerly animal waste management)* Vol. 2, (eds. J. Martinez and M.-N. Mauded), pp. 371–378, Rennes, France, 26–29 May 1998, Ramiran 98, FAO and Cemagref.

Böhm, R. (1999b) Die Bedeutung hygienischer Belange für die Qualität von Produkten der aeroben und anaeroben Bioabfallbehandlung. *Schriftenreihe des Vereins für Wasser-, Boden- und Lufthygiene* **104**, 3–25.

Braumiller, P., Martens, W. and Philipp, W. (2000) Abschlußbericht Forschungsvorhaben 95 HS 059 Aerobe und anaerobe Behandlung von Bio- und Küchenabfällen sowie anderen Rest- und Abfallstoffen tierischer Herkunft unter den Aspekten der Seuchenhygiene, Teil B: Untersuchungen zur Inaktivierung von Indikator-Bakterien und Viren im Kompostierungsprozeß. Bundesanstalt für Landwirtschaft und Ernährung, Frankfurt/M, Verbundvorhaben der BFAV und der Universität Hohenheim.

Breer, C. (1981) Freilandbiologie und Infektzyklen der Salmonellen. *Schweiz. Arch. Tierheilk.* **123**, 89–96.

Breitenfeld, P. (2000) Untersuchungen zur Human- und Veterinärhygiene der Bioabfallkompostierung. Agrarwiss. Diss. Universität Hohenheim.

Bundesgütegemeinschaft Kompost (1994) Methodenhandbuch zur Analyse von Kompost Nr. 222, Methode 8. BGK, Schönauser Str. 3, D-50968 Köln.

Coulson, J.C., Butterfield, J. and Thomas, C. (1983) The herring gull *Loras argentatos* as a likely transmitting agent of *Salmonella montevideo* to sheep and cattle. *J. Hyg. Camb.* **91**, 437–443.

Edel, W. and Kampelmacher, E.H. (1969) *Salmonella* isolation in nine European laboratories using a standardized technique. *Bull. WHO.* **41**, 297–306.

Enigk, K. (1980) Verbreitung freilebender Entwicklungsformen von Parasiten durch Siedlungsabwasser. *Dtsch. Tierärztl. Wschr.* **87**, 365–400.

Foster, J.W. and Spector, M.P. (1995) How *Salmonella* survive against the odds. *Annu. Rev. Microbiol.* **49**, 145–174.

German Biowaste Ordinance (1998) Verordnung über die Verwertung von Bioabfällen auf landwirtschaftlich und gärtnerisch genutzten Böden (Bioabfallverordnung – Bio-AbfV), *Bundesgesetzblatt* Teil I **65**, 2955–2981.

Grüner, R.C. (1996) Gesundheitszustand und Belastung von Beschäftigten im Abfallbereich. Erste Ergebnisse und Schlussfolgerungen für die Praxis. In: *Biologische Abfallbehandlung* III (eds. K. Wiemer and M. Kern), pp. 315–334, Baeza-Verlag: Witzenhausen.

Hellmann, E. (1977) Latente Salmonellen-Infektionen der Tiere und ihre Ursachen. *Wiener Tierärztl. Mschr.* **64**, 173–180.

Hoferer, M. (2001) Seuchenhygienische Untersuchungen zur Inaktivierung ausgewählter Bakterien und Viren bei der mesophilen und thermophilen anaeroben alkalischen Faulung von Bio- und Küchenabfällen sowie anderen Rest- und Abfallstoffen tierischer Herkunft. Diss. Freie Universität Berlin, ISBN 3-8265-9616-1.

Karuniawati, A., Philipp, W. and Böhm, R. (2000) Is enterohaemorrhagic *E. coli* (EHEC) in sewage and slurry an important link in the epidemiology? In: *Proceedings of the Xth International Congress on Animal Hygiene* 2–6 July 2000, Maastricht, Vol. 2 (eds. M.J.M. Tielen and M. Th. Voets), pp. 815–819, Animal Health Series, P.O. Box 4 NL-5280 AA Boxtel.

Köhler, B. (1993) Beispiele für die Anreicherung von Salmonellen in der Umwelt. Dtsch. *tierärztl. Wschr.* **100**, 264–274.

Krauss, P., Krauss, M., Wilke, M. and Wallenhorst, T. (1996) Herkunft und Verbleib organischer und anorganischer Schadstoffe im Bioabfall. In: *Biologische Abfallbehandlung* III (eds. K. Wiemer and M. Kern), pp. 335–365, Baeza-Verlag: Witzenhausen.

Länderarbeitsgemeinschaft Abfall (LAGA) (1995) LAGA-Merkblatt M10: Qualitätskriterien und Anwendungsempfehlungen für Kompost, Stand 15.2.1995. In: *Müll-Handbuch*, Kennzahl 6856, 1–59 (eds. G. Hösel, B. Bilitewski, W. Schenkel, H.E. Schnurer), Erich Schmidt Verlag: Berlin.

LASI-LV 1 (1995) Leitlinien für den Arbeitsschutz in Wertstoffsortieranlagen. Länderausschuss für Arbeitssicherheit, D-Wiesbaden.

LASI-LV 13 (1997) Leitlinien für den Arbeitsschutz in biologischen Abfallbehandlungsanlagen, Länderausschuss für Arbeitssicherheit, D-Wiesbaden.

Malmros, P. (1994) Occupational health problems owing to recycling of waste. In: *Nachweis und Bewertung von Keimemissionen bei der Entsorgung von kommunalen Abfällen sowie spezielle Hygieneprobleme der Bioabfallkompostierung* (ed. R. Böhm), pp. 36–50, Dt. Vet. Med. Ges., Frankfurter Straße 89, D-35392 Giessen.

Martens, W., Böhm, R., Fessel, A., Fricke, K., Haumacher, R., Kellner-Aschenbrenner, K., Köhler, K.K., Mayr, C., Philipp, W., Strauss, R., Witzigmann, R. and Zapirain, R. (1999) Mikrobielle Immissionen bei der Einsammlung von Siedlungsabfällen. *Schriftenreihe des Vereins für Wasser-, Boden- und Lufthygiene* **104**, 503–521.

Mayr, A. (1983) Verbreitung von Infektionserregern über Abfälle durch Haus-, Gemeinde und Freilandungeziefer unter besonderer Berücksichtigung der Gesundheit des Menschen. *Zbl. Bakt.Hyg.*, I. Abt. Orig. 1B, **178**, 53–60.

Menke, G. (1992) Hygienische Aspekte der Bioabfallkompostierung. Teil 1: Phytohygiene, in: Proc. Symposium des Umweltministeriums Baden-Württemberg und der LG-Stiftung "Natur und Umwelt", 26. März 1991 in Stuttgart, "Bioabfallkompostierung – Chance der Abfallverwertung oder Risiko der Bodenbelastung?", pp. 68–73 und Tab 1–5. LG-Stiftung "Natur und Umwelt, Königstraße 3–5, D-70173 Stuttgart.

Metzler, A., Regli, W., Leisinger, M., Heider, H., Schweizer, K. and Tabisch, A. (1996) Viren und Parasiten im Trinkwasser Risiken und Prävention. *Mitt. Gebiete Lebensm. Hyg.* **87**, 55–72.

Moss, A. and Haas, B. (2000) Abschlußbericht Forschungsvorhaben 95 HS 059 "Aerobe und anaerobe Behandlung von Bio- und Küchenabfällen sowie anderen Rest- und Abfallstoffen tierischer Herkunft unter den Aspekten der Seuchenhygiene", Teil A: Tenazität viraler Tierseuchenerreger in biogenen Abfällen in Biogasanlagen bei der Kofermentation mit Gülle sowie in Kompostanlagen. Bundesanstalt für Landwirtschaft und Ernährung, Frankfurt/M, Verbundvorhaben der BFAV und der Universität Hohenheim.

Ottolenghi, A.C. and Hamparian, V.V. (1987) Multiyear study of sludge application to farmland: Prevalence of bacterial enteric pathogens and antibody status of farm families. *Appl. Environ. Microbiol.* **53**(5), 1118–1124.

Pesaro, F., Wellinger, A. and Metzler, A. (1999) Inactivation of animal viruses during thermophilic fermentation of source separated waste in a fall scale biogas plant. In: *IEA Bioenergy Workshop (Proceedings)*, pp. 62–68, Deutsche Veterinärmedizinische Gesellschaft e.V., Frankfurter Str. 89, D-35392 Giessen.

Philipp, W., Gresser, R., Michels, E. and Strauch, D. (1990) Vorkommen von Salmonellen in Gülle, Jauche und Stallmist landwirtschaftlicher Betriebe in einem Wasserschutzgebiet. *Forum Städtehygiene* **41**, 209–212.

Rappaport, F., Konforti, N. and NAVON, B. (1956) A new enrichment medium for certain Salmonellae. J. Clin. Pathol. 9, 261–266.

Scherer, P.A. (1992) Hygienische Aspekte bei der getrennten Abfallsammlung. In: Getrennte Wertstofferfassung und Biokompostierung. In: *Biologische Abfallbehandlung* (eds. K.J. Thomé-Kozmiensky and P.A. Scherer), pp. 135–161, EF-Verlag für Energie und Umwelttechnik GmbH, Berlin.

Strauch, D. (1998) Pathogenic micro-organisms in sludge. Anaerobic digestion and disinfection methods to make sludge usable as fertiliser. *Europ. Water Manage.* **2**(2), 12–26.

Vassiliadis, P. (1983) The rappaport–vassiliadis (RV) enrichment medium for the isolation of *salmonellas*: an overview. *J. Appl. Bacteriol.* **54**, 69–76.

Williams, B.M., Richards, D.W., Stephens, D.P. and Griffiths, T. (1977) The transmission of *S. livingstone* to cattle by the herring gull (*Larus argentatus*). *Vet. Rec.* **100**, 450–451.

23

Remediation technologies for conversion of heavy metal polluted organic wastes into compost

Adrie Veeken

23.1 INTRODUCTION

23.1.1 Recycling of organic wastes

The main strategies for managing solid waste streams are landfilling, incineration and recycling (White *et al.* 1995). Due to the growing volume of solid waste streams management strategies as incineration and landfilling have become problematic in the Netherlands. In 1985, 5 million tons of municipal solid waste was produced in the Netherlands, and this amount is expected to increase to approximately 7.2 million tons in 2010 (RIVM 1988; CBS 1994). The problems related to landfilling are lack of space, the emission of greenhouse gases and the possibility of soil and groundwater contamination due to leaching of hazardous components

 Resource Recovery and Reuse in Organic Solid Waste Management. Edited by Piet Lens, Bert Hamelers, Harry Hoitink, Werner Bidlingmaier. ISBN: 1 84339 054 X. Published by IWA Publishing, London, UK

from landfill sites (Farquhar and Rovers 1973). Recent developments in waste management have led to a decree which forbids the landfilling of solid wastes with an organic matter content higher than 5% (Stax 1995). Incineration can cause air pollution due to emissions (e.g. heavy metals, chlorinated dioxins and dibenzofurans), and thus requires expensive after-treatment (White *et al.* 1995). Incineration also contributes to the greenhouse effect and leads to the production of highly contaminated and hazardous slags and fly ashes, which have to be landfilled (McBean *et al.* 1995). As a result of these problems, the Dutch governmental policy has gradually shifted towards preventing and recycling solid waste streams (RIVM 1988). In this way, a significant reduction of the quantity of solid wastes that eventually have to be landfilled or incinerated can be achieved.

The separate collection of components of municipal solid waste and the development of methods of recycling have been intensified in the Netherlands during the 1990s. The organic fraction of biological origin in municipal solid waste (thus excluding paper and plastics) makes up 50% of the total amount of municipal solid waste (Van Roosmalen and Langerijt 1989). The organic part consists of indoor-collected organic material, such as food remainders and flowers, and outdoor-collected organic matter, such as grass and branches from gardens. In the National Environmental Policy Plan (NMP) of 1989 activities related to separate waste collection were formulated under article A67 following the NMP's philosophy concerning the durable use of the environment (SDU 1991). These plans were further developed in the NMP-plus programme of 1990. The objective for the year 2010 is to recycle 50% of municipal solid waste and to incinerate the remaining 50%. By then, there should be no further landfilling of municipal solid waste. The success of source separation of municipal solid wastes resulted in an increased practice in many European countries (Hogg *et al.* 2002).

The recycling of solid organic wastes is preferred to incineration and landfilling from the view of sustainable development because:

- incineration produces greenhouse gases, contaminated slags and fly ashes,
- landfilling produces greenhouse gases in the long term and requires expensive measures to prevent emissions to soil and groundwater systems,
- recycling of organic matter promotes closed cycles of carbon and nutrients.

23.1.2 Compost application in relation to soil quality

Organic wastes recycling can be achieved by biological conversion into compost. The compost thus produced can be reused as soil improver and natural fertiliser. The biological conversion of organic wastes can be established by aerobic composting (Inbar *et al.* 1990) or anaerobic digestion (Cecchi *et al.* 1988). Advantages of the use of compost on arable land are:

- organic matter has a positive effect on the structure of the soil, thereby improving tilth, aeration and retention of moisture; its capacity to bind soil particles into structural aggregates reduces soil erosion (Huang and Schnitzer 1986)

- organic matter has a nutritive function: it is a source of nutrients and trace metals, and regulates the supply of nutrients from other sources in the soil (Vaughan and Malcolm 1985)
- organic matter has a positive effect on the activity of microbioto and microfaunal organisms (Burns *et al.* 1986).

On the other hand, the application of compost to soil systems is of great concern because the frequent supply of compost may lead to the accumulation of heavy metals in the soil. Increased levels of heavy metals in topsoils due to atmospheric deposition from industrial activities and input via fertilisers, pesticides and animal manure have already been observed (De Haan and Van der Zee 1993). The negative effects of increased heavy metal contents in soil systems are (Chaney and Ryan 1993):

- leaching of heavy metal to groundwater systems,
- heavy metal uptake by plants and animals which introduces more heavy metals into various vital life cycles,
- inhibition of plant growth and of the activity of soil micro-organisms.

The heavy metal content of composts should be limited in order to guarantee the safe use of compost. In the Netherlands, legal criteria for the quality and dosage of compost are laid down in the BOOM law (Dutch abbreviation for Besluit Overige Organische Meststoffen) concerning the quality and use of so-called other organic fertilisers (SDU 1991). The BOOM standards for heavy metals are listed in Table 23.1 and compared with standards for compost as defined by the European Union (EU) and USA. Also other statutory standards are considered to ensure protection of environment and health (e.g. Hogg *et al.* 2002). Table 23.1 shows that the standards for heavy metals in compost are most strict in the Netherlands, somewhat higher in the EU and several orders of magnitude higher than in the USA. The differences in heavy metal standards are due to the fact that different approaches were applied to set up the

Table 23.1 Standards for heavy metals in compost ($mg\,kg^{-1}$ TS)[1].

Parameter	The Netherlands		EU		USA
	Clean compost	Compost	Class 1	Class 2	Biosolids
Cd	0.7	1	0.7	1.5	39
Cr	50	50	100	150	No limit
Cu	25	60	100	150	1500
Hg	0.2	0.3	0.5	1	17
Ni	10	20	50	75	420
Pb	65	100	100	150	300
Zn	75	200	200	400	2800
As	5	15	–	–	41

[1] Detailed information on heavy metal standards and different compost classes can be found in Hogg *et al.* (2002).

heavy metal standards (McGrawth *et al.* 1994). In the Netherlands, the metal balancing approach is applied that tries to match the metal inputs to soil to the small losses of metals due to crop removal, soil erosion and leaching (Moolenaar *et al.* 1997). The USA set up their metal limits based on toxic effects on soil microorganisms under field conditions (Chaney and Ryan 1993). Although both approaches are conceptually valid, there are large discrepancies in the limits between USA and EU countries. An excellent discussion of the different scientific perspectives of heavy metal loading between EU and USA is given by McGrawth *et al.* (1994). Comparison between heavy metal standards in the Netherlands and other European countries shows that Dutch heavy metal limits, especially Zn, are too strict for several heavy metals and should be re-examined (Hogg *et al.* 2002).

Apart from meeting the legal standards for heavy metals as laid down in BOOM, the recycling of organic wastes only makes sense if there is a market for the compost. Possible marketing areas for compost are its application in horticulture, private gardens, public parks and the agricultural sector. However, the level of acceptance of compost by the agricultural sector is rather low. This is mainly because of the quality of the compost, especially with respect to the content of heavy metals and low content of organic matter. Besides that, the compost has to compete with other soil improvers, such as animal manure, inorganic fertilisers and potting substrates (Hauke *et al.* 1996). The recycling of organic wastes can be frustrated if the market does not accept compost. Existing aerobic and anaerobic treatment processes without any pretreatment steps cannot achieve these goals.

There are two ways to guarantee the quality of organic wastes for recycling:

(1) *Prevention*: By changing the collection criteria for biowaste, components which cause the low organic matter content and high heavy metal content of the waste can be avoided. For sewage sludge, the amount of heavy metals in the collected wastewater should be lowered. In most cases prevention is not possible due to economical and social obstacles.
(2) *Technological approach*: Technological processes can be employed to convert organic wastes to compost which can both meet the legal criteria as laid down in BOOM and compete with other fertilising and soil improving products.

This chapter will discuss on technological solutions to reduce the heavy metal content of organic wastes preceeding the biological conversion to compost.

23.2 REMEDIATION TECHNOLOGIES

Up until now, no full-scale processes have been operational for the removal of heavy metals from organic wastes. In the last decades, several studies have been published focusing on the development of technologies for the removal of heavy metals from contaminated soil and sewage sludge (Assink 1986; Wong and Henry

1988). Most of these technologies can be grouped in one of the two following classes:

(1) *Solid-to-liquid extraction*: This process is based on the extraction of metals from the solid waste to an aqueous-liquid phase followed by separation of the solid and liquid phase. To promote solubilisation, an extracting agent (or extractant) is added directly (chemical extraction) or is produced by micro-organisms (microbiological leaching).
(2) *Physical separation*: This process is based on the physical separation of a specific fraction of the solid waste in which the heavy metals are concentrated. Physical separation processes are based on differences in physical properties of different fractions of the solid waste stream.

23.2.1 Extraction processes

The extraction process for the removal of heavy metals from solid wastes normally consists of three steps (Figure 23.1):

(1) The actual extraction process.
(2) Separation of the solid and liquid phase.
(3) Cleaning and recycling of the heavy metal enriched liquid phase.

The possibilities of the extraction process for solid organic waste streams are first of all determined by the extraction efficiency of the applied extracting reagent. The extraction efficiency is defined as the percentage of heavy metals extracted from the solid to the liquid phase. Other factors determining the applicability of the extraction process on a full scale are of economical, technological and environmental concern:

(1) Costs of the process:
    - costs of the extracting reagent,
    - number of process units,
    - wastewater treatment for liquid discharge.

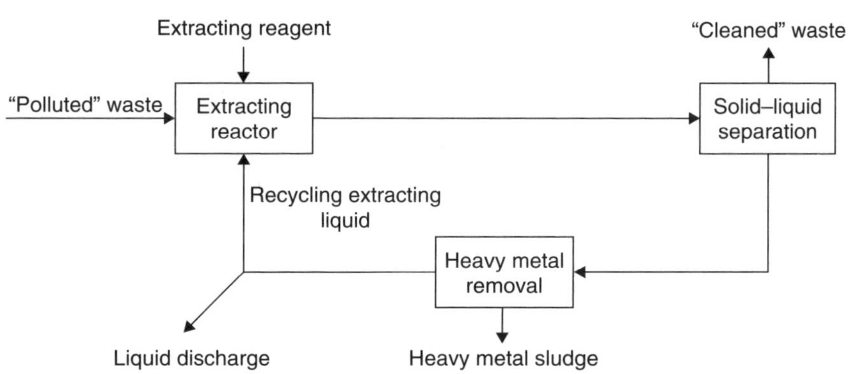

Figure 23.1 Extraction scheme for the removal of heavy metals from solid wastes.

(2) Minimisation of the production of hazardous emissions:
   - air emissions,
   - water emissions,
   - final inert waste.

(3) Possible negative effects of the leaching process of the cleaned solid waste on:
   - the structure of the waste,
   - the bioavailability of essential nutrients in subsequent biological processes,
   - toxicity of heavy metals in subsequent biological processes.

The main mechanisms for the retention of heavy metals in the solid phase are:

(1) adsorption to the organic and inorganic solid phases, e.g. Fe-(hydr)oxides, clay minerals, organic debris, humic substances;
(2) presence as inorganic precipitates, e.g. $CdCO_3$, $Cu_3(PO_4)_2$, PbS.

For chemical extraction of heavy metals from solid wastes, two types of reagents are generally used: acids and complexants. Acid extraction of heavy metals is brought about by the exchange of protons for heavy metals adsorbed to the solid phase and by the solubilisation of heavy metal precipitates. The protons can either be supplied directly by addition of a strong acid (e.g. HCl, HNO_3, H_2SO_4, citric acid) or be produced by micro-organisms. Extraction through the production of acids by bacteria or fungi is referred to as microbiological leaching. For example, H_2SO_4 can be produced by the oxidation of reduced sulphur compounds by *Thiobacilli* strains (Doddema *et al.* 1987; Wong and Henry 1988). Strong organic acids such as oxalic acid and citric acid can be produced by fungi (Burgstaller and Schinner 1993). Weak organic acids such as lactic acid and acetic acid can be produced by various anaerobic consortia (Schlegel 1993).

Heavy metal extraction by complexants is based on the high affinity of the complexant for heavy metals. In this way, heavy metals adsorbed to the solid phase or present as heavy metal precipitates are dissolved. Complexants regularly applied for the extraction of heavy metals are ethylenediaminoacetic acid (EDTA) and nitrotriacetic acid (NTA). Other processes, which are more or less based on these two types of reagents, are as follows:

- *Extraction with lye (e.g. NaOH)*: At high pH values humic substances are solubilised, thus enhancing the solubilisation of heavy metals (Wong and Henry 1988).
- *Ion exchange*: The solid ion exchanger is suspended in the liquid phase and the metal ions are transported from the contaminated solid waste to the complexing ion exchanger. In the ion-exchange particles, the ion-exchanger sites (Wong and Henry 1988) covalently or electrostatically bind the metal ions. This process is very complex and rate limited by mass transport of heavy metal ions from the bulk solution to the surface of the ion-exchange particles and by transport of the metal ions from the solid phase to the liquid phase.
- *Electro-reclamation*: Mostly applied for *in situ* removal of heavy metals from soils by electrically charging the system through the application of several

Table 23.2 Extraction efficiency of heavy metals from sewage sludge (%).

Extracting reagent	Cd	Cu	Pb	Zn	Reference
HCl, pH 1.5	10–90	0–70	5–100	50–90	Oliver and Carey (1976)
0.5 M acetic acid	40	0	5	25	Bloomfield and Pruden (1975)
HCl, pH 1.5	80–100	80–100	40–100	100	Wozniak and Huang (1982)
H_2SO_4	10–70	<2	10–15	35–70	Jenkins *et al.* (1981)
HCl, pH 1	90	50	–	90	Ried (1988)
HCl, pH 1	22–90	2–90	30–100	50–95	Rulkens *et al.* (1989)
H_2SO_4, pH 1.5	–	50–75	50–60	80–95	Tyagi *et al.* (1988)
H_2SO_4, pH 1.5	95–99	8–10	35–65	50–99	Lo and Chen (1990)

electrode arrays; generally the process is promoted by acidification of the soil (Lageman *et al.* 1990).

The extraction of heavy metals from sewage sludge has been studied extensively. Table 23.2 presents a compilation of literature data regarding the extraction efficiencies of Cd, Cu, Pb and Zn from sewage sludge. Table 23.2 shows a very broad range of extraction efficiencies for all heavy metals. The broad range in extraction efficiencies is due to differences in sludge composition, differences in pretreatment of the sludge and differences in extracting conditions.

The variable extraction conditions, such as extracting time, reagent concentration, solid concentration and type of mixing, are important process parameters. As might be expected, the extraction efficiency increases for lower pH values. Theis and Hayes (1978) showed that the metal extractability depends on the redox conditions. The extraction efficiency decreases for most heavy metals when the sludge is anaerobically digested. Especially the Cu extractability decreased drastically, from 61 to 1% (Theis and Hayes 1978). The lower solubility is caused by the reduction of sulphate to sulphide and the subsequent precipitation of heavy metal sulphides. Heavy metal sulphides, especially CuS and PbS, are very difficult to solubilise, even at low pH values (Fletcher and Beckett 1987).

23.2.2 Physical separation processes

If heavy metals are concentrated in specific solid fractions of a solid waste, the selective removal of these particles from the solid waste generally results in a small fraction that is highly contaminated and a large fraction that is relatively clean. The separation of a specific fraction is achieved by means of physical separation processes. The term "physical" is added to distinguish this process from molecular separation processes, such as distillation and dialysis. The processes are widely used in soil remediation, the mining industry and mineral processing (Svarovsky 1985). Until now, adaptations of mining technologies in the field of environmental technology have been applied only in the cleaning of polluted soils and sediments, where "non-polluted" sand is separated from polluted smaller size fractions (Kroenig 1990; Van Dillen and Schotel 1990). The separation of a specific fraction of the solid waste by a solid–solid separation process is established by differences in particle properties. Table 23.3 presents the wet solid–solid separation processes mostly widely used in waste treatment, together with the principal properties on which the separation is based. Descriptions and applications of these processes and adaptations of these processes can be found in various engineering textbooks (Svarovsky 1985; Metcalf and Eddy 1991).

Physical wet-separation processes used in the field of environmental technology can be characterised by a clean solid stream and a contaminated solid stream. Figure 23.2 shows a schematic presentation of a physical wet-separation process where the contaminated waste (influent flow) is separated into a "clean" flow and a "contaminated" flow. The flows are characterised by the flow rate of the liquid (F in $m^3 s^{-1}$), total solids (TS) concentration (S in $kg\,m^{-3}$) and concentration of contaminant in the solids (C in $mg\,kg^{-1}$ dry matter). The separation efficiency

Table 23.3 Waste properties that determine separation in various wet solid–solid separation processes.

Process	Size	Density	Hydrophobicity	Shape	Magnetic properties
Screening (sieving)	+			(+)[1]	
Elutriation	+	+		(+)	
Hydrocyclones	+	+		(+)	
Air flotation			+		
Magnetic separation					+

[1] Brackets indicate secondary importance.

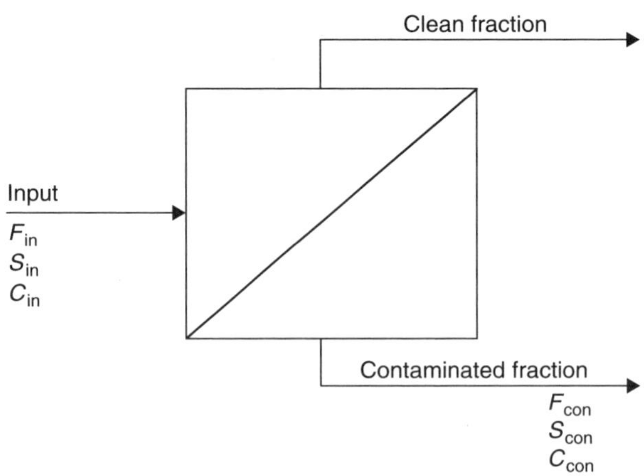

Figure 23.2 Schematic presentation of physical separation process for environmental applications.

(E_S) with respect to the total contaminated flow is defined as the TS mass flow of the contaminated fraction as a fraction of the TS feed mass flow rate:

$$E_S = \frac{F_{con} S_{con}}{F_{in} S_{in}} \tag{23.1}$$

where the subscripts "in" and "con" refer to the incoming flow and contaminated flow, respectively.

The efficiency of the solid–solid separation process is determined not only by the total separation efficiency but also by the distribution of the contaminant over the clean and contaminated flows. The separation efficiency with respect to the contaminants (E_C) is defined as:

$$E_C = \frac{F_{con} S_{con} C_{con}}{F_{in} S_{in} C_{in}} = E_S \frac{C_{con}}{C_{in}} \tag{23.2}$$

The removal efficiency of a solid–solid wet-separation process is high when E_S is low and E_C is high. De Waaij and Van Veen (1990) studied the heavy metal decontamination by hydrocyclones of several types of sediment. Table 23.4 gives the total separation efficiency and contamination efficiencies, recalculated from their data. Hydrocyclone separation gives good results for sandy sediments. However, for muddy sediments, the performance of hydrocyclones is poor. This is caused by the poor separation efficiency and the high contaminant concentrations in the "clean" stream.

The separation efficiency as defined in equation (23.1) includes all the separated solids, irrespective of particle size. However, for most separation processes the separation efficiency varies with particle size. Therefore, the grade efficiency is introduced, i.e. the separation efficiency as a function of particle size:

$$G(x) = E_C \frac{dF_{con}(x)}{dF(x)} = E_S \frac{C_{con}}{C_{in}} \frac{dF_{con}(x)}{dF(x)} \tag{23.3}$$

where $G(x)$ is the grade efficiency, x is the particle size (m), and $F(x)$ and $F_{con}(x)$ are the cumulative particle-size distributions of the feed and contaminated flow, respectively.

Table 23.4 Performance of hydrocyclone separation for different types of sediments with respect to heavy metals (De Waaij and Van Veen 1990).

Origin	Type	Metal	E_S	E_C
River	Mud	Cd	0.50	0.90
River	Sand	Cd	0.10	0.85
Channel	Sand	Cr	0.25	0.85
River	Sand	Cd	0.10	0.70
Channel	Mud	Hg	0.50	0.80
Harbour	Mud	Cd	0.57	0.82
Harbour	Mud	Hg	0.40	0.75
Channel	Mud	Hg	0.46	0.86

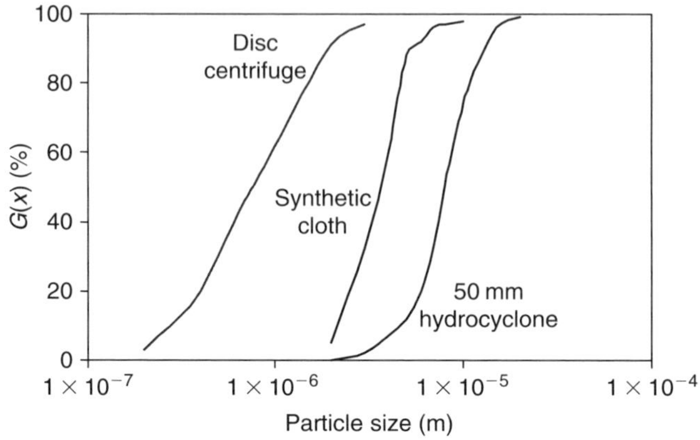

Figure 23.3 Grade efficiency curves for several types of separators (adapted from Svarovsky (1985)).

Figure 23.3 shows the grade efficiency curves of several types of separators. Often a single number is used to replace the grade efficiency curve of a separation device. This number is called the "cut size" (or x_{50}) and corresponds to the place on the curve where the grade efficiency is 50%. A second criterion for the performance of a separation process is the sharpness of the classification, as indicated by the steepness of the grade efficiency curve. An ideal separation would result in a step function, as might be obtained with an ideal screen. In most cases there is, however, some "misplaced" material. The less steep the grade efficiency curve, the greater the amount of misplaced material. The simplest measure of the classification sharpness (H) is the ratio of two sizes corresponding to two different percentages (symmetrical around 50%) on the grade efficiency curve, such as, for example, the quartiles (Svarovksy 1985):

$$H_{(25/75)} = \frac{x_{25}}{x_{75}} \tag{23.4}$$

where x_{25} and x_{75} are the particle sizes corresponding to grade efficiencies of 25% and 75%, respectively.

The solid waste has to be characterised with respect to particle-size distribution and level of contamination before a separation technology is applied on a pilot-plant or practical scale. For this, the solid waste is physically separated by small laboratory-scale devices that differentiate on one of the particle properties given in Table 23.3. After the classification, the heavy metal content of the separated fractions is determined. In this way, it becomes clear to which constituents of the solid waste the heavy metals are bound. On the basis of these results a separation technology has to be selected to separate the most contaminated fractions. Most often a single separation process is insufficient for the separation of a contaminated solid waste into a clean and a contaminated fraction. Instead a cascade of separation units is necessary.

There is still a wide gap between the theory and practice of solid–solid separation processes because the primary properties of a system cannot be easily translated into equipment. Selection and sizing of the equipment can only be done on the basis of small-scale tests either in a laboratory or on a pilot-plant scale.

23.3 CASE STUDIES

Both remediation technologies, extraction and physical separation, were evaluated for sewage sludge and biowaste, respectively. Both studies are extensively described in the following papers:

- Removal of heavy metals from sewage sludge by extraction with organic acids (Veeken and Hamelers 1999).
- Advanced bioconversion of biowaste for production of peat substitute, a renewable energy (Veeken *et al.* 2004).

23.3.1 Extraction from sludge

As shown in Figure 23.1 extraction of heavy metals from sludge typically consists of three steps. In the first step of the extraction process the heavy metals are transferred from the solid phase to the aqueous phase. This requires good contact between solids and liquids that is brought about by intensive mixing. In the second step the aqueous phase is separated from the "cleaned" sludge by a solid–liquid separation process, e.g. decanting centrifuge. In the third step the (heavy) metals are removed from the extracting solution to recover the economical value of the extracting agent and prevent environmental impact associated with discharge of extracting liquid. The most appropriate and economically feasible technologies for removal of heavy metals from solutions are chemical sulphide precipitation and/or selective ion exchange.

Various inorganic acids (HNO_3, HCl, H_2SO_4) and strong complexing agents (NTA, EDTA) are proposed in literature. However, inorganic acids and complexing agents are not applicable on a practical scale as the costs of the chemicals are high and the chemicals have negative environmental impact on the discharged solid and liquid waste streams (Wong and Henry 1988). Organic complexing acids (OCAs) such as citric and oxalic acid can be more promising because of the following reasons:

(1) Heavy metal extraction is partly due to the acidic character but for the greater part to the complexing behaviour of OCAs; therefore extraction can be performed at mildly acidic conditions (pH 3–5).
(2) OCAs are biodegradable under both aerobic and anaerobic conditions. This implies that the "cleaned" sludge does not have to be conditioned, thus leading to a substantial reduction of wastewater. Moreover, wastewater can be treated (an)aerobically.

(3) Heavy metals can be removed from OCAs solution. In this way, the extraction liquid can be recycled, reducing the costs of the process.

Like inorganic acids and strong complexing agents, OCAs are not selective and also metals other than heavy metals are simultaneously extracted from the sludge. The major cations competing with heavy metals for OCAs are Ca, Mg, Fe and Al (here referred to as "competing" metals). Therefore, OCAs have to be dosed to sewage sludge in an equimolar amount to the cumulative content of complexing metals, i.e. heavy metals and competing metals. This last point is often overlooked in relation to heavy metal extraction by complexing agents. Moreover, for recovery of OCAs not only heavy metals but also competing metals have to be removed from the extracting liquid. Otherwise, the competing metals are accumulated in the extraction reactor, in this way lowering the heavy metal extracting capacity of OCAs.

Extractions were performed for oxalic acid and citric acid at pH region 1–6 for Cu, Zn, Ca and Fe (Veeken and Hamelers 1999). Both acids increased the heavy metal extraction efficiency at mildly acidic pH compared to nitric acid. Citric acid had higher extraction efficiencies than oxalic acid as oxalic acid is removed from solution by precipitation as calcium oxalate.

The kinetics of the extraction process were studied for citric acid extraction at various concentrations at pH 3 (Figure 23.4). The course of extraction showed a

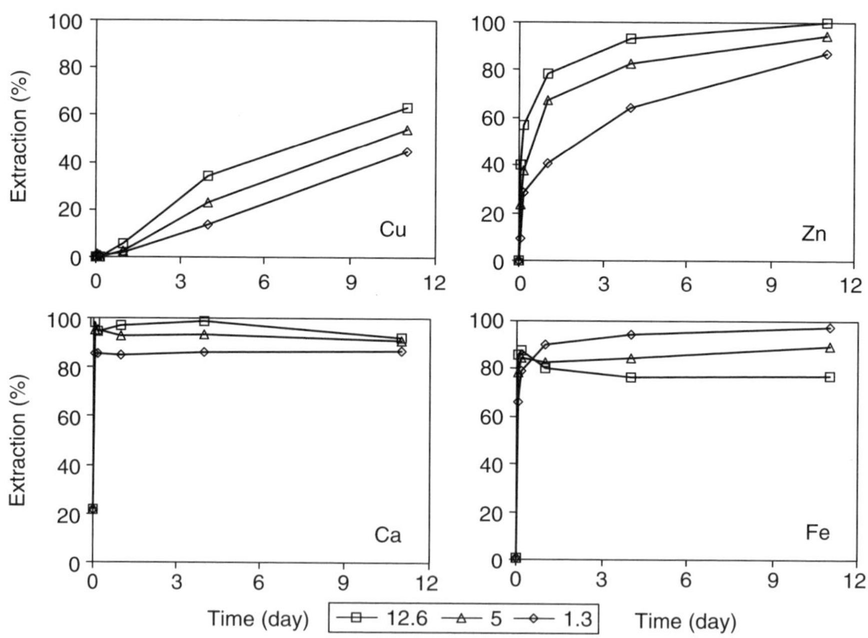

Figure 23.4 Kinetics of extraction of Cu, Zn, Ca and Fe from sewage sludge at various citric acid concentrations (mM).

distinct difference between heavy metals (Cu, Zn) and competing metals (Ca, Fe). The extraction of Cu and Zn as a function of time is typical for the extraction of heavy metals from sewage sludge: less strongly bound heavy metals (Zn) are extracted more rapidly from the sludge than strongly bound heavy metals (Cu). Heavy metals are predominantly incorporated in sludge flocs and the extraction takes time as the heavy metals have to diffuse from the sludge matrix to the bulk solution. The metals Ca and Fe are extracted much more rapidly from sewage sludge because these metals are present as precipitate (e.g. $Fe(OH)_3$, $FePO_4$, $CaCO_3$) or very weakly adsorbed to sludge flocs (Ca). The dissolution of precipitates is a relatively fast process compared to the diffusion of metal ions from within the sludge flocs to the bulk solution. The extraction was also studied at 10°C, 20°C and 30°C for the extraction with 0.1 M citric acid at pH 3.5. The increase in rate of extraction for both Cu and Zn with increasing temperature confirms the hypothesis that the rate of extraction is controlled by diffusion of metal ions from the sludge flocs to the bulk solution.

Heavy metals and competing metals both have to be removed from the extracting liquid to prevent accumulation of these metals in the process. Citric acid is a moderately strong complexing agent and both heavy and competing metals can possibly be removed from the liquid phase by a combination of chemical sulphide precipitation and specific ion exchangers.

Extraction of heavy metals from organic wastes by citric acid could be an attractive option because the extraction can be performed at mildly acidic conditions (pH 3–5). As citric acid is biologically degradable, the extraction process is compatible with composting and wastewater treatment. The extraction was studied for heavy metals Cu and Zn and for competing metals Ca and Fe from polluted sewage sludge. The rate of extraction increases for increasing temperature and citric acid concentration. Cu can be extracted for 60–70% and Zn for 90–100% by citric acid at pH 3–4. However, the process is still critical with respect to the removal of competing and heavy metals from the recycling liquid. Moreover, the process results in a highly toxic metal sludge that has to be landfilled. A first economic valuation of the extraction process showed that the total costs of the treatment process are high and comparable to the costs of incineration, approximately €400 per ton of dry matter.

23.3.2 Physical separation for biowaste

Laboratory-scale particle-size fractionation of biowaste under wet conditions showed that biowaste can be separated in three distinct fractions (Veeken and Hamelers 2002):

(1) a fraction >1 mm mainly composed of organic matter, makes up 40% of the mass of biowaste and the heavy metal content is somewhat lower than for unfractionated biowaste;
(2) inorganic fraction 0.05–1 mm (sand) makes up 30% of the mass and is very low in heavy metals;

(3) fraction composed of organics 0.05–1 mm and the fraction <0.05 mm (30% of mass), which heavy metals exceed the BOOM standards.

The composition of the three fractions is constant over the year as it is composed of the same biowaste components. Only the dimension of the fractions will show variations depending on time of year and place of collection. The compost produced from the organic fraction >1 mm complies with the legal standard for compost and with the exception of Zn with the legal standards for clean compost. Composting of the organic fraction >1 mm after wet physical separation could therefore produce compost for which the quality is independent of the fluctuating composition of the incoming biowaste. In this way, the sales of compost will be facilitated and compost can become competitive with other soil improvers and fertilisers, and possibly partly replace peat in potting mixes. The rest of the biowaste is made up of sand, whereas this fraction contributes to only 5% of the heavy metal content of biowaste. As the sand fraction is very low in heavy metals, it could possibly be reused in road and building construction. On the basis of the results of the laboratory-scale fractionation of biowaste, an advanced bioconversion process was designed that aimed to reach the following goals:

(1) Production of a peat substituting compost from the organic fraction >1–2 mm that complies with the standards for clean compost.
(2) Production of sand with a low organic matter content (<1%) that can be reused in road and building construction.
(3) Production of renewable energy by anaerobic digestion of the organic matter which is solubilised in the washing water to methane.
(4) Residual fraction that can be applied as landfill cover material.

The process was tested on pilot-plant scale (Veeken *et al.* 2004). For this, the performance of the physical separation units was tested with respect to mass balances of dry matter, organic matter and heavy metals (Cu, Pb and Zn). The organic fraction >1–2 mm was subjected to aerobic composting and the water soluble and residual fraction to anaerobic digestion. Figure 23.5 presents the flow sheet of the process together with mass balance for TS, volatile solids (VS), ash, salts, Cu, Pb and Zn on basis of the results of the pilot-plant study.

Washing biowaste results in a quality compost high in organic matter and low in salts and heavy metals; 53% of VS is recovered in the compost. This compost can be an excellent substitute of peat in growing media production systems (Hauke *et al.* 1996). In this way, the mining of peat moss can be reduced which results in a reduction of CO_2 emissions. As the quality of washed biowaste does not depend on the composition of incoming biowaste, compost with constant quality can be produced. The costs of the composting process are significantly reduced, as a smaller part of the total biowaste has to be composted. The major part of the ash (61%) is collected in the sand fraction. As the sand fraction is very low in organic matter and heavy metals, it can be utilised in road and building construction. About 11% of VS are converted to CH_4 by anaerobic digestion. The

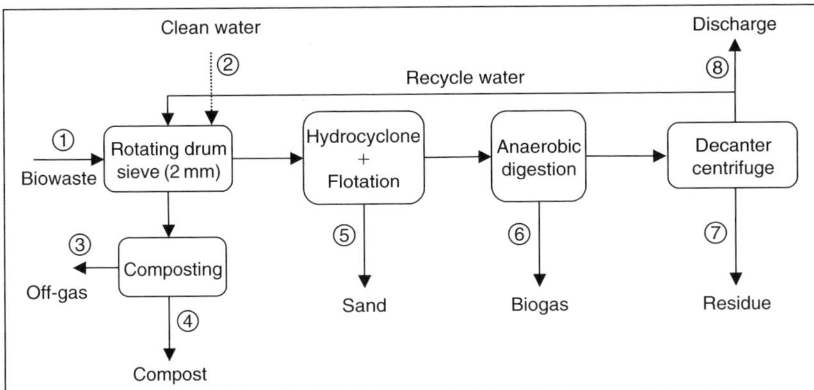

Component	Flow	1 Biowaste	2 Water[2]	3 Off-gas	4 Compost	5 Sand	6 Biogas	7 Residue	8 Discharge
Total flow	(kg)	1000	1250	n.a.[3]	453	293	n.a.	164	736
Water	(kg)	500 (29)[1]	1250 (71)	523 (30)	272 (16)	141 (8)	0	85 (5)	730 (42)
TS	(kg)	500 (100)	0	56 (11)	181 (36)	152 (30)	27 (5)	80 (16)	4.2 (0.8)
VS	(kg)	250 (100)	0	56 (23)	131 (53)	0.7 (0.3)	27 (11)	33 (13)	1.8 (0.7)
Ash	(kg)	250 (100)	0	0	50 (20)	152 (61)	0	46 (18)	2.4 (1.0)
Salts	(kg)	4.4 (100)	0	0	1.8 (40)	n.d.[4]	0	n.d.	2.6 (60)
Cu	(g)	6.0 (100)	0	0	2.2 (36)	0.03 (0.5)	0	3.7 (62)	0.06 (1)
Pb	(g)	10.5 (100)	0	0	6.6 (62)	2.2 (21)	0	1.7 (16)	0.11 (1)
Zn	(g)	31.8 (100)	0	0	19.6 (61)	1.9 (6)	0	10.1 (32)	0.31 (1)

[1]Percentage of total flow given in parenthesis; [2]Contributions of clean water were assumed negligible; [3]Not applicable; [4]Not determined.

Figure 23.5 Flow sheet and mass balances for processing of biowaste.

produced CH_4 can be considered as a renewable energy source. In the composting of unfractionated biowaste energy is lost via the off-gas, i.e. biowaste has to be wetted several times during composting before it is sufficiently stabilised. A substantial fraction of biowaste is recovered as sludge (16% TS) in which heavy metals are accumulated. The sludge cannot be reused as compost as it does not meet the legal standards for compost. As the organic matter originates from the topsoil of gardens, the organic matter is already highly stabilised. The sludge residual fraction can be applied as landfill cover material after dewatering.

An alternative is to combine the flows of sand and the residual fraction and obtain compost as product. In this way the hydrocyclone and flotation units can be omitted, reducing the costs of the process. Combination of the sand and residual fraction results in a product with a VS content of 35% and Cu, Pb and Zn levels of 21, 52 and 85 mg kg^{-1} TS, respectively. This product can be regarded as compost according to the legal demands.

In practice, the water consumption should be limited to minimise the amount of discharged water. Recycling the water can reduce water consumption but clean water has to be used at the end of the drum sieve to prevent the accumulation of salts in the compost. Both the levels of heavy metal and chemical oxygen demand (COD) (and probably also N and P) imply that the discharge water has to be treated in a wastewater treatment plant. The consumption of clean water can be reduced significantly when the water emitted from the composting process is condensed and reused in the sieving step.

Physical separation can be an attractive option when heavy metals are concentrated in a specific fraction of the waste. Moreover, the separation processes are generally cheap and compatible with biological treatment processes such as composting. Physico-chemical separation of biowaste showed that physical separation of biowaste before composting improves the quality of compost. Study on pilot scale showed that the advanced bioconversion process could replace traditional composting systems as it produces a high quality compost and renewable energy. In this way the recycling of biowaste as compost can be guaranteed. Indicative cost calculations including treatment of discharge water show that the costs of the process are comparable to the costs of traditional composting systems, i.e. approximately €50 per ton. These calculations did not yet include the positive contributions of compost sale, renewable energy production and omission of the hydrocyclone and flotation units.

23.4 CONCLUSIONS

Two strategies for remediation of heavy metal polluted organic wastes are available: (1) hydrometallurgical extraction, i.e. extraction with an aqueous phase, can be employed to reduce the heavy metal content of organic wastes and (2) physical separation. In order to assess both technologies, a physico-chemical fractionation scheme can be adopted to (1) gain insight into the distribution of heavy metals over the physical entities of the organic waste and (2) to determine the binding strength of heavy metals to these physical entities (Veeken and Hamelers 2003). Both remediation technologies are discussed and illustrated in this chapter. It was shown that existing extraction processes are complex and expensive and not compatible with succeeding biological treatment steps such as composting, anaerobic digestion or wastewater treatment. Extraction of heavy metals from organic wastes by citric acid could be an option because the extraction can be performed at mildly acidic conditions and citric acid is biologically degradable. The extraction process is still critical with respect to the removal of competing and heavy metals from the recycling liquid. Moreover, a highly toxic metal sludge is produced that has to be landfilled. Physical separation can be an attractive option when heavy metals are concentrated in a specific fraction of the waste. The separation processes are generally cheap and compatible with biological treatment steps.

In order to ensure protection of the environment and health, heavy metal for composts are set up to guarantee soil quality. Comparison of metals limits between EU and USA shows large discrepancies that are due to different approaches to guarantee soil quality. Many organic waste streams, among other sewage sludge and biowaste are regarded as polluted wastes in the EU while they are classified as clean in the USA. Comparison between heavy metal standards in the Netherlands and other European countries shows that Dutch heavy metal limits are very strict and should be re-assessed (Veeken and Hamelers 2002).

REFERENCES

Assink, J.W. (1986) Extractive methods for soil decontamination; a general survey and review of operational treatment installations. In: *Contaminated Soil '90, Third International KfK/TNO Conference on Contaminated Soil* (eds. F. Arendt, M. Hinsenveld and W.J. Brink), Kluwer Academic Publishers, Dordrecht.

Bloomfield, C. and Pruden, G. (1975) The effects of aerobic and anaerobic incubation on the extractabilities of heavy metals in digested sewage sludge. *Environ. Pollut.* **8**, 217–232.

Burgstaller, W. and Schinner, F. (1993) Leaching of metals with fungi, minireview. *J. Biotechnol.* **27**, 91–116.

Burns, R.G., Dell'Agnola, G., Miele, S., Nardi, S., Savoini, G., Schnitzer, M., Sequi, P., Vaughan, D. and Visser, S.A. (1986) *Humic Substances: Effects on Soil and Plants*. Italy, REDA.

CBS (1994) *Environment Statistics of the Netherlands 1994*. SDU, 's-Gravenhage (in Dutch).

Cecchi, F., Traverso, P.G., Mata-Alvarez, J., Clancy, J. and Zaror, C. (1988) State of the art of R&D in the anaerobic digestion process of municipal solid waste in Europe. *Biomass* **16**, 257–284.

Chaney, R. and Ryan, J. (1993) Heavy metals and toxic organic pollutants in MSW composts: research results on phytoavailability, bioavailability, fate. In: *Science and Engineering of Composting: Design, Environmental, Microbiological and Utilization Aspects* (eds. A.J. Hoitink and H.M. Keener), Renaissance Publications, Ohio.

De Haan, F.A.M. and Van der Zee, S.E.A.T.M. (1993) Composts regulations in the Netherlands in view of sustainable soil use. In: *Science and Engineering of Composting: Design, Environmental, Microbiological and Utilization Aspects* (eds. A.J. Hoitink and H.M. Keener), Renaissance Publications, Ohio.

De Waaij, A.C. and Van Veen, H.J. (1990) Processing of contaminated sediments in the Netherlands. In: *Contaminated Soil '90, Third International KfK/TNO Conference on Contaminated Soil* (eds. F. Arendt, M. Hinsenveld and W.J. Brink), Kluwer Academic Publishers, Dordrecht.

Doddema, H.J., Rulkens, W.H., Voornburg, F. van and Annokkee, G.J. (1987) *Verwijdering van Zware Metalen Uit Afvalstromen Door Middel van Bacteriele Uitloging Met Thiobacillus Soorten*. TNO-rapport, ref.nr. 87-333, dossiernr. 8722–18421 (in Dutch).

Farquhar, G.J. and Rovers, F.A. (1973) Gas production during refuse decomposition. *Water, Soil Air Pollut.* **2**, 483.

Fletcher, P. and Beckett, P.H.T. (1987) The chemistry of heavy metals in digested sewage sludge – I. Copper(II) complexation with soluble organic matter. *Water Res.* **21**, 1153–1161.

Hauke, H., Stoeppler-Zimmer, H. and Gottschall, R. (1996) Development of compost products. In: *The Science of Composting* (eds. M. De Bertoldi, P. Sequi, B. Lemmes and T. Papi), Blackie Academic & Professional, London.

Hogg, D., Bart, J., Favioni, E., Centemero, M., Caimi, V., Amlinger, F., Devliegher, W., Brinton, W. and Antler, S. (2002) *Comparison of Compost Standards within the EU,*

North America and Australasia – Main Report. The Waste and Resource Action Programme, Banbury, UK.
Huang, O.M. and Schnitzer, M. (1986) *Interactions of Soil Minerals with Natural Organics and Microbes*. Soil Science of America, Special Publication No. 17.
Inbar, Y., Chen, Y. and Hadar, Y. (1990) Humic substances formed during the composting of organic matter. *Soil Sci. Soc. Am. J.* **13**, 137–148.
Jenkins, R.L., Scheybeler, B.J., Smith, M.L., Baird, R., Lo, M.P. and Haug, R.T. (1981) Metals removal and recovery from municipal sludge. *J. WPCF* **53**, 25–32.
Kroenig, H. (1990) First practical experience with the industrial-scale metha II plant for the mechanical separation of dredged material from the Hamburg harbour. In: *Contaminated Soil '90, Third International KfK/TNO Conference on Contaminated Soil* (eds. F. Arendt, M. Hinsenveld and W.J. Brink), Kluwer Academic Publishers, Dordrecht.
Lageman, R., Pool, W. and Seffinga, G.A. (1990) Electro-reclamation: state-of-art and future developments. In: *Contaminated Soil '90, Third International KfK/TNO Conference on Contaminated Soil* (eds. F. Arendt, M. Hinsenveld and W.J. Brink), Kluwer Academic Publishers, Dordrecht.
Lo, K.S.L. and Chen, Y.H. (1990) Extracting heavy metals from municipal and industrial sources. *Sci. Total Environ.* **90**, 99–116.
McBean, E.A., Rovers, F.A. and Farquhar, G.J. (1995) *Solid Waste Engineering and Design*. Prentice Hall PTR, New Jersey.
McGrawth, S.P., Chang, A.C., Page, A.L. and Witter, E. (1994) Land application of sewage sludge: scientific perspectives of heavy metal loading limits in Europe and the United States. *Environ. Rev.* **2**, 108–118.
Metcalf and Eddy (1991) *Wastewater Engineering – Treatment, Disposal and Reuse*. McGraw-Hill, Inc, New York.
Moolenaar, S.W., Lexmond, T.M. and Van der Zee, S.E.A.T.M. (1997) Calculating heavy metal accumulation in soil: comparison of methods illustrated by a case-study on compost application. *Agr. Ecosyst. Environ.* **66**, 71–82.
Oliver, B.G. and Carey, J.H. (1976) Acid solubilization of sewage sludge and ash constituents for possible recovery. *Water Res.* **10**, 1077.
Ried, M. (1988) Heavy metal removal from sewage sludge: practical experiences with acid treatment, pretreatment in chemical water and wastewater treatment. *Proceedings of the 3rd Gothenburg Symposium*, 1988.
RIVM (1988). *National Environmental Survey 1985–2010*. Samson H.D. Tjeenk Willink, Alphen aan den Rijn (in Dutch).
Rulkens, W.H., Voorneburg, F. van and Joziasse, J. (1989) Removal of heavy metals from sewage sludges. *Sewage Sludge Treatment and Use*. Elsevier Applied Science.
Schegel, H.G. (1993) *General Microbiology*. Cambridge University Press, Cambridge.
SDU (1991) *Besluit Overige Organische Meststoffen (BOOM)*. Staatblad **613**, 1–45 (in Dutch).
Stax, A.B.M. (1995) Landfill guidelines in the Netherlands. In: *Proceedings Sardinia 1995, Fifth International Landfill Symposium*, Calgliari, Italy.
Svarovsky, L. (1985) *Solid–Liquid Separation Processes and Technology*. Elsevier, Amsterdam.
Theis, T.L. and Hayes, T.D. (1978) Chemistry of heavy metals in anaerobic digestion. In: *Chemistry of Wastewater Technology* (ed. A.J. Rubin), Ann Arbor Science Publishers Inc., Michigan.
Tyagi, R.D., Couillard, D. and Tran, F. (1988) Heavy metals removal from anaerobically digested sludge by chemical and microbiological methods. *Environ. Pollut.* **50**, 295–316.
Van Dillen, M.R.B. and Schotel, F.M. (1990) Contaminated harbour and river sediments in the Netherlands: development programme 1989–1990: hydrocyclone separation. In: *Contaminated Soil '90, Third International KfK/TNO Conference on Contaminated Soil* (eds. F. Arendt, M. Hinsenveld and W.J. Brink), Kluwer Academic Publishers, Dordrecht.

Van Roosmalen, G.R.E.M. and Langerijt, J.C. van de (1989) Green waste composting in the Netherlands. *Biocycle* **30**, 32–35.
Vaughan, D. and Malcolm, R.E. (1985) Soil organic matter and biological activity. *Developments in Plant and Soil Sciences Vol. 16*. Martinus Nijhoff/Dr W. Junk Publishers, Dordrecht.
Veeken, A.H.M. and Hamelers, H.V.M. (1999) Removal of heavy metals from sewage sludge by extraction with organic acids. *Water Sci. Technol.* **40**, 129–136.
Veeken, A. and Hamelers, H. (2002) Sources of Cd, Cu, Pb and Zn in biowaste. *Sci. Total Environ.* **300**, 87–89.
Veeken, A. and Hamelers, H. (2003) Assessment of heavy metal removal technologies for biowaste by physico-chemical fractionation. *Environ. Technol.* **24**, 329–337.
Veeken, A., De Wilde, V., Woelders, H. and Hamelers, H. (2004) Advanced bioconversion of biowaste for production of a peat substitute and renewable energy. *Biores. Technol.* **92**, 121–131.
White, P.R., Franke, W. and Hindle, P. (1995) *Integrated Solid Waste Management – A Lifecycle Inventory*. Blackie Academic & Professional, London.
Wong, L.T.K. and Henry, J.G. (1988) Bacterial leaching of heavy metals from anaerobically digested sludge. In: *Biotreatment Systems, Vol. III* (ed. D.L. Wise), CRC Press, Inc., Boca Raton.
Wozniak, D.J. and Huang, J.Y.C. (1982) Variables affecting metal removal from sludge. *J. WPCF* **54**, 1574–1580.

24

Odour monitoring within solid waste management

P. Hobbs

24.1 INTRODUCTION

Of all the odours that are offensive, those from waste processes are probably considered the worst. This is probably because they are associated with decay and something that is discarded. However, for those who deal with the management of wastes, that are increasingly identified as renewable resources, there is a more meaningful context. Odours from waste processes can also be an asset rather than a problem for various reasons. Odours are used to evaluate the maturity of alcoholic beverages and food, so why not also use them to characterise decay processes.

This chapter does not discuss the means of reducing or eliminating odours by chemical or physical techniques as these are adequately described elsewhere for the wastewater industry (see Stuetz and Frechen 2001) and composting process (see Park *et al.* 2002). The approach that is investigated here is the use of odours

 Resource Recovery and Reuse in Organic Solid Waste Management. Edited by Piet Lens, Bert Hamelers, Harry Hoitink, Werner Bidlingmaier. ISBN: 1 84339 054 X. Published by IWA Publishing, London, UK

and other volatile organic compounds (VOCs) to monitor and improve process control of waste degradation. There is good evidence for this approach, the basis of which depends upon the characteristic odour from a substrate that is fermenting.

24.2 ODOUR SOURCES AND PHYSICAL FACTORS

24.2.1 Origin of odours during waste processing

Generally, wastes produce an anaerobic environment because of the high organic content. In an anaerobic environment decay products, normally use the most readily available source of electron acceptors. Therefore, instead of sulphate, carbon dioxide, water and often methane, odourous molecules such as hydrogen sulphide, methanethiol, butanoic acid, phenols and indoles are produced. Some of these odourous compounds can be individually perceived by the olfactory senses at concentrations below 1 ppb and are noted by their malodourous presence. However, they are produced in $mg\,l^{-1}$ quantities in the bulk of organic waste because of the anaerobic environment, or in the case of composting, micro-locations can produce considerable quantities of sulphide based odours.

The majority of odourous compounds originate from the decomposition of protein (Mackie *et al.* 1998; Spoelstra 1980). Protein contains sulphur and nitrogen which decay to odourous molecules. Some are exclusive to specific protein sources (Table 24.1). Of the nitrogen in faeces, up to 50% can be present as undigested protein. Generally, short chain fatty acids originate from a range of sources such as polysaccharides, fibre and lipids as well as amino acids. These are non-specific sources in terms of the fatty acids produced.

Table 24.1 Amino acid as sources of odour after deamination.

Amino acid	Odourous compound(s)	Henry's constant	Odour threshold $mg\,m^{-3}$
Tryptophan	Indole and skatole		0.6
Tyrosine	4-methyl phenol and phenol	2240 and 1870	17 and 2000
Methionine and cysteine	Hydrogen sulphide and methanethiol	0.087 and 0.378	1 and 0.5
Phenyalanine	Phenol and phenylpropionic acid	1870 and unknown	2000 and unknown
Histidine	Butanoic acid and dihydroskatole	2800 and unknown	3500 and unknown
Glycine, alanine and serine	Acetic acid	5700	17,000
Threonine	Propanoic acid	4100	450
Histidine	Butanoic acid	2800	3500
Valine	2-methyl propanoic acid	3400	Unknown
Isoleucine	2-methyl butanoic acid	Unknown	Unknown
Leucine	3-methyl butanoic acid	1200	5

Table 24.1 also shows the Henry's Law constant at standard temperature and pressure and the odour threshold for the major odourous compounds. Henry's Law constants increase as the gas has a greater solubility in water. Standardised values were averaged and are from current literature sources that are summarised at the Max-Planck Institute (website http://www.mpch-mainz.mpg.de/~sander/res/henry.html). Averaged odour thresholds are from Van Gemert and Nettenbreijier (1977).

24.2.2 Determination of emission rates of VOCs

It is necessary to take more exacting measurements to determine the emission rate from a source to obtain data for modelling or to give accurate estimations. This requires the use of a hood or an open or closed chamber system (Cumby *et al.* 1995). Emission rates can be presented as mass emitted per unit area or more often mass emitted per unit area divided by volume of the emitting sample especially in the case of digestive or decaying processes that are normally biogenic. The latter measurement takes into account the amount of VOC produced and the surface area, which will have physical factors that limit the emission rate. Biogenic sources will also have variable ages and starting materials and varying rates at which oxygen diffuses into the storage volume. The surface to be evaluated should have well characterised airflows which can be by diffusion limiting (laminar) or turbulent.

Clean air is introduced at one end for sampling with the open system. However, if a closed system is preferred this was the advantage of the volatile compounds being at higher concentrations. This make analysis easier, but further complexities of mathematics are required to allow for the increase in concentration that suppresses further emissions as saturated vapour pressure (SVP) is approached.

24.2.3 Processes in the odour plume

Each odourous compound has an individual odour character (Dravnieks 1985) which when mixed with others forms a new odour. For example 4-methyl phenol increases the malodour and skatole modifies the nature of the odour (Yasuhara 1980). Whilst odours can be sampled immediately, they will only give a profile of the relative concentrations of components present at the time of measurement.

Odourants are emitted in a multi-chemical plume numerous physical and chemical interactions with the atmosphere are possible (Figure 24.1). Water-soluble components will adsorb into droplets or induce nucleation for water vapour.

They may therefore increase in mass and are affected by gravity. Odourants mix at a greater rate at the edge of the plume. Zahn *et al.* (1997) identified differences in the plume profile from pig manure stored lagoons (Table 24.2). Zahn *et al.* (1997) also recognised that VOCs identified in laboratory studies did not necessarily represent the compounds produced in the field or their emission rate. The determination of emission rates is further complicated by many climatic and management factors such as the site involved, i.e. buildings, yards, storage and

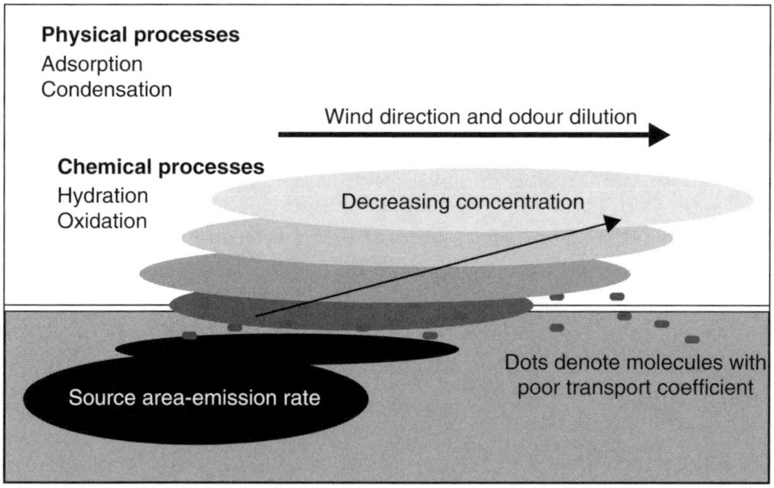

Figure 24.1 Processes in the odour plume affecting VOC concentrations.

Table 24.2 Different concentrations against distance from the source for an odour plume.

	Concentration in air, $mg\,m^{-3}$		
Volatile compound	0 m*	25 m	100 m
Acetic acid	0.52	0.52	0.52
Propionic acid	2.15	1.93	1.14
2-Butanol	0.58	0.57	0.52
Butanoic acid	1.17	1.1	0.89
1-Bromodecane	1.93	1.72	0.99
Pentanoic acid	0.49	0.39	0.18
1-Decanol	0.79	0.65	0.36
1-Bromononane	0.77	0.69	0.32
Hexanoic acid	0.74	0.7	0.33
Bendyl alcohol	1.9	0.86	0.83
Butylated hydroxytoluene	0.72	0.41	0.39
1-Dodecanol	0.63	0.67	0.25
Phenol	1.36	0.79	0.16
Octanoic acid	0.35	0.32	0.21
4-Methyl phenol	4.23	0.88	0.46
Nonanoic acid	0.62	0.51	0.25
4-Ethyl phenol	1.04	0.24	0.09
1,1,3-Trimethyl-Dihydroindene	1.35	1.28	0.78
Hexadecanoic acid methyl ester	0.14	0.12	0.14
Benzoic acid	0.66	0.49	0.1
Dodecanoic acid	1.43	1.06	0.19
2-Amino acetophenone	1.09	0.68	0.08
4-(1,1-dimethylpropyl) phenol	0.26	0.21	0.12
Bis (2-ethylhexyl) phthalate	2.78	2.65	1.37
Total VOC emission, $mg\,m^{-3}$	27.7	19.44	10.67

*Air samples collected at listed distances from odour source. VOCs in the plume may be subject to chemical reactions with non-odourous compounds, such as the hydroxyl radical (Harrison *et al.* 1991) which is an oxidising species. Sunlight will also reduce odour directly by photolytic reactions (Yin *et al.* 1990).

land-spreading. Differing amounts of water and organic matter will also alter degradation and emission rates. Further, as the wastages, emission rates change. For many wastes, emissions of volatile fatty acids (VFAs), phenols and carbon dioxide appear to decrease with increasing storage period (Hobbs *et al.* 1999).

Other sources of error for odour emission rates include those associated with transport of an odour as or on an aerosol particle (Ritter 1989). Such aerosols have been identified and were found to be in the 0.5–3 mm diameter range and at 880–16,000 particles l^{-1} from livestock manure (Hammond *et al.* 1989).

The synergy of odours and the changing chemical and atmospheric conditions combine to make olfactory and instrumental measurement of odours a challenging field of study.

24.3 FACTORS AFFECTING EMISSION RATES

The emissions rates of VOCs, which originate from a liquid surface, depends upon several variables. They mostly depend upon the concentration in the liquid and the chemical or driving energy available for release into the air above the surface. Henry's Law expresses the ratio of the concentration between air and liquid at equilibrium concentrations for a volatile compound, providing no chemical reactions occur and is a thermodynamic expression of the potential for emission. Henry's Law constant can be approximated from the ratio of vapour pressure to solubility. The implications are that those compounds that are virtually insoluble in water, such as the hydrocarbons and the alkyl sulphides have a greater presence in the headspace above the source than those that are soluble (Vitenberg 1991). Those that react with water and are pH dependent, such as the phenols and VFAs make the calculation more complicated but Henry's Law is only concerned with the unionised species. Therefore, the effect of pH needs to be determined before the unionised concentration can be calculated.

Emission rates can be determined in an enclosed chamber from the concentration in the headspace according to ideal Henry's Law behaviour, from an inverse exponential curve with time, limited at the SVP of a volatile compound. The slope is the kinetic rate of emission. However, Henry's Law applies to pure solutions rather than the complex mixture present in wastes. Electrolytes and surfactants in the form of polymeric proteins, saccharides and lipids all cause potential deviations from Henry's law (Thomas 1990).

There are some general observations, such as VOC emission rate decreasing with increasing molecular weight. However, most odourants have a molecular mass from 30 to 150 atomic mass units. When the volatile compounds approach SVP, emission rates are suppressed. If the VP is low then the emission rate is high. The corollary of this argument is that the emission rate from an emitting surface will be dependent on the wind speed. If laminar flows are present at low air speeds then surface emissions will be diffusion limited. At greater wind speeds, turbulence will increase the emission rate because mixing reduces the headspace concentration.

Should this occur then emission rates will be limited by the diffusion of the volatile compound to the liquid surface. Therefore, above a threshold wind speed the emission rate will not increase and, of course, the VOC concentration will decrease in the air. For a liquid or semi-liquid waste, the calculation of emission rates and threshold windspeeds will be complex. The emission profile will change with temperature and airspeed, because of different kinetic rates of release from the waste surface. Different VOCs will produce a changing emission profile which will add to the complexity of determining the olfactory offensiveness and concentration of the odour emitted by introducing a time domain factor. Zahn *et al.* (1997) positively correlated wind velocity with total VOC emissions between 0.2 and $9.4\,m\,s^{-1}$ with a maximum concentration of $9\,\mu g\,l^{-1}$ occurring at $3.6\,m\,s^{-1}$ for odourants from a pig waste lagoon.

24.4 ODOUR MEASUREMENT

24.4.1 Sampling procedures for odours

Sampling can greatly influence the outcome both in terms of quantitative and qualitative measurements. Odours are easy to detect by the olfactory senses, although many odourants are often at concentrations below 10 ppb where there are difficulties for chemical analysis not only because of the volume of the sample required but also because of an increasing ability to disappear. Accurate knowledge of what compounds are present and the ways in which they are sampled are important aspects of identification and determination of the emission rate. Because of the potential loss of VOCs from the measuring system, good quality control methods by using standards and spiking techniques should be used to validate analytical results. Spiking a sample involves adding known quantities of the compound to be measured and identifying if all that which has been added can be detected during analysis. If less than about 70% is present after being introduced as a spike, then the analytical procedure is suspect. Often the sample amount lost will be a greater proportion of the initial concentration for samples at an increasingly lower concentration. At low concentrations, adsorption onto surfaces becomes significant, especially for odourants such as sulphides which can adsorb onto metal surfaces and oxidise or react chemically. Silicone tubing is hydrophobic by nature and will adsorb sulphides and should not be used for connections where odourants are being transferred. Sulphides are difficult to analyse, for example, methanethiol polymerises to produce higher methyl sulphides, *viz.* dimethyl disulphide (CH_3—S—S—CH_3).

24.4.2 Olfactory measurement

Odour is mostly determined as odour concentration (OC). OC is the number of dilutions of odourous air with odourless air at which 50% of an odour panel can just detect an odour using a dynamic dilution olfactometer. The olfactometer enables

Figure 24.2 Odour panellist evaluating an odour from one of the olfactometer ports.

presentation of odourous air to panellists at dilutions with clean air of between 4 and 60,000. An olfactometer normally has two sniffing ports and is of the forced choice type, with odourless air being presented through one port and diluted odourous air through the other (assigned at random) (Figure 24.2). For each presentation, each panellist is required to indicate via a keyboard which port emitted the odourous air and the certainty of their choice. Responses are regarded as being correct if the panellist indicated the correct port and were certain of their choice. To ensure accuracy each sample is presented twice for up to 20 s within a range of five dilution steps of ascending sample concentration with an inter-stimulus period of over 5 min. An experienced olfactometer operator selects the dilution range of five steps between 2^2 and 2^{15} varying by a factor of 2. In addition, each panellist has to respond correctly to the final two (highest) concentrations presented. Individual panellist detection thresholds for each series are calculated as the geometric mean in terms of number of dilutions of the lowest dilution step at which they gave an incorrect response and the next lowest dilution step after which they constantly give a correct response. Panel detection threshold or OC of the sample is calculated at the geometric mean of the individual panellists thresholds.

Equipment and procedures used followed current EU recommendations under guidelines BSEN13725 “Air quality – determination and OC by dynamic dilution olfactometry”. Odour panellists are selected on the basis of sensitivity to butan-1-ol and the calculated variation of the threshold determination from the previous 20 tests with the reference standard.

24.4.3 Chemical analysis of odours

Quantifying odourants can be difficult as some odourants have low odour thresholds of about 1 ppb(v). To have a better indication of the processes occurring in wastes the chemical composition of odour needs to be know. From such data, the odour mixture to build an odour model can be formulated (Hobbs *et al.* 2001).

Gas Chromatography is ideally suited for the rapid separation of complex VOCs that contribute to odour formation, especially since the development of capillary columns that offer a higher resolution. Gas chromatography principally involves a carrier gas that passes over a stationary phase for which the volatile components have a differential affinity that effects separation. Chromatographic efficiency is improved by precision temperature control of the column and the use of a mass flow controller for a constant mass of carrier gas. Generally, a range of detectors can be attached to the end of the column with the mass spectrometer proving the most effective for the identification of unknown components. If, and more commonly, a flame ionisation detector is used, volatile components can be recognised by matching the retention time with compounds that have recognised properties to give a relative retention time. Odour components are introduced into the column through an injector inlet. However, with the introduction of capillary columns small volumes have to be introduced and pre-concentration of an odour is required. Detectors can be used in parallel and the sulphide detector can provide high sensitivities at very low concentrations. Often the flow to the detector can be split and an odour port used to identify the odour signature for a given component.

24.5 CASE STUDY – BETTER QUALITY COMPOST

Compost used to grow mushrooms was used to illustrate that VOCs can be used to diagnose the status of the compost and reduce odourous emissions (Noble *et al.* 2001). Further investigations revealed the importance of the nitrogen sources for compost production (Noble *et al.* 2002).

Odour emissions are a major problem facing compost production for mushroom cultivation in most economically developed countries (Noble and Gaze 1994; Duns *et al.* 1997). Mushroom composting involves wetting and mixing straw and animal manures in heaps (pre-wetting) and leaving in long windrow stacks (Phase I). Malodours originating from the production of compost that uses straw, gypsum and chicken manure (as a nitrogen source) are a problem for the mushroom industry. During these stages, fermentation is uncontrolled, producing anaerobic zones that produce odourous VOCs, causing unacceptable emissions (Figure 24.3). To reduce emissions during pre-wetting and Phase I, new management techniques were necessary. Trace volatile emission profiles are indicators of complex biological processes and were investigated as a tool for making composting management decisions. Composting needs to be well managed, especially with large outputs of about 1000 tons per week.

Understanding the controls over compost quality was improved by using thermal desorption-gas chromatography mass spectrometry in order to detect both odourous and non-odourous VOCs at concentrations as low as parts-per-trillion, 26 VOCs that relate to the processes occurring in different stages of fermentation were analysed. A relatively small number of chemical compounds were identified as responsible for the offensive composting odours and were identified by producing

Figure 24.3 Odour sampling from compost used to grow mushrooms.

the highest value above their odour threshold. Gas detector tubes were used for on-site measurement of gaseous compounds. Odourants that exceeded their published olfactory detection thresholds by the greatest order of magnitude, in decreasing order, were: H_2S, dimethyl sulphide (DMS), butanoic acid, methanethiol and trimethylamine. Concentrations of ammonia were not significantly correlated with OC, and they were not significantly affected by the use of aeration. There was a close correlation ($r = 0.918$, $P < 0.001$) between the OC of air samples from the composting materials and the combined H_2S + DMS concentrations (Figure 24.4), measured on site with detector tubes.

This relationship was unaffected by the ammonia concentration or the type of compost: aerated or un-aerated, pre-wet or Phase I, poultry manure-based or horse and poultry manure-based. However, by using other sources of nitrogen such as chipboard and digester wastes, cocoa shells or ammonium sulphate, odour emissions were reduced but at the cost of mushroom yield (Noble *et al.* 2002).

In many cases, local authorities are threatening to enforce compulsory closures of composting sites because of odour complaints. By adopting different composting turning regimes, composting processes were improved and odour nuisance complaints were reduced by 90% and closure was avoided. Other fermentation processes are likely to benefit from this approach and especially as an alternative to landfill. There is a continuing need to ensure that odour pollution is minimised to achieve legislative requirements and that the finished compost is of the highest quality.

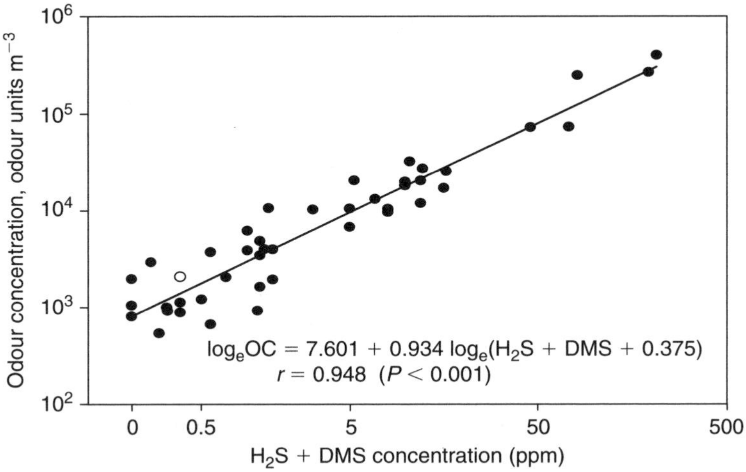

Figure 24.4 Relationship between the combined on-site hydrogen sulphide and dimethyl sulphide concentrations and the 24 h OC on mushroom composting sites. Each point is the mean of two sample determinations.

24.6 CONCLUSION

This chapter does not describe a new approach, but hopefully will enable managers of "waste-streams" to consider odours as providing an insight into the fermentation processes. There is a requirement for improved analytical facilities many of which are likely to be portable. The outcome of odours as a means of process control should lead to the development of a science that enhances sustainability by producing a quality product that can be of use or released into the environment with minimum impact.

REFERENCES

Cumby, T.R., Moses, B. and Nigro, I. (1995) Gases from livestock slurries; Emission kinetics. *Proceedings of the 7th International Conference on Agricultural and Food Wastes*. Chicago, USA.

Dravnieks, A. (1985) Atlas of odor character profiles. *ASTM Committee on Sensory Evaluation of Materials and Products*, ASTM data series, Baltimore MD, USA.

Duns, G.J., Ripley, B.D., Kingsmill, C.A. and Rinker, D.L. (1997) The analysis of mushroom composting odours. *Mushroom World,* June, 58–70.

Hammond, E.G., Heppner, C. and Smith, R. (1989) Odors of swine waste lagoons. *Agricul. Ecosyst. Environ.* **25**, 103–110.

Harrison, R.M., de Mora, S.J., Rapsomanikis, S. and Johnson, W.R. (1991) *Introductory Chemistry for the Environmental Sciences*, pp. 272–275. University Press, Cambridge.

Hobbs, P.J., Misselbrook, T.H. and Cumby, T.R. (1999) Production and emission of odours and gases from ageing pig waste. *J. Agricul. Eng. Res.* **72**, 291–298.

Hobbs, P.J., Misselbrook, T.H., Dhanoa, M.S. and Persaud, K.C. (2001) Development of a relationship between olfactory response and major odorants from organic wastes. *J. Sci. Food Agricul.* **81**, 188–193.

Mackie, R.I., Stroot, P.G. and Varel, V.H. (1998) Biochemical identification and biological origin of key odor components in livestock waste. *J. Animal Sci.* **76**, 1331–1342.

Noble, R. and Gaze, R.H. (1994) Controlled environment composting for mushroom cultivation – substrates based on wheat and barley straw and deep litter poultry manure. *J. Agricul. Sci.* **123**, 71–79.

Noble, R., Hobbs, P.J., Dobrovin-Pennington, A., Misselbrook, T.H. and Mead, A. (2001) Olfactory response to mushroom composting emissions as a function of chemical concentration. *J. Environ. Qual.* **30**, 760–767.

Noble, R., Hobbs, P.J., Mead, A. and Dobrovin-Pennington, A. (2002) Influence of straw types and nitrogen sources on mushroom composting emissions and compost productivity. *J. Ind. Microbiol. Biot.* **29**, 99–110.

Park, K.J., Choi, M.H. and Hong, J.H. (2002) Control of composting odor using biofiltration. *Compost Sci. Util.* **10**, 356–362.

Ritter, W.F. (1989) Odor control of livestock wastes – state-of-the-art in North-America. *J. Agricul. Eng. Res.* **42**, 51–62.

Spoelstra, S.F. (1980) Origin of objectionable odourous components in piggery wastes and the possibility of applying indicator components for the study of odour development. *Agricul. Environ.* **5**, 241–260.

Stuetz, R.M. and Frechen, F.B. (2001) *Odours in Wastewater Treatment.* (eds. R.M. Stuetz, and F.B. Frechen), 1st edn. IWA publishing, London.

Thomas, R.G. (1990) Volatilization from water. In: *Handbook of Chemical Property Estimation Methods* (ed. W.J. Lyman), pp. 15-1–15-34. American Chemical Society, Washington DC.,

Van Gemert, L.J. and Nettenbreijier A.H. (1977) Compilation of odour threshold values in air and water. *National Institute for Water Supply and Central Institute for Food and Nutrition Research*, Netherlands.

Vitenberg, A.G. (1991) Equilibrium accumulation methods in determining volatile trace components by gas-chromatography. *J. Anal. Chem.–USSR* **46**, 1548–1565.

Yasuhara, A. (1980) Relationship between odor and odourous components in solid swine manure. *Chemosphere* **9**, 587–592.

Yin, F., Grosjean, D. and Seinfeld, J.H. (1990) Photooxidation of dimethyl sulphide and dimethyl disulphide I: mechanism development. *J. Atmos. Chem.* **11**, 309–364.

Zahn, J.A., Hatfield, J.L., Do, Y.S., DiSpirito, A.A., Laird, D.A. and Pfeiffer, R.L. (1997) Characterization of volatile organic emissions and wastes from a swine production facility. *J. Environ. Qual.* **26**, 1687–1696.

Index